Reviews in Computational Chemistry Volume 7

Reviews in Computational Chemistry 7

Edited by

Kenny B. Lipkowitz and Donald B. Boyd

Kenny B. Lipkowitz
Department of Chemistry
Indiana University–Purdue University
at Indianapolis
402 North Blackford Street
Indianapolis, IN 46202-3274
USA
lipkowitz@chem.inpui.edu

Donald B. Boyd
Department of Chemistry
Indiana University–Purdue University
at Indianapolis
402 North Blackford Street
Indianapolis, IN 46202-3274
USA
boyd@chem.inpui.edu

This book is printed on acid-free paper ♾

Printed in the United States of America

ISBN 1-56081-915-4 VCH Publishers, Inc.
ISSN-1069-3599

Printing History:
10 9 8 7 6 5 4 3 2 1

Published jointly by

VCH Publishers, Inc.
220 East 23rd Street
New York, New York 10010

VCH Verlagsgesellschaft mbH
P.O. Box 10 11 61
69451 Weinheim, Germany

VCH Publishers (UK) Ltd.
8 Wellington Court
Cambridge CB1 1HZ
United Kingdom

Preface

We begin this seventh volume of *Reviews in Computational Chemistry* by continuing our commentary on employment prospects for computational chemists. As we noted in the Preface of Volume 6, economic and political changes sweeping over the pharmaceutical and chemical industries of the world were having the negative effect of reducing the rate at which new jobs for computational and other chemists were being created. According to the National Science Foundation, the number of research and development scientists employed by the pharmaceutical and chemical industries declined 4% between 1989 and 1992 in the United States. From 1992 through 1994, there was a further drop; the pharmaceutical industry alone had to cut roughly 3500 science jobs, according to the Pharmaceutical Manufacturers Association, and plans for many more cuts have been announced. As also noted in the Preface of Volume 6, computational chemists were not unduly targeted in these cutbacks, fortunately, but neither were they immune to all the macroeconomic changes or governmental influences.

To obtain a perspective on the effects of these changes, we examine a historical measure of the rate of job creation for computational chemists. In Figure 1 we present data on the yearly number of positions for computational chemists advertised in the American Chemical Society's weekly magazine *Chemical and Engineering News* (*C&EN*). Our data go back as far as 1983. Recall that prior to the 1980s the job market in industry for computational chemists (then generally known as theoretical chemists) was minuscule, so the plot covers most of the modern period of computational chemistry.

In Figure 1, the jobs advertised are categorized into postdoctoral research positions, nontenured staff academic positions, tenure-track academic positions, positions at pharmaceutical and chemical companies, or positions at hardware and software companies. It is immediately apparent from the figure that the total number of advertisements has a roughly parabolic shape, with a peak in 1990. The adolescence of the field is seen in the early 1980s. The big increase in advertisements from 1987 to 1990 was due to a combination of all categories. By 1990 the molecular modeling software companies were sustaining the overall high demand for computational chemists. However, by 1991 advertisements in all categories were slumping. The total number of advertisements in 1994 had returned to almost the level in 1983.

Industrial job openings in Figure 1 are mainly at pharmaceutical houses, where the responsibility of the computational chemists is to aid drug discovery. Some chemical and petroleum companies built small in-house computational

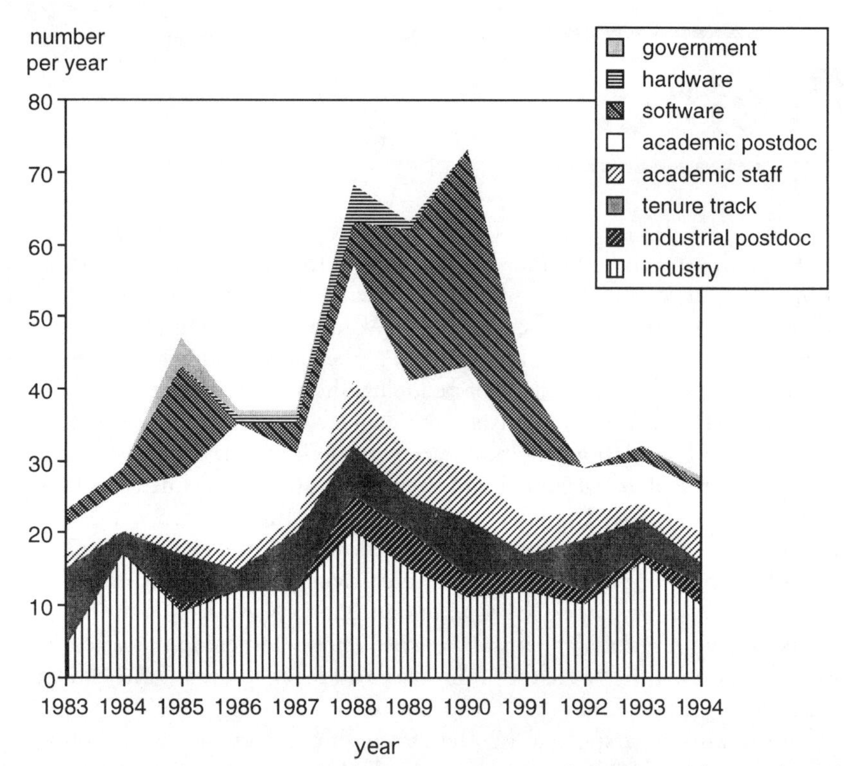

Figure 1 Jobs for scientists in the field of computational chemistry advertised in *Chemical and Engineering News* by year. The available positions are classified as to whether they were for government laboratories, software companies, computer hardware companies, academia (tenure-track professional, nontenured staff, or postdoctoral appointments), or industrial research laboratories (permanent or postdoctoral appointments). The advertised job openings were not divided into B.S.-level, M.S.-level, and Ph.D.-level positions, but because computational chemistry is an advanced science, most of the positions advertised in *C&EN* were for Ph.D.s. In preparing the figure, multiple advertisements for the same job in the same year are counted only once. In some cases, jobs advertised near the end of one year were also advertised early in the following year; in these situations the positions are counted in both years. In cases of advertisements for an unspecified number of positions greater than one, an estimate was made.

chemistry groups for design of other compounds and materials. Postdoctoral situations in private industry were essentially nonexistent prior to 1988, but more recently pharmaceutical management viewed the need for low-cost, highly trained scientists as greater than their concern for the security of proprietary data when a transient labor force has access to it. The hardware and software companies referred to in Figure 1 are those that cater in large part to the molecular modeling and database needs of the pharmaceutical and chemical industries and depend on these industries for much of their revenue.

Surprisingly, the number of job advertisements in the pharmaceutical and chemical industries has been remarkably steady during the 12-year period covered by Figure 1; three crests occurred in 1984, 1988, and 1993. Interestingly, sharp increases in jobs at software companies occurred one or two years after the 1984 and 1988 crests. Many of the openings at the software companies, however, were for temporary, i.e., postdoctoral, positions. The number of jobs advertised in academia lacks any apparent correlation with the other components of the job market. This uncoupling suggests that perceived long-term educational needs and research grants drive the academic job market more than do the immediate job prospects of Ph.D. graduates. Openings at government laboratories have not been advertised in large numbers in *C&EN*.

Does Figure 1 suggest a temporary slowdown in the need for computational chemists, or are we seeing a saturation of the size of the field? Surely the technology of computational chemistry can continue to develop, and the use of the technology can continue to spread to more chemists and other molecular scientists. However, is the field now mature enough that the job market for computational chemists will grow only in proportion to the underlying pharmaceutical, chemical, and materials industries?

To assuage fears that the job market for new graduates specializing in computational chemistry has dried or will dry up, it is instructive to look at the running sum of the number of jobs for all categories of computational chemists advertised in *C&EN*. This sum can be taken as roughly proportional to the overall population of computational chemists with jobs. In Figure 2, we see three legs to the curve. From 1983 through 1987, the number of advertised job openings increased at the robust rate of about 38 per year. From 1987 through 1990, available positions were being advertised at an intoxicating rate of 68 new ones per year. After 1990, the growth rate cooled to 30 per year. Growth has not stopped; it has only slowed. Although we do not have any hard data, it appears that the total number of employed computational chemists has not experienced a shrinkage.

Figure 3 shows the relative sizes of the components of the computational chemistry job market as reflected by advertisements in *C&EN* for the entire 12-year period. The largest component (totaling 45%) is in academia (combining tenure-track, staff, and postdoctoral positions). About one-third of the available positions were in industry, and a fifth were at software companies. Considering the academic, industrial, and software company groups together, about one-third of the advertised jobs were for postdoctoral positions.

It should be pointed out that *C&EN* is only one forum in which jobs for computational chemists are advertised. International science magazines, magazines of the chemical societies of other nations, electronic archives (such as that maintained at the Ohio Supercomputer Center), and employment agencies are other sources of information about available positions. In addition, personal contacts through professors may be the most common way some universities and companies find and hire new scientists.

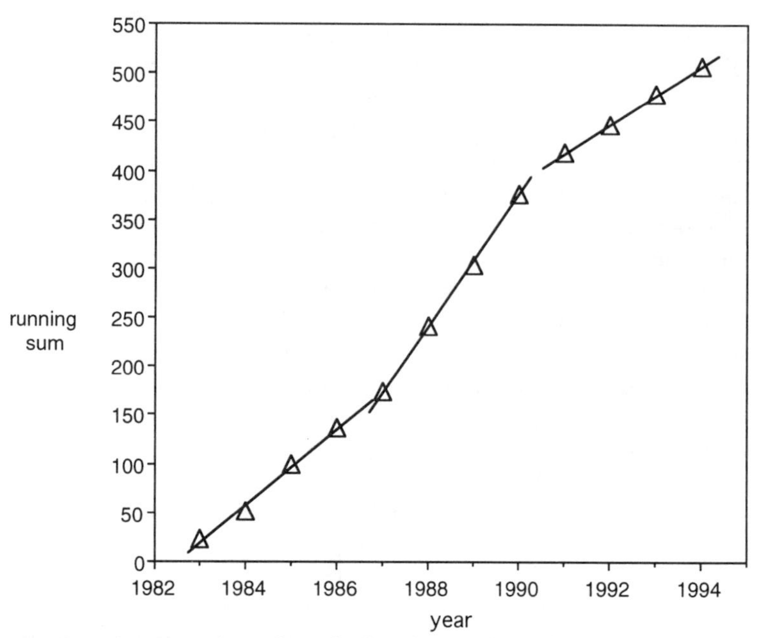

Figure 2 Cumulative total number of jobs advertised in *Chemical and Engineering News* for scientists in the field of computational chemistry. Regression lines are fit to the three legs of the data points. The correlation coefficients are 0.996, 0.999, and 0.999, respectively, for each leg.

Being familiar with the important techniques of computational chemistry is one way chemists can enhance their job qualifications and increase the demand for their expertise at university, company, or government laboratories. Chemists, including experimentalists, are nowadays expected to have a rudimentary working knowledge of computational chemistry. A familiarity with

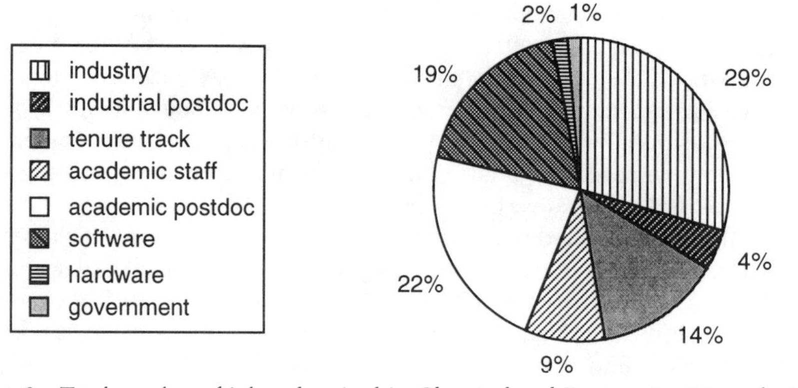

Figure 3 Total number of jobs advertised in *Chemical and Engineering News* during the 12-year period 1983–1994 for scientists in the field of computational chemistry.

molecular modeling has become *de rigueur,* so education in the techniques has reached far beyond just computational chemists. Hardware and software tools for computational chemistry are almost expected to be available in both academic and industrial research settings. With this in mind, we present this latest volume of *Reviews in Computational Chemistry.* Tutorials and reviews on several important topics are featured in this volume.

One of the hottest areas of research in drug discovery today is combinatorial chemistry. The strategy here is to rapidly synthesize tens or hundreds of thousands of peptide, nucleotide, or organic oligomers and congeners for mass screening. Combinatorial chemistry requires the thought process of the organic chemist to change from its traditional custom designing of individual molecules to designing synthetic production lines. The resulting multitude of compounds can then be tested in highly automated, cheap assays for biological or biochemical activities of interest. Most of this century has seen medicinal chemistry move in the direction of more rational approaches to drug discovery. It is therefore especially ironic that as this century comes to a close, we see a resurgence of interest in random screening based on the diversity of structures present in combinatorial libraries. A few more years are needed before it will be clear whether combinatorial chemistry is a fad, a panacea, or just another resource in the armamentarium for drug discovery.

Although the techniques of computational chemistry were obviously involved in rational drug design and discovery, does a rebirth of random searching portend diminished roles for the computational scientist and molecular modeler? We think not. The deluge of data coming from screening hundreds of thousands of compounds can only be effectively handled by computer storage and retrieval techniques. Already some of the more forward-looking software vendors are rolling out new programs to deal with the increased flow of structural and biological data. In addition, molecular modeling will come into play to design druglike molecular frameworks from some of the oligomeric compounds generated by combinatorial chemistry.

Creating and searching databases of molecular structures go hand in hand with combinatorial chemistry. The huge numbers of newly prepared compounds are stored for structure–activity analysis. Old compounds of a pharmaceutical company are searched and reexamined in light of new data and new tests in hopes of formulating libraries of diverse and representative structural types that can be evaluated in the laboratory and docked into target receptor structures on the computer. The traditional methods of drug discovery —synthetic chemistry, pharmacognosy, biological testing, and spectroscopic methods of physical and structural analysis—are strengthened, rather than supplanted, by the new computer-based approaches to discovering therapeutic agents.

This volume has two chapters related to structural databases and how they can be mined for useful information. Chapter 1 by Dr. Geoffrey M. Downs and Professor Peter Willett describes techniques for measuring whether compounds are similar in structure. This is an important concept the chemist uses when trying to retrieve compounds similar to a lead compound or when trying to obtain a diverse set of compounds. In Chapter 2, Drs. Andrew C. Good and Jonathan S. Mason provide a tutorial on approaches to searching databases of three-dimensional structures and illustrate these approaches with practical examples. As typified by these four authors, both academic and industrial scientists are keenly interested in developing methodologies for exploring molecular databases.

Historically, two of the major user groups in computational chemistry were those who implemented quantum mechanics to address electronic effects explicitly and those who used empirical force fields treating electrons implicitly.* Although each approach has its inherent advantages and disadvantages, it was recognized some time ago that in certain situations a hybrid QM/EFF treatment was needed. With such an approach, one could study bond-making and bond-breaking processes of, say, an enzyme using a suitable level of quantum theory, while treating the remainder of the system with molecular mechanics or dynamics. Major efforts have recently been directed toward establishing the hybrid QM/EFF methodology. Chapter 3 is a tour de force by Professor Jiali Gao on the methods and applications of the combined computational approach.

Chapters 4 and 5 teach density functional theory (DFT) and its applications. Drs. Libero J. Bartolotti and Ken Flurchick introduce the theory in Chapter 4. This tutorial rigorously establishes an essential educational foundation for those who intend to use DFT. The chapter is based on courses the authors have taught over the years. Although DFT has been around for many years, it has only recently been used in a routine way by those outside quantum physics to predict molecular structures and properties. Although the methodology is still undergoing evaluation by bench chemists, computational chemists are undertaking crucial computational experiments to validate DFT in diverse applications. In Chapter 5, Professor Alain St-Amant carefully critiques the performance of available functionals to model biologically interesting molecules and interactions relevant to structural biology.

One of the important roles of computational chemistry is to explain and predict electronic structure in conjunction with modern spectroscopic techniques. In Chapter 6, Dr. Danya Yang and Professor Arvi Rauk thoroughly describe the computational determination of intensities of vibrational circular dichroism spectra.

The Appendix by Professor Donald B. Boyd is a broadly updated compendium of software for molecular modeling, computational chemistry, de novo molecular design, quantitative structure–property relationships, synthesis

*J. D. Bolcer and R. B. Hermann, in *Reviews in Computational Chemistry,* Vol. 5, K. B. Lipkowitz and D. B. Boyd, Eds., VCH Publishers, New York, 1994, pp. 1–63. The Development of Computational Chemistry in the United States.

planning, and other facets of computer-aided molecular science. The Appendix is intended to help our readers track down tools of interest. This edition of the Appendix has more than 25% more material than in Volume 6. Addresses, telephone numbers, electronic mailing addresses, and World-Wide Web addresses of suppliers are provided for more than 2500 programs.

We thank the authors of this volume for joining us in this enterprise. We have learned much from these excellent chapters and hope that you too will find them beneficial. We acknowledge Joanne Hequembourg Boyd for invaluable assistance with the editorial processing of this book. Finally, we thank the readers of this series who have found the books useful and have given us encouragement.

Donald B. Boyd and Kenny B. Lipkowitz
Indianapolis
August 1995

Contents

Contributors

Libero J. Bartolotti, Research Institute, North Carolina Supercomputer Center, MCNC, Research Triangle Park, North Carolina 27709, U.S.A. (Electronic mail: bartolot@ncsc.org)

Donald B. Boyd, Department of Chemistry, Indiana University Purdue University at Indianapolis (IUPUI), 402 North Blackford Street, Indianapolis, Indiana 46202-3274, U.S.A. (Electronic mail: boyd@chem.iupui.edu)

Geoffrey M. Downs, Barnard Chemical Information Ltd., 30 Kiveton Lane, Todwick, Sheffield S31 OHL, United Kingdom. (Electronic mail: downs@bci2.demon.co.uk)

Ken Flurchick, Research Institute, North Carolina Supercomputer Center, MCNC, Research Triangle Park, North Carolina 27709, U.S.A. (Electronic mail: kenf@ncsc.org)

Jiali Gao, Department of Chemistry, State University of New York at Buffalo, Buffalo, New York 14260, U.S.A. (Electronic mail: jiali@tams.chem.buffalo.edu)

Andrew C. Good, Dagenham Research Center, Rhône-Poulenc Rorer Ltd., Rainham Road South, Dagenham, Essex RM10 7XS, United Kingdom (Electronic mail: andrew.a.g.good@drc.rp–rorer.rp.fr)

Jonathan S. Mason, Rhône-Poulenc Rorer Central Research, 500 Arcola Road, Collegeville, Pennsylvania 19426-0107 (Electronic mail: jon.mason@rp.fr)

Arvi Rauk, Department of Chemistry, The University of Calgary, Calgary, Alberta, Canada T2N 1N4 (Electronic mail: rauk@acs.ucalgary.ca)

Alain St-Amant, Department of Chemistry, University of Ottawa, 10 Marie Curie, Ottawa, Ontario, Canada K1N 6N5 (Electronic mail: st-amant@theory.chem.uottawa.ca)

Peter Willett, Krebs Institute for Biomolecular Research and Department of Information Studies, University of Sheffield, Western Bank, Sheffield S10 2TN, United Kingdom. (Electronic mail: p.willett@sheffield.ac.uk)

Danya Yang, Department of Chemistry, The University of Calgary, Calgary, Alberta, Canada T2N 1N4 (Electronic mail: dyang@acs.ucalgary.ca)

Contributors to Previous Volumes*

VOLUME I

David Feller and **Ernest R. Davidson,** Basis Sets for Ab Initio Molecular Orbital Calculations and Intermolecular Interactions.

James J. P. Stewart,† Semiempirical Molecular Orbital Methods.

Clifford E. Dykstra,‡ **Joseph D. Augspurger, Bernard Kirtman,** and **David J. Malik,** Properties of Molecules by Direct Calculation.

Ernest L. Plummer, The Application of Quantitative Design Strategies in Pesticide Design.

Peter C. Jurs, Chemometrics and Multivariate Analysis in Analytical Chemistry.

Yvonne C. Martin, Mark G. Bures, and **Peter Willett,** Searching Databases of Three-Dimensional Structures.

Paul G. Mezey, Molecular Surfaces.

Terry P. Lybrand,¶ Computer Simulation of Biomolecular Systems Using Molecular Dynamics and Free Energy Perturbation Methods.

Donald B. Boyd, Aspects of Molecular Modeling.

Donald B. Boyd, Successes of Computer-Assisted Molecular Design.

Ernest R. Davidson, Perspectives on Ab Initio Calculations.

---

*For chapters where no author can be reached at the address given in the original volume, the current affiliation of the senior author is given here.
†Current address: 15210 Paddington Circle, Colorado Springs, CO 80921 (Electronic mail: jstewart@fai.com)
‡Current address: Indiana University-Purdue University at Indianapolis, Indianapolis, IN (Electronic mail: dykstra@chem.iupui.edu)
¶Current address: University of Washington, Seattle, WA (Electronic mail: lybrand@proteus.bioeng.washington.edu)

VOLUME II

Andrew R. Leach,[*] A Survey of Methods for Searching the Conformational Space of Small and Medium-Sized Molecules.

John M. Troyer and **Fred E. Cohen,** Simplified Models for Understanding and Predicting Protein Structure.

J. Phillip Bowen and **Norman L. Allinger,** Molecular Mechanics: The Art and Science of Parameterization.

Uri Dinur and **Arnold T. Hagler,** New Approaches to Empirical Force Fields.

Steve Scheiner, Calculating the Properties of Hydrogen Bonds by Ab Initio Methods.

Donald E. Williams, Net Atomic Charge and Multipole Models for the Ab Initio Molecular Electric Potential.

Peter Politzer and **Jane S. Murray,** Molecular Electrostatic Potentials and Chemical Reactivity.

Michael C. Zerner, Semiempirical Molecular Orbital Methods.

Lowell H. Hall and **Lemont B. Kier,** The Molecular Connectivity Chi Indexes and Kappa Shape Indexes in Structure-Property Modeling.

I. B. Bersuker[†] and **A. S. Dimoglo,** The Electron-Topological Approach to the QSAR Problem.

Donald B. Boyd, The Computational Chemistry Literature.

VOLUME III

Tamar Schlick, Optimization Methods in Computational Chemistry.

Harold A. Scheraga, Predicting Three-Dimensional Structures of Oligopeptides.

---

*Glaxo Wellcome, Greenford, Middlesex, United Kingdom (Electronic mail: arl22958@ggr.co.uk)
†Current address: University of Texas, Austin, TX (Electronic mail: cmao771@charon.cc.utexas.edu)

Andrew E. Torda and **Wilfred F. van Gunsteren,** Molecular Modeling Using NMR Data.

David F. V. Lewis, Computer-Assisted Methods in the Evaluation of Chemical Toxicity.

VOLUME IV

Jerzy Cioslowski, Ab Initio Calculations on Large Molecules: Methodology and Applications.

Michael L. McKee and **Michael Page,** Computing Reaction Pathways on Molecular Potential Energy Surfaces.

Robert M. Whitnell and **Kent R. Wilson,** Computational Molecular Dynamics of Chemical Reactions in Solution.

Roger L. DeKock, Jeffry D. Madura, Frank Rioux, and **Joseph Casanova,** Computational Chemistry in the Undergraduate Curriculum.

VOLUME V

John D. Bolcer and **Robert B. Hermann,** The Development of Computational Chemistry in the United States.

Rodney J. Bartlett and **John F. Stanton,** Applications of Post-Hartree-Fock Methods: A Tutorial.

Steven M. Bachrach, Population Analysis and Electron Densities from Quantum Mechanics.

Jeffry D. Madura, Malcolm E. Davis, Michael K. Gilson, Rebecca C. Wade, Brock A. Luty, and **J. Andrew McCammon,** Biological Applications of Electrostatic Calculations and Brownian Dynamics Simulations.

K. V. Damodaran and **Kenneth M. Merz Jr.,** Computer Simulation of Lipid Systems.

Jeffrey M. Blaney and **J. Scott Dixon,** Distance Geometry in Molecular Modeling.

Lisa M. Balbes, S. Wayne Mascarella, and **Donald B. Boyd,** A Perspective of Modern Methods in Computer-Aided Drug Design.

VOLUME VI

Christopher J. Cramer and **Donald G. Truhlar,** Continuum Solvation Models: Classical and Quantum Mechanical Implementations.

Clark R. Landis, Daniel M. Root, and **Thomas Cleveland,** Molecular Mechanics Force Fields for Modeling Inorganic and Organometallic Compounds.

Vassilios Galiatsatos, Computational Methods for Modeling Polymers: An Introduction.

Rick A. Kendall, Robert J. Harrison, Rik J. Littlefield, and **Martyn F. Guest,** High Performance Computing in Computational Chemistry: Methods and Machines.

Donald B. Boyd, Molecular Modeling Software in Use: Publication Trends.

Eiji Osawa and **Kenny B. Lipkowitz,** Published Force Field Parameters.

CHAPTER 1

Similarity Searching in Databases of Chemical Structures

Geoffrey M. Downs and Peter Willett

Krebs Institute for Biomolecular Research and Department of Information Studies, University of Sheffield, Western Bank, Sheffield S10 2TN, United Kingdom

INTRODUCTION

Chemical structure databases contain machine-readable representations of traditional chemical structure diagrams of molecules or their atomic coordinates: these are normally referred to as two-dimensional (2D) or three-dimensional (3D) databases, respectively. Systems for searching such databases play an increasingly important role in chemical research and development.[1] Structural databases can be publicly available via wide-area telecommunications systems, as with the Chemical Abstracts Service (CAS) Chemical Registry System[2,3] or contain the proprietary compounds of a single organization, with searches being effected with software packages such as MACCS from MDL Information Systems or UNITY from Tripos Associates. In-house applications are of particular importance in the pharmaceutical and agrochemical industries, where database searching is an important component of programs to reveal novel chemical entities with beneficial biological properties.

Computer techniques for searching databases of chemical structures have been the subject of active research and development for more than three decades, and there is now a well-established body of theory and practice underly-

Reviews in Computational Chemistry, Volume 7
Kenny B. Lipkowitz and Donald B. Boyd, Editors
VCH Publishers, Inc. New York, © 1996

ing the design and implementation of chemical-structure retrieval systems. For the first 25 years of their existence, such systems provided access to databases of 2D chemical structure diagrams using techniques derived from graph theory. In the late 1980s, the available techniques were extended to encompass the representation and searching of 3D compounds,[4,5] although some years were to pass before it became possible to provide a full description of the conformational space of flexible molecules.[6–8] Early systems (both in 2D and in 3D) provided two main types of retrieval mechanism: *structure searching* and *substructure searching* (as described in the next section). The last few years have seen the emergence of *similarity searching* as an alternative, and complementary, means of access to structural databases. The algorithms and data structures necessary for similarity searching provide the principal focus of this review.

In a wide-ranging discussion of the role of similarity in the chemical and the physical sciences, Rouvray[9] has noted the ubiquitous nature of the concept of similarity, even when attention is restricted to applications in the area of structural chemistry. Here, we consider one such application, namely, the use of similarity measures to search a database of chemical structures. Specifically, we are interested in techniques that, given a user-defined target structure, identify those molecules in a database that are most similar to the target using some quantitative measure of intermolecular similarity. There are many measures available to quantify the degree of similarity between a pair of structures. The computational requirements (as manifested by their space and time complexities) of these measures vary drastically, the variation being caused in large part by the level of detail that is used to represent the molecules that are being compared. Thus, if a similarity measure is to be employed that is designed for highly complex structural representations requiring a large amount of processing, it will be possible to match a target structure against only a small number of database structures. At the cost of some simplification, we can identify three main types of similarity measure, based on the processing requirements and types of data to which they are to be applied (although the plummeting cost of computer hardware means that the precise nature of each class changes from year to year). In what follows, we shall refer to these as type-A, type-B, and type-C similarity measures.

Type-A measures are appropriate for measuring the similarity between one or, at most, a few pairs of structures that are characterized by quantum-mechanical descriptions of various sorts. Following the pioneering studies of Carbo et al.[10] a large number of similarity measures have been described that fall in this class (see, e.g., Refs. 11–14). Here, a molecule is described by an electron probability density function, and the similarities between pairs of molecules are calculated with measures that describe the overlap of their density functions.[11] Work in this area is exemplified by that of Manaut et al.[14] who describe a procedure to align two molecules so as to maximize the similarity between the sets of molecular electrostatic potential (MEP) values around the molecules, where the MEP values are computed using electron density distributions from either ab initio or semiempirical wavefunctions. Their programs require an initial, manual alignment of the two molecules, which rules out the use of such a procedure for database-searching purposes even if the actual similarity calculation was sufficiently rapid. Analogous comments apply to the OVID program of Hermann and Herron,[15] which seeks to maximize the steric overlap rather than the density-function overlap; these authors also describe another program, called SUPER, which searches all possible structural correspondences, but this typically requires two CPU hours on a Cray-2 supercomputer to compare two structures. As Allan and Cooper[12] note, electron-density measures tend to focus on the dense core electrons, and they have therefore suggested a related approach involving the momentum-space electron densities that emphasizes the chemically interesting features of the valence electron distributions.[13] Other examples of quantum-mechanical approaches to the measurement of similarity are discussed by Bowen-Jenkins et al.[16] and by Ponec and Strnad,[17] inter alia, whereas Mezey reviews topological approaches to the comparison of electron density functions.[18]

Type-B measures are appropriate for measuring the similarities between a few tens or, at most, a few hundreds of structures. Data sets of this size are commonly met with in structure–activity relationship (SAR) studies, which seek to correlate biological activities with structural features and which play a central role in rational approaches to the design of drugs and pesticides (see, e.g., Refs. 19–21). Many of the type-B similarity measures that have been described involve representing a molecule by a 3D grid, each element of which is characterized by the values of the steric, electrostatic, or hydrophobic fields at that point. The similarity between two molecules is obtained by aligning the grids in some way, and then measuring the degree of resemblance of the corresponding sets of grid values. The most extensive examples of this approach are provided by the studies of electrostatic and steric similarity that have been reported by Richards and his associates (see e.g., Refs. 22–25), as well as by several other workers.[26,27] The alignment of grids links this approach to the Comparative Molecular Field Analysis (CoMFA) method for 3D QSAR, which was pioneered by Cramer et al.[28] and which describes each (superimposed) molecule of a set on the basis of its steric and electrostatic fields at intersections of a lattice enclosing the compounds. CoMFA (and approaches that are closely related to it) is now being used very widely for the analysis of sets of a few tens of 3D structures, as is evidenced by even the most cursory scan of the contents pages of journals such as the *Journal of Computer-Aided Molecular Design,* the *Journal of Medicinal Chemistry,* or *Quantitative Structure–Activity Relationships.* Another group of type-B measures derives from the use of superimposition procedures that seek to identify an optimal mapping (in the sense of minimizing some distance metric such as the root mean-squared deviation) of the atoms of one structure onto the atoms of another structure. The studies of Dean and his co-workers provide an excellent introduction to work in this area (see, e.g., Refs. 29–34).

Detailed accounts of many type-A and type-B similarity measures may be found in the books edited by Johnson and Maggiora[35] and by Dean[36] and in the proceedings of a *Symposium on Similarity in Organic Compounds* that was organized by the Beilstein Institute in Bolzano, Northern Italy in June 1992.[37]

We define a type-C similarity measure as one for which the processing requirements are sufficiently small for it to be used to calculate the similarities between a target structure and hundreds of thousands, or even millions, of structures in a database. We shall be primarily concerned with similarity measures of this type in this review, which is structured as follows. The next section reviews the development of chemical information systems, focusing on the representations that are used to describe the 2D and 3D structures of molecules, on the substructure-searching algorithms that have historically provided the main means of access to such data, and the need for similarity searching as a complementary means of access. The section on similarity searching in databases of 2D structures reviews the various types of measure that have been suggested for measuring the similarity between pairs of chemical structure diagrams and then focuses on the fragment-based approaches that are currently most widely used for database searching. Then follows a section on one of the principal applications of 2D similarity searching, the clustering of databases as an aid to the selection of compounds for inclusion in biological screening programs. There is much current interest in developing similarity measures for searching databases of 3D structures, and we give an overview of this research. Finally, the last section summarizes the main points of the review and discusses ways in which we can expect similarity searching to develop over the next few years. In particular, we consider the extent to which it may be possible to use type-B similarity measures in a database context in the future.

CHEMICAL INFORMATION SYSTEMS

In this section, we give a brief introduction to the subject of chemical information systems, with special reference to the graph-theoretic methods that form the basis for current approaches to the representation and searching of both 2D and 3D structures. More detailed accounts of chemical information systems are provided in the book edited by Ash et al.[1] and in the proceedings of the conferences on *Chemical Structures* that are held at Noordwijkerhout, Holland every three years.[38–40]

Representation and Substructure Searching of 2D Chemical Structures

Connection Table Representation of 2D Structures

Many different approaches have been, and continue to be, described for the representation of the 2D structures of chemical compounds.[41,42] Of these approaches, the connection table is now by far the most important. A *connection table* contains a list of all of the (typically nonhydrogen) atoms within a structure, together with bond information that describes the exact manner in which the individual atoms are linked together. An example of a 2D structure diagram and the corresponding connection table are shown in Figure 1.

A connection table provides a complete and explicit description of a

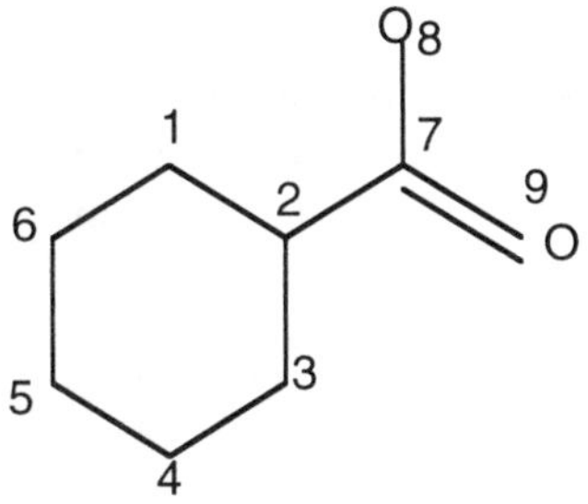

O8
1
7
9
2
6
O
3
5
4

Row	Atom	Nbr	Bond	Nbr	Bond	Nbr	Bond
1	C	2	rs	6	rs		
2	C	1	rs	3	rs	7	cs
3	C	2	rs	4	rs		
4	C	3	rs	5	rs		
5	C	4	rs	6	rs		
6	C	5	rs	1	rs		
7	C	2	cs	8	cs	9	cd
8	O	7	cs				
9	O	7	cd				

Figure 1 Example 2D structure diagram and connection table. r refers to ring; c to acyclic; s to single; and d to double. Nbr refers to neighboring atom.

molecule's topology, and as such can be regarded as a *graph*.[43] This is a mathematical construct that describes a set of objects, called *nodes* or vertices, and the relationships, called *edges* or arcs, that exist between pairs of the objects. Formally, we may say that a graph, *G*, consists of a set of nodes, *V*, together with a set of edges, *E*, connecting pairs of nodes. Two nodes are referred to as *adjacent* if they are connected by an edge. A *labeled graph* is one in which labels are associated with the nodes and/or edges, and a 2D connection table can hence be considered as a labeled graph because the atoms and bonds of the connection table correspond to the nodes and edges, respectively, of a graph. A *subgraph* of *G* is a subset, *P*, of the nodes of *G* together with a subset, *F*, of the edges connecting pairs of nodes in *P*. Two graphs, *A* and *B*, are said to be *isomorphic* if there is a mapping between the nodes of *A* and of *B* such that adjacent pairs of nodes in A are mapped to adjacent pairs of nodes in B. A *common subgraph* of two graphs *A* and *B* consists of a subgraph *a* of *A* and a subgraph *b* of *B* such that *a* is isomorphic to *b;* the *maximal common subgraph* is the largest such common subgraph.

The equivalence between a labeled graph and a connection table means that it is possible to process connection tables, and hence databases of 2D chemical structures, using algorithms that were originally designed for testing isomorphisms between pairs of graphs.[44–46] Thus, *structure search,* which involves the search of a database for the presence or absence of a specified query molecule (e.g., when there is a need to retrieve the physical or chemical data associated with some compound or to determine the novelty of a molecule for purposes of patent protection) is the chemical equivalent of graph isomorphism (i.e., the matching of one graph against another to determine whether there is a one-to-one mapping between them). *Substructure search* involves the search of a database for all molecules that contain some specified query substructure irrespective of the environment in which it occurs (e.g., to identify all molecules that contain a pharmacophoric pattern that has been identified in a molecular modeling study). This is the chemical equivalent of subgraph isomorphism (i.e., the matching of two graphs to determine whether one is contained within the other). Finally, the maximal common subgraph between two graphs is one of the definitions of similarity that are discussed in the main body of this review.

Substructure searching has been one of the main driving forces in the development of chemical information systems over the years: it is described in outline in the remainder of this section to provide a context for the subsequent discussions of similarity-searching methods. A detailed review of current approaches to substructure-searching techniques is provided by Barnard.[47] Subgraph isomorphism is known to belong to a class of computational problems, known as *NP-complete* problems, for which no efficient algorithms are believed to exist[48,49] and its equivalence to substructure searching means that the latter is extremely demanding of computational resources when large databases need to be searched. It is conventionally effected by a two-level search procedure, in which an initial *screening search* precedes the detailed and time-consuming subgraph-isomorphism search, which is normally referred to as *atom-by-atom searching*. That said, the reader should note that other types of substructure-searching system are becoming available that operate in a very different manner, and the two-level approach described here may become less important in the future.[47]

Screening and Atom-by Atom Searches

The screening search identifies those molecules in a database that match the query at the *screen* (or *fragment*) level, where a screen is a substructural feature, the presence of which is necessary, but not sufficient, for a molecule to contain the query substructure. These features are typically small, fragment substructures that are algorithmically generated from a connection table. The screen search involves checking each of the database structures for the presence of those screens which are present in the query substructure; molecules that contain all of the query screens then undergo the detailed atom-by-atom search. The time-consuming nature of the latter search means that the overall efficiency of a substructure-searching system is crucially dependent on the *screenout,* that is, the fraction of the database that is eliminated by the screening search. There has accordingly been considerable interest in the development of algorithmic techniques for the selection of fragment screens that will give high screenout.[50–52] Many different types of fragment screens have been reported for substructure searching, and we now give an overview of these because, as will be seen below, such fragments also form the basis for many operational systems for similarity searching and for clustering. Some examples of typical fragment types are shown in Figure 2.

Much of the early work on automatic generation and selection of structural fragments for use as screens was conducted by Lynch and his coworkers.[50,53] Hierarchies of atom-centered, bond-centered, and ring-centered fragments were investigated, at varying levels of specificity. From these, fixed-size dictionaries of fragments were selected, based on the relative frequencies of occurrence of the fragments in representative subsets of the database that was to be searched. This frequency-based approach to the selection of fragments for substructure searching was adopted and extended, initially by the BASIC group of companies[54] and subsequently by CAS,[3] to produce standard dictionaries containing a few thousand screens that were sufficiently discriminating to permit effective substructure searching in databases containing millions of structures. The fragment types considered included augmented atoms and their derivatives, linear sequences of atoms (with their associated atom types and bond types) three to six atoms in length, ring counts and types, atom counts, element composition, and graph-modifier fragments. These fragment types are all of predefined size; a different approach for substructure searching, based on two concentric levels of successive atoms around a chosen starting atom, is used in the DARC system.[55] Here, the resultant Fragments Reduced to an Environ-

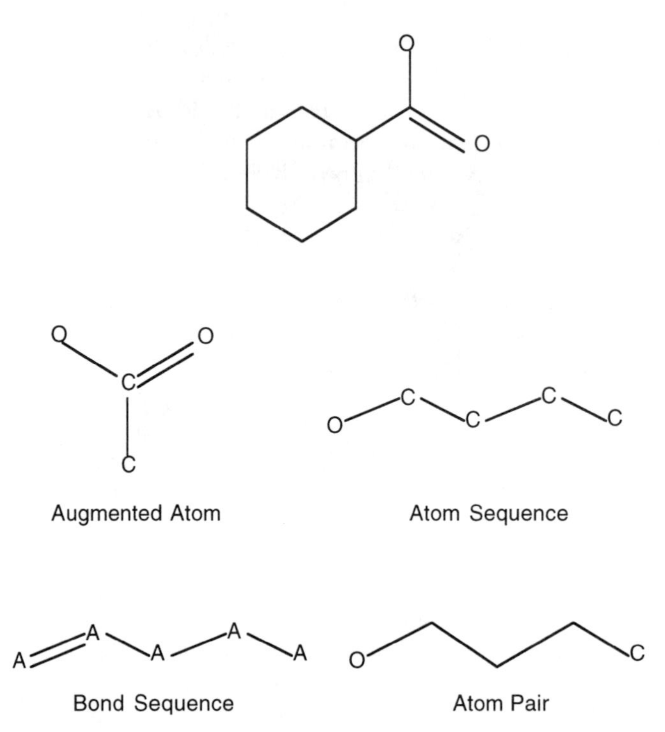

Figure 2 Example 2D structure diagram and some of its associated 2D structural fragments.

ment which is Limited (FRELs) provide a more flexible, open-ended set of descriptors than can be achieved using fixed-size fragment dictionaries.

Several types of fragment have been developed for screening programs associated with large-scale structure–activity analysis, although they can also be used for substructure searching. Thus, instead of a fixed dictionary of relatively small fragments, the system developed at the National Cancer Institute[56,57] involves supplementing single-atom fragments by exhaustively generated larger fragments. These are bond-centered fragments that extend to a further concentric level of bonds if these bonds are other than carbon–carbon single or ring-alternating. Again, Lederle Laboratories have developed two classes of fragment for structure–activity applications. The *atom pair* fragment[58] is an interesting mix of specific and generalized detail. Each pair of atoms in a molecule is given, along with the length of the shortest bond-by-bond path between them. Both atoms have labels to denote their atom type, the number of adjacent nonhydrogen atoms, and the number of incident bonds. The *topological torsion* fragment[59] is a linear sequence of four adjacent nonhydrogen atoms, each labeled in the same way as the atom pairs, so that the fragment is the 2D analog of the conventional 3D torsional angle. In a similar manner, Klopman[60] generates the set of all linear sequences of three to twelve adjacent nonhydrogen atoms, with additional labels for each atom denoting a number of incident bonds and the presence of certain adjacent terminal groups.

Once the required fragment types have been generated and, if necessary, appropriate members selected, the assignment of screen numbers can be conducted in several different ways. The result of such assignment is often represented by a *bit-string,* a binary string with each bit denoting the presence or absence of a specific screen in a structure. The approach adopted by Lynch's research group directly assigns each fragment to its own screen number to create a screen dictionary. Alternatively, the screen dictionary for CAS Online allocates several fragments, which are generally structurally related, to the same screen number. Predefined dictionaries, however, can contain only a limited number of screens and require careful selection procedures to ensure optimal use. A more flexible approach, especially when a database is growing and new classes of molecules need to be accommodated, involves the indirect assignment of fragments via some form of hash-coding or by folding a much larger bit-string into a smaller one of the required length.[61] [*Hashing* is an algorithm that pseudorandomly maps one number into another number that is in a specific range, so as to make the first number more compact and efficient for storage and retrieval. *Folding* is a method for making a shorter bit-string from a long sparse one by combining bits using, e.g., a Boolean combination of the mth and $(n + 1 - m)$th bits in a string of length n until the density of ones exceeds a predefined threshold.] An example of this bit-string superimposition approach is provided by the *fingerprints* that are used for substructure and similarity searching in the Daylight chemical information system.[62] Here, all linear sequences containing up to seven bonds are generated exhaustively and then hashed to four or five bit positions in a bit-string, which may also be folded. The use of indirect assignment means that a given bit in the bit-string can denote the presence of more than one, structurally unrelated, fragment, and that the presence of a given fragment may be represented by more than one bit in the bit-string, whereas the simpler, dedicated bit-string approach involves a one-to-one mapping of fragments to bit locations and vice versa.

All database structures, and query substructures, are analyzed by the same automatic fragmentation and screen-assignment procedures. The screening stage of a substructure search is then effected by matching the bit-string representing a query substructure against the bit-strings representing each of the molecules in the database to identify those molecules that contain all of the fragments that have been specified in the query. The second stage, atom-by-atom search, is carried out for those few molecules that match the query at the screen level. Whereas screen searching can be implemented very rapidly indeed, with many thousands of structures per second being matched against a query, atom-by-atom searching is very time-consuming, with only a few tens of structures per second being processed by the subgraph-isomorphism algorithm that lies at the heart of any substructure-searching system. The NP-complete nature of subgraph isomorphism[5,45] has resulted in the development of sophisticated heuristics to minimize the number of query-atom to database-atom mappings that need to be considered, while still ensuring the retrieval of all of the matching compounds.[63–65]

Representation and Substructure Searching of 3D Chemical Structures

Connection Table Representation of 3D Structures

The last few years have seen substantial interest in the development of database-searching systems for files of 3D structures. The atom coordinate information in such databases is obtained either experimentally, usually from the Cambridge Structural Database, which contains X-ray structures that have been reported in the published literature,[66] or by the use of a structure-builder, which is a computer program that calculates an approximate 3D structure from a 2D connection table.[67,68] The best known of such programs is CONCORD,[69] which uses a knowledge base of rules that describe preferred molecular conformational patterns and a simplified force-field. CONCORD has been widely used for the conversion of both in-house[70] and public[71] databases of 2D structures to 3D form.

It was soon recognized that the graph-theoretic principles outlined above could also be used for the representation and searching of 3D molecules.[72] As we have seen, the nodes and edges of a 2D chemical graph represent the atoms and bonds, respectively, of a molecule; in a 3D chemical graph, the nodes and edges are used to represent the atoms and interatomic distances, respectively, so that the graph is, in fact, an interatomic distance matrix. The edge labels in a 3D chemical graph are thus real numbers, that is, distances in Å, rather than the integer bond labels that denote the edges of a 2D chemical graph. It should be noted at this point that although a connection table forms the basis for all of the searching operations that are discussed below, the actual representation that is stored is usually just the atomic coordinates, since these take up very little space (and thus minimize input/output costs) but can be converted to an interatomic distance matrix extremely rapidly. Similar comments apply to the bounded-distance matrix representation of a flexible structure that is described later.

Screening and Geometric Searching Algorithms

Given a graph-based representation, 3D substructure searching can be effected using screening and subgraph-isomorphism procedures that are analogous to those that we have described previously for searching databases of 2D structures.

Screens for 3D searching typically consist of a pair of atoms, together with an associated *distance range* that represents the separation of the two atoms. For example, a screen might consist of a nitrogen atom separated from an oxygen atom by a distance of between 5.40 and 6.87 Å, and a molecule that contained these two atoms separated by a distance of 5.50 Å would be assigned this screen, inter alia, whereas a distance less than 5.40 Å or greater than 6.87 Å would result in the assignment of an alternative nitrogen-oxygen screen. The screens are identified as the result of a statistical analysis of the distribution of interatomic distances in the molecules in a sample of the database that is to be searched.[52,73,74] Analogous range-based screens can be used to define the distances and/or the angles between sets of three or four atoms.[75–78] Molecules that match a query in the screening search are then passed on for processing in the second-stage, subgraph-isomorphism search, which is normally referred to as *geometric searching* in the 3D case to distinguish it from conventional 2D atom-by-atom searching. It is simple to modify atom-by-atom searching algorithms so that they can be used for geometric searching. The primary modification that is needed is to permit the mapping of edges from the graph representing the query pharmacophore only to those edges from a database structure that represent the same distance, to within any allowed tolerance; in the 2D case, conversely, the condition for two edges to be equivalent is that they represent the same type of bond.

Flexible 3D Substructure Searching

Thus far, we have considered only rigid structures and have taken no account of the conformational flexibility that characterizes very many 3D molecules. The representation of a molecule by an interatomic distance as described above assumes that the molecule exists in just a single low-energy conformation, and it is thus possible for a 3D substructure search to miss many structures that could adopt a conformation containing the query pattern. *Flexible-searching systems* encompass the full conformational space spanned by a flexible molecule and are based on graph models analogous to those used for 2D and for rigid 3D molecules. In a rigid 3D molecule, the distance between each and every pair of atoms is a single, fixed value, whereas the distance between a pair of atoms in a flexible molecule depends on the particular conformation that is adopted. The separation of a pair of atoms here is described by a *distance range,* the lower and upper bounds of which correspond to the minimum possible and maximum possible separations. The set of distance ranges for a molecule will contain all of the geometrically feasible conformations which that molecule can adopt, and a flexible molecule can hence be represented by an interatomic distance matrix in which each element contains the appropriate distance range, rather than just a single distance as in rigid 3D searching.

Martin et al.[4] suggested that sets of distance ranges could be generated using the bounds-smoothing technique that forms one of the main components of distance geometry.[79] Clark et al.[80] developed this idea by noting that the resulting bounded distance matrix can be regarded as a graph in which the nodes are the atoms and the edges are the interatomic distance ranges, and that such matrices can thus be searched using algorithms analogous to those used for rigid 3D substructure searching. Indeed, it is possible to view rigid-searching algorithms merely as a limiting case of the more general algorithms that are required for flexible searching. There is, however, one major difference between flexible 3D and both 2D and rigid 3D substructure searching, in that those molecules that match the query in the geometric search must then undergo a further, and final, check that uses some form of conformational-searching procedure.[81,82] This final check is required because bounds-smoothing is known to overestimate the true range of possible interatomic distances (because it takes no account of correlation effects) and because it gives no information regarding the energies of the possible conformations. There is much current interest in the development of efficient techniques for the implementation of this final stage of a flexible 3D substructure search.[6–8]

Limitations of Substructure Searching

Substructure searching, whether in 2D or in 3D, provides an invaluable tool for accessing databases of chemical structures. It does, however, have several limitations that are inherent in the retrieval criterion that is being used, which is that a database record must contain the entire query substructure in precisely the form that it has been specified by the user (subject to any variations that have been specified, e.g., a distance tolerance or a range of acceptable substituent types for a specific position on a ring system).

Firstly, and most importantly, a substructure search requires that the user who is posing the query must already have acquired a well-defined view of what sorts of structure are expected to be retrieved from the database. For example, a 3D substructure search requires sufficient information about the geometric requirements for activity to be able to specify distance and/or angular constraints to characterize those molecules, and just those molecules, that can fit some putative receptor site. This implies that some form of *pharmacophore map* has been created, using techniques such as those described by Bures et al.,[83] which in turn implies that it has already been possible to identify sufficient active substances to generate the map. Substructure search is much less appropriate at the start of an investigation when perhaps only one or two active structures have been identified and when it is not at all clear which particular feature(s) within them are responsible for the observed activity.

Secondly, there is very little control over the size of the output that is produced by a particular query substructure. Without a detailed knowledge of the contents of the file, the searcher will be unable to predict a priori how many database structures will satisfy the structural constraints defined by a given query. Even in the case of a 2D substructure search of an in-house file, the specification of a common ring system or of several possible substituents at a particular location (or locations) can result in the retrieval of several thousands of structures (unless it is also possible to specify other constraints such as additional functionality or some specific range of values for a physiochemical property). Such problems will be magnified many times over if an analogous search is carried out on CAS Online, which now contains some 12 million 2D structures, or on a 3D database, where search outputs are known to be far larger than in 2D searching, especially when flexible searches are carried out.[6–8] Unless tools are available for processing the large search outputs that may be expected (such as the clustering approaches that are described later), the user will either have to consider only some small sample of the output or reformulate the query in the hope that a second search will result in the retrieval of a more manageable number of structures.

The third point to note is that a substructure search results in a simple partition of the database into two discrete subsets, namely, those structures that contain the query and those that do not. All of the retrieved records are thus presumed to be of equal usefulness to the searcher, and there is no direct mechanism by which they can be ranked in order of decreasing utility, e.g., in order of decreasing probability of activity. In fact, the problem is still worse, in that not only can one not differentiate between retrieved structures, but also one cannot differentiate between different parts of the query, since a substructure search implicitly assumes that all parts are of equal importance. In other words, a substructural query implies that all features contained within it have a weight of one (or some other constant value) and that all other possible features (e.g., the screens in a screening system that have not been assigned to that query) have a weight of zero.

One should not overestimate the scale of these problems, since substructure searching has provided an effective and popular way of accessing chemical databases for many years. That said, these inherent limitations have occasioned interest in similarity searching as a complement to substructure searching. Similarity searching requires the specification of an entire target structure, rather than the partial structure that is required for substructure searching. The structures in a database are then ranked in order of decreasing similarity with the target structure. Specifically, a similarity search compares a set of structural characteristics of the target structure with the corresponding sets of characteristics for each of the database structures, calculates a measure of similarity between the target and each database structure based on the degree of resemblance of these two sets of characteristics, and then sorts the database structures in order of decreasing similarity with the target. The output from the search is a ranked list, where the structures that the system judges to be most similar to the target structure, which are usually referred to as the *nearest neighbors,* are located at the top of the list and are thus displayed first to the user. Accordingly, if an appropriate measure of similarity has been used, the first database structures inspected will be those that have the greatest probability of being of interest to the user.

The use of a ranking mechanism serves to alleviate many of the inherent limitations of substructure searching. Thus, one needs just a single structure to act as the target for a similarity search (e.g., one that has shown activity in a primary biological screen), rather than a query pharmacophore that has been derived from careful analysis of many active structures. Next, the ranking of the database allows complete control over the amount of output that needs to be inspected (since the user can search down the list just as far as is needed). Thirdly, it is very easy to take weighting information into account when calculating the degree of similarity between the target structure and the database structures. Perhaps the main characteristic of similarity searching, and the one that commends itself strongly to end users, is that (given a sufficiently rapid speed of response) it encourages a browsing-like approach to database access, in marked contrast to the tightly defined queries that are required for substructure searching.

SIMILARITY SEARCHING IN DATABASES OF 2D STRUCTURES

Introduction

Having outlined the basic rationale for the development of similarity searching, we shall commence our review of the available approaches by discussing two of the characteristics that can be used to evaluate the merits of a particular similarity measure. They are introduced here with reference to measures of 2D similarity but are equally applicable when considering the 3D measures that are described later.

At the risk of some oversimplification, it is possible to divide all structure-based similarity measures into two classes. *Global similarity measures* are measures that yield a single numeric value for the overall similarity between two molecules without any information being provided as to the parts of the two molecules that are identical, or nearly identical, to each other. *Local similarity measures* provide such locational information and thus result in an *alignment* of one molecule with another, that is, they yield a mapping of features in the target structure to features in a database structure sufficient to superimpose one upon the other. The most obvious basis for such an alignment is the constituent atoms of the two structures, but other types of alignment may also be of importance, such as the overlapping charge clouds that characterize many of the type-A similarity measures that were mentioned in the Introduction. The majority of similarity measures that have been used with 2D structures are global measures.

A further characteristic of a measure that must be considered is its appropriateness, since, as was mentioned in the Introduction, the definition of similarity is application dependent. When very large databases need to be processed, molecular descriptions are required that must not only provide an appropriate characterization of a structure but also be simple enough to be compared rapidly. Thus, a similarity measure must be *effective,* that is, its application must result in measures of similarity that are meaningful and useful to the user, and must also be *efficient,* that is, have acceptable computational requirements, if it is to be used as a similarity-searching tool in databases of nontrivial size. It is often the case that there is an inverse relationship between efficiency and effectiveness, and the designer of a similarity procedure may thus need to consider some degree of tradeoff between the conflicting requirements of speed and utility. The situation is closely related to that encountered by the designers of substructure search systems, where only a small number of all possible descriptors can be used, and careful selection of appropriate descriptors is necessary to minimize the number of structures that need to undergo the atom-by-atom search.

Having introduced these two principal characteristics of similarity measures, we shall now focus upon the three main components of a measure. These are[84] the structural descriptors that are used to characterize the molecules, the weighting scheme that is used to differentiate more important features from less important features, and the similarity coefficient that is used to quantify the degree of similarity between pairs of molecules.

Structural Descriptors

Structural-fragment and pairwise-distance (atom-pair) descriptors have already been reviewed in some detail earlier, since these are the most commonly used descriptors for type-C similarity applications. In many cases, indeed, the same fragment-based, bit-string descriptors are used for both substructure searching and similarity searching. The use of the maximal common substructure (MCS) as a descriptor is outlined in the section on The Upjohn Company (and more fully in the context of 3D similarity searching in MCS methods described later). The two remaining categories of structural descriptors that need to be mentioned are *topological sequences* and *topological indices.*

Topological sequences are exemplified by the path numbers described originally by Randic and Wilkins.[85] The path number sequence for a structure is constructed by generating all self-avoiding interatomic paths and then assigning the total numbers of paths of each length corresponding elements of the sequence. The path number sequence thus starts with element zero (corresponding to the number of atoms in the structure) and, unless arbitrarily truncated, finishes with the element corresponding to the longest path length in the structure. Structures with approximately the same distribution of path lengths tend to be structurally similar. Several improvements to the basic path number sequence have been suggested,[86] the most important being to reduce the effect of different sizes when comparing structures by normalization of the sequences or by weighting in favor of the less frequent path lengths. Weighted path numbers can be used as a sequence or they can be summed, or otherwise processed, to produce a single-valued topological index.

The aim of descriptor selection is one of data reduction: to reduce the infinite variety of potential descriptors to those most appropriate to a given application. The calculation of a topological index can be viewed as the ultimate form of data reduction, since it typically yields a single real or integer value to characterize the structure of a molecule, with the expectation that structurally similar molecules will have similar index values. This is often found to be the case for congeneric series of molecules, where very many high-quality structure–property correlations have been obtained with a range of types of topological indices. A very large number of topological indices have been published, but only a few have been shown to have general applicability. Three such "classical" indices are outlined below.

The first topological index to be published was the Wiener index, *W*, which is half the sum of the bond-by-bond path lengths between each pair of atoms.[87] The Wiener index has some interesting mathematical properties and can correlate well with chemical properties such as melting point and viscosity. It can be calculated from each off-diagonal element of the path length matrix, D_{IJ}, of a structure:

$$\frac{1}{2} \sum_{I,J} D_{IJ}$$

The value is largest for linear structures and smallest for compact structures (e.g., multibranched and/or cyclic molecules).

The Balaban index is also calculated from the distance matrix of a structure and is otherwise known as the average distance sum connectivity index.[88] Each distance sum, D_I, is the sum of the elements of the *I*th row of the distance matrix. The index is normalized by the numbers of bonds, *B*, and rings, *C*, and is calculated from:

$$\frac{B}{C+1} \sum_{I,J} 1/\sqrt{D_I \times D_J}$$

where *I* and *J* are neighboring nonhydrogen atoms. The index increases with structure size and degrees of branching and unsaturation, but decreases with the number of rings.

By far the most widely used indices are the molecular connectivity indices, χ, which were originally formulated by Randic[89] and subsequently generalized and extended by Kier and Hall.[90] Information about size, branching, unsaturation, heteroatoms, and numbers of rings are all included in an extensive series of indices. Each atom is given two descriptors, δ_i (the number of adjacent nonhydrogen atoms) and δ_i^v (the number of valence electrons less the number of adjacent hydrogen atoms). Connectivity indices for a wide variety of path, cluster, path/cluster, and cycle fragments are all calculated by multiplying the descriptor values for each atom in the fragment. The reciprocal square roots are summed over the structure to give the resultant index value. Hall and Kier[91] have also introduced molecular shape indices, κ, which are based on counts of all the 1-, 2-, and 3-bond fragments in a structure, and topological state indices, T_i, which are based on all paths leading from each atom to every other atom.[92] More recently they have added electrotopological state indices, S_i, which are based on the topological state plus intrinsic electronic state differences between atoms.[93] Some examples of topological indices are illustrated in Figure 3.

The large number of topological indices that can be calculated for a structure raises the question as to which should be used in a particular application, especially as it can be difficult to relate many of the higher-order indices to

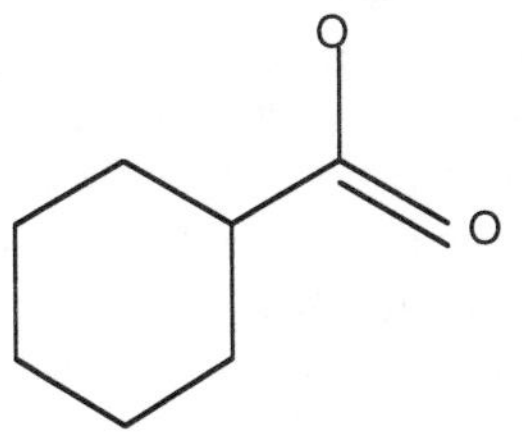

Weiner, W, number = 88

Simple Molecular Connectivity, χ, up to order 4:

	0	1	2	3	4
Path	6.6902	4.3045	3.6421	2.5926	1.8008
Cluster	–	–	–	0.5000	0.0000
P/C	–	–	–	–	1.0404
Chain	–	–	–	0.0000	0.0000

Valence Molecular Connectivity, χ^v, up to Order 4:

	0	1	2	3	4
Path	5.4683	3.5329	2.6732	1.9544	1.3495
Cluster	–	–	–	0.1970	0.0000
P/C	–	–	–	–	0.4021
Chain	–	–	–	0.0000	0.0000

Shape, κ, up to Order 3:

0	1	2	3
–	7.1111	3.2397	2.0000

Topological State, T_i, for the OH group = 8.0989
Electrotopological State, S_i, for the OH group = 8.5382

Figure 3 Example 2D structure diagram and some associated topological indices (calculated using the MOLCONN program of Kier and Hall).

any specific structural features. Structure–activity applications typically consider a number of different indices and then use a multivariate technique such as multiple regression to identify those that are most highly correlated with the activity of interest.[90]

Basak et al.[94] have published one of the few studies on the use of topological indices to determine similarity for large databases. Using the U.S. Environmental Protection Agency's TSCA database of 25,000 structures, they took a subset of 3,692 structures and calculated no fewer than 90 topological indices for each, these encoding size, shape, bonding type, and branching pattern information. Many of these indices were highly correlated, and so principal components analysis was used to reduce them to 10 principal components, which explained 92.6% of the variance and which were used as descriptors in similarity searching of the whole database. The resultant top-five hits for a selection of randomly chosen queries were as good as one might expect from a more traditional structural-fragment similarity search (of the sort that is discussed later). The distinction between functional groups was, however, poorer than one would expect from a structural-fragment approach.

It is important to emphasize that there are very many more types of topological index than those mentioned here and that all of them could be used for similarity searching, either on their own (with two molecules being judged to be structurally similar if they had similar index values) or in combination with other indices (as in the procedure of Basak et al.[94]) However, the great bulk of the work that has been done on 2D similarity searching has been based on the use of fragment substructures, with two molecules being judged to be structurally similar if they have many fragment substructures in common, and we shall focus upon the fragment-similarity approach in the remainder of this section.

Weighting Schemes

It is natural for chemists, when looking at chemical structure diagrams, to regard certain features as more important than others. In so doing, they are, in effect, giving greater weight to those parts that they deem more important, such as functional groups. One of the unresolved questions in implementing similarity searching systems is whether to give greater weight to certain descriptors, so that a match between two compounds on a highly weighted feature makes a greater contribution to the overall similarity than a match on a less important feature.

A trivial example of weighting is provided by the use of similarity measures based on structural-fragment dictionaries, because if a specific fragment that is present in the two molecules is in the dictionary, then it has a weight of one (and a weight of zero if it is not present in the dictionary). More sophisticated approaches to the weighting of fragment substructures have been investigated by Willett and Winterman,[95] who compared three types of statistical information that can be used for the calculation of fragment weights, using simulated property-prediction experiments with a number of small data sets from the structure–activity literature. The types of information studied were as follows: the frequency of occurrence of the fragment in a structure, with greater weight given to fragments occurring more frequently; the frequency of occurrence of the fragment in the database, with greater weight given to fragments occurring less frequently; and the number of fragments present in a structure, with greater weight given to fragments in structures with fewer fragments. Willett and Winterman found that the best results were obtained by taking account of the frequency with which fragments occurred within a molecule, although the effect of including such information was not very large. Conversely, Moock et al.[96] found that search performance was enhanced by using weights inversely proportional to the frequency of occurrence in the database, and this feature was implemented in MDL Information System's REACCS systems. The work of Moock et al. was in the context of reaction similarity searching, whereas structure similarity searching experiments on several large data sets by Downs et al. suggested that such inverse-frequency information was not important.[97] At the National Cancer Institute, the fragments are assigned weights derived from their frequency of occurrence in the molecule, their size, and their frequency of occurrence in the database.[57]

The subsimilarity search method developed at the Upjohn Company,[98] which is discussed later, measures the similarity between a pair of molecules by the *maximal common substructure* (MCS), that is, the largest substructure that is common to the two molecules that are being compared, as illustrated in Figure 4. The system utilizes an inverse-frequency weighting scheme that favors heteroatom–heteroatom pairs over the more frequently occurring carbon–carbon pairs. For each pair of heteroatoms separated by minimum paths of two to six bonds, an additional bond is added to link the atoms directly. This additional link results in an increase in the size of the MCS and thus a greater similarity for heteroatom-containing common substructures.

Closely related to weighting is *standardization,* which involves a rescaling of the variables in a multivariate analysis to ensure that all of them are measured on the same scale and that one, or a few, of them do not dominate the overall similarities.[99] Many different approaches to standardization have been discussed in the literature.[100] Bath et al.[101] evaluated the use of seven of these with fragment-based similarity measures but concluded that their application did little to improve performance in simulated property prediction experiments.

Although several examples of the use of weighting schemes in similarity searching have been given above, the majority of fragment-based similarity searching implementations do not use weighting. The main reasons are the rather equivocal nature of the evidence supporting their use and the increased computational requirements (for storing or calculating the weights when very large numbers of compounds need to be searched).

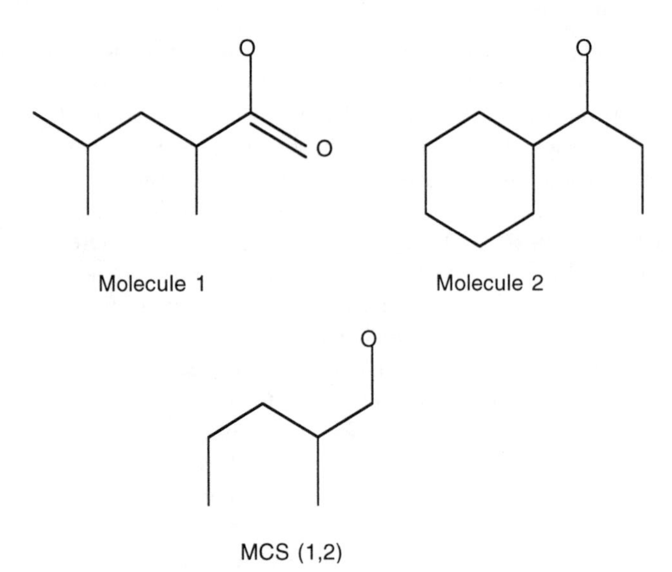

Figure 4 Example of a Maximal Common Subgraph for the two molecules shown above.

Similarity Coefficients

A definitive account of similarity coefficients is provided by Sneath and Sokal.[99] They describe four main classes of similarity coefficients: distance, association, correlation, and probabilistic coefficients. The most common distance measure is the Euclidean distance, and many nearest-neighbor searching algorithms use this to measure the degree of resemblance between pairs of objects. For two points I and J, the Euclidean distance is given by

$$\sqrt{\sum_{K} (x_{IK} - x_{JK})^2}$$

with the summation over all of the descriptors, K (and similar comments apply to the two following formulas in this section). The Euclidean distance is usually the measure of choice when processing real-number data (such as fragment weights or property data).

A study by Willett and Winterman[95] suggested that the cosine, Tanimoto, and correlation coefficients gave better results than distance measures for measuring the similarity using fragment bit-string descriptors and that, of these, the Tanimoto coefficient gave the most intuitively acceptable rankings of molecules. The *Tanimoto coefficient* is an example of an association coefficient[99] and is given by

$$\frac{\sum x_{IK} \times x_{JK}}{\sum x_{IK}^2 + \sum x_{JK}^2 - \sum x_{IK} \times x_{JK}}.$$

This form of the coefficient encompasses both binary and nonbinary variables.[102] Willett et al.[103] suggested that the Tanimoto coefficient was the most appropriate for similarity searching in large 2D databases, and it has now become the coefficient of choice for most operational similarity-searching systems, where the fragment-occurrence data are represented by the bit-strings that are normally used for substructure searching. In this case, a simplified form of the Tanimoto coefficient can be used. Assume that a target molecule and a database molecule have T and D nonzero bits, respectively, in their fragment bit-strings and that C of these are in common. Then the Tanimoto coefficient is given by

$$\frac{C}{T + D - C}$$

The cosine coefficient is also frequently used and is assumed for certain implementations of group-average clustering (as discussed later). The cosine coefficient is given by

$$\frac{\sum x_{IK} \times x_{JK}}{\sqrt{\sum x_{IK}^2 \times \sum x_{JK}^2}}.$$

The product–moment correlation coefficient is widely used in bivariate data analysis to measure the extent of the correlation between two variables, but it is not clear that it is necessarily appropriate to measure the extent of the similarity between two objects.[99] Other sorts of correlation coefficient are available, such as the Spearman rank correlation coefficient which has been used by Manaut et al. as a measure of electrostatic similarity,[14] but these have not found extensive application in similarity searching systems. Similar comments apply to probabilistic coefficients, which are calculated from the frequency distribution of descriptors in a database, and which Adamson and Bush[104] found to give poor results when applied to 2D chemical structures.

Examples of 2D Similarity-Searching Systems

The first two operational 2D similarity-searching systems to be described[58,103] both employed fragment-based, global similarity measures, and this has become the method of choice for the great majority of subsequent systems, as discussed by Bawden[105] and by Barnard and Downs.[106] We shall illustrate this approach here by reference to the facilities for similarity searching provided in the MACCS and REACCS products of MDL Information Systems;[107,108] that said, the reader should note that analogous facilities are provided by other commercial software products (such as those produced by the Cambridge Crystallographic Data Centre, Daylight Chemical Information Systems, and Tripos Associates, inter alia) and by in-house chemical information systems [such as those developed at Lederle Laboratories[58] and at Pfizer Central Research (U.S.)[103]]. We then describe the systems that are under development at CAS[76,109–111] and at the Upjohn Company[98] because they illustrate the application of more sophisticated structural representations and matching algorithms than are required for the simple, fragment-based approach that underlies most current systems for 2D similarity searching.

MDL Information Systems

The MACCS and REACCS database packages produced by MDL Information Systems include similarity-searching routines that are based on atom-pair structural fragments and the Tanimoto coefficient.[108] Specifically, the similarity between a pair of molecules is determined using a set of 933 fragments (called keys), and the similarity between a pair of reactions is determined using the same molecule fragments and an additional 230 fragments that characterize the reaction centers (and their immediate environments) in the reactant and product molecules of a chemical reaction. In addition to conventional similarity searching, these packages also provide two additional types of similarity searching: subsimilarity searching and supersimilarity searching. Assume that a target (query) molecule and a database molecule have T and D non-zero bits, respectively, in their fragment bit-strings and that C of these are in common (as was employed in the definition of the Tanimoto coefficient given earlier). Then the *subsimilarity* is given by C/T, whereas the *supersimilarity* is given by C/D.

By definition, any target structure that is a substructure (or superstructure) of the database molecule will have a subsimilarity (or supersimilarity) of 1.0, and hence a subsimilarity (supersimilarity) search can be viewed as a form of "fuzzy" substructure (superstructure) search.[107] A further difference from the systems described thus far is in the nature of the output. The description of similarity searching that has been given previously assumes that a user browses down a ranked output until sufficient database structures have been identified. Here, conversely, the user specifies a threshold similarity value, and the output consists of an unordered list of all database structures (or database reactions) that have a similarity to the target that exceeds the chosen threshold value.

Grethe and Hounshell[107] provided several examples of the use of the three types of similarity searching. One interesting application involves searches of the MACCS-II Drug Data Report, a database of investigational compounds with known biological activity, for novel structures with potential phosphodiesterase-inhibiting cardiotonic activity. The suggested approach involves carrying out similarity searches with several known drugs as the target compounds and then intersecting the resulting hit-lists to identify molecules that are structurally related to all of the known drugs and that may hence also be expected to exhibit the activity of interest. These authors emphasize the complementary natures of the outputs that are provided by substructure, similarity, subsimilarity, and supersimilarity searching.[108] The reaction similarity-searching facilities in REACCS are exemplified by searches for the well-known Michael reaction, in which donor and acceptor structures can have a variety of functional groups. In a substructure search of the *Journal of Synthetic Methods (JSM)* REACCS database, 155 hits were retrieved, of which less than half were actual Michael reactions. An equivalent similarity search retrieved 65 hits, most of which were Michael reactions, with a variety of differently substituted educts and reaction conditions, and there were also several "Michael-like" reactions.

Chemical Abstracts Service

The CAS Online system has offered facilities for structure and substructure searching of the millions of molecules in the Chemical Abstracts Service (CAS) Registry File for many years.[3] An ongoing project is evaluating techniques that could provide complementary facilities for similarity searching, or *fuzzy-match searching,* as Fisanick et al. refer to their work.[76,109–111] Three types of data have been identified as appropriate for the implementation of similarity searching: the 2D structure, the 3D structure, and molecular property data. The 2D structures are those for the ca. 12 million compounds in the Registry File, and ca. 4.6 million of these have 3D structures generated using CONCORD.[69] The molecular properties are global features that have been calculated using standard computational chemistry routines, principally the semiempirical molecular orbital package MOPAC,[112] and programs written at CAS that implement published property-prediction algorithms. The initial experiments involved a set of approximately 6000 molecules, each of which is characterized by features generated from all three types of data.

Research on 2D similarity searching has concentrated on the use of selected subsets of the existing 2D substructure screen fragments of the CAS Online screen dictionary (as discussed earlier). The idea is to enable a user to get different "similarity views" by selecting the required fragment types. For instance, the selection of augmented atoms and atom sequences gives a general view of the structural relationships between pairs of molecules, whereas the selection of ring composition and type of ring gives a ring-centered view of these relationships. This research suggests that further analysis into mixed descriptor types (e.g., including topological indices in addition to the fragment types) could give users an even more flexible approach to similarity searching. One aspect that is already under investigation is a second stage based on *reduced graphs*[113,114] to calculate a connectivity-based measure of similarity. Unlike substructural fragments, reduced graphs retain the topological relationships between areas of a molecule and are thus more capable of providing a local measure of similarity than the simpler, first-stage fragment approach, which is inherently global in nature. CAS has developed a hierarchical system of reduced graphs[115] for its generic search capability and has now shown their potential for refining the results from a conventional, fragment-based similarity search. Currently, the reduced graphs are used in a direct comparison between query and retrieved database structures, but work is under way to define an appropriate similarity measure to enable refinement based on similar reduced graphs. The work at CAS on 3D and property-based similarity searching is discussed later.

The Upjohn Company

The Upjohn Company's COUSIN database system has a conventional similarity search based on 2D fragment screens; however, this is complemented by a sophisticated routine for subsimilarity searching.[98] A fragment-based definition of subsimilarity searching has been given earlier; a more sophisticated version of this definition has been adopted by the Upjohn group, who employ the MCS, as illustrated in Figure 3, as the similarity measure. The MCS is the largest set of atoms (or bonds) from the target structure that can be superimposed exactly (or within user-defined tolerances if interatomic distances are being employed in a measure of 3D similarity) onto the database structure. In the Upjohn implementation, once the MCS between a target structure and a database structure has been identified (by means that are discussed later), the subsimilarity between these two structures is defined to be the number of bonds in the MCS divided by the number of bonds in the target structure.

The maximal matching set of atoms or bonds in an MCS can be identified by means of a maximal common subgraph isomorphism algorithm. Historically, however, MCS algorithms have been little used for similarity searching in 2D databases, owing to their substantial computational requirements. Specifically, although both the subgraph and maximal common subgraph isomorphism problems are NP-complete, MCS detection is far more time-consuming in practice, because there appear to be few strong constraints available for reducing the number of matches that must be investigated in a subgraph isomorphism search. Moreover, there is no screening procedure that can be used to eliminate the great bulk of the definite non-hits prior to the MCS search, whereas this can be done prior to an atom-by-atom search (as described earlier). The Upjohn system overcomes these efficiency problems by means of a two-level strategy in which an initial, fragment-based search enables the calculation of an upper bound to the size of the MCS that can be obtained in the second-stage MCS search.[98]

In the initial search, the target structure and the database structures are represented by their constituent bonded-atom fragments, for which there are as many per structure as there are bonds in that structure. The number of such fragments in common between a target structure and database structure, together with the number in the target structure, permits the calculation of an upper bound to the size of the MCS between them using the fragment-based subsimilarity formula given earlier. The database structures are then sorted into decreasing order of the subsimilarity upper bounds to reduce the number of matches required (as discussed in detail later); further reductions in the time of the MCS search are achieved by allowing the user to specify threshold values for the minimum and maximum subsimilarity coefficients, the maximum number of database structures that are to be retrieved, or the maximum permissible size of the MCS. The computational requirements of the time-consuming second-stage search are minimized by the use of an approximate MCS algorithm that can occasionally result in the identification of submaximal common substructures, although the results presented by Hagadone[98] suggest that this will occur in less than 1% of the MCS searches that are carried out. Subsimilarity searches on a 57,000-compound database require about 1 CPU minute on average using an IBM 3090-400J processor, although a few searches take considerably longer.

The linking of fragment-based and MCS-based searching provides an extremely powerful mechanism for subsimilarity searching (and could clearly be extended to both similarity and supersimilarity searching); with developments in computing power, it is likely that such systems will become widely used over the next few years.

CLUSTERING DATABASES OF 2D STRUCTURES

Introduction

The general effectiveness and efficiency of the fragment-based approaches to 2D similarity searching have encouraged the development of clustering procedures that group molecules on the basis of the similarities between pairs of molecules. Accordingly, although the principal focus of this review is similarity searching, this section contains an introduction to the use of clustering methods for processing chemical databases. Standard textbooks on cluster analysis include those by Everitt,[116] Gordon,[117] and Sneath and Sokal,[99] while Willett[84] provides an extensive discussion of the application of such methods to databases of 2D chemical structures. Two more recent reviews of this subject are provided by Barnard and Downs[106] and by Downs and Willett,[118] who focus on the use of clustering methods in drug-discovery programs.

There are at least three reasons why one might wish to apply cluster analysis to a large chemical database. The simplest approach is to cluster the output resulting from a substructure search that retrieves too many molecules, to enable the searcher to gain a rapid overview of the extent of the structural heterogeneity in the retrieval list. Second, one may wish to cluster an entire database so that individual clusters can be retrieved and examined by more computationally expensive procedures; this can be particularly appropriate for structure–activity techniques that require homogeneous subsets of molecules from the main database. Finally, there has been substantial interest in the use of techniques to cluster an entire database so that representative molecules can be selected from each cluster for input to biological screening programs (with the aim of maximizing the diversity of the structures that are tested).

Clustering the outputs of 2D substructure searches has been reported by Willett et al.[119] and by Barnard and Downs,[106] but has otherwise not received much attention. This situation may well change with the increased emphasis on 3D substructure searching, which typically results in the retrieval of far more hit structures than does 2D substructure searching, even if attention is restricted to databases of rigid 3D structures. Clustering such output on the basis of 2D descriptors could provide not only a useful overview of the results but also a complementary view to that resulting from the original search.

Clustering an entire database to provide homogeneous subsets suitable for structure–activity studies relies on the *similar property principle,*[35] which states that structurally similar molecules will exhibit similar properties. The extensive studies reported by Willett and co-workers (see, e.g., Refs. 84, 95, 119) have provided substantial evidence of the general validity of this principle when the similarities are measured using 2D fragment bit-strings. This supporting evidence has led to the widespread use of clustering procedures for compound selection in biological screening programs, such as those carried out at the National Cancer Institute,[57] Pfizer Central Research (U.K.),[119] the Upjohn Company,[98,120] and the European Communities Joint Research Centre.[121] Cluster-based compound selection has the following advantages[118]: clustering is a cheap, automatic alternative to the time-consuming and expensive use of highly trained staff; an effective clustering procedure can help to ensure that no classes of compounds are overlooked when selecting structures for testing; and parameterized clustering permits the creation of different sets of structures to suit different screening requirements.

There are many different types of clustering methods,[99,116,117] and there have been several evaluations of these for use in a chemical context, as we now discuss.

Hierarchical Clustering

Agglomerative Methods

Hierarchical clustering methods produce small clusters of very similar molecules nested within successively larger clusters of less similar molecules. Visualization of the resultant hierarchy remains an unsolved problem for databases of more than a few hundred molecules.

The widely used *agglomerative* clustering methods or, more formally, the *Sequential Agglomerative Hierarchical Nonoverlapping* (SAHN) methods[99] generate a classification in a "bottom-up" manner, by a series of agglomerations in which small clusters, initially containing individual molecules, are fused together to form progressively larger clusters. The agglomeration process may be represented by a *dendrogram,* as illustrated in Figure 5.

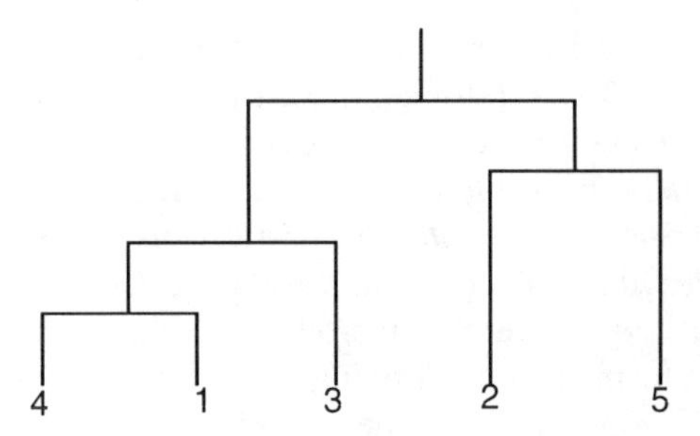

Figure 5 An example of an SAHN clustering dendrogram. Molecules 1 and 4 are the most similar. They fuse, at a lower similarity, with molecule 3. Molecules 2 and 5 then fuse, and finally they all fuse at the top.

Murtagh[122] classifies SAHN techniques into *graph-theoretic* and *geometric* methods. The fusion step in a graph-theoretic method requires the calculation of similarities between individual members of two clusters, whereas a geometric method allows the contents of a cluster to be represented by a single point. Members of the two classes include the *single-linkage, complete-linkage, weighted-average,* and *group-average* methods (graph-theoretic methods), and the *centroid, median,* and *Ward* methods (geometric methods). There is an extensive literature associated with the comparison of SAHN methods (see, e.g., Refs. 123–126), and such comparisons have also been carried out in the context of clustering chemical data sets. For example, Adamson and Bawden[127] studied the groupings resulting from the application of five SAHN methods to a set of 36 benzene derivatives and suggested that the methods were so different that several of them (specifically, the single-linkage, group-average, and Ward methods) should be used in combination; however, although this might be appropriate for the analysis of small data sets, such a procedure would be infeasible if the objective were to cluster an entire database. Accordingly, subsequent comparative studies have focused on the identification of individual methods that provide a reasonable tradeoff between the effectiveness of the clusters resulting from a method and the computational requirements of the algorithm that implements the method.[84]

All of the SAHN methods can be implemented using a standard *stored matrix* approach, in which an initial intermolecule similarity matrix is progressively updated as new clusters are formed by the fusion of pairs of molecules and/or clusters of molecules. The intercluster similarities are calculated using the Lance–Williams matrix-update formula[128]

$$d_{k(i,j)} = \alpha_i d_{ki} + \alpha d_{kj} + \beta d_{ij} + \gamma |d_{ki} - d_{kj}|$$

where $d_{k(i,j)}$ is the similarity measure between point K and a point (I, J) formed by the fusion of the points I and J. The various SAHN methods differ only in the values for the four parameters α_i, α_j, β, and γ, thus facilitating their implementation by a single program. However, for a database of N structures, the stored-matrix approach has space and time complexities of $O(N^2)$ and $O(N^3)$, respectively, which make it totally infeasible for processing large databases.

Murtagh[122] discusses the *reducibility property,* which enables agglomerations to be done in restricted areas of the similarity space and the results then to be amalgamated to form the overall hierarchy of relationships. Such agglomerations can be performed by identifying all *Reciprocal Nearest Neighbors* (RNNs). The RNN approach has time and space complexities of $O(N^2)$ and $O(N)$, respectively, which are sufficiently low to suggest that methods based on the RNN algorithm could be used to cluster chemical databases of nontrivial size. Of the most commonly used geometric methods, Ward's method satisfies the reducibility property. The RNN algorithm is not generally applicable to graph-theoretic methods, but Voorhees[129] has shown that the group-average method also satisfies the reducibility property if the cosine coefficient is used as the similarity coefficient. As a consequence, the group-average and Ward methods can be implemented using the same efficient algorithm. Both of these have been shown to perform well in chemical[84,127] and non-chemical comparisons of SAHN methods,[122–126] and some initial experiments have confirmed that the RNN algorithm allows their use with large chemical databases.[130] It will hence be of interest to see whether this approach will be taken up as a complement, or as an alternative, to the nonhierarchic Jarvis–Patrick method that is discussed later.

Divisive Methods

Hierarchical *divisive* methods generate a classification in a top-down manner, by progressively subdividing the single cluster that represents the entire data set. Most such methods are *monothetic* in character, in that the division at each stage in a classification is based upon the value (or the presence or absence) of a single descriptor from among those that are used to characterize the objects that are being classified.[131] The monothetic divisive methods are generally much faster in operation than the agglomerative methods, which are *polythetic* in nature (i.e., they take account of all of the descriptors when deciding which of the objects should be grouped together within a cluster) and which give noticeably better levels of performance in comparative studies.[84]

One of the few successful polythetic divisive methods is the minimum-diameter hierarchical divisive method of Guenoche et al.,[132] which has time and space requirements of $O(N^2 \log N)$ and $O(N^2)$, respectively. This is rather too high for processing very large databases, such as an in-house corporate structure file, but is feasible for data sets containing a few thousand structures. We have recently shown that this method gives superior results to the Ward and group-average SAHN methods, and particularly to the Jarvis–Patrick non-hierarchical method, when clustering molecules using calculated property descriptors.[130] In this comparative study, 13 of a set of 29 computed molecular properties used at CAS (as discussed further later) were standardized and used as descriptors for the four clustering methods, the effectiveness of which were evaluated by means of simulated property-prediction experiments (as has been the case with most of the previous comparative studies that have been carried out in Sheffield, e.g., Refs. 119, 131).

Nonhierarchical Clustering

Nonhierarchical methods generate individual partitions rather than complete hierarchies of partitions and hence tend to be far less demanding of computational resources.[116,117] The most efficient nonhierarchical methods have time and space complexities of $O(MN)$ (at most) and $O(N)$, respectively, where M is the number of clusters generated. That said, the inverse relationship between efficiency and effectiveness that we have noted previously applies here, and the most effective of the nonhierarchical methods tend to be comparable in efficiency to the hierarchical methods.

Single-Pass Methods

The simplest nonhierarchical methods are the *single-pass* methods. These are very rapid in operation, since they require, as their name suggests, only a single pass through the data set to assign molecules to clusters. The major drawback is that the resultant clusters are often dependent on the order in which the molecules are processed. An example of a single-pass method is that which has been developed for compound selection at the National Cancer Institute.[57] The Leader algorithm method has been optimized for use with the weighted structural-fragment descriptors that were described earlier and utilizes a weighted form of the asymmetric coefficient,[102] which has been found to give better results than the more widely used Tanimoto coefficient. The method is used to cluster many hundreds of thousands of molecules and typically results in very large numbers of small clusters. Many of these are *singletons* (unclustered molecules) that are markedly dissimilar to the other molecules in the database and that are hence worthy of consideration for biological testing. An interesting characteristic of this work is that the algorithm has been implemented so that it produces overlapping clusters, where an individual molecule can be assigned to more than one cluster. All of the other procedures discussed here result in nonoverlapping classifications in which each molecule belongs to just a single cluster.

Relocation Methods

Nonhierarchical *relocation* methods, such as the *k-means* method,[133] start with a user-defined number of *seed* clusters and iteratively reassign molecules between clusters to see if better clusters result, where the success of the solution is measured by some parameter such as the ratio of the mean intercluster similarity to the mean intracluster similarity.[117] The methods can be quite effective in operation[135] but have several inherent limitations, in that they are crucially dependent on the seed clusters that are used to initiate the classification, and they can either fail to converge or terminate in a local minimum that corresponds to a markedly suboptimal solution.

Nearest-Neighbor Methods

Nonhierarchical *nearest-neighbor* methods assign structures to the same cluster as some user-defined number of their nearest neighbors. Of the several available methods, the Jarvis–Patrick[136] method has been highly successful and widely adopted for clustering large databases of molecules represented by structural fragments.[119] The method has two main stages. In the first, each molecule in a database is used in turn as the target structure in a similarity search to identify its K nearest neighbors, that is, the K structures that are most similar to it (typically using a bit-string comparison and a subsequent Tanimoto calculation). In the second stage, molecules are clustered together if they are in each other's nearest-neighbor lists and if they have some minimal number, K_{min}, of their nearest neighbors in common. The groupings resulting from a particular clustering run depend upon user-supplied values for K and K_{min}, this parmeterization providing some degree of control over the precise nature of the classifications that are produced by the method. The generation of the nearest-neighbor lists dominates the computation, which has time and space requirements of $O(N^2)$ and $O(N)$, respectively. The availability of a fast inverted-file algorithm for the calculation of the similarity coefficients[137] means that the method can be used with very large numbers of structures. The method was first tested for compound selection from large databases at Pfizer Central Research (U.K.)[119] and has subsequently been implemented within many other companies and within commercial software systems (as exemplified by the clustering routines in the Daylight[62] and CAVEAT[75] systems). Despite some shortcomings, such as its tendency to produce a highly disparate distribution of cluster sizes unless appropriate parameter values are used, the general acceptance of the Jarvis–Patrick method has enabled users to become familiar with the potential of clustering methods for database processing. This should make it easier for other, more computationally demanding clustering methods to be accepted when the development of appropriate software and/or hardware techniques enables their implementation on a large scale.

Dissimilarity Approaches

A similarity search provides a ranking of the structures in a database in order of decreasing similarity with the target structure; by analogy, a *dissimilarity search* provides a ranking in order of increasing similarity so that the first structures seen by the user are those that are maximally different (as defined by the particular similarity measure that is being used) from the target structure. Both Pfizer[105] and Upjohn[120] have implemented retrieval options that are based on dissimilarity searching, where the dissimilarity measure is simply one minus the Tanimoto coefficient value; an overview of this work is provided by Bawden.[138]

The group at Upjohn[120,139–141] has made extensive use of similarity and dissimilarity methods for the selection of structural representatives from a large database. This work is exemplified in a study by Lajiness,[120] who compared three methods of selecting compounds for biological screening, which were referred to as simple cluster-based selection, dissimilar cluster selection, and maximum dissimilarity selection. The first of these, simple *cluster-based selection,* is the standard way of selecting compounds for screening in which a nonhierarchical clustering method, typically the Jarvis–Patrick method, is used to partition the database into a reasonable number of homogeneous clusters. One compound is then chosen from each cluster, this representative structure either being selected at random or being the closest to the cluster centroid. No account is taken of compounds that might have been used previously in biological screens, and so there is no guarantee that the selected set of compounds will differ substantially from previous selections. *Dissimilar cluster selection* overcomes this problem by simply eliminating clusters that contain any compounds that have been tested previously in the biological screen(s) of interest and selecting one compound from each of the remaining clusters in the same way as simple cluster-based selection. In contrast to these cluster-based methods of selection, the *maximum dissimilarity method* tries to select a set of compounds that are as dissimilar to each other as possible in a single pass of the data set that is to be processed. The identification of the maximally dissimilar set of structures is computationally infeasible; instead, an initial compound is selected at random. Then, at each stage in the processing, that next compound is selected for which the sum of the similarities to all previously selected compounds is a minimum (an analogous algorithm has been described by Bawden).[138] Although this cannot be expected to identify an optimally dissimilar set of compounds, the procedure is found to work reasonably well in practice. Comparison of the results of these three methods, and also random selection, showed that the dissimilar-cluster and maximum-dissimilarity methods are most suitable for the selection of compounds in screening programs. This being so, it will be of interest to see whether other organizations adopt dissimilarity-based, rather than similarity-based, approaches to computer-assisted compound selection.

SIMILARITY SEARCHING IN DATABASES OF 3D STRUCTURES

Introduction

It will be clear from the material presented earlier that 2D similarity searching (either in its basic form or in its subsimilarity or supersimilarity forms) has become a standard retrieval option in chemical information systems and one that provides an effective complement to conventional substructure searching. With the increasing availability of facilities for substructure searching in databases of 3D structures,[142] much research is now under way to develop techniques that will allow similarity searching in such databases. This work is at an early stage, with many different types of 3D similarity measure under active investigation, and there is little sign, thus far, that any one of these will become the standard method of choice, whereas fragment-based measures rapidly established themselves as the basis for nearly all 2D similarity searching systems. In the latter case, the requisite data were already available in the shape of the bit-string screen records that were used for substructure searching, and the rapid development of efficient nearest-neighbor searching algorithms meant that it was simple to move to large-scale operational use once the effectiveness of screen records for similarity searching was demonstrated.[84]

The situation is very different in the case of 3D structures, because there is a divergence of views as to what should constitute a measure of 3D similarity, as is evidenced by the range of measures that have been suggested for the measurement of similarity in QSAR studies. Moreover, as has been noted earlier, many of these measures are type-A or type-B measures, and it will thus not be possible, in the near future at least, to develop implementations that will allow them to be used for searching large databases. The precise natures of these measures differ substantially (as was briefly discussed in the Introduction), but they generally seek to characterize the steric, electrostatic, or hydrophobic fields that are the most important determinants of biological activity.[143] Thus far, the only type-C measures that have been described use distance, angular, or molecular property descriptors, all of which can be generated from a set of 3D atomic coordinates at little computational cost, and all of which permit the use of matching algorithms that are sufficiently rapid to permit the searching of large databases of 3D structures.

We have noted previously that similarity measures can be either local or global in character, depending upon whether they involve the generation of an alignment of the target structure with a database structure, and we shall use this characteristic to provide an initial subdivision of the 3D similarity measures that have been reported to date in the literature. The next section also considers the extension of these measures to encompass flexible 3D molecules and the subject of *docking*, which is probably the most important application, thus far, of 3D similarity measures in the design of novel bioactive compounds.

Global Measures for 3D Similarity Searching

Comparison of Distance-Based Similarity Measures

Most of the 3D similarity measures that have been described involve the use of interatomic distance information and do not take conformational flexibility into account, other than through storing a compound in the database in multiple conformations. The first detailed study of distance-based measures for 3D similarity searching was reported by Pepperrell and Willett.[144]

Given two molecules, A and B, the simplest measure that was tested by Pepperrell and Willett was the *distance-distribution measure.* Here, an N-element frequency distribution is generated for each molecule, in which the Ith element contains the number of interatomic distances in that molecule that lie within the Ith distance range ($1 \leq I \leq N$). The resulting frequency distribution encodes all of the distances within a structure, and hence a comparison of the distributions for two structures provides a measure of their shape similarity. Such an approach takes no account of the elemental types involved in each distance; this information is contained in the second measure suggested by Pepperrell and Willett, which they call the *individual-distances measure.* This is, in essence, the 3D analog of the fragment-based measures described earlier, but with the screen occurrence data being replaced by interatomic distance information. The individual-distances method involves generating the constituent sets of interatomic distances, in the form of the pair of atoms and the distance separating them, for a target structure and a database structure, and then carrying out a Tanimoto calculation based on the numbers of distances that are, and are not, common to the two molecules. Given two atoms of elemental types a_1 and a_2, separated by a distance of d, a common distance is one that involves the same two elemental types and has a distance of $d \pm t$, where t is some user-defined tolerance (such as 0.5 Å).

Both the distance-distribution and individual-distances measures are examples of global similarity measures in that they give a figure for the overall similarity between two molecules without any information being provided as to the parts of the two molecules that are identical, or nearly identical, to each other. The remaining two measures discussed by Pepperrell and Willett were local similarity measures: the *atom-mapping* measure and the *3D MCS* measure. These are discussed in a later section, which focuses on local measures for 3D similarity searching.

The effectiveness of the four 3D similarity measures was compared by means of simulated property prediction using data sets from the QSAR literature for which both 3D coordinates (specifically, CONCORD structures[69]) and biological activity data were available.[144] The experiments suggested that the atom-mapping and MCS measures gave the best correlations between structure and property and that they were of about the same level of effectiveness. However, they differ substantially in terms of computational efficiency. MCS detection is known to belong to the class of NP-complete computational problems,[48,49] whereas a complexity analysis of the atom-mapping method shows it to have an expected time scaling of $O(N^3)$ for the matching of two structures, each containing N atoms.[145] It was accordingly concluded that atom-mapping was the most cost-effective of the methods that had been tested and was thus chosen for the operational implementation that is discussed in detail later.

Bemis–Kuntz and Lederle Algorithms

The measure reported by Bemis and Kuntz[146] involves the decomposition of a structure into all possible three-atom substructures [so that a molecule containing N heavy atoms yields $N(N-1)(N-2)/6$ such substructures]. The procedure represents a molecule by a frequency distribution, the 64 elements of which are initially set to zero. A three-atom substructure is selected, and the interatomic distances are calculated over all of the pairs of atoms in the chosen substructure. Let these three distances be n_1, n_2 and n_3; then an increment of one is made to that element of the distribution that corresponds to the sum $n_1^2 + n_2^2 + n_3^2$. This summation is repeated for all of the possible three-atom substructures, and the resulting frequency distribution acts as the input to a hash-coding routine that produces a single numeric descriptor characterizing the overall shape of a molecule. The similarities between pairs of molecules can then be determined with great efficiency by the similarity between the corresponding hash codes. Bemis and Kuntz note that an entirely comparable approach can be used to calculate measures of 2D similarity using path lengths, rather than Euclidean distances, in the construction of the frequency distributions (which results in a similarity measure that is closely related to the topological-sequence and topological-index procedures that have been described earlier).[85–94] They report an extended series of experiments to compare the two types of measure with each other and with an RMS-fitting procedure derived from the DOCK program,[147] which is used to dock ligands into macromolecular receptor sites and which is described in detail in the section on docking.

Unfortunately, the use of a hashing procedure results in meaningful similarities only for pairs of molecules that have the same (or very similar) molecular formulas,[146] which reduces the attractiveness of this approach as a general mechanism for 3D similarity searching. Bath et al. noted that this restriction can be removed if the original frequency distributions, rather than the hash codes that are generated from them, are used to calculate intermolecular similarities.[148] If the distributions are used, then a rapid similarity search can be effected by matching the distribution for a target structure against the distribution for each of the molecules in a database, using a goodness-of-fit criterion to determine the similarity of the two distributions (and hence of the corresponding structures). Such an approach is clearly analogous to the distance-distribution measure described in the previous section, with the obvious difference that the latter method considered only pairs of atoms, rather than the more discriminating three-atom substructures studied by Bemis and Kuntz. Bath et al. also noted that Bemis and Kuntz's work could be generalized to accommodate n-atom substructures, rather than just three-atom substructures, and they then described experiments with n set to 2, 3, and 4. They found that the best results were given by the most discriminating substructures, that is, those containing four atoms.[148]

Nilakantan et al. at Lederle Laboratories have described a similarity measure that is closely related to the Bemis–Kuntz measure, in that it is based on the three interatomic distances for a set of three nonhydrogen atoms.[149] Their procedure is in two stages. In the first stage, each set of three atoms, which they refer to as an *atom triplet,* is used to generate an integer that characterizes the interatomic distances within the triplet. Specifically, the three interatomic distances, n_1, n_2, and n_3, for a given atom triplet are calculated and sorted into increasing order. Assume that the distances are such that $n_1 \leq n_2 \leq n_3$; then the integer code is calculated as $n_1 + 1{,}000 \times n_2 + 1{,}000{,}000 \times n_3$. The integers resulting from the first stage are used as the seeds for a hashing procedure that addresses individual bits in a bit-string representation of molecular shape. The similarity between a target structure and a data set structure is obtained by comparing their bit-strings and then calculating a similarity coefficient based on the number of bits in common, in a manner that is comparable to that used for fragment-based 2D similarity searching.

Nilakantan et al. discuss several applications of their approach. The first involved comparing the rankings resulting from an atom-triplet search of a 22,495-compound subset of the Cambridge Structural Database with the rankings from a DOCK search[147] of the same subset using a complex of HIV-protease with the inhibitor MVT-101[150] as the basis for the two searches. This comparison revealed that the simple atom-triplet program had been able to encompass at least some of the detailed shape information that is manipulated in DOCK. The second approach involved merely taking the top-ranking atom-triplet compounds and then docking them, either manually or automatically, onto a target receptor site. This was investigated using a data set published by Grootenhuis et al.,[151] who have discussed techniques for the discovery of drugs that bind to the minor groove of DNA using the netropsin-B-DNA crystal structure as a prototype.[152] Similarity-searching experiments with netropsin as the target structure showed that many of the high-ranking compounds identified by Grootenhuis et al.[151] were also identified by the atom-triplet searches. Modifications to, and other applications of, the atom-triplet approach to 3D similarity searching are described by Nilakantan et al.[149]

Use of Angular Information

The 3D similarity measures described thus far are all based on interatomic distance information. Bath et al.[148] have described two measures for 3D similarity searching that are based on angular information and have compared the effectiveness of searches using these measures with those of the atom-triplet measures described in the previous section. It should be noted that all of these angular and atom-triplet measures make use only of geometrical information, that is, distances and angles between the constituent atoms of the pairs of molecules that are being compared, and do not involve chemical information such as atomic type, polarity, hydrophobicity, etc.

Given a set of four atoms, *ABCD,* a *torsional angle,* ω, is the angle between the two three-atom planes, *ABC* and *BCD*. It describes the twist of the vector *AB* relative to the vector *CD* when viewed along the vector *BC*. A torsional angle is very well suited to 3D searching, because it provides 3D information, unlike a valence angle, which is 2D, or a bond, which is 1D. Indeed, the importance of torsional angles has led to attempts to characterize them by means of 2D similarity descriptors.[59]

A torsional angle is normally defined in terms of sets of four connected atoms. Poirrette et al.[78] have discussed the concept of a *generalized torsional angle,* in which it is not mandatory for the atoms to be connected. Let the presence or absence of a bond between two of the four atoms comprising a torsional angle be characterized by the letters "B" or "N," respectively. A conventional, fully bonded torsional angle can then be described as a BBB torsion. The 3D similarity measures of Bath et al.[148] are based on BNB torsions and NBN torsions. A BNB torsion is one in which there are bonds between the two outer pairs of atoms forming the four-atom set, that is, *AB* and *CD,* but in which there is no bond between the central pair of atoms, *BC* (or, for that matter, between *A* and *D*). An NBN torsion is one in which only the two central atoms of the four-atom set are bonded, that is, *BC* in our example. These, and other, types of generalized torsional angle have been evaluated for the creation of force fields by Weiner et al.,[153] and for 3D substructure searching by Poirrette et al.,[78] Lauri and Bartlett,[75] and Fisanick et al.[111]

Bath et al.[148] describe a BNB torsion, *ABCD,* by an integer code that encompasses the arithmetic mean of four quantities: the two intervector angles *ABC* and *BCD,* the absolute value of the torsional angle, ω, and the interatomic distance for the nonbonded central atoms, *B* and *C.* No account is taken of the different units or magnitudes of these quantities. A molecule can be characterized by generating the integer codes for all of the BNB fragments in that molecule, and the sets of codes for two molecules can then be used for the calculation of a Tanimoto similarity analogous to that defined for 2D fragment data. The NBN measure is calculated in a similar manner, by generating integer codes for all of the possible NBN fragments in a molecule. The code here encompasses the absolute value of ω and the sums of the distances of the triangle sides *ABD* and *ACD*. No account is taken of the central separation in a fragment, that is, the distance *B–C,* because this bonded distance varies little in comparison with the other distances in the quadrilateral, *ABCD,* that defines an NBN torsional angle.

The effectiveness of the BNB and NBN measures was assessed by simulated property-prediction experiments. These experiments involved the QSAR data sets studied previously by Pepperrell and Willett[144] for the evaluation of distance-based similarity measures and a large set of 6-deoxyhexopyranose carbohydrates, which had previously been classified into 14 shape classes using numerical clustering methods based on torsional dissimilarity coefficients. The comparison encompassed the Bemis–Kuntz and Lederle measures, including not just the atom-triplet but also the atom-pair and atom-quadruplet versions of the former measure. The results were equivocal, in that it was impossible to identify any one measure as consistently giving the best level of performance, although the four-atom measures tended to be superior to the others.[148]

Research at Chemical Abstracts Service

In the earlier section on the Chemical Abstracts Service (CAS), we gave a brief overview of the range of similarity measures that are under investigation by Fisanick et al. and described the novel descriptors that these workers have developed for similarity searching in databases of 2D structures. Their studies have also involved both geometric and property-based 3D similarity measures: in all, a molecule is characterized by a maximum of no fewer than 3172 different features.

Fisanick et al.[76] identify eight classes of 3D features, some of which are shown in Figure 6. For each class, the features are generated for all of the structures in a sample file containing about 6000 substances in the shape of CONCORD structures. The frequencies are cumulated and the range of values for the chosen feature is identified; for example, the range of distances corresponding to the separation, in Ångstroms, between a pair of heteroatoms in a structure. A binning procedure is used to identify appropriate subranges within the total range for a feature type, and a molecule is then characterized by the number of occurrences of a feature of that type within each such range. With the sole exception of the 3-Bonded Atoms Angle Bin Counts (see below), all the bin ranges are identified using equifrequency procedures analogous to those used to generate screen sets for 2D and 3D substructure searching.[52,53,73,74]

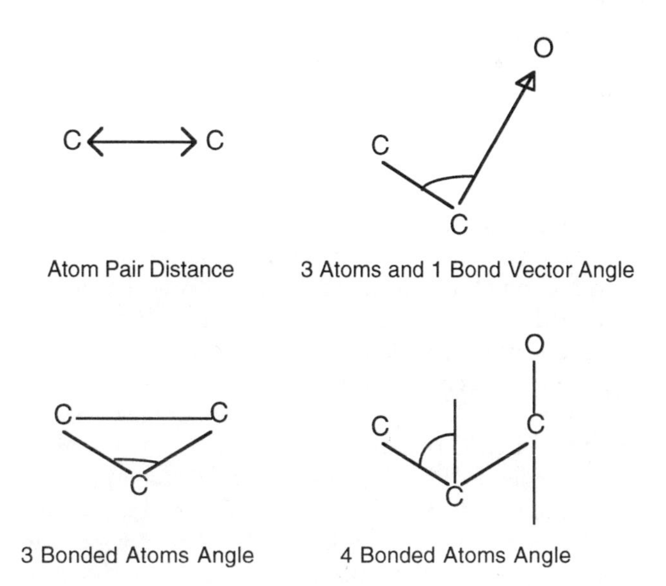

Figure 6 Examples of the classes of 3D fragments studied by CAS (Reference 76).

The features describe interatomic distances, valence, and torsional angles (both normal and generalized as discussed by Poirrette et al.[77,78]), and various geometric characteristics of triangles of three atoms.

There is one class of Distance Bin Count, this being the Atom-Pair counts for X–X, C–C, C–Het, and Het–Het pairs of atoms (where C, Het, and X denote carbon, any heteroatom, and any atom, respectively). Next, there are four classes of Angle Bin Counts. The first, 3-Bonded Atoms, are counts for bonded atom triplets involving C, Het, and X atoms. The bins have been selected to encompass the common valence angles (i.e., 109°, 120°, and 180°). 4-Bonded Atoms are counts for the torsional angle between sets of four atoms (C, Het, and X) that are bonded in a chain, that is, BBB torsions in the nomenclature of Bath et al.[148] 3 Atoms and 1 Bond Vector are counts for atom triplets (C, Het, X, and H, where H is hydrogen) where two atoms are bonded and the third atom can be any other atom in the structure. Only one hydrogen is permitted. 4 Atoms and 2 Bond Vectors are counts for BNB torsional angles (with C, Het, X, and H). Finally, there are three classes of Atom Triangle Bin Counts. Atom Triangles and Related Metric Bin Counts are counts for triangle atoms (C, H-donor, H-acceptor, Het, Nitrogen, Oxygen, Het other than nitrogen or oxygen, and X) and related metrics such as the area and the perimeter of the triangle comprising the three atoms. Atom Triangle 3-Slot Bin Counts are counts for the first two ordered atoms in a triangle and their binned interatomic distance. Atom Triangle 5-Slot Bin Counts are counts of the first two ordered atoms in a triangle and the ordered set of three binned interatomic distances in the triangle.

It should be clear from this brief description that the geometric arrangement of the components of an individual 3D molecule is described in very great detail. An example of the features identified in a typical structure is discussed by Fisanick et al.[76] Two measures are available to determine the degree of similarity between the target structure and a database structure. In the "presence/absence" method, a feature is deemed to be in common if it occurs with a nonzero frequency in both molecules, and the similarity is then the percentage overlap of the database-structure features with the target-structure features. The more discriminating "statistical" method involves calculating the standard deviation for a particular bin count across the whole database. The similarity between the target structure and a database structure is incremented by 1, 2, 3, or 4 if the count for a specific feature in the database structure is within 1.0, 0.75, 0.5, or 0.25 standard deviations, respectively, of the corresponding count for the target structure. This matching procedure is repeated for all of the counts to give the overall similarity. Fisanick et al.[76] have also described an alternative, but rather more complex, version of this similarity measure.

Visual inspection of the top-ranking structures from similarity searches using the triangle counts suggests that these features are sufficiently discriminating to permit the detection of both shape and size similarity and that they result in the retrieval of substances that can be very different from those retrieved using 2D fragment similarities or similarities based on the interatomic distance counts. The shape and size of a molecule are related to its volume, and Fisanick et al. report some limited experiments in which an attempt is made to predict the molecular volume of the target structure. However, as the authors themselves note, further testing is required to identify the most appropriate ways of utilizing the wealth of information that is encoded in the many features that they describe.

The CAS workers have also calculated intermolecular similarities using global molecular properties. Here, each structure is represented by a total of 29 computed properties, including molar refractivity, log P, HOMO, LUMO, van der Waals surface area, dipole moment, heat of formation, and the mean atom charge, inter alia. The similarity between a target structure and a database structure is calculated using the "statistical" measure described above. The search can use all of the properties, a reduced set in which only the less highly correlated are considered, individual properties, or sets that are thought to reflect some particular aspect of molecular reactivity, such as the use of the various charge and electron-density features to allow electrostatic similarity searching. Initial experiments suggest that the most similar structures resulting from a database search do, indeed, closely resemble the target molecule.[76,109]

Local Measures for Distance-Based 3D Similarity Searching

The Atom-Mapping Method

The atom-mapping method provides a quantitative measure of the similarity between a pair of 3D molecules, A and B, that are represented by their interatomic distance matrices.[144,145,154] The measure is calculated in two stages. First, the geometric environment of each atom in A is compared with the corresponding environment of each atom in B to determine the similarity between each possible pair of atoms. The resulting interatomic similarities are then used to identify pairs of geometrically related atoms, and these equivalences allow the calculation of the overall intermolecular similarity.

Assume that the molecule A contains $N(A)$ nonhydrogen atoms. The distance matrix for A, $\mathbf{DA}$, is then an $N(A) \times N(A)$ matrix such that the IJth element, $DA(I, J)$, contains the distance between the Ith and Jth atoms in A (and similarly for the matrix $\mathbf{DB}$ representing the $N(B)$-atom molecule B). The first stage of the procedure compares each atom from A with each atom from B. Consider the Ith atom in A, $A(I)$; then the Ith row of the distance matrix $\mathbf{DA}$ contains the distances from I to all of the other atoms in A. This set of distances is compared with each of the rows, J, from the distance matrix $\mathbf{DB}$ to identify the matching distances. These common distances are used to calculate a Tanimoto coefficient, $S(I, J)$, that measures the similarity between $A(I)$ and $B(J)$. $S(I, J)$ is the IJth element of an $N(A) \times N(B)$ *atom-match matrix*, $\mathbf{S}$, that contains the similarities between all pairs of atoms, $A(I)$ and $B(J)$, from A and B. The interatomic similarities in *S* are then employed to establish a set of equivalences of the form $A(I) \Leftrightarrow B(J)$, where $B(J)$ now represents that atom in *B* that has the maximum value of $S(I, J)$ for all of the coefficients involving $A(I)$. A note is made of this maximal similarity for each such equivalence, and the overall similarity between *A* and *B* is then the mean of these maximal similarities when averaged over all of the atoms in *A*. This is clearly a global similarity measure, but the atom-mapping method also provides a local measure of 3D similarity, since it permits mappings to be made from atoms in *A* to atoms in *B*. Modifications of the basic procedure are discussed by Pepperrell et al.[154]

The computational requirements of atom-mapping are dominated by the calculation of the elements of the atom-match matrix. Rapid similarity searching is hence possible only if it is possible to eliminate many of the structures in a database from the detailed atom-mapping search, and Pepperrell et al.[145] have discussed the use of upper bound strategies for this purpose. Specifically, these workers showed that it was possible to calculate an upper bound to the atom-mapping similarity between a target structure and an individual database structure by means of a comparison of the corresponding molecular formulas, and that this procedure could eliminate from further searching up to ca. 80% of the database in a typical atom-mapping search. An implementation of these ideas is now used for similarity searches of the Zeneca Agrochemicals corporate database of ca. 180,000 structures. The algorithm is coded in C and runs on a UNIX workstation, with a response time for a typical search in the range of 20–60 minutes (the precise time depending primarily upon the size of the target structure). It is also possible to use hardware, rather than software, means to increase the speed of the algorithm, and a recent paper has discussed several ways in which atom-mapping can be implemented on massively parallel processors.[155]

The description thus far has assumed that atoms can be mapped to each other if, and only if, they are of the same elemental type. However, this restriction is not a necessary part of the algorithm, and it is hence equally possible for them to be mapped if they have comparable atomic properties.[145] Specifically, an atom can be described in terms of one or more of its hydrogen-bonding characteristics, its partial charge, and its van der Waals radius, with each atom in a molecule being assigned an integer class number signifying the particular value(s) of the chosen characteristic(s) that is (are) associated with that atom. It is also possible to specify weights so that the user can designate certain atoms as being of greater importance than others; for example, a greater degree of importance might be assigned to those atoms that can form hydrogen bonds than to those that cannot. If very high weights are assigned to some atoms, it is possible to obtain an output in which the top-ranked structures will contain all (or most) of the highly weighted atoms, so that one can execute a form of ranked 3D substructure search.

The SPERM Method

The SPERM (Superpositioning by PERMutations) program was developed by a group at Organon International[156,157] to provide a computationally efficient means of quantifying the degree of shape similarity between pairs of 3D molecules. SPERM derives from studies by Dean and co-workers[30,31] that position a molecule at the center of a sphere and characterize the molecule by a set of points on the surface of the sphere; the similarity between a pair of molecules is then determined by a comparison of the corresponding sphere surfaces. The shape of a molecule in SPERM is described by mapping a specified property onto the vertices of a tessellated icosahedron; this property is either the minimum distance or the radial distance from the vertex to the surface of the molecule. The precision with which the shape is described is determined by the extent of the tessellation. Perry and van Geerestein[157] report experiments that suggest that acceptable results will be obtained only if at least 162 vertices are used (whereas earlier work had suggested that as few as 32 vertices would be sufficient, these being the 12 vertices of an icosahedron and the 20 vertices of a dodecahedron oriented such that its vertices lie on the vectors from the center of the sphere through the midpoints of the icosahedral faces).[156]

A database structure is aligned with the target structure by translating and rotating the sphere representing the former molecule, and the dissimilarity between the two structures for this particular alignment is calculated from the root mean squared difference (RMSD) of the distances at the matched vertices of the two tessellated icosahedra. An improved version of an algorithm originally due to Bladon[158] is used to minimize the number of alignments that need to be evaluated in this way, and the output of the match of the target structure with a database structure is some number of the best RMSD values and the corresponding alignments. Optimization of the translational component of the alignment is approximated by aligning the geometric centers of the molecules that are being compared, this procedure being justified by a restriction of interest only to those database structures that are very similar to the target structure. The molecules in a database are ranked in decreasing order of the resulting similarities, and some number of the top-ranked molecules are then input to the similarity measure described by Hodgkin and Richards.[159] This is based on the superposition of a pair of electron-density grids and is about three orders of magnitude slower than the sphere-matching procedure; it is thus typically applied to just the top structures in the ranking resulting from the SPERM procedure. The orientation with the best Hodgkin–Richards score for a particular molecule is then taken to be the shape similarity for that molecule with the target structure.

Van Geerestein et al.[156] discuss the use of SPERM to search a 30,000-structure subset of the Cambridge Structural Database for molecules that are similar to the antitumor antibiotics netropsin and daunomycin, which bind to DNA[152] and which have also been studied by Nilakantan et al. as discussed in the earlier section on Lederle algorithms. The best-matching molecules from the search represented a wide range of structural types, included some known active compounds, and also suggested novel DNA-binding molecules. There was very little overlap either with 2D similarity searches using fragment bit-strings or with the outputs of DOCK[147] searches for these two target molecules on a smaller subset of the Cambridge Structural Database. The search of the 30,000 Cambridge structures took about 24 hours on a VAXstation 3100, although van Geerestein et al. noted that this could be much reduced by the use of screening and other heuristics with little effect on the quality of the final output. An evaluation of several such optimization procedures was reported by Perry and van Geerestein.[157] Thus, upper bound techniques were used to enable the termination of a match with a database structure at an early stage in the calculation if the dissimilarity became greater than that for the least similar of the nearest neighbors identified thus far in a search. In addition, a bit-string screening procedure was developed in which matches were only attempted if there was a high degree of match between the molecular volumes and the first three principal moments of inertia of the target structure and of a database structure.

The description above has assumed that the property that is stored at the vertices of the tessellated icosahedron describing a molecule is either the minimal or the radial distance to the molecular surface. However, it is equally possible to store at each vertex any sort of property that can be calculated for that point in 3D space, such as the molecular electrostatic potential.[156] Thus SPERM, like atom-mapping, allows local, property-based 3D similarity searching.

MCS Methods

We have already made several references to the use of the maximal common substructure, or MCS, as a measure of structural similarity. The computational requirements of MCS detection, and the efficiency and effectiveness of simple, bit-string approaches, have meant that there have been few studies of the use of the MCS as a similarity measure for 2D database systems. The only obvious exception to this is the subsimilarity system at Upjohn, which has been described earlier, and even this makes use of an approximate procedure that may result in the identification of less-than-maximal common substructures.[98] Efficient algorithms for 2D MCS detection do exist, such as that described by McGregor,[160] but their use has principally been for the generation of reaction databases, where there is a need to identify the largest structures common to the reactants and the products in a chemical reaction.[161] We have stated previously that 3D similarity searching is more demanding of computational resources than is 2D similarity searching; that being so, it is rather surprising that a fair amount of interest has been shown in the use of the MCS as a measure for 3D similarity searching, where the 3D MCS is defined as the largest set of atoms that have matching interatomic distances (to within user-defined tolerance values) in the two molecules that are being compared. The MCS thus represents the maximal geometric superimposition of one molecule onto the other.

Pepperrell and Willett[144] used the MCS as a measure of intermolecular similarity in the comparison of distance-based measures for 3D similarity searching that has been described in the earlier subsection on distance-based similarity measures. Assume that two molecules have *A* and *B* atoms and that the MCS contains *C* atoms; then Pepperrell and Willett were able to use the simpler form of the Tanimoto coefficient that has been defined previously. In their work, Pepperrell and Willett drew on an earlier study by Brint and Willett[162] that presented an upper bound algorithm to maximize the speed of 3D similarity searching when the MCS is used as the similarity measure. On the basis of simulated property-prediction experiments, Pepperrell and Willett suggested that the MCS provided a less cost-effective measure of 3D similarity than the atom-mapping measure that has been described earlier (and hence used the latter measure in an operational system for 3D similarity searching at Zeneca Agrochemicals).[145] But this has not stopped several other workers from investigating the merits of what is, without any doubt, an intuitively attractive measure of structural resemblance. Thus, Ho and Marshall have described a program, called FOUNDATION, that retrieves all structures from a database that contain any combination of a user-specified minimum number of matching atoms, for example, all structures that contain at least seven atoms in the same geometric arrangement (to within any allowed distance tolerances), that occur in the target structure.[163] FOUNDATION is thus rather different from most similarity-searching programs, whether in 2D or in 3D, since these tend to retrieve some number (either fixed or user-defined) of the top-ranked structures. An example of the latter approach is provided by Moon and Howe[164] in their program MOSAIC, which provides facilities for MCS-based 3D similarity searching and for both substructure and superstructure searching, using an approach that is closely related to the subsimilarity system described by Hagadone.[98]

All of the systems that have been described here use the graph-theoretic technique known as *clique detection* for the identification of the 3D MCS. A *clique* is a subgraph of a graph in which every node is connected to every other node and which is not contained in any larger subgraph with this property. The clique detection approach to the identification of 3D MCSs involves the identification of cliques in a *correspondence graph,* a data structure that contains all of the possible equivalences between the two graphs that are being compared. Specifically, given a pair of graphs *A* and *B*, a correspondence graph, *C*, can be formed by creating the set of all pairs of nodes, one from each of the two graphs, such that the nodes of each pair are of the same type: *C* is then the graph whose nodes are these pairs of nodes. Two correspondence graph nodes $(A(I), B(X))$ and $(A(J), B(Y))$ are connected in *C* if the values of the edges from $A(I)$ to $A(J)$ and $B(X)$ to $B(Y)$ are the same. The maximal common subgraphs

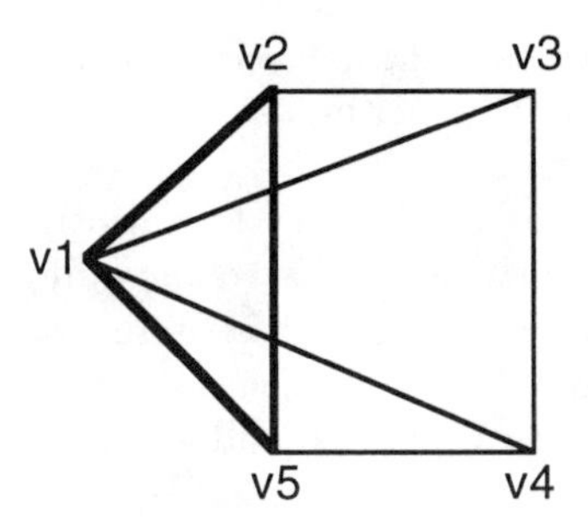

Figure 7 An example of a clique. Assume that comparing the atoms and interatomic distances of two molecules yields the above graph in which vertices v1–v5 are the resulting correspondence. The maximally connected subgraph v1 v2 v3 denotes the presence of a 3-atom common substructure.

then correspond to the cliques of C, so that the identification of the MCS for a pair of 3D structures, where the nodes and edges correspond to the atoms and to the interatomic distances, respectively, is equivalent to the identification of the largest clique in the correspondence graph linking together the two structures. An example of a clique in a 2D graph is shown in Figure 7. A range of different clique-detection algorithms may be used for MCS-based similarity searching. Ho and Marshall[163] use an exhaustive enumeration procedure that derives from an early 3D substructure searching algorithm described by Jakes et al.[165] Pepperrell and Willett[144] and Brint and Willett[162] use a much more efficient algorithm due to Bron and Kerbosch[166] following a detailed comparison of several 3D MCS algorithms by Brint and Willett,[167] whereas Moon and Howe[164] use the routines developed by Kuntz for the DOCK program.[147] In view of the extensive use that is made of 3D superimposition procedures in molecular modeling (as exemplified by the docking and pharmacophore mapping procedures that are described later, and by the integration with de novo design tools that has been achieved in MOSAIC),[164] we may expect that MCS-based similarity searching will develop rapidly as a general-purpose mechanism for 3D database access.

Flexible Similarity Searching

Most of the early systems for 3D substructure searching and those described earlier assumed that all of the molecules in the database that was to be searched were completely rigid or could be represented by a single (presumably low-energy) conformation, that is, the systems took little account of the conformational flexibility that characterized most 3D molecules. Such an approach was appropriate because the only databases that were generally available were the X-ray structures in the Cambridge Structural Database[66] and the single conformations resulting from structure-generation programs such as CONCORD[69] and because appropriate techniques had not been developed for the representation and searching of flexible structures. However, it is now clear that a substructure search in a database of rigid structures often results in the retrieval of only a small fraction of the total number of molecules that could adopt a conformation that contains the query pattern, and this has led to the development of sophisticated data structures and matching algorithms that allow the retrieval of all such molecules.[6,7]

It is a common approach to investigate a special case of some scientific problem before considering how the initial solution can be generalized to encompass all of the occurrences of that problem: it is hence hardly surprising that the 3D similarity measures that have been described thus far have involved only rigid structures. However, even with such a constraint, the results that have been obtained to date are sufficiently encouraging to suggest that the next major step is to consider how the measures can be extended to encompass conformational flexibility.

As discussed earlier, the initial stage of a flexible 3D substructure search involves representing a flexible database structure by a bounded distance matrix, in which each element specifies the maximum-possible and minimum-possible distances between each pair of atoms in the molecule (rather than just a single distance, as in the case of a rigid molecule.[80] An extremely simple way of implementing a flexible 3D similarity search would be to use interatomic distance ranges, rather than interatomic distances, as the inputs to suitably modified versions of some of the procedures described thus far. For example, there is, in principle at least, no obvious reason why one could not implement a flexible version of atom-mapping, given an appropriate definition of when two distance ranges were to be regarded as being the same. A simpler approach has been adopted in the UNITY chemical-information system developed by Tripos Associates, which provides facilities for similarity searching based on 2D fragment bit-screens, and on both rigid and flexible 3D distance-range bit-screens. In all three cases, the bit-string representing the target structure is matched against the bit-string describing a database structure, and the Tanimoto coefficient is calculated in the normal way. However, there are substantial differences in the structures that result from such a procedure, as befits the very disparate structural representations that are employed.

Docking Procedures

We now describe an important application of 3D similarity searching, one that is rather different in concept from those discussed thus far. Specifically, we discuss the use of similarity-based techniques to identify molecules in a database of 3D structures that are *complementary* to a biological receptor site and can thus exhibit activity at it, rather than seeking to identify molecules that are structurally similar to a target structure that is known to exhibit activity at the site of interest.

The principles whereby macromolecular biological receptors can recognize small-molecule substrates or inhibitors are of paramount importance in rational approaches to drug design where the binding-site geometry of the receptor is known, typically from X-ray or nuclear magnetic resonance (NMR) analysis.[168] Given such information, the *docking problem* involves the identification of molecules that are complementary to the site and that might thus be putative ligands for it.[169]

The first computer program to be described for the docking problem was that of Kuntz et al.[147] In their program, called DOCK, the binding site is described by spheres that are complementary to the grooves and ridges in the receptor's surface and that fill the available binding site. The atoms comprising a putative ligand are represented by a similar set of spheres, and the shape similarity of the ligand to the site is then determined by the extent to which it is possible to overlap, or to dock, the two sets of spheres. The docking is effected by means of an approximate clique-detection procedure that matches subsets of the ligand interatomic distances with subsets of the receptor intersphere distances until the best fit has been obtained. A least-squares superimposition is then carried out for the sphere-to-sphere equivalences that have been identified. The resulting orientations of the ligand in the site can be checked by comparison with crystallographic studies of the bound ligand. More recent versions of the program augment the steric matching scores with electrostatic and molecular mechanics interaction energies for the ligand–receptor complex,[170] and consider the use of atomic hydrophobicity descriptors in scoring docked orientations.[171]

The original DOCK program considered the docking of a single ligand into a site, but the procedure has since been extended to permit the searching of a 3D database by means of a scoring function that ranks the molecules in order of decreasing steric fit to the receptor site.[172] A detailed evaluation of the use of DOCK for this purpose has been reported by Stewart et al., who docked 103 ligands that had been previously tested as inhibitors of α-chymotrypsin catalysis into the active site of the enzyme.[173] A statistically significant relationship was found between the DOCK goodness-of-fit scores of the docked ligands and the observed inhibition strengths, with eight of the ten most active inhibitors appearing at the top of the DOCK ranking.

The demonstration that DOCK scores are positively correlated with binding affinities has led to widespread use of the program to identify those structures in a database that are most likely to fit the receptor site on steric grounds.[151,168,169,174–177] Usage is exemplified by a recent paper that discusses the design of potential HIV-1 protease inhibitors.[177] The search used the CONCORD structures for 9561 molecules from the SmithKline Beecham database that contained two oxygen atoms within 5.87 ± 1.0 Å of each other, this requirement being based on the ligand–complex crystal structures for two known inhibitors (MVT-101 and JG-365) with HIV-1 protease. Spheres were generated to characterize the shape of the active site within 8 Å of these two bound inhibitors, and two further spheres defined the presence of water in the MVT-101 complex and of a hydroxyethylamine oxygen in the JG-365 complex. DOCK was then used to search the database, with a match being obtained for each database structure only if it was possible to map the two specified oxygens to the two added spheres. Inspection of the top-ranking hits showed that they shared a common six-membered ring with *para*-hydroxyl groups overlaying the two added spheres such that the hydroxyls could interact with pairs of Asp and of Ile residues in the active site: this finding was used to guide the design of a series of micromolar inhibitors of HIV-1 protease. A similar set of structures was independently designed by the group at DuPont,[178] using 3D database searching and molecular modeling.

Lawrence and Davis[179] describe a program, called CLIX, that performs a similar function to DOCK. CLIX utilizes information from Goodford's GRID program,[180] which identifies regions of high affinity for chemical probes on the molecular surface of a binding site. CLIX takes a 3D structure from the Cambridge Structural Database[66] and then exhaustively tests whether it is possible to superimpose a pair of the candidate's substituent chemical groups with a pair of corresponding favorable interaction sites proposed by GRID (only non-hydrogen atoms are considered). All possible combinations of ligand pairs and GRID binding-site pairs are tested; if a match is obtained, then the candidate ligand is rotated about the two pairs of groups and checked for steric hindrance and coincidence of other candidate atomic groups with appropriate GRID sites. Lawrence and Davis demonstrate that the program is capable of predicting the correct binding geometry of sialic acid to a mutant influenza-virus hemagglutin and also report the best-matching potential ligands resulting from a search of 27,720 3D structures from the Cambridge Structural Database. The search took 33 CPU hours on a Silicon Graphics 4D/240 workstation.

Both DOCK and CLIX assume that the ligand molecules are completely rigid and that no account need be taken of the inherent flexibility of potential ligands during docking. There have been several attempts to develop modified versions of DOCK that overcome this limitation,[181–183] but these are all far too slow for database searching. Similar comments apply to the AUTODOCK program of Goodsell and Olson,[184] which permits the docking of fully flexible ligands by means of simulated annealing but which takes several CPU hours for the processing of a single ligand. Such approaches can hence be regarded as involving type-A similarity measures. Two approaches, which are described next, permit the use of flexible docking in a database context.

Smellie et al.[185] replaced the sphere-matching algorithm in DOCK with a clique-detection procedure in which the hydrogen-bond donor and acceptor atoms of a ligand are docked onto corresponding atoms in the protein. The flexibility of the ligand is described by distance ranges that include all of the allowed distances, in all conformations, between the particular points. If the distances between ligand atoms fall within hydrogen-bonding distance of the complementary distances in the protein, then the potential matches are tested using distance-geometry embedding. Smellie et al. suggest that the distance-range procedure can be sufficiently fast to act as a screening mechanism for the search of a full database. However, the NP-complete nature of clique detection means that this will only be the case if small numbers of points need to be considered by the matching algorithm; moreover, the final 3D embedding stage is also slow.

The FLOG program described by Miller et al.[176] represents the full conformational space of a flexible ligand by a small number of representative, low-energy conformations; specifically, the conformation generator, which is based on distance geometry,[79] results in an average of 8 and a maximum of 25 conformers per molecule, the minimum root-mean-squared distance between any two conformers being at least 1.2 Å. Such a set of individual conformations is clearly an approximation to the true conformational space of a flexible ligand (and one that was also used in early approaches to flexible 3D substructure searching),[5] but it does provide a representation that is sufficiently simple to allow database searches to be carried out in an acceptable amount of time. FLOG draws heavily on the sphere-matching approach of DOCK but augments it by labeling each of the nonhydrogen ligand and receptor atoms with atom-type classes. Precomputed energy grids allow the rapid calculation of the energy of interaction between each possible pair of ligand and receptor atom classes that can arise from an alignment of one of the chosen ligand conformations in the receptor site. In addition, a modified sphere-matching algorithm is described that achieves efficiencies in operation by the specification of *essential points,* locations within the active site that must be mapped to ligand atoms if an acceptable docking is to be achieved, and by the use of an heuristic clique-detection procedure that constrains the number of mappings of site spheres to ligand atoms that need to be investigated. Miller et al. exemplify the use of FLOG by searching 57,531 conformations from 7636 structures in the *Merck Index* for potential ligands for dihydrofolate reductase (DHFR) from *L. casei.* The search required about 12 CPU hours on a four-processor Cray YMP and succeeded in identifying several novel classes of compound that appeared able to bind to DHFR.

The discussion of docking has focused on the docking of small-molecule ligands with proteins. There is also interest in the related *protein docking problem,*[186] which involves modeling the interactions of proteins, such as those that occur at the antibody–antigen interface. Most approaches to protein docking have adopted a geometric approach, in which docking is initially effected by means of descriptors that characterize the shapes of the interacting surfaces, and other factors (such as electrostatic complementarity, hydrogen bonding, and molecular surface shape and size) are considered once steric complementarity has been achieved (in a manner that mirrors the development of the DOCK program). Examples of this two-level approach are described by Walls and Sternberg,[187] who check for an electrostatic fit once the complementarity of the van der Waals surfaces has been confirmed, by Jiang and Kim,[188] and by Shoichet and Kuntz,[189] who describe energy calculations that are carried out after an initial geometric match. Other methods have been described by Connolly[190] and by Cherfils et al.[191] However, it is not clear that such approaches are applicable to similarity searching in small-molecule databases.

CONCLUSIONS

Overview

In this chapter, we have reviewed the development and current status of similarity-searching systems for databases of 2D and 3D chemical structures. Although there are many ways in which the similarity can be calculated for a pair of 2D structures, the great majority of current software systems employ similarity measures that are based on the fragment bit-string representations of molecules that had been developed previously for the screening stage of 2D substructure searching. Despite their great simplicity, and consequent efficiency of operation, such representations have been shown to provide an effective way of accessing large chemical databases, as is exemplified by the many in-house and commercial systems that utilize this approach. It has also proved relatively simple to extend the techniques to encompass the clustering of large databases. Currently, the main focus of research in the area of similarity searching is on development of analogous methods for databases of 3D structures, and we have sought to provide a detailed overview of this research. At present, the situation is rather confusing in that although many different similarity measures are being described, there have been few detailed comparative studies to date that provide insights into their relative strengths and limitations. It is to be hoped that such studies will be forthcoming in the not-too-distant future.

A characteristic of many of the 3D similarity measures that have been described is that they are much more demanding of computational resources than the fragment-based measures that are used for 2D similarity searching. Accordingly, many of the 3D measures will not be used on a large scale until new software and hardware techniques become available, and their extension to encompass databases of flexible 3D molecules (as has recently been achieved in the case of substructure searching)[6–8] will require still further computational resources. This being so, it may be some time before it is possible to apply clustering methods to databases of 3D structures and thence to determine whether the resulting classifications are superior to those resulting from the current generation of 2D clustering approaches. In view of this need for increased performance, it is appropriate briefly to consider some of the algorithms that have been described for the implementation of similarity searching in large databases.

Efficiency of Searching

The user of a similarity-searching system may know little about the structures that are to be retrieved, other than that they are likely to be related, in some imprecisely defined way, to the target (query) structure. Such an imprecise approach to query formulation necessitates a browsing-like retrieval mechanism in which the user is able to inspect output (i.e., the top-ranked structures) and then to refine the search in an iterative manner until an appropriate set of structures has been obtained. A rapid response is required if the browsing is to be effective, and it is thus important that efficient algorithms be available if one wishes to implement similarity searching on a large scale. This is true of any type of nearest-neighbor search procedure, whatever the application area, and there is an extensive literature that discusses a range of approaches to this very general problem. The reader is referred to the recent review by Murtagh[192] for an excellent, wide-ranging introduction to this subject.

Efficient algorithms for database searching are available in some cases. For example, Willett[193] has reviewed several algorithms that can be used for fragment-based, 2D similarity searching. These typically involve the creation of an inverted index to the fragment bit-strings so as to provide rapid access to those structures that contain fragments in common with the target structure. Instead of a list of structures with their associated screen numbers, the list is inverted to form a matrix, with each row representing a screen number and each column a structure number. A nonzero entry for the *IJ*th matrix element indicates that the *I*th screen is present in the *J*th structure. When comparing a target structure, *T*, with all of the structures in a database, consideration need be given only to those rows in the matrix that represent the screens that are present in *T*. All such rows are added together to yield a row of summations representing the number of screens common to *T* and each database structure. This information is then used for the calculation of the final set of similarity coefficients.

In other cases, *upperbound* strategies may be used to reduce the number of database structures that need to be matched against *T*. Such strategies are based on the assumptions that *T* is to be matched against each of the *N* structures in a database, that there is a need to minimize the number of such matches, and that the search is to retrieve the *M* nearest neighbors $(M << N)$ for *T*. At a given point in the search of the set of *N* structures, let the similarity of the current *M*th nearest neighbor (i.e., the database record that is the *M*th most similar to *T* of those that have been tested thus far) be $Sim(M)$. If this is the case, then none of the remaining, untested records need be considered further for inclusion among the set of *M* nearest neighbors unless there is a nonzero probability that their similarity could be greater than *Sim(M)*. An upperbound strategy involves calculating the maximum possible similarity, *MaxPossSim,* for each record that has not yet been matched against *T*. Each such match then needs to carried out if, and only if, *MaxPossSim* > *Sim(M),* because the full match cannot affect the current list of *M* nearest neighbors if this inequality is not satisfied. Such a strategy will typically eliminate far fewer molecules than can the screening stage of a substructure search. That said, an appropriate upperbound calculation may be sufficient to bring about quite substantial reductions in the computational cost of a nearest-neighbor search if the calculation can be carried out more cheaply than can the full similarity match; if this is not the case, then one might just as well carry out the full match for each of the *N* structures in the database. There are several published examples, some of which have been discussed previously, that have made successful use of an upperbound strategy to increase the efficiency of similarity searching.[98,130,145,162,193]

Upperbound strategies, such as those described above, provide an example of the use of software technology to increase the speed of searching. Alternatively, hardware techniques can be employed to lessen the computational requirements, and there has already been some interest in the use of parallel hardware for similarity-searching applications. For example, Whaley and Hodes[194] and Rasmussen et al.[195] have described the use of massively parallel processors for clustering large 2D databases, Wild and Willett[155] have compared three different massively parallel implementations of the atom-mapping approach to 3D similarity searching, Miller et al.[176] have discussed the use of a Cray-2 for docking flexible ligands, and Brint and Willett[196] have simulated the use of transputer networks for 3D MCS detection. We can expect substantial developments in this area with the falling costs of parallel machines and, perhaps more importantly, the ability to build powerful distributed systems from large numbers of basic workstations and/or personal computers.

Applications of Similarity Searching

It was emphasized previously that there are many possible applications of similarity in the general area of chemical structure handling. In this chapter, we have focused on similarity searching in large structural databases; even so, we have managed to provide at least some discussion of two important applications, viz clustering and docking, and there is much further material that could have been included in a more wide-ranging review. One obvious example, which has already been mentioned in passing earlier, is the use of similarity as a retrieval mechanism for reaction databases, rather than the structure databases that are the principal focus of this review.

Reaction similarity searching differs from conventional similarity searching in that it involves calculating the similarities between pairs of reactions, rather than pairs of molecules. Even so, the basic searching algorithm is the same, in that a target record (i.e., a structure or a reaction) is matched against a database of comparable records to find those that are most similar to it (or that have a similarity greater than some predefined match value). MDL Information System's implementation of similarity searching in the REACCS system has been discussed earlier. An alternative implementation, based on the Hierarchical Organization of Reactions through Attribute and Condition Eduction (HORACE) algorithm, is described by Rose and Gasteiger.[197] This method classifies a reaction database using both topological and physicochemical descriptors. A whole molecule is classified in terms of physicochemical parameters, such as atomic charges and bond dissociation energies. A single-linkage type clustering method is used to group the reactions on the basis of the physicochemical descriptors. The highest level of the hierarchy comprises generalized descriptions of the reactions in each of these clusters. The molecules in each cluster then undergo topological analysis. The reaction centers, plus up to one bond beyond, are classified on the basis of 114 substructural descriptors. A hybrid clustering algorithm, based on single-pass but extended to produce a hierarchy in a polythetic divisive manner, is used to classify further each physicochemically-based cluster on the basis of the topological descriptors. By this method, similar types of reaction are found within the same region of the hierarchy, with the different levels giving varying degrees of specificity, thus improving the efficiency of reaction searching.

Closely related to similarity searching in reaction databases is similarity searching in synthesis design. Here, similarity comparisons are made between the target structure and a database of potential starting structures. Comparison between the carbon skeletons is particularly important, because bonds to heteroatoms are more likely to change during the course of a reaction. The similarity-searching capability with the SST system of Wipke and Rogers[198,199] and the WODCA systems of Gasteiger et al.[200] is based on an initial 2D MCS comparison of the carbon skeletons, with an analogous set reduction comparison used in the LHASA system.[201] In the SST system, the database of potential starting compounds is sorted into a hierarchy based on their MCSs. Bonds in the target structure are then matched against bonds in the MCS tree to build up the largest MCS between the two. On completion, the starting compounds associated with this MCS are retrieved. In the WODCA system, comparison between MCSs is just one of the similarity measures used. Complementary measures include one based on a refinement of the path numbers advocated by Randic and Wilkins[85] (as discussed earlier), and one based on transformation similarity as discussed next.

Using a series of reaction type transforms, such as substitution, elimination, oxidation, and reduction, both target and starting structure are transformed into some new structure. These new structures are tested for identity using a hash-coding algorithm. Identical hash codes imply similar original structures and also give an indication of the reaction type necessary to transform starting structure to target structure. Comparison of carbon skeletons in the LHASA system uses a set reduction method to find the central atoms of the *Largest Identical Regions* (LIRs) between target and starting compounds. A second stage then considers the position and nature of any functional groups and heteroatoms. The LIR central atoms are used as root atoms to build up carbon skeleton "partmaps" via a breadth-first search. A backtrack algorithm is used in the second stage to try to expand the partmaps by including unmapped areas, such as heteroatoms. Many potential mappings can result, and so they are scored using Heuristic Estimate of Synthetic Proximity (HESP) values,[202] and the top 20 mappings are displayed to the user.

Finally, the discussion of 3D MCS algorithms is closely related to the emerging technique of *pharmacophore mapping*, which seeks to provide automated tools for the identification and subsequent representation of pharmacophore models for sets of biologically related compounds, that is, the 3D chemical and spatial requirements for bioactivity (as reviewed by Bures et al.).[142] An example of work in this area is the DISCO (DIStance COmparisons) program that has been developed by Martin et al.[203] DISCO identifies the maximum overlap (to within allowed interatomic distance tolerances) between a set of structures, which can be either a set of different molecules or a set of different conformations for the same molecule. The points for the resulting superimposition are typically the ligand hydrogen-bond donor, hydrogen-bond acceptor, and charged atoms; centers of ligand hydrophobic regions; and extensions of the hydrogen bonds and electrostatic interactions to binding sites in the receptor. These various types of points are identified using the ALADDIN package.[204] DISCO uses the clique-detection approach to MCS detection to identify patterns of points that are present in all of the bioactive molecules and/or conformations that are under investigation, so that the MCS algorithm is applied not just to pairs of 3D structures but to some or many 3D structures. Martin et al. achieve this by means of a technique first described by Brint and Willett.[167] A more efficient, and highly elegant, approach to the identification of the MCS when large numbers of structures are involved has been described by Bayada et al.[205] In addition to DISCO, the group at Abbott has developed a further MCS-based program, called FAMILY, that provides an effective way of grouping the hits from a 3D substructure search by means of the sets of points that they have in common.[83] This can be visualized as a 3D equivalent to the procedures for clustering 2D substructure-search outputs that are described by Barnard and Downs.[106]

Future Developments

Given the speed with which similarity searching has developed as a general approach to accessing databases of chemical structures, it is of interest to consider the future evolution of similarity-searching systems. To answer this question, we must return to the introduction to this chapter, where we identified three types of measure that can be used to quantify the degree of resemblance between a target structure and each of the molecules in a database. The previous sections have provided a detailed account of type-C measures, that is, those that are sufficiently rapid in execution to enable a search to be carried out on a large database in an acceptable amount of time. Different people will have different views of what is an acceptable response time. However, its browsing-like capability is one of the most attractive features of similarity searching, and this capability will be severely affected if it takes more than a few hours to identify the nearest neighbors for a user's target structure.

Type-A and type-B similarity measures involve much more detailed representations of molecular structure than are provided by the connectivity, coordinate, and property information that forms the basis for current type-C similarity-searching algorithms. It is to be expected that the use of a more detailed representation will allow the calculation of similarity measures that reflect more closely the structural relationships between pairs of molecules, and hence that type-A and type-B measures will provide a still more effective means of accessing structural databases than does the present generation of type-C measures. However, the implementation of type-A and type-B measures for database searching will be feasible if, and only if, it is possible to develop efficient implementations of such measures. Despite the massive improvements that are currently being seen in CPU speeds and in the performance of parallel-processing systems, it will not be possible to use quantum-mechanical and other type-A approaches for similarity searching for a considerable time. The final part of this chapter hence considers the extent to which type-B similarity measures might be used to search large databases, rather than merely to calculate the similarities between small numbers of pairs of structures, as at present.

The most important class of type-B measures is the grid-based approaches first developed by Carbo et al.[10] These workers suggested that the similarity between a pair of molecules could be estimated by a similarity coefficient based on the overlap of the molecules' electron charge clouds. A molecule is positioned at the center of a 3D grid, and the electrostatic potential is calculated at each point in the grid. The similarity between a pair of molecules is then estimated by comparing the potentials at each grid point and summing over the entire grid, with a suitable normalizing factor being used to bring the similarities into the range -1.0 to $+1.0$. This numerical approach can also be used for the measurement of shape similarity,[23] but it necessarily involves the matching of very large numbers of grid points (unless extremely coarse grids are used in the calculation). Good et al.[24] have recently reported an alternative approach in which the potential distribution is approximated by a series of Gaussian functions that can be processed analytically, with a substantial increase in the speed of the similarity calculation and with only a minimal effect on its accuracy. A similar approach can be used for the calculation of shape similarity.[206] This elegant idea removes one of the main limitations of field-based approaches to 3D similarity searching, but still requires searching for the alignments of the two molecules that are being compared to ensure that analogous grid points are matched.

The most obvious way of obtaining an alignment is by a superimposition procedure, and the MCS methods described earlier provide a direct and reasonably efficient means by which this can be achieved (at least when rigid molecules are being considered). Alternative approaches based on atom matching are exemplified by the work of Dean and his collaborators, who have discussed the matching of pairs of atoms, one from a target structure and one from a database structure, so as to minimize the sum of the squared distance errors using a range of matching algorithms.[29–34] At least some of these measures may be sufficiently fast for implementation in a database-searching program. Meyer and Richards[23] measure the similarity of a pair of molecules by the extent to which their volumes overlap and describe an efficient algorithm to obtain good, but not necessarily optimal, overlaps for pairs of structurally related molecules.

However, it is not clear that an atom-based superimposition procedure will result in the correct alignment of molecular fields, and a usable method for field-based similarity searching in large databases thus requires the identification of an appropriate alignment procedure. Exhaustive searching for alignments is completely infeasible unless extremely coarse grids are used, in which case the calculated similarities are unlikely to reflect accurately the true degree of resemblance of the molecules that are being compared. Even stochastic searching procedures can run for extended periods if one wishes to conduct a detailed examination of the alignment space. For example, Kearsley and Smith have described a method, called SEAL (Steric and Electrostatic ALignment), which uses a Monte Carlo procedure to optimize the alignment of two rigid 3D molecules and which bases the matching on atomic partial charges and steric volumes.[207] A simpler, but more efficient, alignment procedure based on a genetic algorithm has been described recently by Payne and Glen,[208] but there have been no reports to date on the use of this approach for similarity searching.

The absence of an appropriate alignment procedure has thus meant that field-based processing of large databases is not practicable, unless an initial screening procedure is used to eliminate the great majority of the structures from the time-consuming calculation. Examples of the latter approach have been suggested by Good et al.,[209] who consider only the output of a conventional 3D pharmacophoric-pattern search, and by van Geerestein et al.,[156] who consider only the top-ranked molecules resulting from a distance-based SPERM search. Unfortunately, such approaches are limited, in that the screening procedures use angular and distance information rather than field information; accordingly, it is possible for a molecule to be eliminated from the second stage of the search even though it has a high degree of similarity with the target structure at the field level. Such retrieval failures are inevitable unless the screening stage also uses field information, and we hence believe that field-based similarity searching will only become of practical value if it is possible to identify some mechanism to eliminate alignments that cannot possibly yield a large final similarity value. The identification of an appropriate mechanism for this purpose is now the subject of detailed study in our laboratory and, we would expect, in many others.

There are several further topics relating to similarity searching that merit urgent investigation. For example, work to date has tended to try to identify a single, hopefully optimal (in some sense) technique for measuring the similarities between pairs of molecules. However, molecular recognition is known to involve several factors (electrostatic, steric, and hydrophobic), and there is thus some basis for believing that more effective measures might be achieved by the use of approaches that consider more than one type of similarity criterion (in a manner analogous to the inclusion of various types of parameter in the classical multiple-regression approach to the identification of QSARs). This could be done by the development of new measures that take account of multiple similarity criteria, as is exemplified by the SEAL algorithm.[207] Such an approach is likely to be very time-consuming, and a more efficient similarity-searching algorithm might be obtained by combining sets of rankings, each of which resulted from the use of a single-criterion measure. As a simple example of this, Bath et al.[148] have discussed simulated property-prediction experiments that involved several of the 3D similarity measures that were discussed earlier; their combined rankings showed little improvement in predictive performance over the rankings resulting from the single-criterion measures that they studied, but this may have been because all of the measures were based on simple geometric criteria. Other areas that have already been mentioned include the development of 3D clustering methods, the need to integrate database searching and molecular modeling more closely than is currently the case, and the development of new software techniques that can make full use of new hardware technologies as they become available.

At the time of writing, less than a decade has passed since the first reports of similarity searching.[58,103] In the intervening years, similarity searching has evolved rapidly so that it now plays a central role in chemical database systems, acting as a complement to the long-established substructure-search facilities. As the developments that have been reviewed here come to fruition, we believe that similarity searching will increase further in importance, as chemists discover that its inherent browsing-like search capability provides a powerful tool for exploring the multitude of structural data that are available in both corporate and public chemical databases.[210]

ACKNOWLEDGMENTS

We gratefully acknowledge the contributions of Frank Allen, Peter Bath, David Bawden, Andrew Brint, Bill Fisanick, Bobby Glen, Trevor Heritage, Gareth Jones, Helen King, Carol Morris, Catherine Pepperell, Andrew Poirrette, Robin Taylor, David Thorner, David Turner, Peter Walsh,

David Wild, Vivienne Winterman, and Matt Wright to the work that has been carried out in Sheffield on similarity searching and clustering. We thank Andrew Poirrette for providing some of the references and for carefully reading an initial draft of this chapter. Funding for our work in this area has been provided by the Cambridge Crystallographic Data Centre, Chemical Abstracts Service, the European Communities Joint Research Centre, the James Black Foundation, Pfizer Central Research (U.K.), the Science and Engineering Research Council, Shell Research Centre, Tripos Associates, Wellcome Research Laboratories, and Zeneca Agrochemicals. This paper is a contribution from the Krebs Institute for Biomolecular Research, which is a designated center for biomolecular sciences of the Science and Engineering Research Council. Correspondence should be addressed to P. W.

REFERENCES

1. J. E. Ash, W. A. Warr, and P. Willett, Eds., *Chemical Structure Systems,* Ellis Horwood, Chichester, 1991.
2. P. G. Dittmar, R. E. Stobaugh, and C. E. Watson, *J. Chem. Inf. Comput. Sci.,* **16,** 111 (1976). The Chemical Abstracts Service Chemical Registry System. I. General Design. (And subsequent papers in this long-running series).
3. P. G. Dittmar, N. A. Farmer, W. Fisanick, R. C. Haines, and J. Mockus, *J. Chem. Inf. Comput. Sci.,* **23,** 93 (1983). The CAS ONLINE Search System. I. General System Design and Selection, Generation and Use of Search Screens.
4. Y. C. Martin, M. G. Bures, and P. Willett, in *Reviews in Computational Chemistry,* Vol.1, K. B. Lipkowitz and D. B. Boyd, Eds., VCH Publishers, New York, 1990, pp. 213–263. Searching Databases of Three-Dimensional Structures.
5. P. Willett, *Three-Dimensional Chemical Structure Handling,* Research Studies Press, Taunton, U.K., 1991.
6. T. E. Moock, D. R. Henry, A. G. Ozkabak, and M. Alamgir, *J. Chem. Inf. Comput. Sci.,* **34,** 184 (1994). Conformational Searching in ISIS/3D Databases.
7. T. Hurst. *J. Chem. Inf. Comput. Sci.,* **34,** 190 (1994). Flexible 3D Searching: The Directed Tweak Technique.
8. D. E. Clark, G. Jones, P. Willett, P. W. Kenny, and R. C. Glen, *J. Chem. Inf. Comput. Sci.,* **34,** 197 (1994). Pharmacophoric Pattern Matching in Files of Three-Dimensional Chemical Structures: Comparison of Conformational-Searching Algorithms for Flexible Searching.
9. D. H. Rouvray, *J. Chem. Inf. Comput. Sci.,* **32,** 580 (1992). Definition and Role of Similarity Concepts in the Chemical and Physical Sciences.
10. R. Carbo, M. Arnau, and L. Leyda, *Int. J. Quantum Chem.* **17,** 1185 (1980). How Similar Is a Molecule to Another? An Electron Density Measure of Similarity Between Two Molecular Structures.
11. R. Carbo and B. Calabuig, *J. Chem. Inf. Comput. Sci.* **32,** 600 (1992). Quantum Similarity Measures, Molecular Cloud Description and Structure-Property Relationships.
12. N. L. Allan and D. L. Cooper, *J. Chem. Inf. Comput. Sci.* **32,** 587 (1992). A Momentum-Space Approach to Molecular Similarity.
13. D. L. Cooper and N. L. Allan, *J. Am. Chem. Soc.* **114,** 4773 (1992). Molecular Dissimilarity: A Momentum Space Criterion.
14. F. Manaut, F. Sanz, J. Jose, and M. Milesi, *J. Comput.-Aided Mol. Design,* **5,** 371 (1991). Automatic Search for Maximum Similarity Between Molecular Electrostatic Potential Distributions.
15. R. B. Hermann and D. K. Herron, *J. Comput.-Aided Mol. Design,* **5,** 511 (1991). OVID and SUPER: Two Overlap Programs for Drug Design.
16. P. E. Bowen-Jenkins, D. L. Cooper, and W. G. Richards, *J. Phys. Chem.,* **89,** 2195 (1985). *Ab Initio* Computation of Molecular Similarity.

17. R. Ponec and M. Strnad, *Int. J. Quantum Chem.*, **42**, 501 (1992). Electron Correlation in Pericyclic Reactivity: A Similarity Approach.
18. P. G. Mezey, *J. Chem. Inf. Comput. Sci.*, **32**, 650 (1992). Shape-Similarity Measures for Molecular Bodies: A Three-Dimensional Topological Approach to Quantitative Shape-Activity Relations.
19. C. A. Ramsden, Ed., *Comprehensive Medicinal Chemistry. Volume 4. Quantitative Drug Design,* Pergamon Press, Oxford, 1990.
20. H. Kubinyi, Ed., *3D QSAR in Drug Design,* ESCOM, Leiden, 1993.
21. C. G. Wermuth, Ed., *Trends in QSAR and Molecular Modelling '92,* ESCOM, Leiden, 1993.
22. C. Burt, W. G. Richards, and P. Huxley, *J. Comput. Chem.*, **11**, 1139 (1990). The Application of Molecular Similarity Calculations.
23. A. Y. Meyer and W. G. Richards, *J. Comput.-Aided Mol. Design,* **5**, 427 (1991). Similarity of Molecular Shape.
24. A. C. Good, E. E. Hodgkin, and W. G. Richards, *J. Chem. Inf. Comput. Sci.*, **32**, 188 (1992). The Utilisation of Gaussian Functions for the Rapid Evaluation of Molecular Similarity.
25. A. C. Good, S. J. Peterson, and W. G. Richards, *J. Med. Chem.*, **36**, 2929 (1993). QSARs from Similarity Matrices. Technique Validation and Application in the Comparison of Different Similarity Evaluation Methods.
26. A. M. Richard, *J. Comput. Chem.*, **12**, 959 (1991). Quantitative Comparison of Molecular Electrostatic Potentials for Structure-Activity Studies.
27. J. D. Petke, *J. Comput. Chem.*, **14**, 928 (1993). Cumulative and Discrete Similarity Analysis of Electrostatic Potentials and Fields.
28. R. D. Cramer, D. E. Patterson, and J. D. Bunce, *J. Am. Chem. Soc.*, **110**, 5959 (1988). Comparative Molecular Field Analysis (CoMFA). 1. Effect of Shape on Binding of Steroids to Carrier Proteins.
29. D. J. Danziger and P. M. Dean, *J. Theoret. Biol.* **116**, 215 (1985). The Search for Functional Correspondences in Molecular Structure Between Two Dissimilar Molecules.
30. P. L. Chau and P. M. Dean, *J. Mol. Graphics,* **5**, 97 (1987). Molecular Recognition: 3D Structure Comparison by Gnomonic Projection.
31. P. M. Dean, P. Callow, and P. L. Chau, *J. Mol. Graphics,* **6**, 28 (1988). Molecular Recognition: Blind Searching for Regions of Strong Structural Match on the Surfaces of Two Dissimilar Molecules.
32. M. T. Bakarat and P. M. Dean. *J. Comput.-Aided Mol. Design,* **4**, 295 (1990). Molecular Structure Matching by Simulated Annealing. 1. A Comparison of Different Cooling Schedules. (And subsequent papers in this series).
33. P. M. Dean, P. L. Chau, and M. T. Bakarat, *J. Mol. Structure (THEOCHEM),* **265**, 75 (1992). Development of Quantitative Methods for Studying Electrostatic Complementarity in Molecular Recognition and Drug Design.
34. T. D. J. Perkins and P. M. Dean. *J. Comput.-Aided Mol. Design,* **7**, 155 (1993). An Exploration of a Novel Strategy for Superposing Several Flexible Molecules.
35. M. A. Johnson and G. M. Maggiora, Eds., *Concepts and Applications of Molecular Similarity,* John Wiley & Sons, New York, 1990.
36. P. M. Dean, Ed., *Molecular Similarity in Drug Design,* Chapman and Hall, Glasgow, 1995, and Blackie Academic & Professional, London, 1995.
37. *J. Chem. Inf. Comput. Sci.*, **32**, 577–752 (1992).
38. W. A. Warr, Ed., *Chemical Structures. The International Language of Chemistry,* Springer-Verlag, Heidelberg, 1988.
39. W. A. Warr, Ed., *Chemical Structures 2. The International Language of Chemistry,* Springer-Verlag, Heidelberg, 1993.
40. *J. Chem. Inf. Comput. Sci.*, **34** (*1*), 1–231 (1994). Proceedings of the Third International Conference on Chemical Structures, Noordwijkerhout, The Netherlands, June 6–10, 1993.

41. P. Willett, *J. Chemometrics,* **1,** 139 (1987). A Review of Chemical Structure Retrieval Systems.

42. J. E. Ash, P. A. Chubb, S. E. Ward, S. M. Welford, and P. Willett, *Communication, Storage and Retrieval of Chemical Information,* Ellis Horwood, Chichester, 1985.

43. R. Wilson, *Introduction to Graph Theory,* Oliver and Boyd, Edinburgh, 1972.

44. N. Trinajstic, Ed., *Chemical Graph Theory,* CRC Press, Boca Raton, 1983.

45. N. A. B. Gray, *Computer-Assisted Structure Elucidation,* John Wiley & Sons, New York, 1986.

46. R. C. Read and D. G. Corneil, *J. Graph Theory,* **1,** 339 (1977). The Graph Isomorphism Disease.

47. J. M. Barnard, *J. Chem. Inf. Comput. Sci.* **33,** 532 (1993). Substructure Searching Methods: Old and New.

48. M. R. Garey and D. S. Johnson, *Computers and Intractability: A Guide to the Theory of NP-Completeness,* WH Freeman, San Francisco, CA, 1977.

49. D. Harel, *Algorithmics: The Spirit of Computing,* Addison-Wesley, Reading, MA, 1987.

50. M. F. Lynch, in *Chemical Information Systems,* J. E. Ash and E. Hyde, Eds., Ellis Horwood, Chichester, 1975, pp. 177–194. Screening Large Chemical Files.

51. L. Hodes, *J. Chem. Inf. Comput. Sci.,* **16,** 88 (1976). Selection of Descriptors According to Discrimination and Redundancy. Application to Chemical Substructure Searching.

52. J. K. Cringean, C. A. Pepperrell, A. R. Poirrette, and P. Willett, *Tetrahedron Comput. Methodol.,* **3,** 37 (1990). Selection of Screens for Three-Dimensional Substructure Searching.

53. G. W. Adamson, J. Cowell, M. F. Lynch, A. H. W. McLure, W. G. Town, and A. M. Yapp, *J. Chem. Document.,* **13,** 153 (1973). Strategic Considerations in the Design of a Screening System for Substructure Searches of Chemical Structure Files.

54. W. Graf, H. K. Kaindl, H. Kneiss, and R. Warszawski, *J. Chem. Inf. Comput. Sci.,* **22,** 177 (1982). The Third BASIC Fragment Search Dictionary.

55. R. Attias, *J. Chem. Inf. Comput. Sci.,* **23,** 102 (1983). DARC Substructure Search System: A New Approach to Chemical Information.

56. A. Feldman and L. Hodes, *J. Chem. Inf. Comput. Sci.,* **15,** 147 (1975). An Efficient Design for Chemical Structure Searching. 1. The Screens.

57. L. Hodes, *J. Chem. Inf. Comput. Sci.,* **29,** 66 (1989). Clustering a Large Number of Compounds. 1. Establishing the Method on an Initial Sample.

58. R. E. Carhart, D. H. Smith, and R. Venkataraghavan. *J. Chem. Inf. Comput. Sci.,* **25,** 64 (1985). Atom Pairs as Molecular Features in Structure-Activity Studies: Definitions and Applications.

59. R. Nilakantan, N. Bauman, J. S. Dixon, and R. Venkataraghavan, *J. Chem. Inf. Comput. Sci.,* **27,** 82 (1987). Topological Torsion: A New Molecular Descriptor for SAR Applications. Comparison with Other Descriptors.

60. G. Klopman, *J. Am. Chem. Soc.,* **106,** 7315 (1984). Artificial Intelligence Approach to Structure-Activity Studies. Computer Automated Structure Evaluation of Biological Activity of Organic Molecules.

61. C. N. Mooers, *Aslib Proc.,* **8,** (1956). Zatocoding and Developments in Information Retrieval.

62. C. A. James and A. Weininger, *Daylight Chemical Information Systems Theory Manual,* Daylight CIS Inc., Irvine, CA and Santa Fe, NM.

63. E. H. Sussenguth, *J. Chem. Doc.,* **5,** 36 (1965). A Graph-Theoretic Algorithm for Matching Chemical Structures.

64. J. Figueras, *J. Chem. Doc.,* **12,** 237 (1972). Substructure Search by Set Reduction.

65. J. R. Ullmann, *J. Assoc. Comput. Machinery,* **16,** 31 (1976). An Algorithm for Subgraph Isomorphism.

66. F. H. Allen, J. E. Davies, J. J. Galloy, O. Johnson, O. Kennard, C. Macrae, E. M. Mitchell, G. F. Mitchell, J. M. Smith, and D. G. Watson, *J. Chem. Inf. Comput. Sci.*, **31**, 187 (1991). The Development of Versions 3 and 4 of the Cambridge Structural Database System.

67. J. Sadowski and J. Gasteiger, *Chem. Rev.*, **93**, 2567 (1993). From Atoms and Bonds to Three-Dimensional Atomic Coordinates: Automatic Model Builders.

68. E. M. Ricketts, J. Bradshaw, M. Hann, F. Hayes, N. Tanna, and D. M. Ricketts, *J. Chem. Inf. Comput. Sci.*, **33**, 905 (1993). Comparison of Conformations of Small Molecules from the Protein Data Bank with Those Generated by Concord, Cobra, ChemDBS-3D and Converter, and Those Extracted from the Cambridge Structural Database.

69. A. Rusinko, J. M. Skell, R. Balducci, and R. S. Pearlman, *CONCORD User's Manual*, Tripos Associates, St. Louis, MO.

70. A. Rusinko, R. P. Sheridan, R. Nilakantan, K. S. Haraki, N. Bauman, and R. Venkataraghavan, *J. Chem. Inf. Comput. Sci.*, **29**, 251 (1989). Using CONCORD to Construct a Large Database of Three-Dimensional Coordinates from Connection Tables.

71. D. R. Henry, P. J. McHale, B. D. Christie, and D. Hillman, *Tetrahedron Comput. Methodol.*, **3**, 531 (1990). Building 3D Structural Databases: Experiences with MDDR-3D and FCD-3D.

72. P. Gund, *Progr. Mol. Subcell. Biol.*, **5**, 117 (1977). Three-Dimensional Pharmacophoric Pattern Searching.

73. S. E. Jakes and P. Willett, *J. Mol. Graphics*, **5**, 12 (1986). Pharmacophoric Pattern Matching in Files of Three-Dimensional Chemical Structures. Selection of Inter-Atomic Distance Screens.

74. R. P. Sheridan, R. Nilakantan, A. Rusinko, N. Bauman, K. S. Haraki, and R. Venkataraghavan, *J. Chem. Inf. Comput. Sci.*, **29**, 255 (1979). 3DSEARCH: A System for Three-Dimensional Substructure Searching.

75. G. Lauri and P. A. Bartlett, *J. Comput.-Aided Mol. Design*, **8**, 51 (1994). CAVEAT: A Program to Facilitate the Design of Organic Molecules.

76. W. Fisanick, K. P. Cross, and A. Rusinko, *J. Chem. Inf. Comput. Sci.*, **32**, 664 (1992). Similarity Searching on CAS Registry Substances. 1. Global Molecular Property and Generic Atom Triangle Geometric Searching.

77. A. R. Poirrette, P. Willett, and F. H. Allen, *J. Mol. Graphics*, **9**, 203 (1991). Pharmacophoric Pattern Matching in Files of Three-Dimensional Chemical Structures. Characterisation and Use of Generalised Valence Angle Screens.

78. A. R. Poirrette, P. Willett, and F. H. Allen, *J. Mol. Graphics*, **11**, 2 (1993). Pharmacophoric Pattern Matching in Files of Three-Dimensional Chemical Structures. Characterisation and Use of Generalised Torsion Angle Screens.

79. G. M. Crippen and T. F. Havel, *Distance Geometry and Molecular Conformation*, Research Studies Press, Letchworth, 1988.

80. D. E. Clark, P. Willett, and P. W. Kenny, *J. Chem. Inf. Comput. Sci.*, **32**, 197 (1992). Pharmacophoric Pattern Matching in Files of Three-Dimensional Chemical Structures. Use of Smoothed Bounded-Distance Matrices for the Representation and Searching of Conformationally-Flexible Molecules.

81. A. E. Howard and P. A. Kollman, *J. Med. Chem.*, **31**, 1669 (1988). An Analysis of Current Methodologies for Conformational Searching of Complex Molecules.

82. A. R. Leach, in *Reviews in Computational Chemistry*, Vol. 2, K. B. Lipkowitz and D. B. Boyd, Eds., VCH Publishers, New York, 1991, pp. 1–55. A Survey of Methods for Searching the Conformational Space of Small and Medium-Sized Molecules.

83. M. G. Bures, E. Danaher, J. DeLazzer, and Y. C. Martin, *J. Chem. Inf. Comput. Sci.*, **34**, 218 (1994). New Molecular Modelling Tools Using Three-Dimensional Chemical Substructures.

84. P. Willett, *Similarity and Clustering in Chemical Information Systems*, Research Studies Press, Letchworth, 1987.

85. M. Randic and C. L. Wilkins, *J. Chem. Inf. Comput. Sci.*, **19**, 31 (1979). Graph Theoretical Approach to Recognition of Structural Similarity in Molecules.

86. M. Randic, in *Concepts and Applicatons of Molecular Similarity,* M. A. Johnson and G. M. Maggiora, Eds., John Wiley & Sons, New York, 1990, pp. 77–145. Design of Molecules with Desired Properties.

87. H. Wiener, *J. Am. Chem. Soc.,* **69,** 17 (1947). Structural Determination of Paraffin Boiling Point.

88. A. T. Balaban, Ed., *Chemical Applications of Graph Theory,* Academic Press, London, 1976.

89. M. Randic, *J. Am. Chem. Soc.,* **97,** 6609 (1975). On Characterisation of Molecular Branching.

90. L. B. Kier and L. H. Hall, *Molecular Connectivity in Structure-Activity Analysis,* John Wiley & Sons, New York, 1986.

91. L. H. Hall and L. B. Kier, in *Reviews in Computational Chemistry,* Vol. 2, K. B. Lipkowitz and D. B. Boyd, Eds., VCH Publishers, New York, 1991, pp. 367–422. The Molecular Connectivity Chi Indexes and Kappa Shape Indexes in Structure-Property Modeling.

92. L. H. Hall and L. B. Kier, *Quant. Struct.-Act. Relat.,* **9,** 115 (1990). Determination of Toplogical Equivalence in Molecular Graphs from the Toplogical State.

93. L. H. Hall, B. Mohney, and L. B. Kier, *J. Chem. Inf. Comput. Sci.,* **31,** 76 (1991). The Electrotopological State: Structure Information at the Atomic Level for Molecular Graphs.

94. S. C. Basak, V. R. Magnuson, G. J. Niemi, and R. R. Regal, *Discrete Appl. Math.,* **19,** 17 (1988). Determining Structural Similarity of Chemicals Using Graph-Theoretic Indices.

95. P. Willett and V. Winterman, *Quant. Struct.-Act. Relat.,* **5,** 18 (1986). A Comparison of Some Measures for the Determination of Inter-Molecular Structural Similarity.

96. T. E. Moock, D. L. Grier, W. D. Hounshell, G. Grethe, K. Cronin, and J. G. Nourse, *Tetrahedron Comput. Methodol.,* **1,** 117 (1988). Similiarity Searching in the Organic Reaction Domain.

97. G. M. Downs, A. R. Poirrette, P. Walsh, and P. Willett, in *Chemical Structures 2. The International Language of Chemistry,* W. A. Warr, Ed., Springer Verlag, Heidelberg, 1993, pp. 409–421. Evaluation of Similarity Searching Methods Using Activity and Toxicity Data.

98. T. R. Hagadone, *J. Chem. Inf. Comput. Sci.,* **32,** 515 (1992). Molecular Substructure Similarity Searching: Efficient Retrieval in Two-Dimensional Structure Databases.

99. P. H. A. Sneath and R. R. Sokal, *Numerical Taxonomy,* WH Freeman, San Francisco, CA, 1973.

100. G. W. Milligan and M. C. Cooper, *J. Classif.,* **5,** 181 (1988). A Study of Standardisation of Variables in Cluster Analysis.

101. P. A. Bath, C. A. Morris, and P. Willett, *J. Chemometrics,* **7,** 543 (1993). Effect of Standardisation of Fragment-Based Measures of Structural Similarity.

102. G. Salton and M. J. McGill, *Introduction to Modern Information Retrieval,* McGraw-Hill, Englewood Cliffs, NJ, 1983.

103. P. Willett, V. Winterman, and D. Bawden, *J. Chem. Inf. Comput. Sci.,* **36,** 36 (1986). Implementation of Nearest-Neighbour Searching in an Online Chemical Structure Search System.

104. G. W. Adamson and J. A. Bush, *J. Chem. Inf. Comput. Sci.,* **15,** (1975). A Comparison of the Performance of Some Similarity and Dissimilarity Measures in the Automatic Classification of Chemical Structures.

105. D. Bawden, in *Concepts and Applications of Molecular Similarity,* M. A. Johnson and G. M. Maggiora, Eds., John Wiley & Sons, New York, 1990, pp. 65–76. Applications of 2-D Chemical Similarity Measures to Database Analysis and Querying.

106. J. M. Barnard and G. M. Downs, *J. Chem. Inf. Comput. Sci.,* **32,** 644 (1992). Clustering of Chemical Structures on the Basis of Two-Dimensional Similarity Measures.

107. G. Grethe and T. E. Moock, in *Chemical Structures 2. The International Language of Chemistry,* W. A. Warr, Ed., Springer-Verlag, Heidelberg, 1993, pp. 399–407. Similarity Searching in the Development of New Bioactive Compounds. An Application.

108. G. Grethe and T. E. Moock, *J. Chem. Inf. Comput. Sci.,* **30,** 511 (1990). Similarity Searching in REACCS. A New Tool for the Synthetic Chemist.

109. W. Fisanick, K. P. Cross, and A. Rusinko, *Tetrahedron Comput. Methodol.*, **3**, 635 (1990). Characteristics of Computer-Generated 3D and Related Molecular Property Data for CAS Registry Substances.

110. W. Fisanick, A. H. Lipkus, and A. Rusinko, *J. Chem. Inf. Comput. Sci.*, **34**, 130 (1994). Similarity Searching on CAS Registry Substances. 2. 2D Structural Similarity.

111. W. Fisanick, K. P. Cross, J. C. Forman, and A. Rusinko, *J. Chem. Inf. Comput. Sci.*, **33**, 548 (1993). An Experimental System for Similarity and 3D Substructure Searching of CAS Registry Substances. 1. 3D Substructure Searching.

112. J. J. P. Stewart in *Reviews in Computational Chemistry,* Vol. 1, K. B. Lipkowitz and D. B. Boyd, Eds., VCH Publishers, New York, 1990, pp. 45–81. Semiempirical Molecular Orbital Methods.

113. V. J. Gillett, G. M. Downs, A. Ling, M. J. Lynch, P. Venkataram, J. V. Wood, and W. Dethlefsen, *J. Chem. Inf. Comput. Sci.*, **27**, 126 (1987). Computer Storage and Retrieval of Generic Chemical Structures in Patents. Part 8. Reduced Chemical Graphs and Their Applications in Generic Chemical Structure Retrieval.

114. Y. Takahashi, M. Sukekawa, and S. Sasaki, *J. Chem. Inf. Comput. Sci.*, **32**, 639 (1992). Automatic Identification of Molecular Similarity Using Reduced-Graph Representation of Chemical Structure.

115. W. Fisanick, *J. Chem. Inf. Comput. Sci.*, **30**, 145 (1990). The Chemical Abstracts Service Generic Chemical (Markush) Structure Storage and Retrieval Capability: 1. Basic Concepts.

116. B. S. Everitt, *Cluster Analysis,* 3rd ed., Edward Arnold, London, 1993.

117. A. D. Gordon, *Classification,* Chapman and Hall, London, 1981.

118. G. M. Downs and P. Willett, in *Chemometric Methods in Molecular Design,* H. van de Waterbeemd, Ed., VCH Publishers, New York, 1995, pp. 111–130. Clustering of Chemical-Structure Databases for Compound Selection.

119. P. Willett, V. Winterman, and D. Bawden, *J. Chem. Inf. Comput. Sci.*, **26**, 109 (1986). Implementation of Non-Hierarchic Cluster Analysis Methods in Chemical Information Systems: Selection of Compounds for Biological Testing and Clustering of Substructure Search Output.

120. M. S. Lajiness, in *Computational Chemical Graph Theory,* D. H. Rouvray, Ed., Nova Science, New York, 1990, pp. 299–315. Molecular Similarity-Based Methods for Selecting Compounds for Screening.

121. G. M. Downs and P. Willett, in *Applied Multivariate Analysis in SAR and Environmental Studies,* J. Devillers and W. Karcher, Eds., European Communities, Brussels, 1991, pp. 247–279. The Use of Similarity and Clustering Techniques for the Prediction of Molecular Properties.

122. F. Murtagh, *Multidimensional Clustering Algorithms,* Physica-Verlag, Vienna, 1985.

123. R. Dubes and A. K. Jain, *Adv. Comput.*, **19**, 113 (1980). Clustering Methodologies in Exploratory Data Analysis.

124. G. W. Milligan, *Multivariate Behav. Res.*, **16**, 379 (1981). A Review of Monte Carlo Tests of Cluster Analysis.

125. C. K. Bayne, J. J. Beauchamp, C. L. Begovich, and V. E. Kane, *Pattern Recogn.*, **12**, 51 (1980). Monte Carlo Comparisons of Selected Clustering Procedures.

126. L. C. Morey, R. K. Blashfield, and H. A. Skinner, *Multivariate Behav. Res.*, **18**, 309 (1983). A Comparison of Cluster Analysis Techniques Within a Sequential Validation Framework.

127. G. W. Adamson and D. Bawden, *J. Chem. Inf. Comput. Sci.*, **21**, 204 (1981). Comparison of Hierarchical Analysis Techniques for Automatic Classification of Chemical Structures.

128. G. N. Lance and W. T. A. Williams, *Comput. J.*, **9**, 373 (1967). A General Theory of Classificatory Sorting Strategies. I. Hierarchical Systems.

129. E. M. Voorhees, *Inf. Proc. Manage.*, **22**, 465 (1986). Implementing Agglomerative Hierarchical Clustering Algorithms for Use in Document Retrieval.

130. G. M. Downs, P. Willett, and W. Fisanick, *J. Chem. Inf. Comput. Sci.,* **34,** 1094 (1994). Similarity Searching and Clustering of Chemical-Structure Databases Using Molecular Property Data.

131. V. Rubin and P. Willett, *Analyt. Chim. Acta,* **151,** 161 (1983). A Comparison of Some Hierarchical Monothetic Divisive Clustering Algorithms for Structure Property Correlation.

132. A. Guenoche, P. Hansen, and B. Jaumard, *J. Classif.,* **8,** 5 (1991). Efficient Algorithms for Divisive Hierarchical Clustering with the Diameter Criterion.

133. J. A. Hartigan, *Clustering Algorithms,* John Wiley & Sons, New York, 1975.

134. E. Forgy, *Biometrics,* **21,** 768 (1965). Cluster Analysis of Multivariate Data: Efficiency Versus Interpretability of Classifications.

135. P. Willett, *J. Chem. Inf. Comput. Sci.,* **24,** 29 (1984). An Evaluation of Relocation Clustering Algorithms for the Automatic Classification of Chemical Structures.

136. R. A. Jarvis and E. A. Patrick, *IEEE Trans. Comput.,* **C-22,** 1025 (1973). Clustering Using a Similarity Measure Based on Shared Nearest Neighbours.

137. P. Willett, *Analyt. Chim. Acta,* **138,** 339 (1982). The Calculation of Inter-Molecular Similarity Coefficients Using an Inverted File Algorithm.

138. D. Bawden, in *Chemical Structures 2. The International Language of Chemistry,* W. A. Warr, Ed., Springer-Verlag, Heidelberg, 1993, pp. 383–388. Molecular Dissimilarity in Chemical Information Systems.

139. M. A. Johnson, M. S. Lajiness, and G. Maggiora, in *QSAR: Quantitative Structure-Activity Relationships in Drug Design,* J. L. Fauchere, Ed., Alan R. Liss Inc., New York, 1989, pp. 167–171. Molecular Similarity: A Basis for Designing Drug Screening Programs.

140. M. S. Lajiness, M. A. Johnson, and G. Maggiora, in *QSAR: Quantitative Structure-Activity Relationships in Drug Design,* J. L. Fauchere, Ed., Alan R. Liss Inc., New York, 1989, pp. 173–176. Implementing Drug Screening Programs Using Molecular Similarity Methods.

141. M. S. Lajiness, in *QSAR: Rational Approaches to the Design of Bioactive Compounds,* C. Silipo and A. Vittoria, Eds., Elsevier Science Publishers, Amsterdam, 1991, pp. 201–204. An Evaluation of the Performance of Dissimilarity Selection.

142. M. G. Bures, Y. C. Martin, and P. Willett, *Top. Stereochem.,* **21,** 467 (1994). Searching Techniques for Databases of Three-Dimensional Chemical Structures.

143. P. M. Dean, in *Concepts and Applications of Molecular Similarity,* M. A. Johnson and G. M. Maggiora, Eds., John Wiley & Sons, New York, 1990, pp. 211–238. Molecular Recognition: The Measurement and Search for Molecular Similarity in Ligand–Receptor Interaction.

144. C. A. Pepperrell and P. Willett, *J. Comput.-Aided Mol. Design,* **5,** 455 (1991). Techniques for the Calculation of Three-Dimensional Structural Similarity Using Inter-Atomic Distances.

145. C. A. Pepperrell, P. Willett, and R. Taylor, *Tetrahedron Comput. Methodol.,* **3,** 575 (1990). Implementation and Use of an Atom-Mapping Procedure for Similarity Searching in Databases of 3-D Chemical Structures.

146. G. W. Bemis and I. D. Kuntz, *J. Comput.-Aided Mol. Design,* **6,** 607 (1992). A Fast and Efficient Method for 2D and 3D Molecular Shape Description.

147. I. D. Kuntz, J. M. Blaney, S. J. Oatley, R. Langridge, and T. E. Ferrin, *J. Mol. Biol.,* **161,** 269 (1982). A Geometric Approach to Macromolecule–Ligand Interactions.

148. P. A. Bath, A. R. Poirrette, P. Willett, and F. H. Allen, *J. Chem. Inf. Comput. Sci.,* **34,** 141 (1994). Similarity Searching in Files of Three-Dimensional Chemical Structures: Comparison of Fragment-Based Measures of Shape Similarity.

149. R. Nilakantan, N. Bauman, and R. Venkataraghavan, *J. Chem. Inf. Comput. Sci.,* **33,** 79 (1993). A New Method for Rapid Characterisation of Molecular Shape: Applications in Drug Design.

150. M. Miller, J. Schneider, B. K. Sathyanarayana, M. V. Toth, G. R. L. Marshall, S. B. H. Kent, and A. Wlodawer, *Science,* **246,** 1149 (1989). Structure of a Complex of Synthetic HIV-1 Protease with a Substrate-Based Inhibitor at 2.3-Å Resolution.

151. P. D. J. Grootenhuis, P. A. Kollman, G. L. Seibel, R. L. DesJarlais, and I. D. Kuntz, *Anti-Cancer Drug Design,* **5**, 237 (1990). Computerised Selection of Potential DNA Binding Compounds.
152. M. L. Kopka, C. Yoon, D. Goodsell, P. Pjura, and R. E. Dickerson, *Proc. Natl. Acad. Sci. U.S.A.,* **82**, 1376 (1985). The Molecular Origin of DNA-Drug Specificity in Netropsin and Distamycin.
153. S. J. Weiner, P. A. Kollman, D. A. Case, U. C. Singh, C. Ghio, G. Alagona, S. Profeta, and P. Weiner, *J. Am. Chem. Soc.,* **106**, 765 (1984). A New Force Field for Molecular Mechanical Simulation of Nucleic Acids and Proteins.
154. C. A. Pepperrell, A. R. Poirrette, P. Willett, and R. Taylor, *Pesticide Sci.,* **33**, 97 (1991). Development of an Atom-Mapping Procedure for Similarity Searching in Databases of Three-Dimensional Chemical Structures.
155. D. J. Wild and P. Willett, *J. Chem. Inf. Comput. Sci.,* **34**, 224 (1994). Similarity Searching in Files of Three-Dimensional Chemical Structures: Implementation of Atom Mapping on the Distributed Array Processor DAP-610, the MasPar MP-1104 and the Connection Machine CM-200.
156. V. J. van Geerestein, N. C. Perry, P. G. Grootenhuis, and C. A. G. Haasnoot, *Tetrahedron Comput. Methodol.,* **3**, 595 (1990). 3D Database Searching on the Basis of Ligand Shape Using the SPERM Prototype Method.
157. N. C. Perry and V. J. van Geerestein, *J. Chem. Inf. Comput. Sci.,* **32**, 607 (1992). Database Searching on the Basis of Three-Dimensional Molecular Similarity Using the SPERM Program.
158. P. Bladon, *J. Mol. Graphics,* **7**, 130 (1989). A Rapid Method for Comparing and Matching the Spherical Parameter Surfaces of Molecules and Other Irregular Objects.
159. E. E. Hodgkin and W. G. Richards, *Int. J. Quantum Chem., Quantum Biol. Symp.,* **14**, 105 (1987). Molecular Similarity Based on Electrostatic Potential and Electric Field.
160. J. J. McGregor, *Software Pract. Exp.,* **12**, 23 (1982). Backtrack Search Algorithms and the Maximal Common Subgraph Problem.
161. P. Willett, Ed., *Modern Approaches to Chemical Reaction Searching,* Gower, Aldershot, U.K., 1986.
162. A. T. Brint and P. Willett, *J. Comput.-Aided Mol. Design,* **2**, 311 (1988). Upperbound Procedures for the Identification of Similar Three-Dimensional Chemical Structures.
163. C. M. W. Ho and G. R. Marshall, *J. Comput.-Aided Mol. Design,* **7**, 3 (1993). FOUNDATION: A Program to Retrieve All Possible Structures Containing a User-Defined Minimum Number of Matching Query Elements from Three-Dimensional Databases.
164. J. B. Moon and W. J. Howe, *Tetrahedron Comput. Methodol.,* **3**, 697 (1990). 3D Database Searching and *De Novo* Construction Methods in Molecular Design.
165. S. E. Jakes, N. J. Watts, P. Willett, D. Bawden, and J. D. Fisher, *J. Mol. Graphics,* **5**, 41 (1987). Pharmacophoric Pattern Matching in Files of Three-Dimensional Chemical Structures: Comparison of Geometric Searching Algorithms.
166. C. Bron and J. Kerbosch, *Commun. ACM,* **16**, 575 (1973). Algorithm 457. Finding All Cliques of an Undirected Graph.
167. A. T. Brint and P. Willett, *J. Chem. Inf. Comput. Sci.,* **27**, 152 (1987). Algorithms for the Identification of Three-Dimensional Maximal Common Substructures.
168. I. D. Kuntz, *Science,* **257**, 1078 (1992). Structure-Based Strategies for Drug Design and Discovery.
169. J. M. Blaney and J. S. Dixon, *Perspect. Drug Discovery Design,* **1**, 301 (1993). A Good Ligand Is Hard to Find: Automated Docking Methods.
170. E. C. Meng, B. K. Shoichet, and I. D. Kuntz, *J. Comput. Chem.,* **13**, 505 (1992). Automatic Docking with Grid-Based Energy Evaluation.
171. E. C. Meng, I. D. Kuntz, D. J. Abraham, and G. E. Kellogg, *J. Comput.-Aided Mol. Design,* **8**, 299 (1994). Evaluating Docked Complexes with the HINT Exponential Function and Empirical Atomic Hydrophobicities.

172. R. L. DesJarlais, R. P. Sheridan, G. L. Seibel, J. S. Dixon, I. D. Kuntz, and R. Venkataraghavan, *J. Med. Chem.,* **31,** 722 (1988). Using Shape Complementarity as an Initial Screen in Designing Ligands for a Receptor Binding Site of Known Three-Dimensional Structure.

173. K. D. Stewart, J. A. Bentley, and M. Cory, *Tetrahedron Comput. Methodol.,* **3,** 713 (1990). DOCKing Ligands into Receptors: The Test Case of α-Chymotrypsin.

174. R. L. DesJarlais, G. L. Seibel, I. D. Kuntz, P. S. Furth, J. C. Alvarez, P. R. Ortiz de Montellano, D. L. DeCamp, L. M. Babe, and C. S. Craik, *Proc. Natl. Acad. Sci. U.S.A.,* **87,** 6644 (1990). Structure-Based Design of Nonpeptide Inhibitors Specific for the Human Immunodeficiency Virus 1 Protease.

175. B. K. Shoichet, R. M. Stroud, D. V. Santi, I. D. Kuntz, and K. M. Perry, *Science,* **259,** 1445 (1993). Structure-Based Discovery of Inhibitors of Thymidylate Synthase.

176. M. D. Miller, S. K. Kearsley, D. J. Underwood, and R. P. Sheridan, *J. Comput.-Aided Mol. Design,* **8,** 153 (1994). FLOG: A System to Select 'Quasi-Flexible' Ligands Complementary to a Receptor of Known Three-Dimensional Structure.

177. R. L. DesJarlais and J. S. Dixon, *J. Comput.-Aided Mol. Design,* **8,** 231 (1994). A Shape- and Chemistry-Based Docking Method and Its Use in the Design of HIV-1 Protease Inhibitors.

178. P. Y. S. Lam, P. K. Jadhav, C. J. Eyermann, C. N. Hodge, Y. Ru, L. T. Bacheler, J. L. Meek, M. J. Otto, M. M. Rayner, Y. N. Wong, C.-H. Chang, P. C. Weber, D. A. Jackson, T. R. Sharpe, and S. Erickson-Vitanen, *Science,* **263,** 380 (1994). Rational Design of Potent, Bioavailable, Nonpeptide Cyclic Ureas as HIV Protease Inhibitors.

179. M. C. Lawrence and P. C. Davis, *Proteins: Structure, Function and Genetics,* **12,** 31 (1992). CLIX–A Search Algorithm for Finding Novel Ligands Capable of Binding Proteins of Known 3-Dimensional Structure.

180. P. J. Goodford, *J. Med. Chem.,* **28,** 849 (1985). A Computational Procedure for Determining Energetically Favourable Binding Sites on Biologically Important Macromolecules.

181. R. L. DesJarlais, R. P. Sheridan, J. S. Dixon, I. D. Kuntz, and R. Venkataraghavan, *J. Med. Chem.,* **29,** 2149 (1986). Docking Flexible Ligands to Macromolecular Receptors by Molecular Shape.

182. A. R. Leach and I. D. Kuntz, *J. Comput. Chem.,* **13,** 730 (1992). Conformational Analysis of Flexible Ligands in Macromolecular Receptor Sites.

183. A. R. Leach, *J. Mol. Biol.,* **235,** 345 (1994). Ligand Docking to Proteins with Discrete Side-Chain Flexibility.

184. D. S. Goodsell and A. J. Olson, *Proteins: Structure, Function, and Genetics,* **8,** 195 (1990). Automated Docking of Substrates to Proteins by Simulated Annealing.

185. A. S. Smellie, G. M. Crippen, and W. G. Richards, *J. Chem. Inf. Comput. Sci.,* **31,** 386 (1991). Fast Drug-Receptor Mapping by Site-Directed Distances: A Novel Method of Predicting New Pharmacological Leads.

186. M. L. Connolly, *Biopolymers,* **25,** 1229 (1986). Shape Complementarity at the Haemoglobin$_{11}$ Subunit Interface.

187. P. H. Walls and M. J. E. Sternberg, *J. Mol. Biol.,* **228,** 277 (1992). New Algorithm to Model Protein–Protein Recognition Based on Surface Complementarity. Applications to Antibody–Antigen Docking.

188. F. Jiang and S.-H. Kim, *J. Mol. Biol.,* **219,** 79 (1991). "Soft Docking": Matching of Molecular Surface Cubes.

189. B. K. Shoichet and I. D. Kuntz, *J. Mol. Biol.,* **221,** 327 (1991). Protein Docking and Complementarity.

190. M. L. Connolly, *Biopolymers,* **32,** 1215 (1992). Shape Distributions of Protein Topography.

191. J. C. Cherfils, S. Duquerroy, and J. Janin, *Proteins,* **11,** 271 (1991). Protein–Protein Recognition Analysed by Docking Simulation.

192. F. Murtagh, in *Intelligent Information Retrieval: The Case of Astronomy and Related Space Sciences,* A. Heck and F. Murtagh, Eds., Kluwer, The Netherlands, 1993, pp. 29–48. Search Algorithms for Numeric and Quantitative Data.

193. P. Willett, *J. Chem. Inf. Comput. Sci.,* **23,** 22 (1983). Some Heuristics for Nearest Neighbour Searching in Chemical Structure Files.

194. R. Whaley and L. Hodes, *J. Chem. Inf. Comput. Sci.,* **31,** 345 (1991). Clustering a Large Number of Compounds. 2. Using the Connection Machine.

195. E. M. Rasmussen, G. M. Downs, and P. Willett, *J. Comput. Chem.,* **9,** 378 (1988). Automatic Classification of Chemical Structure Databases Using a Highly Parallel Array Processor.

196. A. T. Brint and P. Willett, *J. Mol. Graphics,* **5,** 200 (1987). Identifiction of 3-D Maximal Common Substructures Using Transputer Networks.

197. R. R. Rose and J. Gasteiger, *J. Chem. Inf. Comput. Sci.,* **34,** 74 (1994). HORACE: An Automatic System for the Hierarchical Classification of Chemical Reactions.

198. W. T. Wipke and D. Rogers, *J. Chem. Inf. Comput. Sci.,* **24,** 71 (1984). Artificial Intelligence in Organic Synthesis. SST: Starting Material Selection Strategies. An Application of Superstructure Search.

199. W. T. Wipke and D. Rogers, *Tetrahedron Comput. Methodol.,* **2,** 177 (1989). Tree-Structured Maximal Common Subgraph Searching. An Example of Parallel Computation with a Single Sequential Processor.

200. J. Gasteiger, W. D. Ihlenfeldt, and P. Rose, *Recl. Trav. Chim. Pays-Bas,* **111,** 270 (1992). A Collection of Computer Methods for Synthesis Design and Reaction Prediction.

201. A. P. Johnson and C. Marshall, *J. Chem. Inf. Comput. Sci.,* **32,** 418 (1992). Starting Material Oriented Retrosynthetic Analysis in the LHASA Program. 2. Mapping the Starting Material and Target Structures.

202. A. P. Johnson and C. Marshall, *J. Chem. Inf. Computer. Sci.,* **32,** 418 (1992). Starting Material Oriented Retrosynthetic Analysis in the LHASA Program. 3. Heuristic Estimation of Synthetic Proximity.

203. Y. C. Martin, M. G. Bures, E. A. Danaher, J. DeLazzer, I. Lico, and P. A. Pavlik, *J. Comput.-Aided Mol. Design,* **7,** 83 (1993). A Fast New Approach to Pharmacophore Mapping and Its Application to Dopaminergic and Benzodiazepine Agonists.

204. J. H. van Drie, D. Weininger, and Y. C. Martin, *J. Comput.-Aided Mol. Design,* **3,** 225 (1989). ALADDIN: An Integrated Tool for Computer-Assisted Molecular Design and Pharmacophoric Recognition from Geometric, Steric and Substructure Searching of Three-Dimensional Molecular Structures.

205. D. M. Bayada, R. W. Simpson, A. P. Johnson, and C. Laurenco, *J. Chem. Inf. Comput. Sci.,* **32,** 680 (1992). An Algorithm for the Multiple Common Subgraph Problem.

206. A. C. Good and W. G. Richards, *J. Chem. Inf. Comput. Sci.,* **33,** 112 (1993). Rapid Evaluation of Shape Similarity Using Gaussian Functions.

207. S. K. Kearsley and G. M. Smith, *Tetrahedron Comput. Methodol.,* **3,** 615 (1990). An Alternative Method for the Alignment of Molecular Structures: Maximising Electrostatic and Steric Overlap.

208. W. Payne and R. C. Glen, *J. Mol. Graphics,* **11,** 74 (1993). Molecular Recognition Using a Binary Genetic Search Algorithm.

209. A. C. Good, E. E. Hodgkin, and W. G. Richards, *J. Comput.-Aided Mol. Design,* **6,** 513 (1992). Similarity Screening of Molecular Data Sets.

210. A. C. Good and J. S. Mason, this volume.

CHAPTER 2

Three-Dimensional Structure Database Searches

Andrew C. Good[*†] and Jonathan S. Mason[‡]

[*]Department of Pharmaceutical Chemistry, University of California, San Francisco, 513 Parnassus Avenue, San Francisco, California 94143-0446, [†](permanent address) Dagenham Research Centre, Rhône-Poulenc Rorer Ltd., Rainham Road South, Dagenham, Essex RM10 7XS, United Kingdom, and [‡]Rhône-Poulenc Rorer Central Research, 500 Arcola Road, Collegeville, Pennsylvania 19426-0107

It is now five years since a chapter on three-dimensional (3D) structure searching was chronicled within *Reviews in Computational Chemistry*.[1] Although the original review still provides a comprehensive grounding in the principles of computational screening, a new treatise is necessary to detail the recent advances made in methodology and application. This chapter therefore focuses on the new techniques available for use with 3D structure searches and their application within an integrated computer-aided drug design (CADD) framework.

Reviews in Computational Chemistry, Volume 7
Kenny B. Lipkowitz and Donald B. Boyd, Editors
VCH Publishers, Inc. New York, © 1996

3D DATABASE SEARCHING AND THE DRUG DESIGN PROCESS

The methods that encompass computational screening of 3D databases have advanced to such an extent that they now form an integral part of the CADD process.[2,3] These search techniques attempt to exploit the structural information provided by experimental procedures such as nuclear magnetic resonance (NMR), X-ray crystallography, and random screening, so as to expedite the lead discovery process. An important supposition of the paradigm is that good agonists/antagonists must possess significant geometric and electronic complementarity to their target receptor in order to bind. Based on this assumption, binding mode models are elucidated through the development of structure-activity relationships (SARs). The resulting models are then used as constraints to search the huge number of 3D structures available through in-house and commercial chemical databases, in the quest for new lead compounds for biological evaluation.

All the elements of modern drug design, from the gathering of structural data, through the development of SAR models, to the synthesis and testing of potential leads, form part of an iterative design process. From the outset, the integrated manner in which 3D screening techniques (and all the other elements of CADD) are accommodated within the paradigm needs to be emphasized (see Figure 1).

ADVANCES IN 3D SEARCH METHODOLOGY

In the 1970s, Gund described how binding mode models made up of 3D arrangements of functional groups (pharmacophores) could be used to search databases for molecules sharing the same structural features.[4] Limitations, such as the small databases then available for screening, restricted technique utility. For 3D screening to be successful, it must be possible to convert large, available 2D chemical databases into 3D. A number of techniques have been devised[5–10] that allow the generation of good-quality 3D structures from two-dimensional (2D) connection tables. This has allowed both pharmaceutical and chemical information companies to convert their compound databases for use in 3D screening.[11–13] Techniques are also required that allow the resulting 3D databases to be searched rapidly. Many pharmaceutical companies developed in-house searching systems in the late 1980s to answer their immediate requirements for 3D screening.[14–17] Molecular Design Ltd. also created MACCS-3D,[18] which provided a bridge between its widely used 2D database system[19] and 3D search functionality. None of these systems in their early

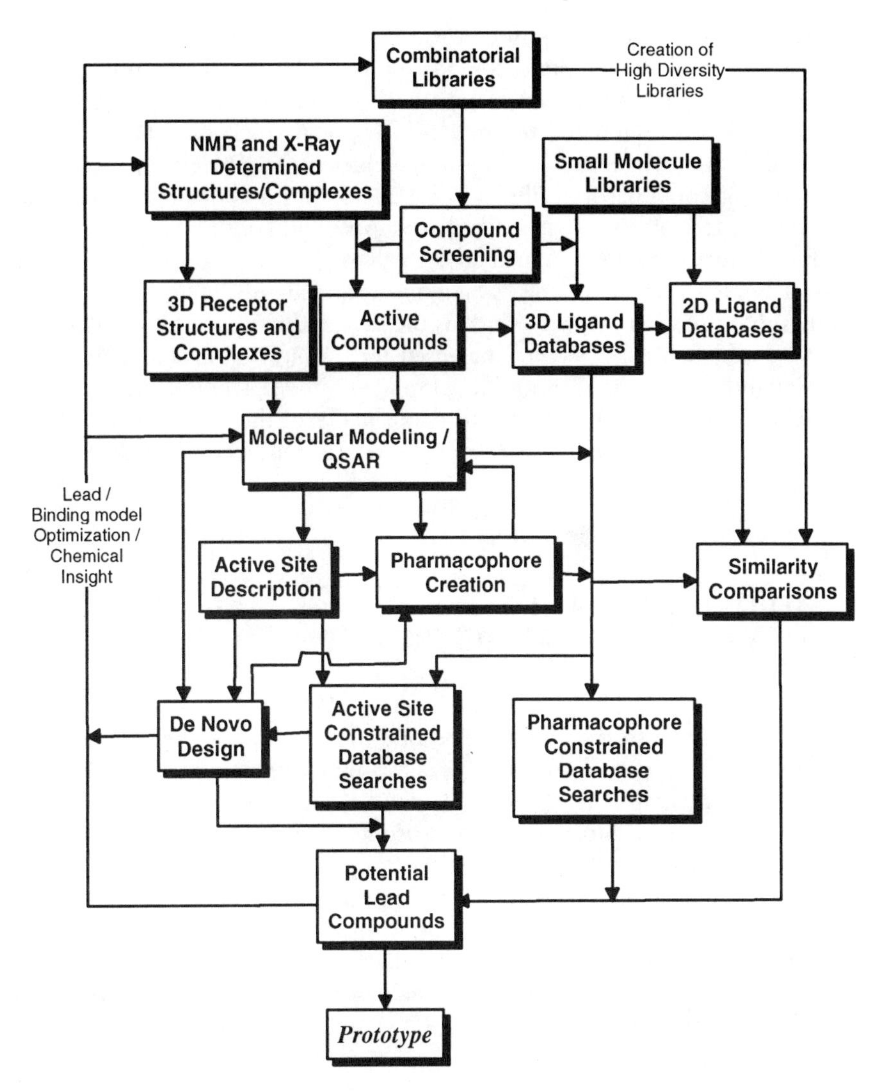

Figure 1 The iterative and integrated process of rational drug design.

incarnations explicitly addressed the important problem of conformational flexibility. It is only recently that commercial software packages have been developed to deal with this problem,[20–23] using a variety of innovative techniques.[24–27]

A 3D database cannot be screened without a binding model hypothesis to define the search (although many searches are based solely on potentially interesting molecular features, with little thought given to a specific binding mode). To this end, a host of techniques have been devised to create, define, and validate potential pharmacophores.[20,24,28–36] New methodologies have also been devised to allow the rapid comparison of molecular shape, thus permitting the use of ligand shape constraints in 3D database screens.[37–43] Continuing advances in experimental techniques have greatly increased the amount of structural data available for binding mode elucidation. For example, the new technology of combinatorial library screening allows millions of potential ligands to be tested for activity in a relatively short period of time.[44,45] The rate at which structures are being elucidated by X-ray crystallography and NMR analysis is also ever-increasing. This has expanded the interest in database searches that use receptor active sites as the search constraints for 3D database screening calculations,[46–60] and increased the amount of structural information available for pharmacophore development and CADD in general.[61,62]

The aims of this chapter are to detail many of these advances and give examples of the success of 3D searching and related CADD tools in drug design problems.

3D DATABASE AND PHARMACOPHORE QUERY CONSTRUCTION

A prerequisite to the construction of a 3D database is the generation of representative 3D structures. This step is often undertaken independently of the database being created (e.g., using a stand-alone program such as CONCORD[5]), although scientists sometimes use features of the database program they intend to work with (e.g., Chem-X/ChemDBS-3D[20]). A reasonable 3D structure is thus usually generated first, and this single conformation is registered in the database. Either during or after this registration, keys (descriptors to flag the presence or absence of a molecular feature and some possible inter–feature distances) can be generated to enable “unsuitable” compounds to be rapidly screened out before the slower 3D geometry search.

Structure Generation

The conversion of a database of chemical compounds to 3D structures first involves the creation of appropriate 2D connectivity table files with atom and bonding descriptions for each structure. This is used as input for an automated 3D structure generation program. Most techniques use SMILES string connection tables as their starting point for the conversion process,[9,10] though some use 2D MOL (molecule) and SD (standard data) files from MACCS[19] (the original 3D database chapter in *Reviews in Computational Chemistry* contains detailed examples of the use and syntax of SMILES strings[1]).

Many of the conversion programs apply a knowledge base to construct the 3D structure. For example, the CONCORD program combines rules with energy estimation procedures in an attempt to produce the lowest-energy 3D conformation for each structure.[5] Cyclic and acyclic portions of each structure are constructed separately, and the resulting 3D sections are then pieced together to form the complete molecule. Acyclic bond lengths and angles are extracted from a table of published values. Torsional angles are assigned so as to generate optimum 1–4 interactions. Similarly, bond lengths and torsional angles of single cyclic portions are constructed using precalculated rules. Ring systems are constructed through the assignment of gross conformations of each ring. Each constituent ring is then fused into the system, and a strain minimization function is employed to produce a reasonable geometry. Once the molecule has been constructed, it can be further optimized through energy minimization by other programs. CONCORD is able to produce reasonable low-energy conformations for most small organic structures, though large ($>$10) rings and peptides are not well handled, and most organometallic compounds cannot be converted. The program is rapid, gives useful structure quality information (e.g., a close-contact ratio), and allows the output of many file formats.

The 3D builder within the Chem-X[7] modeling program (Chem-Model) employs a knowledge base of ring fragments that are fused together as required by the SMILES string construction. Acyclic portions of the molecule are then added using standard molecular mechanics parameters. Whereas regular torsiional angles are used in molecular construction, no attempt is made to ensure a low energy structure, because it is assumed that the molecule will be used as a starting point for 3D database conformational searches.

The program CORINA is another rule-based generation paradigm gaining acceptance with the CADD community.[66,70] Features of CORINA include advanced ring handling techniques, the ability to handle organometallics, and a high successful conversion rate.

An alternative approach has been used in the Converter program,[8] which applies distance geometry[63] in conjunction with upper and lower distance bounds (interatomic distance ranges) based on published tables of bond lengths. These bounds are used together with topological rules to generate a 3D distance geometry description of the 2D connection data. Complex ring systems are readily handled, but 3D structure generation is slow, and often the resulting structures need verification and lack uniformity in bond geometries, possibly needing some further minimization. In spite of this shortcoming, the technique is able to cope with most structures, frequently generating structures successfully in cases where CONCORD has problems.

Attempts have been made to compare the quality of structures produced by the various 2D–3D building programs.[64–66] These studies measured the differences between experimental Cambridge Structural Database[67] (CSD) and Protein Data Bank[68] crystal structures and the same structures constructed using the 2D–3D conversion programs, to compare algorithm speed and reliability. The basic conclusion of these studies is that techniques accounting for conformational flexibility through the creation of multiple models tend to be better at generating structures similar to those found in the crystals. This is intuitively what one would expect, since the more models one creates, the better the chance that a conformation close to the experimental structure will be obtained. Combine this with the fact the global minimum conformation of a molecule in isolation is not necessarily the same as when the molecule is bound to a receptor,[69] and it is clear that multiple model generation is preferable.

The more recently developed 3D screening algorithms are able to consider conformational flexibility at run time.[24–26] Consequently, if one is undertaking a search that includes conformational flexibility, the torsional angle choices made for the parent structure should be of lesser importance, as long as the 3D builder is capable of generating structures with reasonable bond lengths and angles. If one wishes to undertake a rigid search on a database, in contrast, then the quality of the stored 3D structures becomes an important factor because it is then the explicit geometries of these structures that determine search results.

As we have intimated, one problem for rule-based, 3D structure generation systems is that they are not able to create some of the more exotic structures present in a given database.[64] If, for example, an 11-membered ring with complex functionality were required, it is quite possible that the rules for such a ring system would not have been stored in the rules tables or fragment database. Some programs (e.g., Chem-Model) add fragments as required. Generating correct stereochemistry plus the type of database search paradigm being used must all be considered when deciding upon the builder for the problem at hand. For a more complete review of these and other automated 3D builder programs, see Reference 70 (though because of the fast-moving nature of the field, reading should be supplemented with References 64–66).

Atom Typing on Compound Registration

When 3D structures are read into a database system, different actions can occur. For example, if atom typing is used, then a parameterization database of special fragments is applied to assign the desired ring centroids, hydrophobic centers, and atom types to each structure (this is the method used in the ChemDBS-3D module of Chem-X). The process is user-definable, allowing extensive customization to suit particular needs and structures. The ability to differentiate and associate different atom types and environments can greatly alter 3D search efficiency and results. For many pharmacophore searches, it is important to be able to search for centers with a particular property not limited to a single group. For example, one may search for a basic nitrogen (which, for example, can be an sp^3 nitrogen of aliphatic amines and hydrazines, an sp^2 nitrogen of amidines, guanidines, and 2- or 4-amino pyridines but not other sp^2 and sp^3 nitrogens), an acidic center (such as both oxygens of a carboxylic acid and all nitrogens of a tetrazole), and aromatic ring centroids and/or hydrophobes (5- or 6-membered aromatic rings, isopropyl, butyl, cyclohexyl, and so on).

Chem-X allows the inclusion of user-defined atom types and centers (e.g., hydrophobes) through the addition of new fragments in a substructure database, which is automatically applied when a structure is read into the program. In the definition of a query, different atom types can be readily equivalenced or a general element type used. In a modified parameterization, many different atom types have been assigned during registration (see Table 1 for an example). With Chem-X software, hydrogen bond donor and/or acceptor center types can be selectively given only to unsubstituted groups (e.g., amides, aromatic N) and changed upon substitution (e.g., an aniline type N can change from a hydrogen donor to an acceptor upon di-substitution). By this method, a hydrogen bond donor can be identified solely by its atom type; it is not necessary to have an attached hydrogen. This allows better use of tautomeric nitrogens (e.g., in imidazole) and oxygens (e.g., in 2-pyridone) where only one of the atoms of a tautomeric pair will be connected to a hydrogen. Tautomeric atoms can thus be coded identically, allowing the simulation of physiological conditions for acidic and basic centers.

Two types of fragment parameterization are used in the ChemDBS-3D method, one requiring an exact connectivity match but ignoring bond order, the other defining a minimum connectivity count and requiring an exact bond order match (this definition is also used to define "dummy" centroid atoms, e.g., aromatic ring centroids). Atom modifications are applied in the order of the fragment's appearance in the molecule, using a special parameterization database, making the order very important and allowing for selective modifications after a generic modification. A fragment atom may force a modification of the corresponding atom in the molecule or exist purely to identify the environment of another, depending on the identity of its name.

Each atom type can be associated with one or more center types, i.e., hydrogen bond donor, hydrogen bond acceptor, positively charged center,

Table 1 Customized ChemDBS-3D Nitrogen Atom Types[87]

Nsp nitrile	14A	
General sp^2 (imine)	14M (NH)	14B (no H)
Quarternary	14C	
Nsp^3 (basic, amine)	14D	
Nsp^3 (basic, hydrazine)	14N (NH)	14P (no H)
Nsp^2 (basic, amidine, . . .)	14U	
Nsp^3 (tertiary amine)	14W	
Amide NH	14E (NH)	14Q (no H)
Pyrrole	14F	
Azo	14G	
Aromatic	14H	14Z (substituted)
Nitro	14J	
Sulphonamide	14S (NH)	14R (no H)
Tetrazole-H	14T	
Ar-NH_2	14V (NH)	14X (disubstituted)
Tautomeric, e.g., imidazole	14Y	

aromatic ring center, or hydrophobe (optionally combined with an aromatic center).

A query point can thus be defined intuitively, simply by allowing the drawn atom type to be equivalent to specified (additionally required) atom types.

Search-Time-Defined Atom Environment

Rather than defining atom environments during compound registration, the search-time-defined atom typing technique only stores the element type within the database, and the query must define the desired atom environment. This can be simple, for example, with an amide nitrogen, but it becomes complicated when the environment definition requires the exclusion of unwanted chemical groups or the linking of different groups (e.g., a carboxylic acid oxygen and a tetrazole nitrogen, or the centroid for 5- and 6-membered aromatic rings). This method is used by UNITY[21] and ISIS/3D.[23] In UNITY the atom environment is defined at search time using SYBYL Line Notation (SLN: an extended SMILES-like language). Limited environments can also be defined using graphical substructures. The SLN language is powerful and flexible but is rather complicated for some atom environments. On the other hand, ISIS/3D uses a powerful graphical definition technique as the principal method for definition. This is an advantage for simple cases, but the ISIS approach is more complicated for environments such as basic and acidic centers, though the use of predefined environment templates should make the ISIS methodology somewhat easier. The search-time method, although complicated, is potentially flexible, as there are no predetermined environment restrictions for a particular search. Additional complications can arise as fragments must often be used in environment definition, and the correct attachment of the 3D constraint for, say, the distance definition must be assured. For example, a carboxylic acid in SLN, Aci{Aci:OHC=O} would calculate the distance from the centroid of all atoms defining the query; to use only the oxygen atoms one would need Oacid{Oacid:O&!(!OHC=O)|(O&!(!O=COH)}.

DEALING WITH CONFORMATIONAL FLEXIBILITY

The first implementations of 3D database searching software relied on the use of a single low-energy conformation to represent each molecule within a given database.[14–18] This was recognized as a serious deficiency, and new methodology was developed to obviate the problem. Attempts were made to construct pharmacophore search queries that implicitly took account of conformational flexibility.[71] More generally applicable approaches have recently been developed, which employ explicit conformational information regarding database structures, storing the flexibility information within the database, and/or determining conformational possibilities at search time.[24–27,64]

Explicit Conformation Storage

At the simplest level, conformational flexibility has been represented through the explicit storage of multiple molecular conformations.[27,64,133,134] The problem with such an approach is that, if one wishes to keep the storage and search times down to a reasonable level, only a small number of conformers may be stored per molecule. Thus, if this technique is to be successful, careful attention must be paid to which molecular conformations should be retained in the database. A general approach has been to apply conformational sampling methods[63,72–74] to search conformational space and then use a difference metric, for example, the root mean squared (RMS) difference in interatomic distances, to allow the clustering of conformations into families.[75]

This form of postprocessing methodology suffers from a number of drawbacks. A major problem is obtaining good coverage of conformational space. Sampling methods are likely to miss regions of space unless they are run for prohibitively long times. This leads to the second drawback, which is conformational redundancy. All these techniques tend to produce many thousands of conformations, which are subsequently pared down to a just a few conformational families. The problem is further exacerbated if one requires each conformation to be a local minimum, since all conformations lying in the same energy well will collapse down into a single family. As a consequence, conformational sampling can be inefficient in terms of both the keying time required and the amount of conformational space covered.

To mitigate these problems, a new technique for conformational sampling called poling has been developed.[27] This method is designed to promote conformational variation, while being easy to calculate and being compatible with existing conformer generation techniques. This has been accomplished through the addition of a "poling function" to a standard molecular mechanics force field (see Eqs. [1]–[3]).

$$\text{Energy} = F_{\text{bond}} + F_{\text{angle}} + F_{\text{dihed}} + F_{\text{VDW}} + F_{\text{coulomb}} + F_{\text{pole}} \quad [1]$$

where

$$F_{\text{pole}} = W_{\text{pole}} \sum_i \frac{1}{(D_i)^N} \quad [2]$$

and

$$D_i = \left[\frac{\sum_j (d_j - d_{ij})^2}{N_d} \right]^{1/2} \quad [3]$$

In these equations, D_i = RMS difference between the poling distances of the current conformation being poled (d_j) and the corresponding distances of the ith preceding conformation (d_{ij}) for all N_d poling differences. The values for d are user-defined and are usually based upon distances between chemical features that would be used in subsequent pharmacophore searches. Pole gradient is adjusted through alteration to D_i to an arbitrary power N and adjustment of the pole weighting W_{pole}. The function has the effect of changing the energy surface being minimized, as shown in Figure 2. By penalizing conformational space around any previously accepted conformers, the technique increases conformational variation and eliminates redundancy within the limits imposed by the function. Poling has been applied within the Catalyst 3D searching software.[22]

3D Screens Incorporating Conformational Flexibility

An alternative approach to the explicit storage of multiple conformations is the use of 3D distance screens. The ChemDBS-3D program[20,24] provides a good example of such a technique and deals with the problem of conformational flexibility in the following manner:

1. Define ligands in terms of chemical centers as described in the earlier section on atom typing on compound registration.
2. Determine the location of all these centers within the molecule.
3. Identify the rotatable bonds within the structure.
4. Initiate a conformational analysis of the molecule. Conformations are usually generated through the application of rule-based techniques arising from the ideas of Dolata et al.[76] In this way, it is possible to undertake crude systematic conformational searches in a reasonably short period of time. During the search, each conformer is checked for the distances between all the centers present in the molecule. These distances are stored within a one-word (32-bit) screen set. Each bit within the word represents a distance range bin. Therefore, if a particular conformer were found to have a donor–donor distance of 5.5 Å, using the bin settings shown in Table 2, the 20th bit of the donor–donor word screen would be switched on. Bin sizes are assigned to reflect the general size of uncertainties in known pharmacophores. The bin ranges are also designed to provide the most accuracy (i.e., smaller range settings) for the most commonly occurring distances, with larger range bins used for distances that occur with a lesser frequency. For example, previous studies[77] showed that distances greater than 15 Å rarely occur, so the 31st bin is reserved for all these distances. A typical bin distribution is shown in Table 2.
5. Store the following parameters within the 3D database:
   a. The combined 3D screens.

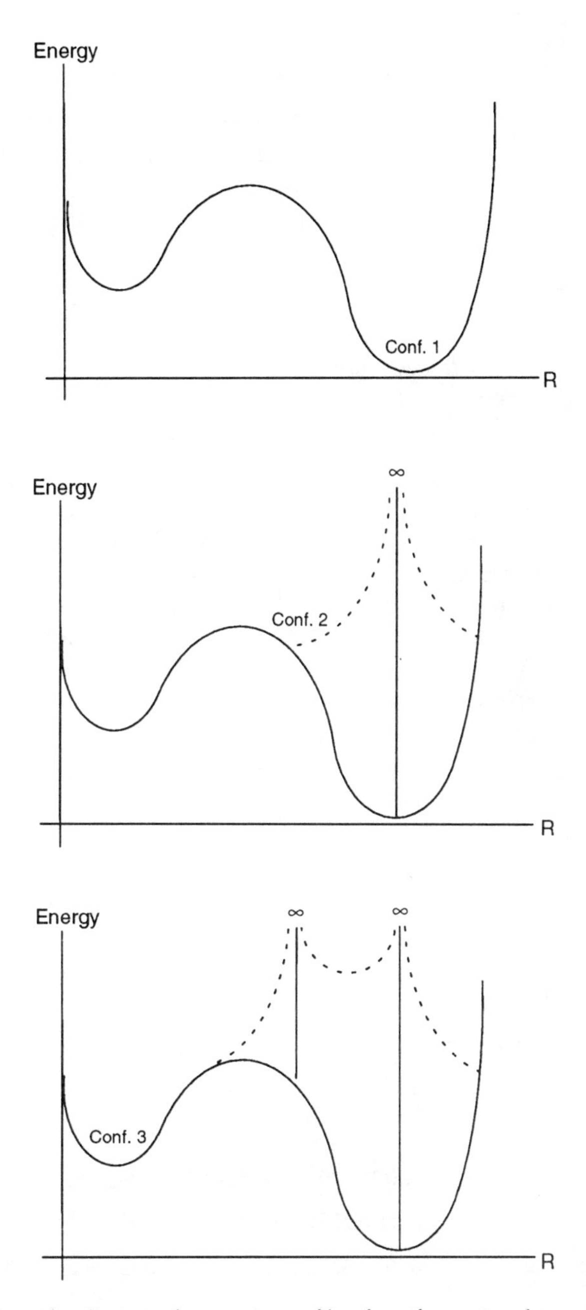

Figure 2 Effect of poling on the energy profile of conformational space.[27] The solid line represents the standard energy surface of a given molecule. Once the first conformation has been determined, the poling function has the effect of modifying the energy surface as shown by the broken lines. Conformation 2 is then determined using the modified energy profile. The poling function then alters the energy surface at two points as shown, and this profile is used to determine conformation 3.

Table 2 Typical Bin Distribution Used for ChemDBS-3D 3D Screens[24]

Distance range	Bins assigned	Bin step
<1.7 Å	1	
1.7 Å–3.0 Å	2–14	0.1 Å
3.0 Å–7.0 Å	15–22	0.5 Å
7.0 Å–15.0 Å	23–30	1.0 Å
>15.0 Å	31	

b. A formula key, which describes the number of each type of center present within the molecule.
c. The energy (if calculated) of the most stable conformation.

To search the resulting 3D screens, a query can be constructed using the molecular fragments that make up the required pharmacophore. As with the screens keyed during database construction, the pharmacophore center types and locations are determined together with the distances between them. User-defined distance tolerances are applied, and then a separate search key is generated for each center–center distance, together with a formula key for the pharmacophore. A sample query and its associated keys are shown in Figure 3.[78] The 3D search screens are used in database searches using a six-step procedure.

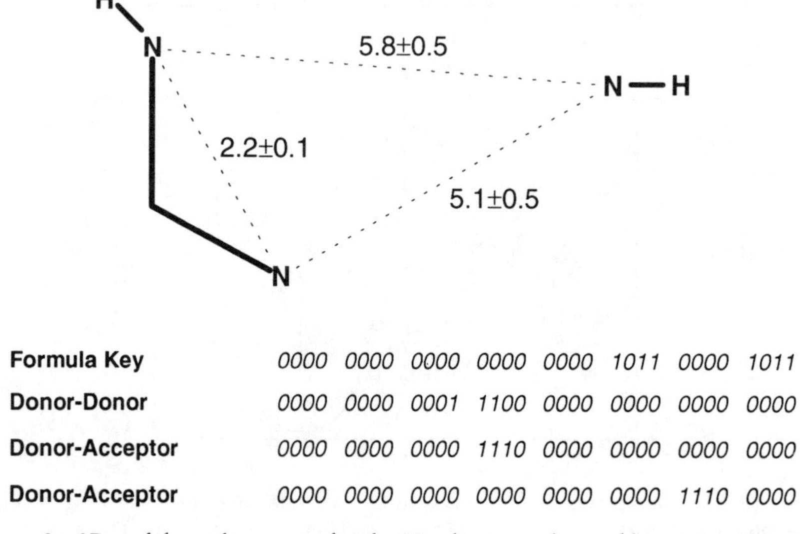

Figure 3 3D and formula screens for the H_1 pharmacophore of histamine.[24,78] The binary 1 and 0 values signal the presence or absence of a particular center type and center–center distance.

1. Formula keys are compared with a logical AND. A result equal to the pharmacophore formula indicates that all the pharmacophore features are present.
2. For those molecules with matching formula key, compare each pharmacophore screen key with the corresponding key of the database molecule using a logical AND. A nonzero result indicates that the keys have at least one bit in common, and so the screen criterion is satisfied.
3. If all pharmacophore keys are satisfied, the molecule is stored in an answer set.
4. Once the key search is complete, a substructure search is undertaken, where connectivity and atom type information between pharmacophore and answer set molecule are compared.
5. Property field comparisons are then undertaken. Property field data can include any user-defined numerical information regarding the structures within the database, for example, log P or LD_{50} data.
6. If all these tests are satisfied, the conformations of the passing molecules are regenerated, with the query pattern acting as a further restraint to speed up the calculation. Each low energy conformer that conforms to the pharmacophore is then superimposed on the query using a least-squares fit[79] and is written out to a results database. The relative speeds of each part of the search process are shown in Table 3.

Tests have been carried out using ChemDBS-3D to show how flexible 3D database searches are able to find hits that would be missed by searches against a single parent conformation database of the same compounds.[80] In one such test, a 3D database built using Chem-Model[7] from the Derwent Standard Drug File was searched using a published pharmacophore for inhibition of thromboxane synthetase.[81] Using a pharmacophore consisting of a nitrogen to acid oxygen distance of 9.6 ± 1.5 Å, inclusion of the conformational regeneration stage of the database search was found to yield an extra 30% of hits over a straight search of the single stored conformation. In all, 53 of the 63 thromboxane inhibitors within the database were found with the flexible search. Of the remaining 10 compounds, 9 had an unsuitable nitrogen atom, while the other missing compound had too short a distance between a possible nitrogen and the acid oxygen.

Table 3 Average Search Speeds for Each Component of a ChemDBS-3D Search.[24] Speeds Are in Compounds per CPU Second on a VAX 8600

Search technique	Speed
Screen	100,000
Substructure	10
Field	150
Final matching	3

Torsional Fitting

This is the latest technique devised to deal with the conformational searching problem.[25,26] The technique involves undertaking torsional optimization of a candidate molecule, in an attempt to tweak the rotatable bonds of the structure so that it meets the 3D search query constraints. Variants of this approach are being or have been implemented within MACCS-II/3D,[18] ChemDBS-3D,[20] SYBYL/3DB[21] UNITY, and ISIS/3D.[23]

The screening procedure employed by ISIS/3D illustrates how the torsional optimization methodology can be incorporated into 3D searches[26]:

1. Execute a topological screen to ensure that the functional groups present in the search query can also be found on the candidate structure.
2. Match search query atoms to atoms in the candidate, using a rapidly calculable approximation of the 3D screens described previously. Rather than using a complete systematic comformational analysis to determine the attainable atom pair distances, the technique used here determines values for the upper and lower distances possible between a given atom pair. The procedure is illustrated in Figure 4.
3. For searches containing only one distance constraint, molecules passing step 2 are considered to have satisfied the constraint and are added to the hit list. If, however, two or more distance constraints exist, specific molecular torsions are tweaked to verify that all the constraints can be met simultaneously. To do this, the rotatable bonds between matched atoms are determined (here defined as all acyclic bonds, excluding amide bonds and bonds between two unsaturated atoms), and the molecule is subjected to torsional optimization to minimize the RMS deviation between the constraint and fitted model values, according to the following equation:

$P =$

$$\sum_i w_d|d_i^2 - d_o^2| + \qquad \text{Distance}$$

$$\sum_i w_{a3}|\cos\theta_i - \cos\theta_{oi}| + \qquad \text{Angles (3-point)}$$

$$\sum_i w_{a4}|\cos\Omega_i - \cos\Omega_{oi}| + \qquad \text{Dihedral angles (4 point)}$$

$$\sum_i w_l(RMS_{\text{linei}}) + \qquad \text{Line constraint}$$

$$\sum_i w_p(RMS_{\text{planei}}) + \qquad \text{Plane constraint}$$

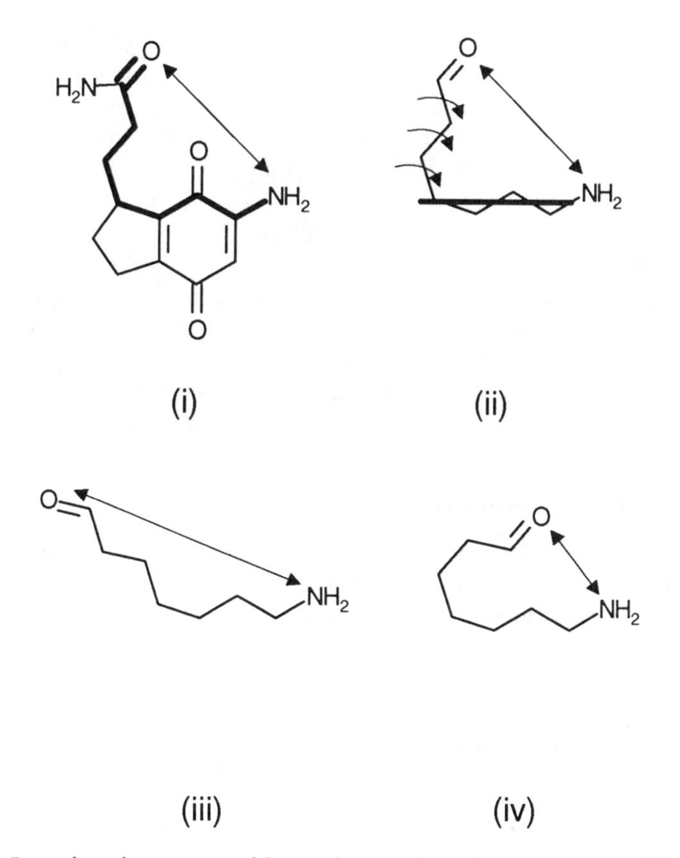

Figure 4 Procedure for upper and lower distance bound calculation in ISIS/3D[23]: (i) Determine the topological construction of the molecule ring systems and the rigid and flexible regions. (ii) Calculate the shortest path (fewest rotatable bonds) between a given pair of atoms and reduce the contiguous nonrotatable bonds within the path to fixed distances. (iii) Determine the maximum possible distance by setting all rotatable bonds to 180°. (iv) Calculate the minimum distance using a depth first conformational analysis. Stop if any distance found is less than the sum of the atom pair VDW radii.

$\sum_i w_e(N_i) +$	Exclusion sphere (N = no. of atoms in sphere i)
$\sum_i w_f(RMS_{\text{fixed atoms}}) +$	Fixed atoms
$\sum_m w_m \dfrac{r_{o,i} + r_{o,j}}{r_{i,j}}$	Bump check [4]

$w_{d,a3,a4...}$ are empirical scaling factors, $r_{o,i}$ is the default VDW radius of atom *i*, and $r_{i,j}$ is the measured distance between atoms *i* and *j*. Simplex optimization is used by default,[82,83] with derivative-based optimization provided for queries containing only distances and angles.[83,84] Both fast and slow searches excluding and including van der Waals (VDW) bump checking are possible.

The system was tested through searches of the Comprehensive Medicinal Chemistry (CMC) database[19] (5374 structures of known pharmacological interest) using six different pharmacophores. The results showed that searches including a torsional optimization with no bump check took 3 to 20 times as long as a static 3D search and yielded 2 to 14 times as many hits. Inclusion of a bump check caused the optimization procedure to take up to 7 times as long and screened out approximately 20% of the hits. Sample timings for a search of the CMC using an antihistamine pharmacophore are shown in Table 4.

Choice of Technique

In outlining the torsional tweak methodology,[26] it was described as "providing the best compromise of resources, using commonly available computing hardware." Other studies have corroborated this general statement,[25,85] concluding that the directed tweak approach provides the most rapidly calculable method for scanning the full range of conformational space. Our experiences with ChemDBS-3D, UNITY, and ISIS/3D also find torsional minimization to be the fastest search technique, able to handle very flexible molecules rapidly

Table 4 Timing Tests for Search of 5374 Models of the CMC Database, Using the Antihistamine Pharmacophore Shown. CPU Times Are in Seconds and Were Carried Out on a VAX 6610 Computer[26]

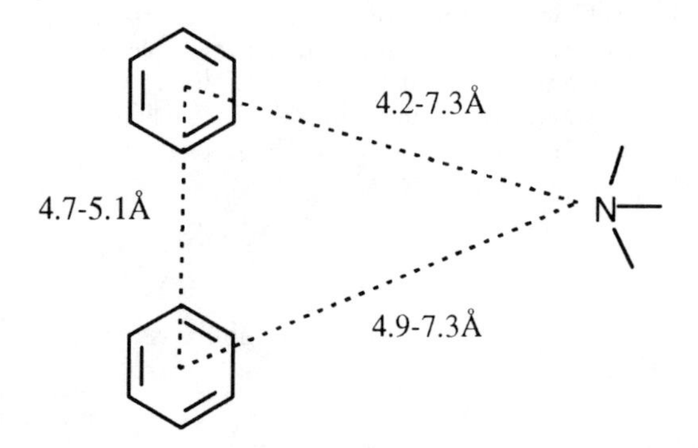

Search method	Number of hits found	CPU search time
Static 3D substructure search	208	13
Torsional tweak with no bump check	556	45
Torsional tweak with bump check	463	162

and thoroughly using small query distance tolerances.[86,87] The method is limited, however, in that it is not systematic and can identify only a very limited number of possible conformational solutions; some implementations are even more restrictive in returning only one substructure mapping. This limitation is not important when the goal of a search is to identify all compounds in a database that can fit a particular pharmacophore, but it can be restrictive for applications such as pharmacophore identification and idea evaluation.

The fast, rule-based, systematic conformational analysis used in the ChemDBS-3D system is not suitable for very flexible molecules (e.g., >14 rotatable bonds, ~5% of compounds in a 62,000 corporate structure database) or for use in conjunction with small tolerance distances (a minimum of 0.5 Å or 20% of the target distance is recommended). It does, however, have the advantage of allowing the systematic evaluation of all conformational and substructure mappings of a molecule to a particular 3D query, within a reasonable time scale for many structures. A full energy calculation is not feasible, except for less flexible molecules (e.g., up to 20,000 conformations), but the use of the rules or a fast steric bump check is usually adequate. We have no published information regarding the nature of answer sets obtained from databases created using the poling methodology, but in principle, it is possible that multiple hits for a single molecule could also be obtained using this approach, though the number of conformations sampled per molecule using this technique is somewhat lower.

Overall, it is clear that each method of conformational regeneration has distinct advantages and disadvantages, depending on the application. A comparative study was made using 58,000 3D structures of our company's registry compounds, generated by CONCORD, built into 3D databases for ChemDBS-3D, ISIS/3D, and UNITY.[87] The results of some simple 3D pharmacophore distance searches are given in Table 5. It was clear from these studies that for very flexible molecules, torsional minimization techniques are the most effective and that tolerance values need to be adjusted according to the method of conformational exploration used. The effect of allowing for ring flexibility can be seen with the UNITY results. Considerations such as atom types versus element types or speed of search versus thoroughness of conformational and substructure mapping search, together with visualization and analysis possibilities, have an influence on the effectiveness of any particular implementation or algorithm. For example, the ability to instantly evaluate hit compounds in 3D superimposed on the query centers, as implemented in ChemDBS-3D, was found to be extremely useful. The pharmacophore searches using rule-based conformational analysis in ChemDBS-3D normally proceed at acceptable rates, but this is generally slower than torsional fitting methods. The torsional fitting methods do, however, use less refined 3D distance keys, leading to a less rapid screen-out of compounds from the database prior to the much slower substructure and 3D searching. Although the yield of hits is also generally higher with torsional fitting, not all the matching compounds from

Table 5 Comparison of 3D Database Search Results

Technique	Hits from conformationally flexible 3D search
	Pharmacophoric distance search: Aromatic nitrogen to oxygen of carboxylic acid N ↔ O 9.25 ± 0.75 Å:
UNITY	1016 with ring flexibility
	1009 with no ring flexibility
	990 when VDW bump check & ring flexibility included
ISIS/3D	961 with no ring flexibility
	954 with explicit VDW bump check, no ring flexibility
ChemDBS-3D[a]	919 rule-based conformational analysis, no ring flexibility
	Pharmacophoric distance search: Nitrogen to carboxylic acid oxygen to aromatic ring (6-membered): N ↔ O 14 ± 0.5 Å; N ↔ AR6 14 ± 0.5 Å; O ↔ AR6 7 ± 0.5 Å:
UNITY	151 with no ring flexibility
	160 with ring flexibility
	122 with N = basic nitrogen
ISIS/3D	152
ChemDBS-3D[a]	*Systematic rule-based analysis:*
	Most suitable compounds too flexibile (>14 rotatable bonds) for systematic rule-based analysis (118 of the 122 hits from UNITY)
	Searching for the 122 hits of UNITY search by
	Random conformational analysis:
	60/122 (tol = ±0.5 Å)
	90/122 (tol = ±1.0 Å)
	Flexifit (torsional minimization):
	122/122 (tol = ±0.5 Å)

[a]ChemDBS-3D uses rule-based systematic conformational analysis (single = 3, alpha = 4, double/conjugated = 2 points) or random rule-based (max. time = 2 min), or flexifit (no VDW bump check); ISIS/3D and UNITY use torsional minimization (conformationally flexible searching and directed tweak, respectively.

the conformational search are necessarily found. Indeed, for handling cis-trans double bond isomerism or mixtures, a conformational method may be necessary.

It was found that structures with up to 14 rotatable bonds are readily handled using the rule-based conformational analysis; 2 (conjugated and double bonds), 3 (single bonds), and 4 (sp^2-sp^3 bonds) rotamers per bond were calculated in the databases for such structures.[86] The default for "alpha" (sp^2-sp^3) bonds in Chem-X is a 6-point analysis, but 4 points can offer a large gain in time without necessarily a reduction in yield. The number of points is based on the increment of rotation about a particular bond and equals 360° per rotational increment. Thus a 60° increment would equal a 6-point bond. For a corporate database of 62,000 company compounds, a total of 29 million conformations were accepted by the rules from a total of 912 million considered (on average, there were ~500 conformations accepted per compound, requiring about 3 seconds CPU for the calculation on a Silicon Graphics workstation; a maximum time of 60 seconds CPU was allowed). About 6000 compounds had no suitable bonds for rotation, and 1600 were too flexible (>14 bonds for rotation); for the latter class of compounds, other analysis strategies, such as torsional fitting or random conformation generation are more effective.

The yield of compounds from a 3D search can also be affected by how a potential hit compound is fitted to the query. The requirements in some implementations are that the final distances are within the tolerances defined in the query; only other factors such as unacceptable steric bumps could stop the compound from being a valid hit (e.g., in ISIS/3D or UNITY). However, if the query is in 3D, rather than being a 2D sketch with distance constraints, the hit conformation is fitted to the query, using the corresponding atoms to define the superposition (e.g., in ChemDBS-3D or UNITY). In this latter method, a second fitting tolerance may come into play, and the acceptance of a particular molecule may be affected by the chirality of the query and whether this final fitting uses just the atoms or centroids defining the distances or all atoms in the query (i.e., including those defining a particular substructure such as an amide NH or an aromatic ring centroid). The most striking difference is with a nonplanar 4- or more point query; since this is chiral in 3D, the searches may only identify structures where the "chirality" defined by the corresponding points is correct in addition to the interpoint distances; it may be possible to vary the fitting tolerance in addition to the distance tolerances to mitigate these effects.

PHARMACOPHORE GENERATION AND VALIDATION

In order to undertake database searches for a biomolecular system of interest, the creation of some form of binding model is essential. Generally, whereas a number of ligands for a given receptor may be known, the receptor structure itself may not be. In this instance, one must infer the critical small molecule–receptor interactions from the data provided by the ligand structures. We now consider some of the many techniques used to solve this pattern recognition problem.

Molecular Graphics

The classic approach to pharmacophore development has been through the visual analysis of low-energy ligand conformer superpositions using molecular modeling packages.[88] The work undertaken by Bures et al.[89] to design new inhibitors of auxin transport provides an excellent example of the use of this approach. A summary of the study is shown in Figure 5. Using compound 5 as a template, the seven compounds were overlaid, using low-energy conformations (determined from conformational analysis and energy minimization) and ensuring maximum overlap of the pharmacophore points. The rotatable bonds in each structure were then adjusted manually to give the best overall overlap for the full compound set, while ensuring that all resulting conformers were within 5 kcal/mol of the determined global energy minimum. The resulting overlay was used to determine the interpoint distances required for activity. Three separate 3D database searches were then undertaken in ALADDIN[14] using a number of pharmacophore variants. The searches produced 467 hits, of which 77 were taken forward for testing. Nineteen of these showed significant activity (a much higher hit rate than would generally be achieved through random screening).

Automating the Search for Pharmacophores

Molecular modeling and graphics provide powerful tools in the search for potential binding models. It has been recognized for some time, however, that attempting to manually overlay a series of known ligands based on a postulated binding mode can prove a daunting assignment, especially when the ligands contain a large number of functional groups and significant conformational flexibility. As a consequence, techniques have been developed to automate all or part of the pharmacophore elucidation process. The active analog approach[90–92] was the first automated methodology to be developed for such a task. The technique looks for interpoint distances within user-supplied superposition rules (defined pharmacophore centers) common to all active molecules, using systematic conformational analysis to determine possible ligand superpositions. A number of variants of this approach have been developed, utilizing alternative techniques to systematic bond rotation in the search of conformational space. These include the application of distance geometry in the generation of molecular conformational ensembles with matching pharmacophore center–center distances[28] and the use of pharmacophore center constrained Monte Carlo, high-temperature annealing searches.[31]

An alternative approach to automated rule creation and ligand overlay has been devised by Martin et al. with the program DISCO[30] (DIStance COmparisons). DISCO takes as its input a series of low energy conformations for each active molecule. Potential pharmacophore site points are generated automatically for each conformer using ALADDIN.[14] An interesting feature of DISCO is the ability to define the locations of potential hydrogen bond donors and acceptors, thus allowing the directionality of hydrogen bonds to be taken into account. This can be important if ligands are found to approach the same polar site point from different directions, something that is not easily accounted for when the atoms themselves are superimposed. The program then uses a

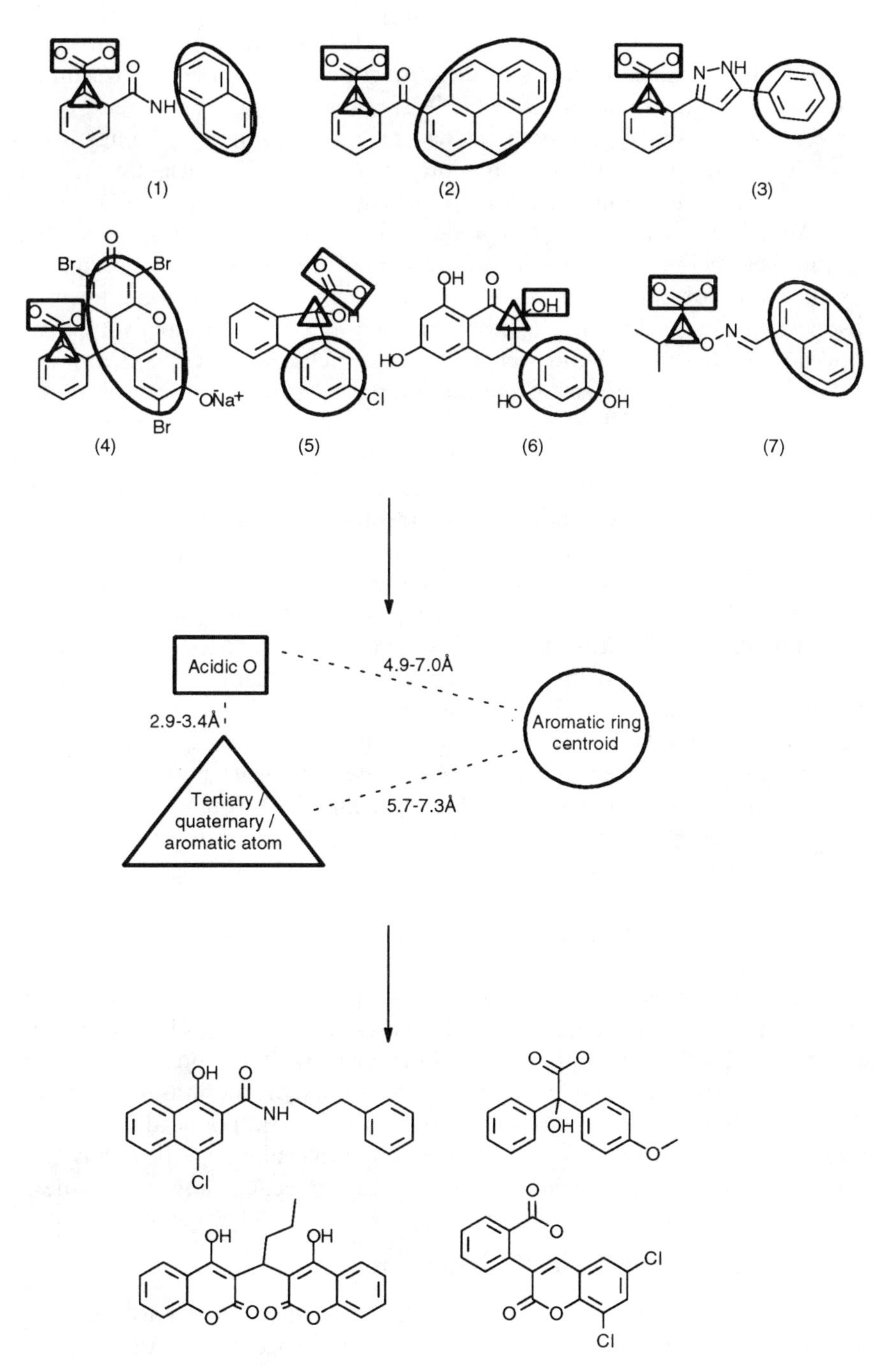

Figure 5 Summary of drug design process used to discover novel inhibitors of auxin transport.[89]

clique-detection algorithm[93] to determine matching groups of distances between site points on conformers of a specified number of molecules within the study set, using the molecule with the least conformations (most rigid) as the reference. The resulting site point matches are used to overlay the ligands. A summary of the DISCO approach is depicted in Figure 6. A schematic representation of clique detection is shown later in Figure 11.

An alternative to the use of a predefined set of conformations is to analyze distance keys derived from a full conformational analysis of the active ligands. Distance keys are calculated in the Chem-X/ChemDBS-3D software from the observed distances between all the pharmacophore centers (normally hydrogen bond donors, hydrogen bond acceptors, positive charge centers, and aromatic ring centroids) in the molecule, and the composite result for all conformations generated is stored.[24] The resulting binding models can then be validated through their use as queries in 3D searches on the database of active ligands. Many, if not all, of the known active ligands should be retrieved before the model can be considered valid. This step identifies any conformations of compounds in the ligand training set that can fit a particular proposed pharmacophore model. By this method, models are tested, many of which will not be valid for all compounds. However, a proposed model can easily be modified. For example, where multiple potential pharmacophore centers exist in a structure in close proximity to a proposed model point, the center type in the model can be modified (e.g., changing an aromatic ring centroid to a hydrogen bond acceptor for an imidazole ring) and the new model evaluated.

An investigation of this method was made using eight nonpeptidic structurally distinct angiotensin II antagonists (see Figure 7). Potential hydrophobe centers (e.g., ethyl or butyl groups) were explicitly coded together with the aromatic ring centers, and the 3D keys were generated through a systematic conformational analysis of all the compounds. Depending on the defined tolerances, between 3 and 44 possible 4-point pharmacophores were proposed. Different combinations of 4 possible center types appeared in the 4-point models, mainly involving hydrogen bond acceptors (all carboxyl oxygens and tetrazole nitrogens were coded in the databases equivalently as hydrogen bond acceptors only) and aromatic ring or other hydrophobic centers. There were many false positives, which failed to validate with all compounds when used as 3D pharmacophore queries to analyze the initial 8-compound database, but interesting models were determined, including molecular superpositions close to those derived from time-consuming classical molecular modeling studies.

Molecular Similarity Calculations

The ligand overlap techniques we have considered so far use distances between ligand centers to determine potential pharmacophores. An alternative technique is to use overlap or difference measures of more continuous molecular properties, such as molecular electrostatic potential and molecular volume,

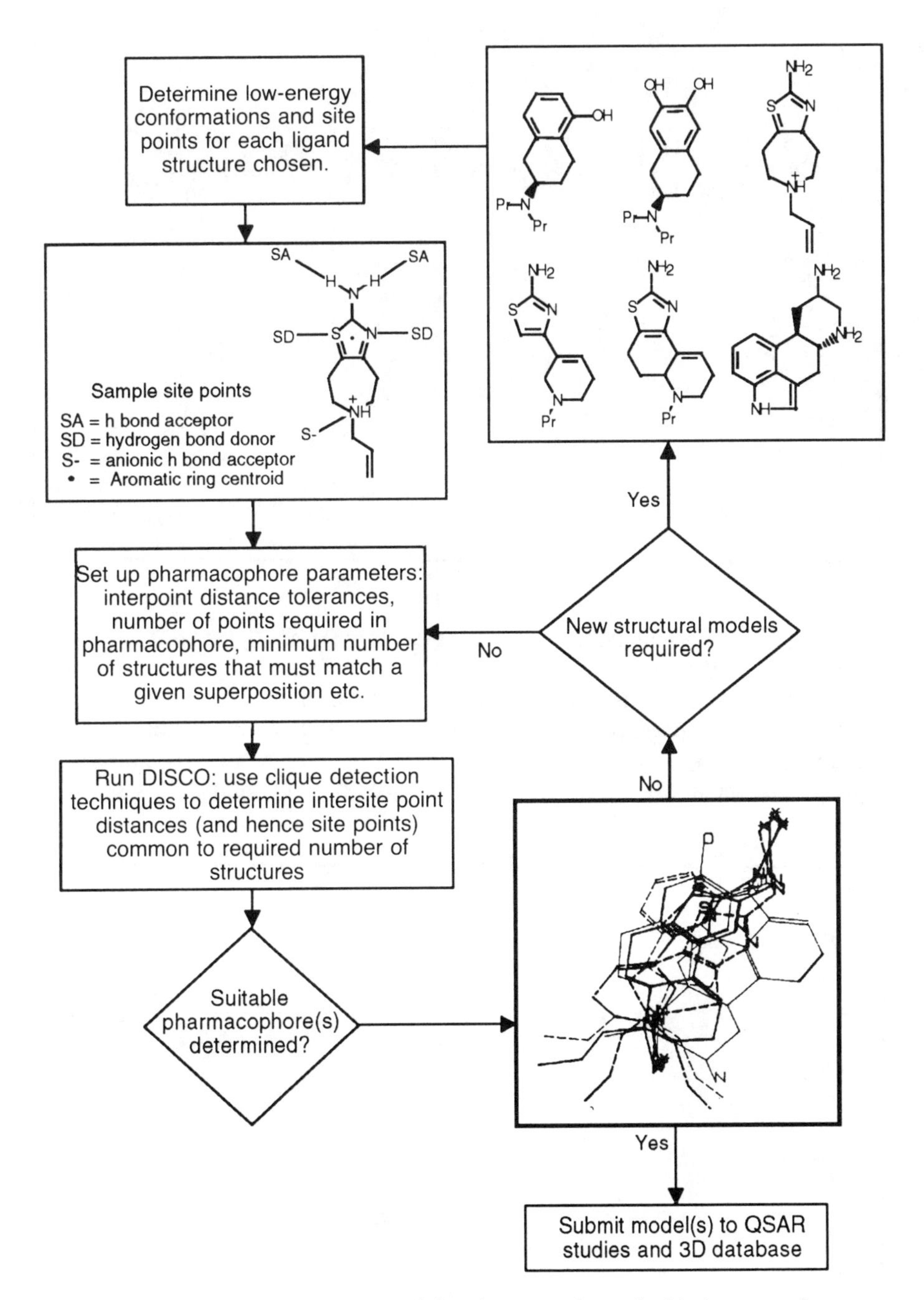

Figure 6 Schematic representation of the pharmacophore elucidation procedure employed by DISCO.[30]

SKB

(i) Define potential pharmacophore centers (H bond donors / acceptors, hydrophobes etc.) for each ligand test set.
(ii) Generate 3D keys using systematic conformational analysis.
(iii) Use resulting pharmacophore models to analyze the initial ligand database.
(iv) Visualize queries which all ligands are able to match.

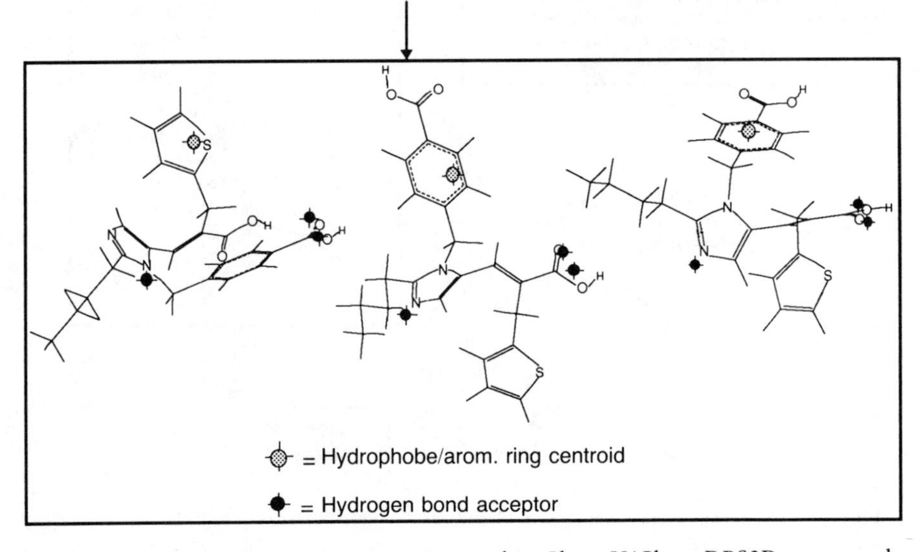

Figure 7 Eight angiotensin II antagonists used in Chem-X/Chem-DBS3D automated pharmacophore identification, together with a number of superpositions of the SKB antagonist onto one of the proposed binding models.

to determine possible ligand superpositions. The quantitative measurements of such property overlaps are known as molecular similarity calculations. These techniques generally use 3D molecular similarity indices to equate the properties of two molecules. A 3D molecular similarity index is essentially a mathematical function of the molecular property being compared. By overlaying two ligands and applying such an index over property space, it is possible to determine a numerical value for the molecular similarity. The similarity value can be used as a maximizable function in molecular alignment optimizations, allowing the index to help determine which ligand superpositions provide the best overlap of molecular properties. In this way it is possible to use such calculations in binding model development.

The basic requirement for a successful molecular (dis)similarity index is a function able to describe the product or difference of a property for two overlaid molecules over all (relevant) property space. Perhaps the most widely applied index for the calculation of 3D molecular similarity was pioneered by Carbo et al.[94]:

$$C_{AB} = \frac{\int P_A P_B dv}{(\int P_A^2 dv)^{1/2} (\int P_B^2 dv)^{1/2}} \qquad [5]$$

Using the Carbo index, molecular similarity C_{AB} is determined from the structural properties P_A and P_B of molecules A and B compared over all space. The numerator measures property overlap, while the denominator normalizes the similarity result. Many alternative formalisms have been proposed for the calculation of molecular similarity between two overlaid molecules.[95–99]

A wide variety of molecular similarity calculation techniques have been developed to aid in the determination of optimal ligand superposition.[33,96–97,99–105] Hodgkin and Richards[96] used variants of the Carbo index to optimize the electrostatic potential similarity between two superimposed molecules. Numerical approximations of index integrals were determined through comparisons of electrostatic potential determined at the intersections of a rectilinear grid constructed around the two molecules. Burt and Richards[100] extended the technique to incorporate flexible fitting during similarity optimization, and Good et al.[101,102] provided further enhancements through the addition of Gaussian function approximations, which allow the rapid analytical comparisons of both electrostatic potential and shape. Dean and Chau[103,104] developed similar optimization techniques, using both gnomonic projections[106] and matching accessible surface areas to describe electrostatic property and shape similarity. Blaney et al.[107] have extended the use of gnomonic projections to encompass interactive graphical comparisons of electrostatic similarity. (An explanation of gnomonic projection is provided in the section on shape-constrained 3D database searches.) Hermann and Herron[105] developed two programs, OVID and SUPER, for pharmacophore elucidation. OVID measures and optimizes the overlap volume between ligand atoms prespecified as being key to activity. SUPER is used to undertake more exhaustive pharmacophore determination searches. The program compares the molecular dot surfaces of overlaid molecules, with surface points considered as matching when they are within a predefined distance of each other. Electrostatic potential data can also be taken into account by discarding matches for points with an electrostatic potential difference greater than a predefined cutoff. (See References 108 and 109 for more complete reviews of molecular similarity calculation methodology.)

We will use a technique recently devised by Masek et al.[33] to illustrate a molecular similarity application. Their methodology employs techniques similar to OVID, but rather than looking at specific atom overlaps, the analytical volume calculations of Connolly[110] are used to determine optimum molecular shape comparisons between entire molecules. Their procedure essentially works as follows:

1. Determine the low energy conformations of each structure.
2. Representing each molecule as a series of spheres of VDW radii corresponding to the constituent atoms, align molecules according to their principal axes of rotation and centers of geometry. The molecular volume overlap of this orientation is optimized using a Newton–Raphson minimizer.[83,84] Three further orientations are then produced by 180° rotations of the mobile molecule about the three principal axes. These superpositions are optimized, and then the mobile molecule undergoes multiple random rotations followed by optimization to probe volume overlap space. Each unique superposition is stored, together with the resulting molecular overlap value. The optimizations can be refined by preferentially weighting overlaps between atoms of matching chemistry. In this way, overlap of both shape and electrostatic properties may be optimized simultaneously.
3. The procedure is repeated for all combinations of molecular conformations, and the top overlap scoring superpositions are analyzed as possible pharmacophores.

Figure 8 illustrates the procedure as undertaken for a pair of angiotensin II antagonists.[33]

Pharmacophore Validation

The procedures we have described are often able to generate many potential pharmacophores for consideration. How can we discriminate between binding models to find the right one? From the perspective of our initial choice of technique for pharmacophore generation, there are a number of factors to consider. Molecular graphics packages allow maximum user intervention (control) within the model-building process. Although this may seem attractive, manual pharmacophore creation can be a time-consuming process prone to user bias. Molecular graphics and modeling tools are thus best left to the validation phase, where visual analysis of computer-generated pharmacophores can prove valuable.

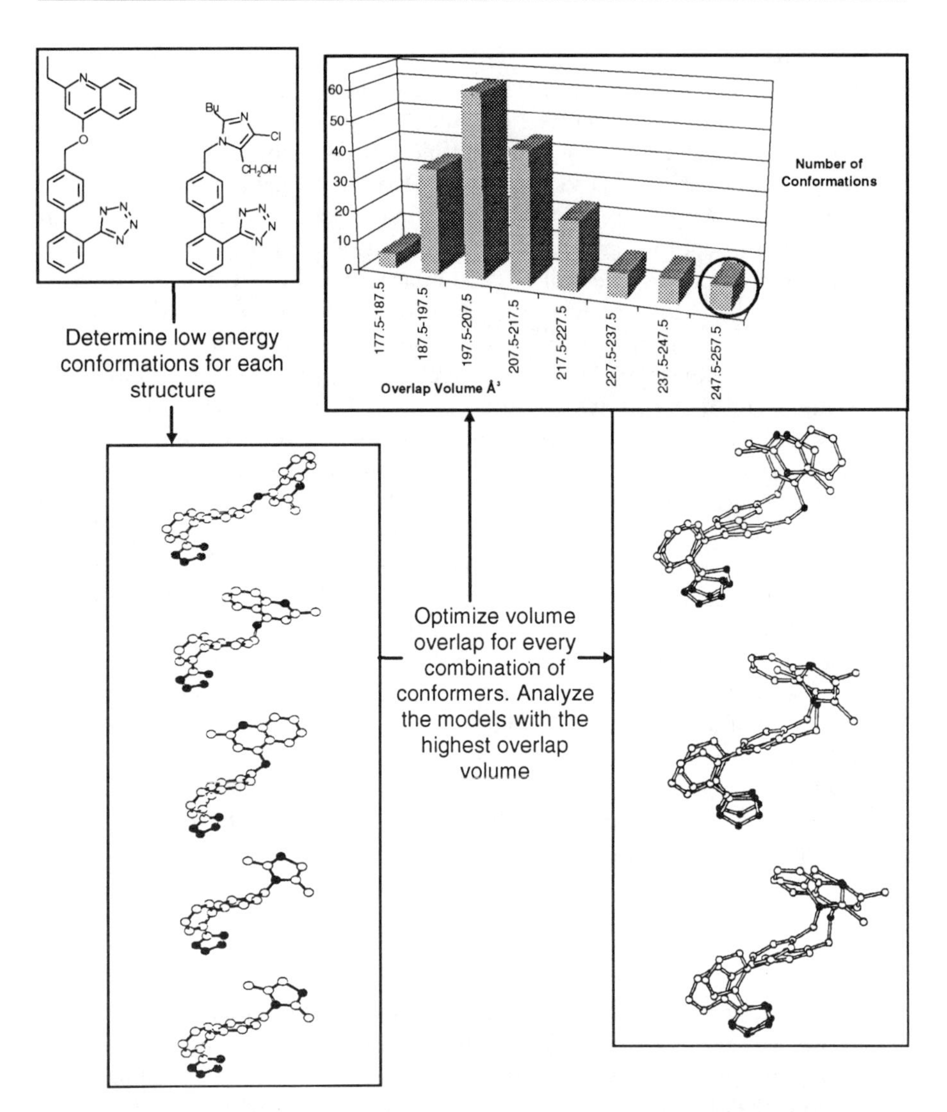

Figure 8 Molecular volume overlap programs of Masek et al. as applied to a pair of angiotensin II antagonists.[33] Note that not all the conformers considered in the overlap calculations are shown.

Each of the automated techniques we have described has its own strengths and weaknesses. The 3D key approach provided within ChemDBS-3D[20] is attractive, because all stages of the process, from conformational analysis through model building, are handled automatically. The technique tends to produce many models, however. This is not necessarily a problem, but it does mean that more false positives will need to be weeded out during the validation process. The approach taken by the DISCO[30] program is interesting, in that the methodology uses a more sophisticated site-point-based approach in defining its ligand pharmacophore centers. Calculations are also rapid, typically completing in a matter of minutes, and multiple superposition rules are generated automatically, as in the 3D key approach. Conformational analysis must be dealt with separately, however, and the technique does not really allow for the large number of conformations that can be considered using the active analog and 3D key techniques. The molecular similarity methods, such as the molecular overlap calculations of Masek and co-workers,[33] have the advantage of being more thorough in their molecular comparisons because they deal with more complete molecular descriptions such as shape, rather than simple subsets (cliques) of interatom (site point) distances. As a consequence, these techniques are able to provide a more quantitative measure of model quality (as the volume overlap values show in Figure 8), generally reducing the number of false positives that need to be dealt with. The problem with these approaches, however, is that in dealing with more complex model elucidation criterion, there is a corresponding increase in the required calculation time. For example, 96 CPU hours on a Silicon Graphics workstation were required to complete the superposition calculations for 192 angiotensin II antagonist conformational pairs considered in the Figure 8 example. A hefty CPU investment is thus often required to undertake thorough searches using such techniques.

Overall, it is clear that a good degree of variation exists in the techniques used to automatically determine potential pharmacophores. As a consequence, many of the techniques have the potential to extract models that could easily be missed using a different approach. The use of a variety of techniques is thus recommended if one is to maximize the chance of finding the best binding model.

Given this, suppose we have a number of potential binding models at our disposal. How do we determine which one is most promising? As has already been mentioned, molecular graphics and molecular modeling provide excellent tools for examining ligand superpositions. The ability to visualize the overlaid molecules can allow nonsensical models to be rapidly weeded out, an example being superpositions between multiple ligand centers, where the bulk of the remaining ligand atoms are found to occupy different regions of Cartesian space. Questions such as, "Can we explain the differences between active and inactive molecules based on this model?" need to be considered. Techniques like volume difference mapping[92] provide tools that can help answer these questions.

Whereas such qualitative approaches are valuable, it would be useful if they could be augmented with techniques that quantitatively link the binding models with ligand binding data. Recent progress in QSAR methodology has provided tools to allow us to do just that. With the advances in statistical techniques applied to QSAR problems in the late 1980s,[111] the technique of Comparative Molecular Field Analysis (CoMFA) was developed.[34] The

CoMFA method relies on the basic premise that noncovalent forces dominate receptor–drug interactions, and that these forces can be described in terms of steric and electrostatic effects. CoMFA attempts to describe these forces in terms of electrostatic and steric (or sometimes another related property) grids calculated around a given series of overlaid ligands. The properties at each grid point are then used to generate a 3D QSAR for the system. As with most QSAR techniques, the primary purpose of CoMFA was to allow the computational chemist to exploit pharmacophores to predict the activity of new ligands of unknown activity.[112] However, the ability of CoMFA to correlate the properties of pharmacophore-aligned ligands with ligand activity allows the technique to be used as a method for testing binding model quality.

Recent studies have shown how CoMFA may be used to determine correct ligand alignment.[113,114] Waller et al.[113] studied a set of 59 HIV-1 protease inhibitors containing five different classes of transition state isosteres. The backbones of HIV-1 protease structures taken from known X-ray complexes containing members from each class of inhibitor were aligned. The relative positions of the inhibitors were extracted from the superposition to provide an initial set of ligand alignment rules. The remaining congeners were fitted onto this template, using the field fit procedure provided within the SYBYL modeling package to optimize the alignment.[115] The field fit routine is essentially a molecular similarity calculation tool, which applies a penalty to the energy minimization of a structure according to the function:

$$E_{\text{pen}} = \frac{Sff}{\sum_{i=1,n} W_i} \sum_{i=1,n} [W_i(P_{A,i} - P_{B,i})^2] \qquad [6]$$

For each point in space considered, the squared difference in field values is determined between molecules A and B. Sff and W_i are user-supplied weights for the overall function and individual lattice point i, respectively. The property P can refer to Lennard Jones and/or Coulombic fields. The function value is added to the standard molecular mechanics force field (akin to poling: see Eqs. [1]–[3]) to alter the potential energy surface and thus force the mobile molecule to adopt a conformation with steric and electrostatic properties similar to the static structure. Additional alignment rules were generated through minimization of the inhibitors into the HIV-1 active site. The five alignment rules produced were tested using CoMFA on both the 59-compound training set and an 18-compound test set. The resulting QSAR models were used to explain the relative merits of the ligand superpositions. This procedure has recently been extended to semiautomate the procedure of conformer selection for a given alignment rule.[114]

A number of alternative 3D QSAR techniques have been advanced that also show potential in the area of pharmacophore development. Good et al.[35] have developed a technique that utilizes molecular similarity matrices to derive

3D QSAR models. Each matrix contains electrostatic and/or shape similarity values calculated between all combinations of molecule pairs involved in the *N* ligand alignment. The resulting *N* by *N* similarity matrix is then used as the variable metric in a 3D QSAR model. The difference between the techniques is that whereas CoMFA will describe a region around a group of molecules using a large number of grid points, the similarity matrix will attempt to describe the same region using just a few numbers. A problem of this methodology is that whereas CoMFA is able to display the coefficients of its QSAR equations as maps of favorable and unfavorable structural interactions, no such method exists for extracting chemical meaning from similarity matrix equations. Nevertheless, the compact nature of the QSAR data means that many models can be tested in a short time. The technique is thus attractive as a potential means for pharmacophore validation.

Another new 3D QSAR technique has recently been published by Jain and co-workers.[36] Their methodology, known as COMPASS, measures steric and electrostatic ligand properties using sampling points scattered around the surface of the aligned molecules. Surface points are used in place of grids to determine ligand properties more efficiently and quickly. COMPASS uses a neural network to execute the statistical model generation, and the method is specifically designed to allow multiple ligand superpositions to be processed. COMPASS is also able to adjust ligand positions to optimize the QSAR model and can provide a graphical interpretation of the QSAR models generated. The technique thus seems to hold much promise as a method both for 3D QSAR generation and pharmacophore validation. For a more complete discussion on general issues relevant to both CoMFA and 3D QSAR in general, see Reference 116.

PHARMACOPHORE SEARCH SUCCESSES

An example of the power of pharmacophore search techniques arose in new leads generation for an LDL (low density lipoprotein) receptor upregulator screen.[117] The screen formed part of a cardiovascular program and was designed to identify compounds that may reduce blood LDL concentrations by a novel mechanism. Low-activity (EC_{50} 1.7 μM) compounds based on the dibenzamide structure shown in Figure 9 were already known, but this series was not showing promise for optimization to more potent compounds. A three-point 3D pharmacophore model was derived from these compounds (illustrated in Figure 9), and 3D searches of the Rhône-Poulenc Rorer corporate databases using this query in ChemDBS-3D yielded two new potent lead series (see Figure 9). One of the new active leads, a diaminoquinazoline (EC_{50} = 0.8 μM), was identified following a 3D search of the stored CONCORD generated

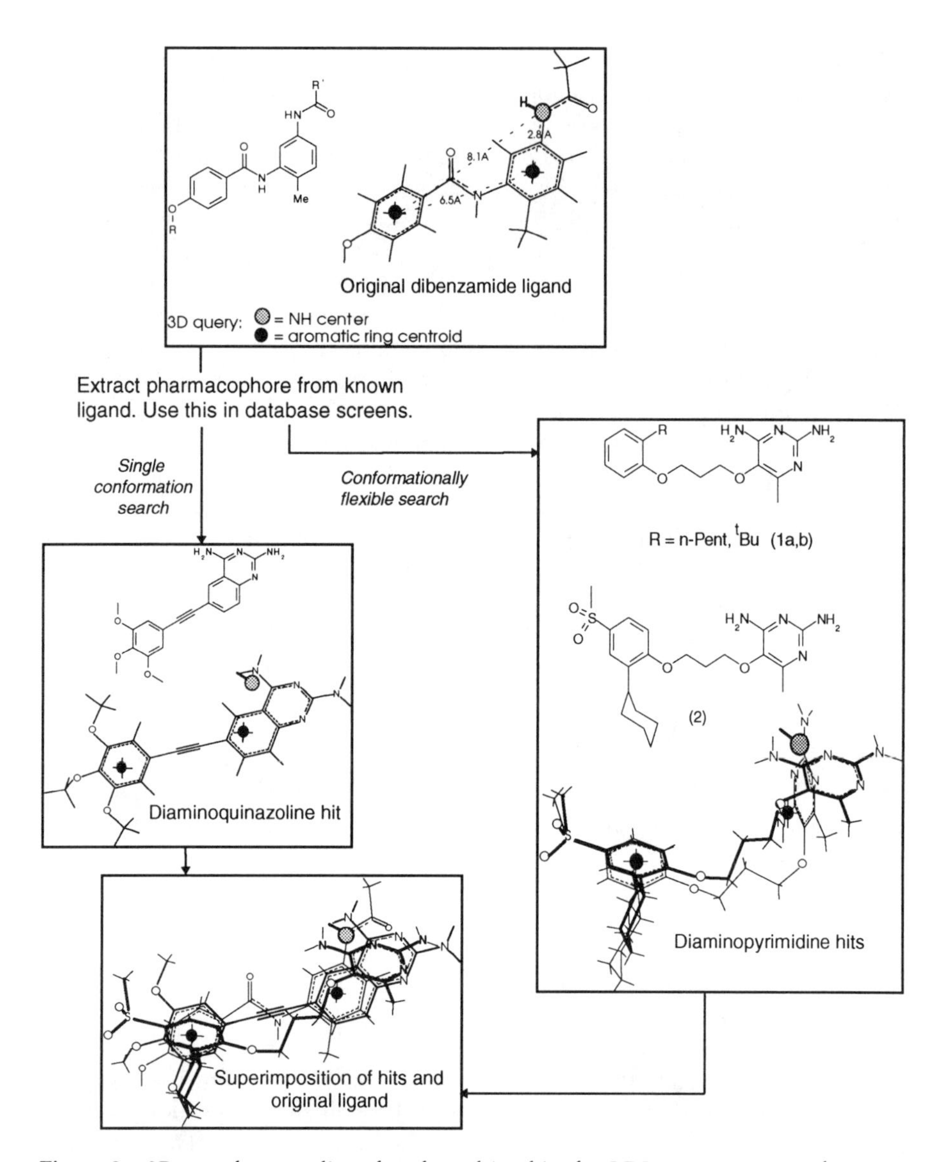

Figure 9 3D search query ligand and resulting hits for LDL receptor upregulator lead screens.

conformation (i.e., a single conformation search). When conformational flexibility was allowed in the search, using the rule-based conformer regeneration in ChemDBS-3D, a second, more potent diaminopyrimidine series was discovered. From this series, compounds already existing in the corporate registry, but made for a very different reason, were identified with activities as low as 5 nM. As well as using different conformations, the compounds can fit the query using different substructure mappings. For example, both substructure mappings of the amino groups are found when the second aromatic ring distance constraint is relaxed or modified based on the diaminoquinazoline lead. These quite different ways of being able to fit the query are illustrated in Figure 9 for diaminopyrimidine compound **2** (EC_{50} = 6 nM). The default extended CONCORD conformation for this structure clearly would not fit the query, showing the importance of allowing for conformational flexibility in the 3D searches.

Another illustration of the capacity of 3D pharmacophore searches to determine new leads was found in an endothelin receptor-binding assay.[118] The assay was part of a general new leads screening program. After approximately 4000 general compounds had been screened, no leads of interest (IC_{50} less than 10 μM) had been found. When a pharmacophore model derived from molecular modeling studies on a cyclic pentapeptide structure and another active compound were used to generate a directed screening set, a much improved yield of hits was obtained. 3D searches were performed on the corporate databases to identify 700 compounds of potential interest; screening of this directed set of compounds identified 10 new leads at less than 10 μM. The most potent compound (~500 nM) was identified only when the 3D search allowed for conformational flexibility. It has been possible to turn these leads into very potent endothelin receptor antagonists through a medicinal chemistry program with the development of improved 3D models for activity. Screening a further 400 compounds related to the 3D hits by their 2D structural characteristics did not yield any further hits. This indicates that the activity was indeed due to the correct 3D disposition of key groups, rather than the chemical family itself.

RECEPTOR-CONSTRAINED 3D SCREENING

Although it is still the case that the structures of the majority of receptors are unknown, the rate of their elucidation through the application of X-ray crystallography and NMR analysis is continually accelerating. This has led to the creation of an area of CADD known as structure-based design. Structure-based design techniques use structural information taken directly from the receptor active site to discover new ligands or improve known ones.[61,62] Among the battery of structure-based design tools at the computational chemist's disposal are 3D screening techniques that use receptor active sites as the constraints in a database search.

The most widely used form of this 3D search paradigm is the DOCK program originally developed by Kuntz et al.[46] DOCK describes the receptor active site using sets of overlapping spheres, each touching two points on the

Connolly molecular surface[119] of the receptor.[46] The centers of these spheres are used to define potential ligand atom positions and effectively define a negative image of the active site. The general structure of a DOCK run is shown in Figure 10. During a database search, ligand atoms and sphere centers are matched using a form of clique detection[46,51,109] (see Figure 11). Each time the requisite number of nodes (ligand atom–receptor sphere center pairs) are

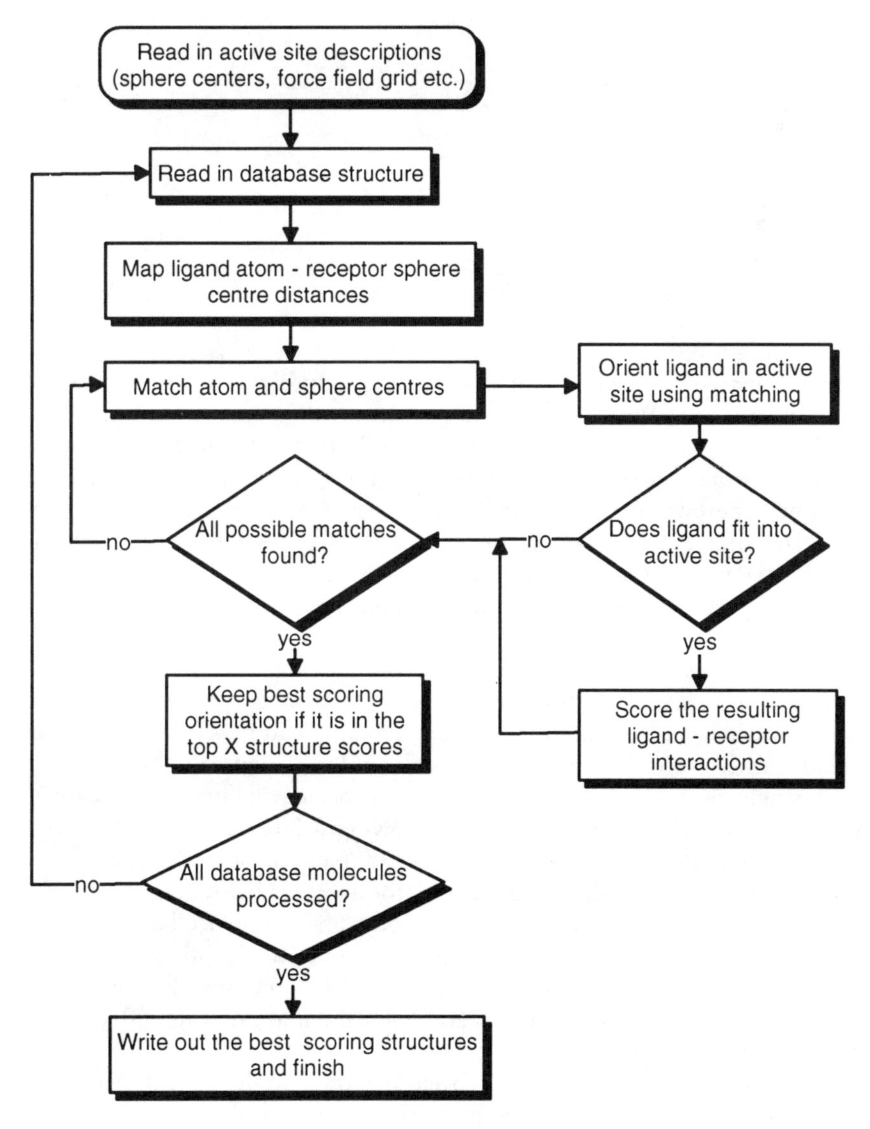

Figure 10 Overview of the DOCK search procedure.[46,47,51,52]

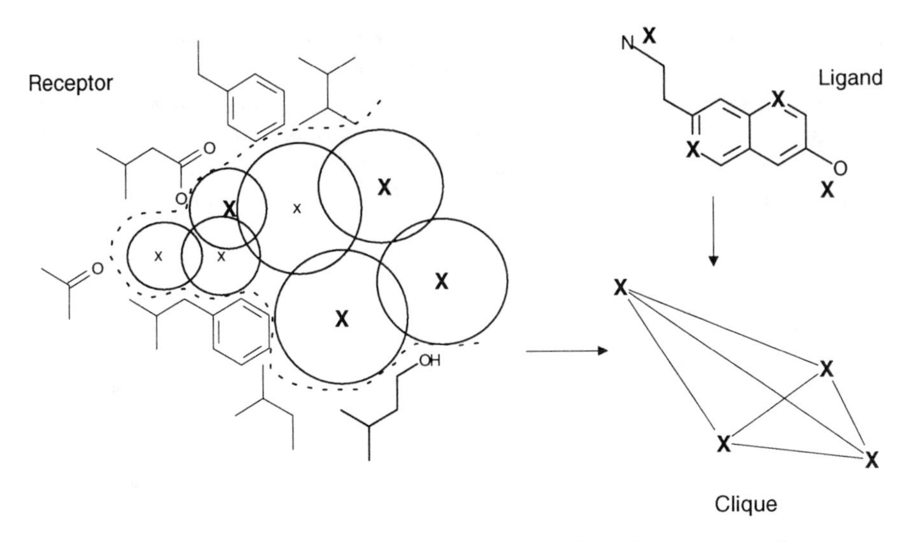

Figure 11 Schematic overview of the receptor sphere/ligand atom clique detection paradigm used in DOCK.[51]

found to match, the ligand atoms are mapped onto the sphere centers by a least-squares fit.[79] Where the resulting fit leaves the ligand positioned within the active site so as to be under the bump limit (the bump limit equals the number of times a ligand and receptor atom clash), the complementarity of the resulting superposition is measured. This measurement (score) is based either on steric complementarity[47] or full intermolecular force field calculations.[52] Recently, rigid body minimization functions were also added to DOCK,[57] to further improve the accuracy of the resulting ligand–receptor superposition scores.

In this way, DOCK explores the six degrees of freedom of the ligand with respect to the active site, allowing the examination of a large number of putative ligand–receptor orientations. An outstanding feature of such an approach is that, rather than trying to find molecules able to adopt suitable geometries for a single postulated binding mode (i.e., pharmacophore-based searches), whole receptor sites can be explored, allowing a large diversity of potential receptor–ligand interactions to be considered. New classes of ligands can thus be discovered that may bind with the receptor in hitherto unforeseen ways. This advantage is also the source of a major problem for structure-based searching tools, such as DOCK, because the thousands of putative ligand–receptor orientations tested make the searching process very CPU-intensive. It is not uncommon to spend one or two weeks searching a database of 100,000 structures (single conformers only).

To combat this, methods have been developed to decrease the number of nonproductive (high-energy) ligand–receptor orientations tested by DOCK, which is where the program spends much of its time.[53,59] The basic process behind the filtering techniques used is to prune the cliques formed in the DOCK matching process. This dramatically increases DOCK speed because the major program bottleneck is the time spent orienting and scoring the potential ligand within the active site. Two techniques are currently used to achieve this goal. The first method is to "color" each ligand atom and sphere center according to its chemistry (hydrogen bond donor, hydrophobe, etc.).[53] Only cliques in which each matched atom and colored sphere center have the same chemistry are allowed to pass through to the orientation routine. The second technique allows the user to designate receptor sphere centers as being critical. In this way, searches can be forced to focus on regions of the receptor considered crucial to binding (e.g., a binding-site metal ion). Only cliques that include the critical sphere centers are allowed to pass through to the ligand orientation routine. Combining these approaches can lead to an increase in search speed greater than two orders of magnitude, while allowing the user to exercise more control over the search.

DOCK has been used to determine inhibitors for a number of systems.[3,48,54–56,59] One recently published design project provides a model example of how DOCK can be used for lead generation.[54] The central aim of the project was to design novel inhibitors of thymidylate synthase (TS). The active site of the protein was explored using ligands taken from a 3D version of the Fine Chemicals Database (FCD). The 25 top-scoring novel ligands, structures not known for TS inhibition and with significant structural dissimilarity to the TS substrate dUMP, were tested for activity. Three of the molecules were found to have IC_{50} values in the high micromolar range. One of these compounds, sulisobenzone (SB), was crystallized with TS and found to bind outside the nucleotide binding region predicted by DOCK, the molecule being displaced by a buffer-derived anion. The resulting binding site was in a previously unexplored receptor region of significant functional importance. For this reason, SB was used as a template for shape similarity searches within MACCS-3D[18] of the FCD database. DOCK was used to fit and score resulting hits in the SB binding region. The most commonly occurring hits that scored highly were found to be derivatives of phenytoin. Five of these derivatives had IC_{50} activity in the 600–1200 μM range. The shape similarity search was repeated, this time combining the SB and phenytoin substructures, and 992 compounds were retrieved. Of the 33 compounds tested as a result of their high DOCK scores, 3 had IC_{50} values under 3 μM and 26 others showed significant activity. The most common hits were phenolphthalein derivatives. Several of these compounds were crystallized and their binding positions bore close resemblance to those determined by DOCK. This is an excellent example of how multiple CADD searching tools can combine with techniques such as X-ray crystallography, forming a drug design process akin to the paradigm illustrated in Figure 1. A summary of this study is illustrated in Figure 12.

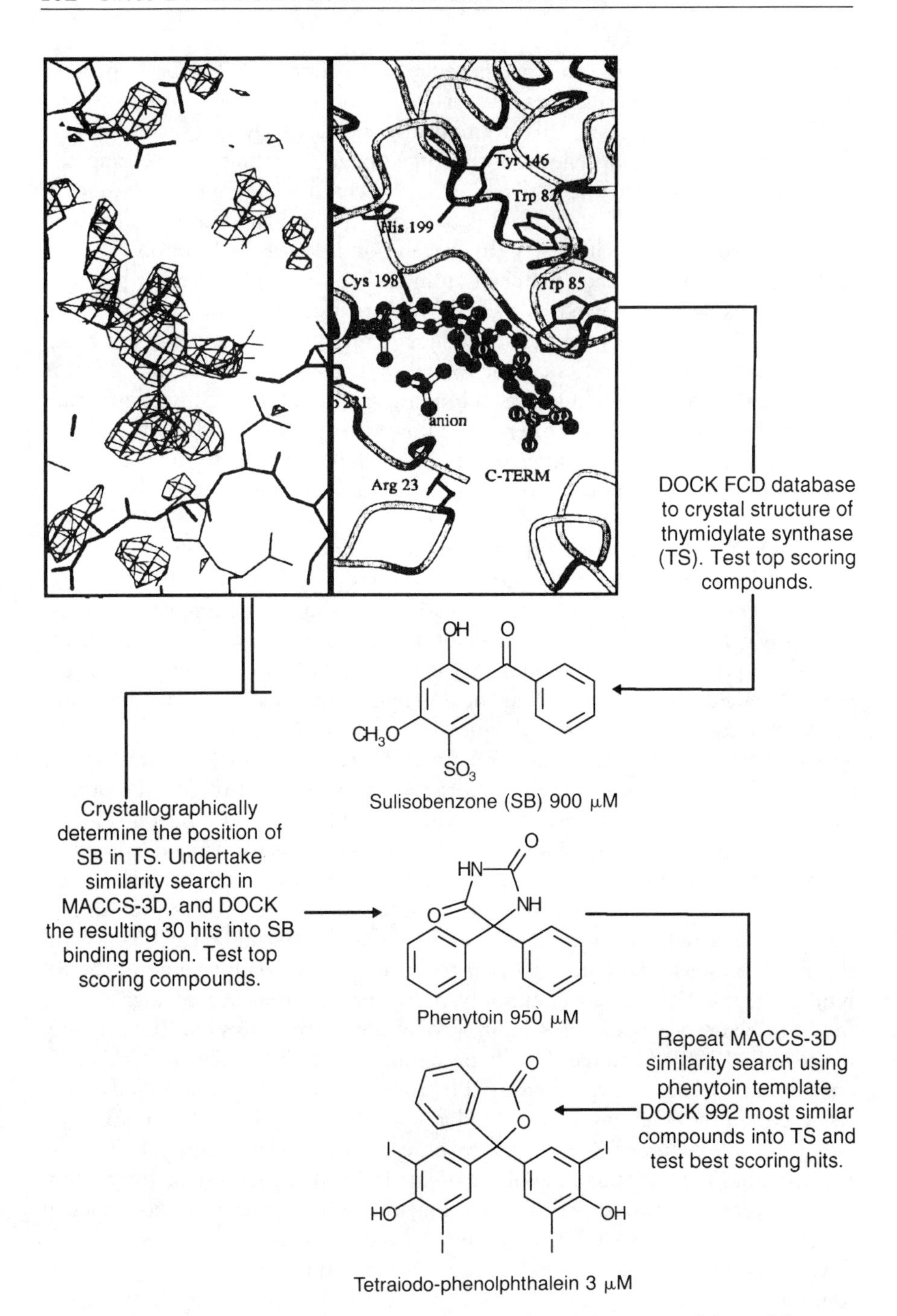

Tyr 146
Trp 82
His 199
Cys 198
Trp 85
anion
Arg 23
C-TERM
DOCK FCD database to crystal structure of thymidylate synthase (TS). Test top scoring compounds.
OH
O
CH3O
SO3
Sulisobenzone (SB) 900 μM
Crystallographically determine the position of SB in TS. Undertake similarity search in MACCS-3D, and DOCK the resulting 30 hits into SB binding region. Test top scoring compounds.
HN
NH
Phenytoin 950 μM
Repeat MACCS-3D similarity search using phenytoin template. DOCK 992 most similar compounds into TS and test best scoring hits.
HO
OH
Tetraiodo-phenolphthalein 3 μM

Figure 12 Design paradigm used in the structure-based discovery of thymidylate synthase inhibitors.[54]

A number of other approaches have been used to tackle the problem of receptor-constained 3D database searches. Smellie et al.[49] have developed a technique that uses a variant of clique-based detection. Hydrogen bond donor and acceptor atoms are used to describe the receptor site points, which are matched to corresponding ligand atoms. An interesting feature of the method is the fact that upper and lower bound information is included with the ligand atom–atom distances (analogous, though not identical, to the approach shown in Figure 4). In this way, conformational freedom is implicitly included in the matching process. Once matching cliques have been determined, distance geometry is used to generate ligand structures conforming to the clique constraints. Lawrence and Davis[50] use a technique that generates receptor site points using GRID[120] (a program for determining interaction energies between 3D receptor structures and small fragment probes). During database searching, only ligand atoms of the same chemistry may match a given receptor site point, analogous to DOCK site point "coloring." Miller et al.[58] have created a program FLOG, which contains a number of alterations to the DOCK paradigm, including a more restrictive matching algorithm, to maximize database search speed.

Another interesting method has recently been introduced by Norel et al.,[121] who employ triangle descriptors to calculate molecular shape. For each molecule in a database, every pair of atoms that are within a defined distance range of each other is extracted, and triangles are constructed by adding a third vertex in the form of a molecular surface point. Triangle data for all the atom pair-surface triangle descriptions of a molecular database are stored in a hash table. The hash table is then used to rapidly screen for potentially complementary ligand–receptor interactions, through comparison with equivalent receptor triangle data. An elegant aspect of this approach is that the triangle information is used to determine potentially complementary ligand receptor shapes before the ligand is oriented into the site. As a consequence, many potentially nonproductive ligand superpositions can be filtered out early in the docking process, thus speeding up the search procedure considerably.

SHAPE-CONSTRAINED 3D DATABASE SEARCHES

Although it is most common to search a 3D database for structures matching a functional group-based pharmacophore, a number of techniques have recently been introduced in which searches are undertaken for molecules matching the shape of known ligands. These searches can be used as an additional screen to standard pharmacophore searches, or the skeletons can be studied using molecular modeling and the knowledge of the synthetic chemist, to see if the structure may be altered so as to present the correct functionality at the required pharmacophore positions. Many of these approaches bear a close resemblance to the molecular similarity calculations we described earlier, the major difference being that the methods used with these procedures have been optimized for speed, thus permitting their use in large database searches.

One of the more innovative techniques advanced to accomplish this task is the SPERM program developed by van Geerestein and Perry.[37,122] SPERM uses gnomonic projection techniques to describe the shape of a molecule in terms of numerical values positioned on the surface of a sphere (see Figure 13). A major problem of this approach is obtaining an even distribution of points across the surface of the sphere used in the projection. In the original incarnation of the program, the vertices of a superimposed dodecahedron and icosahedron were used to define the points.[105,122] This was later modified to the points obtained by passing vectors of sphere radius length (centered at the geometric center of each molecule) through the points generated in tessellating an icosahedron (Figure 14).[37] During a database search, structures undergo initial rapid shape screen comparisons with the template structure (e.g., molecular volume or principal moments of inertia). Molecules passing these tests are then overlaid with the template so that their centers and principal inertial axes coincide. The molecular shape (dis)similarity is calculated based on the root mean square difference (RMSD) of the sphere surface point values:

$$\mathrm{RMSD}_{AB} = \left[\sum_{i=1,n} \{(P_{A,i} - P_{B,i})^2\}/n \right]^{1/2} \qquad [7]$$

The template sphere is then rotated with respect to the molecule, searching for the orientation with minimum dissimilarity. A lot of effort has been taken to optimize the search procedure, for example, through the prior storage of all possible template orientations, and the use of icosahedral symmetry to mini-

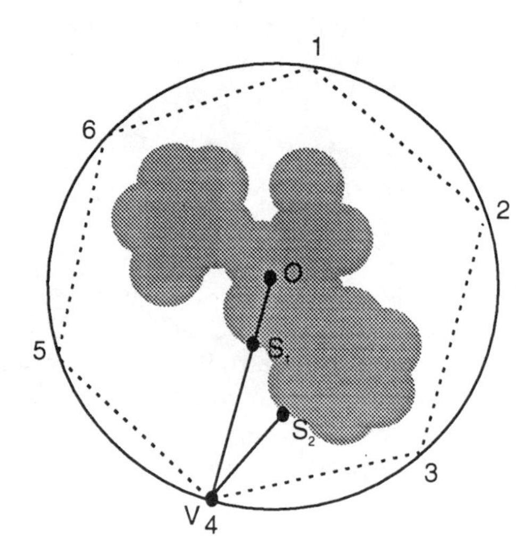

Figure 13 2D representation of gnomonic projection. The two vectors shown refer to the radial distance (VS_1), and the minimum distance from the vertex to the molecular surface (VS_2).[37,122]

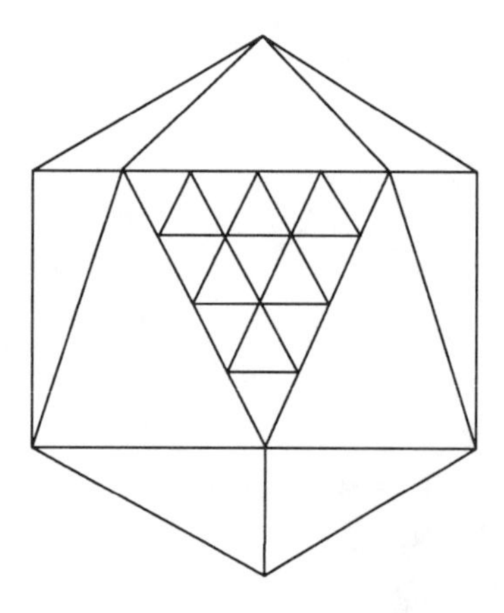

Figure 14 Tessellated icosahedron of the kind applied to define the position of sphere surface points used in SPERM gnomonic projection calculations.[37]

mize the consequent storage needs. The result is a rapid search procedure able to process thousands of molecules in a matter of hours. The top 10 hits of a search of the Cambridge Structural Database (CSD) overlaid on the tetrodotoxin template are shown in Figure 15.

The SPERM program uses an orientation-dependent measure of shape similarity in its search calculations. A number of alternative shape measures have been proposed that use orientation-independent measures of similarity.[38–43] Pepperell and Willett[38] considered a number of similarity-searching techniques involving comparisons of molecular interatomic distance matrices. Fisanick et al.[39,40] have developed descriptors that employ the geometric features of atom triangles, for example, describing perimeter and area to measure molecular shape. These descriptors (along with many others) have been applied to searches of the Chemical Abstracts Service (CAS) registry substances.

Bemis and Kuntz[41] calculated molecular shape utilizing histograms of atom triangle perimeter data and applied the resulting shape descriptions to database clustering. Nilakantan et al.[42] have utilized the distribution of atom triplet distances within a molecule to describe molecular shape. Triangle side lengths are calculated for each atom triplet. These lengths are sorted and scaled by size, then assigned a triplet value according to the equation

$$nt = n1 + 100(n2) + 1{,}000{,}000(n3) \qquad [8]$$

where nt is the packed integer, and $n1$, $n2$, and $n3$ are the three digitized and sorted sides of the triplet ($n1$ shortest, $n3$ longest). The triplet value is digitized and packed into a 2048-bit signature to save space. Signature similarity is determined using formulas of the type

$$s = 2c/(na + nb) \qquad [9]$$

Here s is the similarity, c is the number of "on"(1) bits in common, and na and nb are the number of on bits in the template and database molecules. Triplets

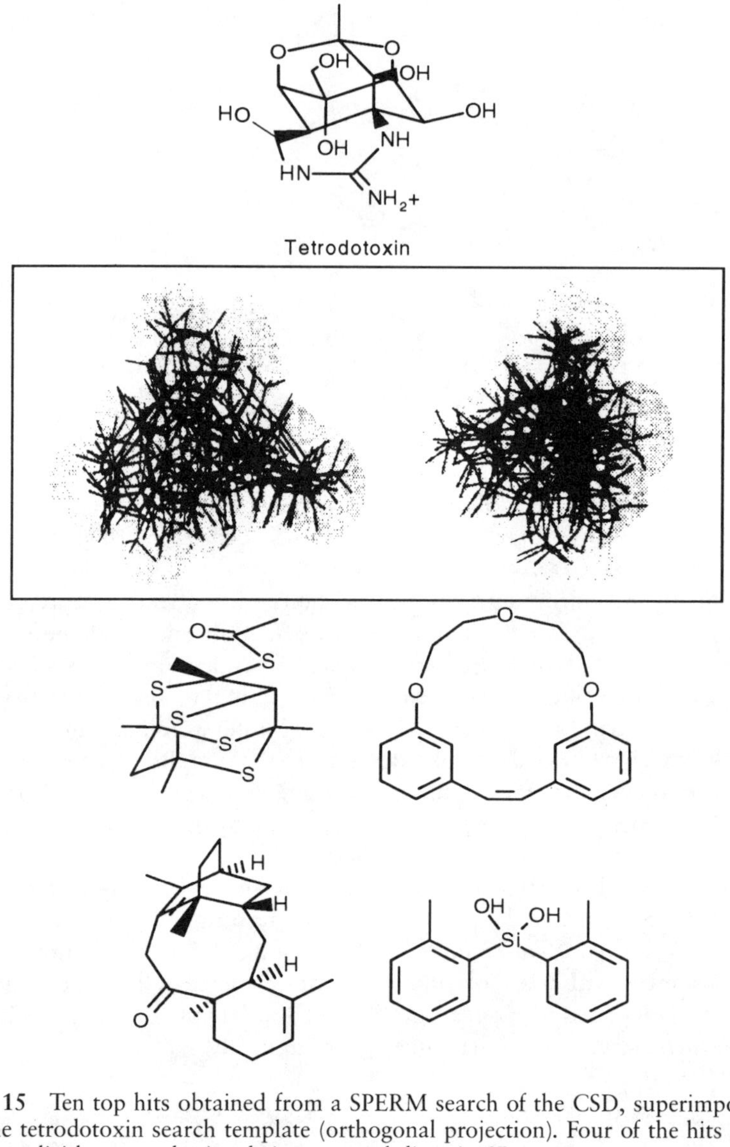

Tetrodotoxin

Figure 15 Ten top hits obtained from a SPERM search of the CSD, superimposed over the tetrodotoxin search template (orthogonal projection). Four of the hits are shown explicitly to emphasize their structural diversity.[37]

are recalculated on the fly for database molecules deemed similar enough to the template based on signature. A more accurate similarity value is then determined using the resulting triplet data and equations analogous to Equation [8].

Good et al.[43] have extended the use of simple ligand triplet descriptors to encompass molecular surface comparisons and have made it possible to store the shape descriptions directly onto hard disk, thus facilitating faster searches. Both Nilakantan et al.[42] and Good et al. have used their systems to prescreen DOCK searches as well as in stand-alone searches of 3D databases. (It should be noted that DOCK can also be used to undertake shape searches using receptor models defined through overlaid active ligands.) Screens undertaken using these orientation-independent procedures are not as accurate as those performed using programs such as SPERM, but they can prove extremely rapid. For example, searches by Good et al.[43] were found to proceed at rates of up to 2000 ligands per second on a Silicon Graphics workstation.

VECTOR-CONSTRAINED 3D SEARCHES

Another technique deserving mention is the program CAVEAT,[123] which provides a method for searching molecular databases for bonds matching a vector-based 3D search query (see Figure 16). Each molecule in the database is represented using a pairwise specification of selected bonds as unit vectors, with bond atoms defined as base and tip. The tip atom is defined as suitable for replacement. Vector pairs are stored only if they are connected by a path through the molecule containing no tip atoms. The user is given control over which bonds may be treated as vectors, and a screening and clustering program CLASS is used[123] to prune CAVEAT search answer sets. Search times are extremely rapid, with a search of 43,550 CSD structures using the four-vector search shown in Figure 16, requiring only 142 seconds on an SGI Indigo R3000 computer. These programs can be used to find potential template molecules in applications such as mimetic synthesis.

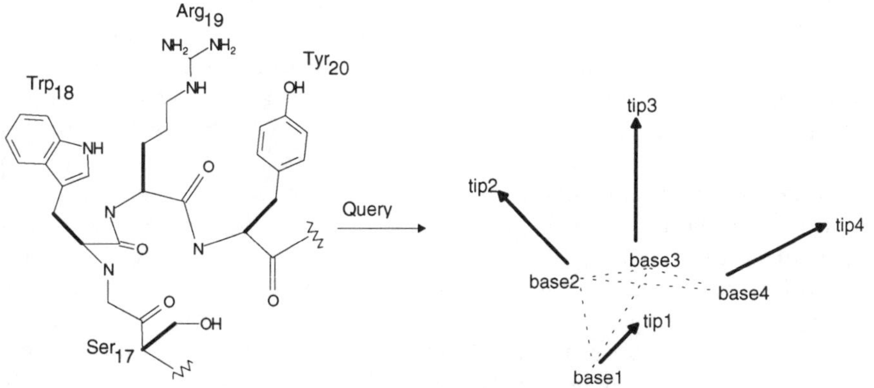

Figure 16 Four-vector CAVEAT query extracted from the C_a–C_b side chain bonds of the α-amylase inhibiting tendamistat beta turn.[123]

DE NOVO DESIGN

Since Goodford developed the GRID program to determine interaction energies between 3D receptor structures and small fragment probes,[120] the de novo construction of molecules has become a popular method of lead generation. Initially, design was accomplished by linking favorable probe sites together to form a complete molecule using molecular graphics. This methodology has been successfully applied to create novel inhibitors of thymidylate synthase.[61] More recently, a variety of techniques have been devised to further exploit de novo design through procedure automation.[124–132]

Automated receptor-based de novo design is generally accomplished in one of three ways. Molecules are commonly built from linked fragments that have been matched to receptor site points[127,128,132] or grown from a seed point inside a receptor site using some form of potential function.[125,126,129–131] The third way is to undertake de novo design without receptor knowledge through automatic lead modification and/or random fragment linking and constraints derived from SAR studies.[124]

Although de novo design techniques are not in themselves database screening methods, they are closely related and complementary. The two techniques have been likened to having a dress designed to your exact requirements, rather than going to your local clothes store to obtain one that is to your general taste.[130] This is due to the fact that, rather than restricting the search for leads to structures already contained within a corporate database, molecules are designed from the ground up based on user-defined binding mode constraints. De novo design methodology thus lends itself to the generation of artificial diversity tailored directly towards a specific target. Most of the published techniques have shown how structures can be built that closely resemble or match known inhibitors. The LUDI program was shown to construct known HIV inhibitors when building onto a known ligand.[128]

In addition to direct lead generation, molecules produced from de novo design packages allow the construction of pharmacophore models from the resulting structural motifs. Molecules generated using the CONCEPTS program were used to create pharmacophores that yielded known inhibitors among the hits generated using 3D database searches.[130] An extension to pharmacophore creation arises from the fact that the structures generated are built to fit to the active site or pharmacophore used in their construction. As a consequence, the conformation of any de novo built structure is already optimized for that particular binding model. Combine this with the fact that de novo design techniques can be used to build thousands of complementary structures[131] in a short time, and it becomes possible to apply de novo design techniques directly to molecular database searches. This can be achieved simply by converting de novo answer set structures into 2D topological queries and then applying the queries to search for molecules with similar topologies within 2D databases of known structures.

A nice example of the potential integration of database screening and de novo design is the use of conformationally flexible 3D database searching to classify and optimize a set of "idea" structures for a particular target. Given a pharmacophore, thorough systematic analysis of a 3D database of designed structures enables the rapid selection of a small set of the most favorable compounds for synthesis. The automation achievable in the construction and evaluation of the 3D structures using different pharmacophore models means that all substitution and isomeric possibilities can be easily evaluated, in what would normally be a time-consuming effort. These "idea structures" can be readily coded into SMILES line notation for 2D structure representation. Simple modification of these SMILES codes (where there is a single line per structure), before 3D structure generation with CONCORD, enables a thorough, systematic, and automated investigation of structural variations for the idea structures containing, for example, different chain length, substitution, and isomeric patterns. From 3D searches with full conformer regeneration, superimposed best-fit conformations for each structure can be obtained. This facilitates the fine-tuning of the "idea" and the incorporation of other features not readily defined within the 3D query.

The need to be able to view multiple conformations and substructure mappings found during the search becomes apparent during this type of project because conformer regeneration can identify many different and interesting matches. The easy and rapid visualization of these multiple matches (different conformations and substructure mappings) for each structure fitted onto the query is a key feature of the technique, from which new ideas are suggested and easily evaluated.

This concept has been extended and implemented in the form of "virtual" databases. These are databases of hypothetical structures that can be used to select structures for synthesis or for designing combinatorial libraries with structural diversity. The type of structures that can be readily generated, by methods such as using SMILES in CONCORD, can be oligomers (e.g., peptides) or combinations of a scaffold with a set of substituents (groups for interaction). The full power of conformationally flexible 3D database searching can then be applied to these virtual databases. The combinatorial generation of hypothetical structures can be achieved through simple user-written programs using structure codes, such as SMILES, or in some of the commercial modeling programs (e.g., Chem-X), based on a scaffold with a set of substituents.

The capability of using idea databases was investigated in the design of angiotensin II antagonists.[87] Using pharmacophore models already proposed from molecular modeling studies of various nonpeptidic angiotensin II antagonists, 3D searches using ChemDBS-3D were used to classify a database of about 1000 idea structures. One of the compounds identified as being of interest was a secondary amide that fitted both the 3D search pharmacophores applied in the study using different low-energy conformations, none of which was the stored conformation. Shortly after this analysis was carried out, a very similar compound appeared in a patent as a potent angiotensin II antagonist, an encouraging result for the effectiveness of the technique.[87]

CONCLUSIONS

The technical advances detailed here have allowed 3D database searching to mature into a powerful technique for lead generation. Perhaps the most important single development has been the ability to account for conformational flexibility, thereby greatly increasing the potential of searches to yield fruitful hits. The whole of CADD is maturing hand in hand with database search technology, and we have emphasized throughout this chapter how all the computational tools available can be integrated to optimize the rational drug design process.

Future developments will further enhance the efficacy of database searching. These include coupling pharmacophore and receptor constraint search techniques, the use of more sophisticated binding models incorporating shape and molecular fields, and improved techniques for scoring hit quality, e.g., through closer integration with molecular similarity and clustering calculations.[109] Combine these advances with the ever-increasing speed of computers, and the future of 3D database searching looks bright indeed.

ACKNOWLEDGMENTS

A.C.G. thanks Professor I. D. Kuntz (University of California, San Francisco) for his help with the creation of this chapter. J.S.M. thanks Dr. Iain McLay (Dagenham Research Center) for his contribution to the development and application of 3D searching at Rhône-Poulenc Rorer.

REFERENCES

1. Y. C. Martin, M. G. Bures, and P. Willett, in *Reviews in Computational Chemistry,* Vol. 1, K. B. Lipkowitz and D. B. Boyd, Eds., VCH Publishers, New York, 1990, pp. 213–263. Searching Databases of Three-Dimensional Structures.
2. Y. C. Martin, *J. Med. Chem.,* **35,** 2145 (1992). 3D Database Searching in Drug Design.
3. I. D. Kuntz, *Science,* **257,** 1078 (1992). Structure-Based Strategies for Drug Design and Discovery.
4. P. Gund, in *Progress in Molecular and Subcellular Biology,* Vol. 5, F. E. Hahn, Ed., Springer-Verlag, Berlin, 1977, pp. 117–143. Three-Dimensional Pharmacophore Pattern Searching.

5. A. Rusinko III, J. M. Skell, R. Balducci, C. M. McGarity, and R. S. Pearlman, University of Texas, Austin. CONCORD, A Program for the Rapid Generation of High Quality Approximate 3D Molecular Structures. Distributed by Tripos Associates, St. Louis, MO.

6. A. R. Leach and K. Prout, *J. Comput. Chem.*, **11**, 1193 (1990). Automated Conformational Analysis: Directed Conformational Search Using the A* Approach. Program known as COBRA, distributed by Oxford Molecular Ltd., Oxford, U.K.

7. Chem/Model, developed and distributed as part of the Chem-X modeling package by Chemical Design Ltd., Chipping Norton, Oxon, U.K.

8. Converter, developed and distributed by BIOSYM Technologies, San Diego, CA.

9. D. Weininger, *J. Chem. Inf. Comput. Sci.*, **28**, 31 (1988). SMILES, a Chemical Language and Information System. 1. Introduction to Methodology and Encoding Rules.

10. D. Weininger, A. Weininger, and J. L. Weininger, *J. Chem. Inf. Comput. Sci.* **29**, 97 (1989). Algorithm for Generation of Unique SMILES Notation.

11. J. Elks and C. R. Ganellin, *Chapman and Hall Dictionary of Drugs 3D. Chemical Data, Structure and Bibliography.* Cambridge University Press, 1990. Distributed by Chemical Design Ltd., Chipping Norton, Oxon, U.K.

12. Available Chemicals Directory, developed and distributed by MDL Information Systems, San Leandro, CA. Formerly distributed as the Fine Chemicals Directory (FCD), which was smaller.

13. CAST 3D, developed and distributed by Chemical Abstracts Service, Columbus, OH.

14. J. H. Van Drie, D. Weininger, and Y. C. Martin, *J. Comput.-Aided Mol. Design,* **3**, 225 (1989). ALADDIN: An Integrated Tool for Computer-Assisted Drug Design and Pharmacophore Recognition from Geometric, Steric and Substructure Searching of 3D Molecular Structures. Distributed by Daylight Chemical Information Systems, Inc., Sante Fe, NM.

15. R. P. Sheridan, A. Rusinko III, R. Nilakantan, and R. Venkataraghavan, *Proc. Natl. Acad. Sci. USA,* **86**, 8165 (1989). Searching for Pharmacophores in Large Co-Ordinate Databases and Its Use in Drug Design.

16. R. P. Sheridan, R. Nilakantan, A. Rusinko III, N. Bauman, K. S. Haraki, and R. Venkataraghavan, *J. Chem. Inf. Comput. Sci.,* **29**, 255 (1989). 3DSEARCH, a System for Three-Dimensional Substructure Searching.

17. J. B. Moon and W. J. Howe, *Tetrahedron Comput. Methodol.,* **3**, 697 (1990). 3D Database Searching and De Novo Construction Methods in Molecular Design.

18. O. F. Guner, D. W. Hughes, and L. M. Dumont, *J. Chem. Inf. Comput. Sci.,* **31**, 408 (1991). An Integrated Approach to 3D Information Management with MACCS-3D.

19. MACCS (Molecular ACCess System), developed and distributed by MDL Information Systems, San Leandro, CA. This older system is being augmented with ISIS (Ref. 23).

20. ChemDBS-3D, developed and distributed as part of the Chem-X modeling package by Chemical Design Ltd., Chipping Norton, Oxon, U.K.

21. UNITY, developed and distributed by Tripos Associates, St. Louis, MO.

22. P. W. Sprague, in *Recent Advances in Chemical Information—Proceedings of the 1991 Chemical Information Conference,* H. Collier, Ed., Royal Society of Chemistry, Cambridge, U.K., 1992, pp. 107–111. Catalyst: A Computer Aided Drug Design System Specifically Designed for Medicinal Chemists. Catalyst, developed by BioCAD and now distributed by Molecular Simulations Inc., Burlington, MA.

23. ISIS/3D, developed and distributed by MDL Information Systems, San Leandro, CA.

24. N. W. Murrall and E. K. Davies, *J. Chem. Inf. Comput. Sci.* **30**, 312 (1990). Conformational Freedom in 3D Databases. 1. Techniques.

25. T. Hurst, *J. Chem. Inf. Comput. Sci.,* **34**, 190 (1994). Flexible 3D Searching: The Directed Tweak Technique.

26. T. E. Moock, D. R. Henry, A. G. Ozkabak, and M. Alamgir, *J. Chem. Inf. Comput. Sci.,* **34**, 184 (1994). Conformational Searching in ISIS/3D Databases.

27. A. Smellie, S. L. Teig, and P. Towbin, *J. Comput. Chem.*, **16**, 171 (1995). Poling: Promoting Conformational Variation.
28. R. P. Sheridan, R. Nilakantan, J. S. Dixon, and R. Venkataraghavan, *J. Med. Chem.*, **29**, 899 (1986). The Ensemble Approach to Distance Geometry: Application to the Nicotinic Pharmacophore.
29. U. Cosentino, G. Moro, D. Pitea, S. Scolastico, R. Todeschini, and C. Scolastico, *J. Comput.-Aided Mol. Design*, **6**, 47 (1992). Pharmacophore Identification by Molecular Modeling and Chemometrics: The Case of HMG-CoA Reductase Inhibitors.
30. Y. C. Martin, M. G. Bures, E. A. Danaher, J. DeLazzer, I. Lico, and P. A. Pavlik, *J. Comput.-Aided Mol. Design*, **7**, 83 (1993). A Fast New Approach to Pharmacophore Mapping and Its Application to Dopaminergic and Benzodiazepine Agonists. (The program known as DISCO is distributed by Tripos Associates, St. Louis, MO.)
31. E. E. Hodgkin, A. Miller, and M. Whittaker, *J. Comput.-Aided Mol. Design*, **7**, 515 (1993). A Monte Carlo Pharmacophore Generation Procedure: Application to the Human PAF Receptor.
32. J. Greene, S. Kahn, H. Savoj, P. Sprague, and S. L. Teig, *J. Chem. Inf. Comput. Sci.*, submitted. Chemical Function Queries for 3D Database Search.
33. B. B. Masek, A. Merchant, and J. B. Matthew, *J. Med. Chem.*, **36**, 1230 (1993). Molecular Shape Comparison of Angiotensin II Receptor Antagonists.
34. R. D. Cramer III, D. E. Patterson, and J. D. Bunce, *J. Am. Chem. Soc.*, **110**, 5959 (1988). Comparative Molecular Field Analysis (CoMFA). Effect of Shape on Binding of Steroids to Carrier Proteins.
35. A. C. Good, S. J. Peterson, and W. G. Richards, *J. Med. Chem.*, **36**, 2928 (1993). QSARs from Similarity Matrices. Technique Validation and Application in the Comparison of Different Similarity Evaluation Methods.
36. A. N. Jain, K. Koile, and D. Chapman, *J. Med Chem.*, **37**, 2315 (1994). COMPASS—Predicting Biological Activities from Molecular Surface Properties—Performance Comparisons on a Steroid Benchmark.
37. N. C. Perry and V. J. van Geerestein, *J. Chem. Inf. Comput. Sci.*, **32**, 607 (1992). Database Searching on the Basis of Three-Dimensional Molecular Similarity Using the SPERM Program.
38. C. A. Pepperrell and P. Willett, *J. Comput.-Aided Mol. Design*, **5**, 455 (1991). Techniques for the Calculation of Three-Dimensional Structural Similarity Using Inter-Atomic Distances.
39. W. Fisanick, K. P. Cross, and A. Rusinko III, *J. Chem. Inf. Comput. Sci.*, **32**, 664 (1992). Similarity Searching on CAS Registry Substances. 1. Global Molecular Properties and Generic Atom Triangle Searching.
40. W. Fisanick, K. P. Cross, and A. Rusinko III, *J. Chem. Inf. Comput. Sci.*, **33**, 548 (1993). Experimental System for Similarity and 3D Searching of CAS Registry Substances. 1. 3D Substructure Searching.
41. G. W. Bemis and I. D. Kuntz, *J. Comput.-Aided Mol. Design*, **6**, 607 (1992). A Fast and Efficient Method for 2D and 3D Molecular Shape Description.
42. R. Nilakantan, N. Bauman, and R. Venkataraghavan, *J. Chem. Inf. Comput. Sci.*, **33**, 79 (1993). New Method for Rapid Characterization of Molecular Shapes: Application in Drug Design.
43. A. C. Good, T. J. A. Ewing, D. A. Gschwend, and I. D. Kuntz. *J. Comput.-Aided Mol. Design*, **9**, 1 (1995). New Molecular Shape Descriptors: Application in Database Searching.
44. M. A. Gallop, R. W. Barrett, W. J. Dower, S. P. A. Foder, and E. M. Gordon, *J. Med. Chem.*, **37**, 1233 (1994). Application of Combinatorial Technologies to Drug Discovery. 1. Background and Peptide Combinatorial Libraries.
45. D. J. Keenan, D. E. Tsai, and J. D. Keene, *T.I.B.S.*, **19**, 57 (1994). Exploring Molecular Diversity with Combinatorial Shape Libraries.

46. I. D. Kuntz, J. M. Blaney, S. J. Oatley, R. Langridge, and T. E. Ferrin, *J. Mol. Biol.*, **161**, 269 (1982). A Geometric Approach to Macromolecule–Ligand Interactions.
47. R. L. Desjarlais, R. P. Sheridan, G. L. Seibel, J. S. Dixon, I. D. Kuntz, and R. Venkataraghavan, *J. Med. Chem.*, **31**, 722 (1988). Using Shape Complementarity as an Initial Screen in Designing Ligands for a Receptor Binding Site of Known 3D Structure.
48. R. L. Dejarlais, G. L. Seibel, I. D. Kuntz, P. S. Furth, J. C. Alvarez, P. R. Ortiz de Montellano, D. L. DeCamp, L. M. Babé, and C. S. Craik, *Proc. Natl. Acad. Sci. USA*, **87**, 6644 (1990). Structure-Based Design of Nonpeptide Inhibitors Specific for HIV-1 Protease.
49. A. S. Smellie, G. M. Crippen, and W. G. Richards, *J. Chem. Inf. Comput. Sci.*, **31**, 386 1991). Fast Drug Receptor Mapping by Site Directed Distances.
50. M. C. Lawrence and P. C. Davis, *Proteins*, **12**, 31 (1992). CLIX: A Search Algorithm for Finding Novel Ligands Capable of Binding to Proteins of Known 3D Structure.
51. B. K. Shoichet, D. L. Bodian, and I. D. Kuntz, *J. Comput. Chem.*, **13**, 380 (1992). Molecular Docking Using Shape Descriptors.
52. E. C. Meng, B. K. Shoichet, and I. D. Kuntz, *J. Comput. Chem.*, **13**, 505 (1992). Automated Docking with Grid Based Energy Evaluation.
53. B. K. Shoichet and I. D. Kuntz, *Protein Eng.*, **6**, 723 (1993). Matching Chemistry and Shape in Molecular Docking.
54. B. K. Shoichet, R. M. Stroud, D. V. Santi, I. D. Kuntz, and K. M. Perry, *Science*, **259**, 1445 (1993). Structure-Based Discovery of Inhibitors of Thymidylate Synthase.
55. C. S. Ring, E. Sun, J. H. McKerrow, G. K. Lee, P. J. Rosenthal, I. D. Kuntz, and F. E. Cohen. *Proc. Natl. Acad. Sci. U.S.A.*, **90**, 3853 (1993). Structure-Based Design by Using Protein Models for the Development of Antiparasitic Agents.
56. D. L. Bodian, R. B. Yamasaki, R. L. Buswell, J. F. Stearns, J. M. White, and I. D. Kuntz, *Biochemistry*, **32**, 2967 (1993). Inhibition of the Fusion-Inducing Conformational Change of Influenza Hemagglutinin by Benzoquinones and Hydroquinones.
57. E. C. Meng, D. A. Gschwend, J. M. Blaney, and I. D. Kuntz, *Proteins*, **17**, 266 (1993). Orientational Sampling and Rigid Body Minimization in Molecular Docking.
58. M. D. Miller, S. K. Kearsley, D. J. Underwood, and R. P. Sheridan, *J. Comput.-Aided Mol. Design*, **8**, 153 (1994). FLOG: A System to Select "Quasi-Flexible" Ligands Complementary to a Receptor of Known 3D Structure.
59. R. L. Desjarlais and J. S. Dixon, *J. Comput.-Aided Mol. Design*, **8**, 231 (1994). A Shape and Chemistry Based Docking Method and Its Use in the Design of HIV-1 Protease Inhibitors.
60. I. D. Kuntz, E. C. Meng, and B. K. Shoichet, *Acc. Chem. Res.*, **27**, 117 (1994). Structure-Based Molecular Design.
61. K. Appelt, R. J. Baquet, C. A. Bartlett, C. L. J. Booth, S. T. Freer, M. A. N. Fuhry, M. R. Gehring, S. M. Herrmann, E. F. Howland, C. A. Janson, T. R. Jones, C.-C. Kan, V. Kathardekar, K. K. Lewis, G. P. Marzoni, D. A. Matthews, C. Mohr, E. W. Moomaw, C. A. Morse, S. J. Oatley, R. C. Ogden, M. R. Reddy, S. H. Reich, W. S. Schoettlin, W. W. Smith, M. D. Varney, J. E. Villafranca, R. W. Ward, S. Webber, S. E. Webber, K. M. Welsh, and J. White. *J. Med. Chem.*, **34**, 1925 (1991). Design of Enzyme Inhibitors Using Iterative Protein Crystallographic Analysis.
62. J. Greer, J. W. Erikson, J. J. Baldwin, and M. D. Varney, *J. Med. Chem.*, **37**, 1035 (1994). Application of the Three-Dimensional Structures of Protein Target Molecules in Structure-Based Drug Design.
63. T. Havel, *Progr. Mol. Biol. Biophys.*, **56**, 43 (1991). An Evaluation of Computational Strategies for Use in the Determination of Protein Structure from Distance Constraints by NMR. For a tutorial on distance geometry, see J. M. Blaney and J. S. Dixon, in *Reviews in Computational Chemistry*, Vol. 5, K. B. Lipkowitz and D. B. Boyd, Eds., VCH Publishers, New York, 1994, pp. 299–335. Distance Geometry in Molecular Modeling.

64. E. M. Ricketts, J. Bradshaw, M. Hann, F. Hayes, N. Tanna, and D. M. Ricketts, *J. Chem. Inf. Comput. Sci.* **33,** 905 (1993). Comparison of Conformations of Small Molecule Structures from the Protein Data Bank with Those Generated by CONCORD, COBRA, ChemDBS-3D, and CONVERTOR and Those Extracted from the Cambridge Database.
65. M. C. Nicklaus, G. W. A. Milne, and D. Zaharevitz, *J. Chem. Inf. Comput. Sci.,* **33,** 639 (1993). Chem-X and CAMBRIDGE. Comparison of Computer Generated Chemical Structures with X-ray Crystallographic Data.
66. J. Sadowski, J. Gasteiger, and G. Klebe, *J. Chem. Inf. Comput. Sci.,* **34,** 1000 (1994). Comparison of Automatic Three-Dimensional Model Builders Using X-ray Structures.
67. F. H. Allen, S. Bellard, M. D. Brice, B. A. Cartwright, A. Doubleday, H. Higgs, T. Hummelink, B. G. Hummelink-Peters, O. Kennard, W. D. S. Motherwell, J. R. Rodgers, and D. G. Watson, *J. Chem. Inf. Comput. Sci.,* **31,** 187 (1992). The Development of Versions 3 and 4 of the Cambridge Structural Database System.
68. F. C. Bernstein, T. F. Koetzle, G. J. B. Williams, E. F. Meyer Jr., M. D. Brice, J. R. Rodgers, O. Kennard, T. Shimanouchi, and M. Tasumi, *J. Mol. Biol.,* **112,** 535 (1977). The Protein Data Bank: A Computer Based Archival File for Macromolecular Structures.
69. W. L. Jorgenson, *Science,* **254,** 954 (1991). Rusting of the Lock and Key Model for Protein Ligand Binding.
70. J. Sadowski and J. Gasteiger, *Chem. Rev.,* **93,** 2567 (1993). From Atoms and Bonds to Three-Dimensional Atomic Co-Ordinates: Automatic Model Builders.
71. O. F. Guner, D. R. Henry, and R. S. Pearlman, *J. Chem. Inf. Comput. Sci.,* **32,** 101 (1992). Use of Flexible Queries for Searching Conformationally Flexible Molecules in Databases of 3D Structures.
72. M. Saunders, *J. Am. Chem. Soc.,* **109,** 3051 (1987). Stochastic Exploration of Molecular Mechanics Energy: Hunting for the Global Minimum.
73. J. A. McCammon and S. C. Harvey, *Dynamics of Proteins and Nucleic Acids,* Cambridge University Press, Cambridge, U.K., 1987.
74. N. Gō and H. A. Scheraga, *Macromolecules,* **11,** 552 (1978). Calculation of the Conformation of Cyclo-Hexaglycil.
75. H. C. Romesburg, *Cluster Analysis for Researchers,* Lifetime Learning Pub., Wadsworth Inc., Belmont, CA, 1984.
76. P. D. Dolata, A. R. Leach, and K. Prout, *J. Comput.-Aided Mol. Design,* **1,** 73 (1987). AI in Conformational Analysis.
77. S. E. Jakes and P. Willett, *J. Mol. Graphics,* **4,** 12 (1986). Pharmacophore Pattern Matching in Files of 3D Chemical Structures: Selection of Inter-Atomic Distance Screens.
78. C. R. Ganellin, in *Pharmacology of Histamine Receptors,* C. R. Ganellin and M. E. Parsons, Eds., Wright-PSG, London, 1982, pp. 21–77. Chemistry and Structure Activity Relationships of Drugs Acting at Histamine Receptors.
79. D. R. Ferro and J. A. Hermans, *Acta Crystallogr.,* **A33,** 345 (1977). A Difference Based Rigid Body Molecular Fit Routine.
80. J. S. Mason, in *Trends in QSAR and Molecular Modelling 92,* C. G. Wermuth, Ed., ESCOM, Leiden, 1992, pp. 252–255. Experiences with Conformationally Flexible Databases.
81. K. Kato, S. Ohkawa, S. Terao, Z. Terashita, and K. Nishikawa, *J. Med. Chem.,* **28,** 287 (1985). Thromboxane Synthetase Inhibitors (TXSI). Design, Synthesis and Evaluation of a Novel Series of ω-Pyridylalkenoic Acids.
82. J. A. Nelder and R. Mead, *Comput. J.,* **7,** 308 (1965). A Simplex Method for Function Optimization.
83. W. H. Press, B. P. Flannery, S. A. Teukolsky, and W. T. Vetterling, *Numerical Recipes—Fortran Version,* Cambridge University Press, Cambridge, U.K., 1989.
84. T. Schlick, in *Reviews in Computational Chemistry,* K. B. Lipkowitz and D. B. Boyd, Eds., VCH Publishers, New York, 1992, pp. 1–30. Optimization Methods in Computational Chemistry.

85. D. E. Clark, G. Jones, and P. Willett, *J. Chem. Inf. Comput. Sci.*, **34**, 197 (1994). Pharmacophoric Pattern Matching in Files of Three-Dimensional Chemical Structures: Comparison of Conformational Searching Algorithms for Flexible Searching.

86. J. S. Mason, in *Trends in Drug Research: Proceedings of the 9th Noordwijkhout-Camerino Symposium,* Noordwijkhout, The Netherlands, 23–28th May 1993, Series title: Pharmacochemistry Library, 20, V. Claassen, Ed., Elsevier, Amsterdam, 1994, pp. 147–156. Drug Design Using Conformationally Flexible Molecules in 3D Databases.

87. J. S. Mason, in *Molecular Similarity in Drug Design,* P. M. Dean, Ed., Blackie Academic and Professional, Glasgow, 1995, pp. 138–162. Chapter 6. Experiences with Searching for Molecular Similarity in Flexible 3D Databases.

88. N. C. Cohen, J. M. Blaney, C. Humblet, P. Gund, and D. C. Barry, *J. Med. Chem.*, **33**, 883 (1990). Molecular Modeling Software and Methods for Medicinal Chemistry. There are too many molecular modeling packages to mention individually. The reference given here is a perspective on molecular modeling packages and the functionality they provide. Its vintage is such that many new programs have since become accessible. See D. B. Boyd, this volume's appendix for a current, comprehensive listing of molecular modeling packages.

89. M. G. Bures, C. Black-Schaefer, and G. Gardner, *J. Comput.-Aided Mol. Design,* **5**, 323 (1991). The Discovery of Novel Auxin Transport Inhibitors by Molecular Modeling and Three-Dimensional Pattern Analysis.

90. G. R. Marshall, C. D. Barry, H. E. Bosshard, R. A. Dammkoehler, and D. A. Dunn, in *Computer-Assisted Drug Design,* American Chemical Society Symposium, No. 112, E. C. Olsen and R. E. Christoffersen, Eds., ACS, Washington, 1979, pp. 205–226. The Conformational Parameter in Drug Design: The Active Analog Approach.

91. D. Mayer, C. B. Naylor, I. Motoc, and G. R. Marshall, *J. Comput.-Aided Mol. Design,* **1**, 3 (1987). A Unique Geometry of the Active Site of Angiotensin-Converting Enzyme Consistent with Structure Activity Studies.

92. D. F. Ortwine, T. C. Malone, C. F. Bigge, J. T. Drummond, C. Humblet, G. Johnson, and G. W. Pinter, *J. Med. Chem.*, **35**, 1345 (1992). Generation of *N*-Methyl-D-Aspartate Agonist and Competitive Antagonist Pharmacophore Models. Design and Synthesis of Phosphonalkyl-Substituted Tetrahydroisoquinolines as Novel Antagonists.

93. A. T. Brint and P. Willett, *J. Chem. Inf. Comput. Sci.*, **27**, 152 (1987). Algorithm for the Identification of Three-Dimensional Maximal Common Substructures.

94. R. Carbo and L. Domingo, *Int. J. Quantum Chem.*, **32**, 517 (1987). LCAO-MO Similarity Measures and Taxonomy.

95. A. J. Hopfinger, *J. Med. Chem.*, **26**, 990 (1983). Theory and Analysis of Molecular Potential Energy Fields in Molecular Shape Analysis: A QSAR Study of 2,4-Diamino-5-Benzylpyrimidines as DHFR Inhibitors.

96. E. E. Hodgkin and W. G. Richards, *Int. J. Quantum Chem., Quantum Biol. Symp.*, **14**, 105 (1987). Molecular Similarity Based on Electrostatic Potential and Electric Field.

97. M. Manaut, F. Sanz, J. Jose, and M. Milesi, *J. Comput.-Aided Mol. Design,* **5**, 371 (1991). Automatic Search for Maximum Similarity between Molecular Electrostatic Potential Distributions.

98. J. D. Petke, *J. Comput. Chem.*, **14**, 928 (1993). Cumulative and Discrete Similarity Analysis of Electrostatic Potentials and Fields.

99. A. C. Good, *J. Mol. Graph.*, **10**, 144 (1992). The Calculation of Molecular Similarity: Alternative Formulas, Data Manipulation and Graphical Display.

100. C. Burt and W. G. Richards, *J. Comput.-Aided Mol. Design.*, **4**, 231 (1990). Molecular Similarity: The Introduction of Flexible Fitting.

101. A. C. Good, E. E. Hodgkin, and W. G. Richards, *J. Chem. Inf. Comput. Sci.*, **32**, 188 (1992). The Utilization of Gaussian Functions for the Rapid Evaluation of Molecular Similarity.

102. A. C. Good and W. G. Richards, *J. Chem. Inf. Comput. Sci.*, **33**, 112 (1993). Rapid Evaluation of Shape Similarity Using Gaussian Functions.

103. P.-L. Chau and P. M. Dean, *J. Mol. Graphics,* **5,** 97 (1987). Molecular Recognition: 3D Surface Structure Comparison by Gnomonic Projection.

104. P. M. Dean and P.-L. Chau, *J. Mol. Graph.* **5,** 152 (1987). Molecular Recognition: Optimized Searching Through Molecular 3-Space for Pattern Matches on Molecular Surfaces.

105. R. B. Hermann and D. K. Herron, *J. Comput.-Aided Mol. Design,* **5,** 511 (1991). OVID and SUPER: Two Overlap Programs for Drug Design.

106. P. Bladen, *J. Mol. Graphics,* **7,** 130 (1989). A Rapid Method for Comparing and Matching the Spherical Parameter Surfaces of Molecules and Other Irregular Objects.

107. F. E. Blaney, D. Naylor, and J. Woods, *J. Mol. Graphics,* **11,** 157 (1993). MAMBAS: A Real Time Graphics Environment for QSAR.

108. A. C. Good, in *Molecular Similarity in Drug Design,* P. M. Dean, Ed., Blackie Academic and Professional, Glasgow, 1995, pp. 25–56. Chapter 2. 3D Molecular Similarity Indices and Their Application in QSAR Studies.

109. G. M. Downs and P. Willett, this volume. Similarity Searching in Databases of Chemical Structures.

110. M. L. Connolly, *J. Am. Chem. Soc.,* **107,** 1118 (1985). Computation of Molecular Volume.

111. R. D. Cramer III, J. D. Bunce, and D. E. Patterson, *Quant. Struct. Act. Relat.,* **7,** 18 (1988). Cross-Validation, Bootstrapping, and PLS Compared with Multiple Linear Regression in Conventional QSAR Studies.

112. J. W. McFarland, *J. Med. Chem.,* **35,** 2543 (1992). Comparative Molecular Field Analysis of Anticoccidial Triazines. This paper contains a number of references to other interesting articles that utilize the CoMFA technique.

113. C. L. Waller, T. I. Oprea, A. Giolitti, and G. R. Marshall, *J. Med. Chem.,* **36,** 4152 (1993). Three-Dimensional QSAR of HIV-1 Protease Inhibitors. 1. A CoMFA Study Employing Experimentally Determined Alignment Rules.

114. T. I. Oprea, C. L. Waller, and G. R. Marshall, *J. Med. Chem.,* **37,** 2206 (1994). Three-Dimensional QSAR of HIV-1 Protease Inhibitors. 2. Predictive Power using Limited Exploration of Alternative Binding Modes.

115. SYBYL, a molecular modeling package developed and distributed by Tripos Associates, St. Louis, MO.

116. H. Kubinyi, Ed., *3D QSAR in Drug Design, Theory, Methods and Applications,* ESCOM, Leiden, 1993.

117. J. S. Mason, unpublished results. For more information on the LDL receptor system, see M. Huettinger, M. Herrman, H. Goldenberg, E. Granzer, and M. Leineweber, *Arteriosclerosis and Thrombosis,* **13,** 1005 (1993). Hypolipidemic Activity of HOE-402 Is Mediated by Stimulation of the LDL Receptor Pathway.

118. J. S. Mason, unpublished results. For more information on the endothelin receptor system, see A. M. Doherty, *J. Med. Chem.,* **35,** 1493 (1992). Endothelin: A New Challenge.

119. M. L. Connolly, *J. Mol. Graphics,* **11,** 139 (1993). The Molecular Surface Package.

120. P. J. Goodford, *J. Med. Chem.,* **28,** 849 (1985). Computational Procedure for Determining Energetically Favorable Binding Sites on Biologically Important Molecules. Program known as GRID, developed and distributed by Molecular Discovery Ltd., Oxford, U.K.

121. R. Norel, D. Fischer, H. J. Wolfson, and R. Nussinov, *Protein. Eng.,* 7, 39 (1994). Molecular Surface Recognition by a Computer Vision Based Technique.

122. V. J. van Geerestein, N. C. Perry, P. D. J. Grootenhuis, and C. A. G. Haasnoot, *Tetrahedron Comput. Methodol.,* **3,** 595 (1992). 3D Database Searching on the Basis of Ligand Shape Using the SPERM Prototype Method.

123. G. Lauri and P. Bartlett, *J. Comput.-Aided Mol. Design,* **8,** 51 (1994). CAVEAT: A Program to Facilitate the Design of Organic Molecules.

124. R. Nilakantan, N. Bauman, and R. Venkataraghavan, *J. Chem. Inf. Comput. Sci.,* **31,** 527 (1991). A Method for Automatic Generation of Novel Chemical Structures and Its Potential Applications to Drug Design.

125. J. B. Moon and W. J. Howe, *Proteins,* **11,** 314 (1991). Computer Design of Bioactive Molecules: A Method for Receptor Based De Novo Design.

126. S. H. Rotstein and M. A. Murcko, *J. Comput.-Aided Mol. Design,* 7, 23 (1992). GENSTAR: A Method for De Novo Drug Design.

127. V. J. Gillet, W. Newell, P. Mata, G. Myatt, S. Sike, Z. Zsoldos, and P. Johnson, *J. Chem. Inf. Comput. Sci.,* **34,** 207 (1994). SPROUT: Recent Developments in the De Novo Design of Molecules.

128. H. J. Bohm, *J. Comput.-Aided Mol. Design,* **6,** 593 (1992). LUDI: Rule Based Automatic Design of New Substituents for Enzyme Inhibitor Leads.

129. D. A. Pearlman and M. A. Murcko, *J. Comput. Chem.,* **14,** 1184 (1993). CONCEPTS: A New Dynamic Algorithm for De Novo Drug Suggestion.

130. Y. Nishibata and A. Itai, *J. Med. Chem.,* **36,** 2921 (1993). Confirmation of the Usefulness of a Structure Construction Program Based on 3D Receptor Structure for Rational Lead Generation.

131. R. S. Bohacek and C. McMartin, *J. Am. Chem. Soc.,* **116,** 5560 (1994). Multiple Highly Diverse Structures Complementary to Enzyme Binding Sites—Results of Extensive Application of a De Novo Design Method Incorporating Combinatorial Growth.

132. R. A. Lewis, D. C. Roe, C. Huang, T. E. Ferrin, R. Langridge, and I. D. Kuntz, *J. Mol. Graphics,* **10,** 66 (1992). Automated Site Directed Drug Design Using Molecular Lattices.

133. A. Smellie, S. D. Kahn, and S. L. Teig, *J. Chem. Inf. Comput. Sci.,* **35,** 285 (1995). Analysis of Conformational Coverage. 1. Validation and Estimation of Coverage.

134. A. Smellie, S. D. Kahn, and S. L. Teig, *J. Chem. Inf. Comput. Sci.,* **35,** 295 (1995). Analysis of Conformational Coverage. 2. Application of Conformational Models.

CHAPTER 3

Methods and Applications of Combined Quantum Mechanical and Molecular Mechanical Potentials

Jiali Gao

Department of Chemistry, State University of New York at Buffalo, Buffalo, New York 14260

INTRODUCTION

Solvent effects can significantly influence the structure and reactivity of organic compounds in solution.[1] This is illustrated by the well-studied S_N2 reactions, for which the reaction rates can vary up to 20 orders of magnitude on transfer from the gas phase into aqueous solution.[2–6] Consequently, it has been a major goal in theoretical chemistry to develop computational methods that can provide quantitative prediction of the behavior of chemical species in solution. Statistical mechanical Monte Carlo and molecular dynamics methods are well suited for investigations of this problem at the molecular level.[7–12] In these studies, one or more solute molecules are placed in a periodic unit cell consisting of a few hundred or perhaps a few thousand solvent molecules. Thermodynamic properties are then averaged over configurations sampled randomly by the Metropolis Monte Carlo simulation, or over a time-dependent trajectory in molecular dynamics calculation.[7] The molecular dynamics simulation method has the advantage of yielding dynamic properties of the solution. Nevertheless, in both calculations the key element that ultimately determines the

Reviews in Computational Chemistry, Volume 7
Kenny B. Lipkowitz and Donald B. Boyd, Editors
VCH Publishers, Inc. New York, © 1996

success of computer simulations of condensed phase systems is the availability of potential energy functions that can accurately describe intermolecular interactions.

In principle, ab initio quantum mechanical (QM) methods can be used to provide a systematic and reliable treatment of the intermolecular interactions in solution and to generate the necessary structural and energetic data.[13–15] Indeed, quantum mechanical potential surfaces have been used in classical trajectory calculations in the gas phase[16,17] and, remarkably, in a molecular dynamics simulation of liquid water using the Car–Parrinello density functional algorithm.[18–20] Similar methods for performing molecular dynamics simulations and simulated-annealing calculations have been extended to Hartree–Fock wavefunctions.[21–24] However, it is generally impossible to treat the entire condensed phase system quantum mechanically because the computer time required in these calculations, even with the fastest supercomputers available today, will rapidly become intractable as the number of atoms increases. Thus, approximations must be made. For this reason, empirical molecular mechanics-type potential functions are typically used in Monte Carlo and molecular dynamics simulation of solutions and proteins.[8–11,25–30] Despite the approximations, empirical potential functions have been extremely successful and can provide valuable and meaningful insights about intermolecular interactions for biomolecules in solution.[8–11] However, there are several well-known difficulties associated with the molecular mechanics (MM) force field. Because electronic structures are not explicitly represented, MM force fields are generally not suitable, and the form of empirical potential functions is not appropriate for chemical reactions that involve bond forming and breaking. For cases in which analytical potential forms are fitted to reproduce quantum mechanical results,[6] the parameterization process is laborious and difficult due to insufficient experimental data along the entire reaction coordinate. As a result, this has limited the application of computer simulations to only a few well-defined systems in solution. The problem is further exacerbated by the need for specific consideration of the solvent polarization effects,[31] which have been treated in an average sense in the past.[26–30]

An alternative approach is to combine QM and MM methods such that the reacting system (or the active site in an enzyme) is treated explicitly by a quantum mechanical method, while the surrounding "environmental" solvent molecules (or amino acids), which constitute the most time-consuming part in the evaluation of the potential energy surface, are approximated by a standard MM force field.[32–37] Such a method takes advantage of the accuracy and generality of the QM treatment for chemical reactions[13–15] and of the computational efficiency of the MM calculation.[25–30] Because the reactant electronic structure and solute–solvent interactions are determined quantum mechanically, the procedure is appropriate for studying chemical reactions, and there is no need to parameterize potential functions for every new reaction. Furthermore, the solvent polarization effects on the solute are naturally included in the

QM calculations. Consequently, the method may be conveniently used by organic chemists, who will be able to concentrate on the chemical problems rather than the technical issues.

The idea of incorporating solvent effects into quantum mechanical calculations can be traced back to the early days of quantum mechanics. In a classical paper dealing with the structure and properties of liquid water, Bernal and Fowler proposed the partitioning of a condensed phase system into discrete solute and solvent regions, although the solvent was treated as a dielectric continuum.[38] Since then, significant efforts have gone into the QM treatment of solvation energies using the Onsager reaction field or generalized Born models.[39] In these calculations, the solute charges polarize the surrounding medium, generating a reaction field, which is then introduced into the solute Hamiltonian. In turn, the reaction field influences the solute charge distribution. The new charges are then used to generate a new reaction field, which gives another modification to the Hamiltonian. This procedure continues until there is no further change in the reaction field. In recent implementations, the crude approximation of a spherical cavity has been extended to a more realistic molecular-shaped cavity. The continuum solvation models have recently been reviewed in this series by Cramer and Truhlar.[39]

The continuum models have the advantage of effectively modeling the long-range electrostatic contributions in solvation; however, specific interactions between the solute and solvent are not considered. In many cases, it is essential to include these interactions explicitly to obtain quantitative results on solvation.[40,41] Consequently, the supermolecular approach has been introduced to treat the solute along with a few solvent molecules in quantum mechanical calculations. The incorporation of explicit classical solvent models in quantum mechanical calculations for reactions in solution and enzymes was studied by Warshel and co-workers.[11,32,33,42] In a seminal paper,[32] Warshel and Levitt laid out the basic algorithms and subsequently investigated numerous chemical and biological systems in solution using, in particular, the empirical valence bond (EVB) theory.[42]

Several groups, including Singh and Kollman, have also developed programs that combine quantum mechanical and molecular mechanical potentials.[34,43–50] In the past, these calculations were primarily performed to determine energies of the reacting molecules in a fixed-charge environment, or to determine the reaction path in enzymes via energy minimization methods.[51] Although the implementation of the combined QM/MM potential appears to be straightforward, full molecular dynamics or Monte Carlo simulation was not possible until recently, thanks to advances in both computer technology and QM algorithms. In 1987, Bash, Field, and Karplus carried out a molecular dynamics simulation of the S_N2 reaction, $Cl^- + CH_3Cl \rightarrow ClCH_3 + Cl^-$, in water using a combined semiempirical AM1 (MNDO) and TIP3P potential.[52] Employing the EVB method, Warshel and co-workers studied the same reaction as well as other reactions, including heterolytic bond cleavage in solution.[11,53] More recently, the combined AM1/classical approach has been extended to Monte Carlo simulations at Buffalo to determine the solvent polarization effects and the potential of mean force for chemical reactions in solution.[54]

In this review, we shall focus on the development and application of combined QM/MM potentials in condensed phase simulations where solute and solvent molecules are explicitly treated, in both aqueous and organic environments. The emphasis of this chapter is the use of molecular orbital (MO) theory in the QM treatment because it is well documented and familiar to chemists. Details of the Monte Carlo and molecular dynamics simulation techniques are available in an excellent book by Allen and Tildesley.[7] In a recent paper, Åqvist and Warshel provided a number of additional details of the methodology, particularly on the use of the EVB method.[42] We have also benefited both intellectually and technically from the thorough paper by Field, Bash, and Karplus,[35] which gives additional insights on the implementation of combining molecular orbital and molecular mechanics programs.

The review is organized as follows. The second section describes some of the technical details concerning partitioning of a condensed phase system, implementation of the method, and computational procedures. In the third section, the performance and suitability of combined QM/MM models are verified through a series of studies, including comparison with high-level ab initio MO or experimental results on bimolecular complexes, free energies of hydration of organic compounds, and solvent effects on chemical equilibria. The performance is further tested by comparing with results obtained from continuum self-consistent reaction field (SCRF) calculations. The fourth section contains results on chemical reactions in solution. Comparison is also made with results obtained using parameterized empirical potential functions, and the inherent advantages and disadvantages of both approaches are described. Finally, the present status of the field and possible future developments are summarized.

INCORPORATION OF EXPLICIT SOLVENT EFFECTS IN QUANTUM MECHANICAL CALCULATIONS

Combined Quantum Mechanical and Molecular Mechanical Potentials

The difficulties associated with the study of chemical reactions in solution and enzymes are the complexity of the calculations and the large number of particles in the system. Although a full quantum mechanical treatment would avoid all the problems inherent in the empirical MM force field, it is practically impossible, and unnecessary, to solve the Schrödinger equation for the entire condensed phase system along with statistical ensemble averaging. What is most interesting to chemists is the chemical reactions that are taking place in the condensed phase, where typically only a small number of atoms participate in the bond formation and bond breaking processes. The surrounding "solvent" molecules generally serve as a "bath" which influences the properties and reactivities of the reactants. It is not essential to treat these environmental molecules explicitly by QM calculations, although the effect of the solvent charges on the solute should be included.

Since only a small number of solvents are widely used in organic chemistry and there are only 20 common amino acids in proteins,[1,55] well-tested empirical potential functions may be developed and utilized to represent polypeptides and the usual solvents. Indeed, Jorgensen and co-workers have created a series of optimized potentials for liquid simulations (OPLS) for organic solvents and water,[56,57] and a number of excellent force fields are available for protein simulations.[26–30] Thus, it appears to be natural to divide a condensed phase system into quantum mechanical and classical regions according to the priority of interest or of importance.

In the combined QM/MM treatment, the molecular system (either a dilute solution or an enzyme–substrate system) is partitioned into (1) a quantum mechanical region consisting of the solute molecules or the active site residues, and (2) a molecular mechanical region containing the surrounding solvent monomers (Figure 1).[32–37] To overcome problems arising from the finite system size used in Monte Carlo and molecular dynamics simulations, boundary conditions are imposed using periodic/stochastic approximations or continuum models.[7,33] These features will be addressed later. For the QM region, which is assumed to be a closed shell system, the solute is represented by nuclei and electrons with a restricted Hartree–Fock wavefunction Φ. The wavefunction is written as a single Slater determinant of all doubly occupied molecular orbitals, $\{\psi_i\}$ (configuration interaction calculations will be described later):

$$\Phi(\mathbf{r},\mathbf{R}_X,\mathbf{R}_s) = |\psi_1\alpha(1)\psi_1\beta(2)\ldots\psi_N\beta(2N)\rangle \qquad [1]$$

where $2N$ is the total number of electrons, α and β are the electron spin eigenfunctions, and the MOs are linear combinations of an atomic basis set, $\{\phi_\mu\}$. The wavefunction of the electrons of the QM atoms is a function of the coordinates of the electrons, $\mathbf{r}$, and positions of both the nuclei of the QM atoms $\mathbf{R}_X$ and the solvent atoms $\mathbf{R}_s$. In the following, we simply use Φ to represent the solute wavefunction at a given configuration sampled in the simulation. On the other hand, solvent atoms in the environmental MM region are described by interaction sites bearing partial atomic charges, which may be augmented with atomic polarizabilities. Both the QM and MM atoms are treated explicitly in fluid simulations.

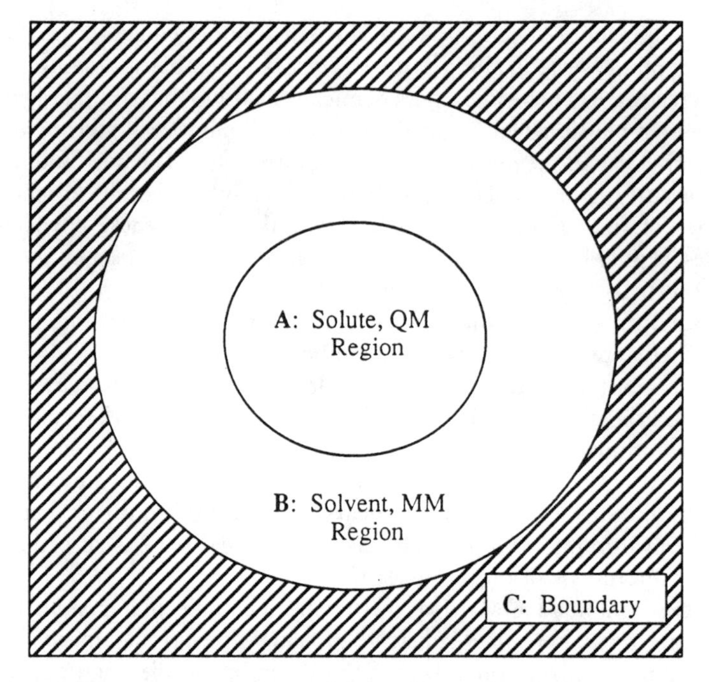

Figure 1 Schematic representation of the partition of a condensed phase system.

With the Born–Oppenheimer approximation and assuming that there is no charge transfer between regions, the effective Hamiltonian of the system can be separated into three terms:

$$\hat{H}_{\text{eff}} = \hat{H}_{\text{X}}^{\text{o}} + H_{\text{Xs}} + \hat{H}_{\text{ss}} \qquad [2]$$

where $\hat{H}_{\text{X}}^{\text{o}}$ is the Hamiltonian of the solute molecule in the QM region, $\hat{H}_{\text{ss}}$ is the interaction energy between solvent molecules in the MM area, and $\hat{H}_{\text{Xs}}$ represents the QM/MM solute–solvent interaction Hamiltonian, which couples the solvent effects into quantum mechanical calculations. In general, a boundary term should also be included in Eq. [2].[35] However, for simplicity, it is assumed to be treated as part of the simulation process or in a way similar to the SCRF approach.[39] Throughout this chapter, the subscript X represents the solute/reactant molecule in the QM region, while s designates the classical solvent particles.

The first term in Eq. [2] is the Hamiltonian of the QM particles in vacuo, which is given as follows in atomic units:

$$\hat{H}_{\text{X}}^{\text{o}} = -\frac{1}{2}\sum_{i=1}^{2N} \nabla_i^2 - \sum_{m=1}^{M}\sum_{i=1}^{2N} \frac{Z_m}{r_{im}} + \sum_{i<j} \frac{1}{r_{ij}} + \sum_{m<n} \frac{Z_m Z_n}{R_{mn}} \qquad [3]$$

where i,j and m,n indicate electrons and nuclei, respectively, r is a distance between two electrons or between an electron and a nucleus, R is a nucleus–nucleus distance, M is the total number of QM atoms, and Z is an effective nuclear charge.

The second term in Eq. [2] describes the interaction between the QM particles and the MM solvent sites, which consists of electrostatic $\hat{H}_{Xs}^{el}$, van der Waals $\hat{H}_{Xs}^{vdW}$, and polarization $\hat{H}_{Xs}^{pol}$ terms:

$$\hat{H}_{Xs} = \hat{H}_{Xs}^{el} + \hat{H}_{Xs}^{vdW} + \hat{H}_{Xs}^{pol}$$
$$= \left(- \sum_{s=1}^{S} \sum_{i=1}^{2N} \frac{q_s}{r_{is}} + \sum_{s=1}^{S} \sum_{m=1}^{M} \frac{q_s Z_m}{R_{ms}} \right)$$
$$+ \sum_{s=1}^{S} \sum_{m=1}^{M} 4\epsilon_{ms} \left[\left(\frac{\sigma_{ms}}{R_{ms}} \right)^{12} - \left(\frac{\sigma_{ms}}{R_{ms}} \right)^{6} \right]$$
$$- \frac{1}{2} \sum_{s=1}^{S} \sum_{i=1}^{2N} \frac{\boldsymbol{\mu}_s^{ind}}{r^3_{is}} \mathbf{r}_{is} + \frac{1}{2} \sum_{s=1}^{S} \sum_{m=1}^{M} \frac{Z_m \boldsymbol{\mu}_s^{ind}}{R_{ms}^3} \mathbf{R}_{ms} \qquad [4]$$

where q_s is MM partial charge on a solvent atom, S is the total number of solvent interaction sites, **r** and **R** are distances of the QM electrons and nuclei from the solvent sites, respectively, and $\boldsymbol{\mu}_s^{ind}$ is the induced dipole of solvent atom s. The induced dipoles are calculated by

$$\boldsymbol{\mu}_s^{ind} = \boldsymbol{\alpha}_s \mathbf{E}_s^{tot} \qquad [5]$$

where $\boldsymbol{\alpha}_s$ is the atomic polarizability and $\mathbf{E}_s^{tot}$ is the total electric field on atom s. $\mathbf{E}_s^{tot}$ includes contributions from the permanent charges (q_s) and induced dipoles $(\boldsymbol{\mu}_s^{ind})$ of all other MM atoms as well as from the QM nucleus charges (Z_m) and wavefunction (Φ). Thus, $\boldsymbol{\mu}_s^{ind}$ and Φ are coupled and should be determined self-consistently. Although the implementation is straightforward, the computational process is very time-consuming. Consequently, pairwise, effective potentials are typically adopted, as in the empirical approach, such that the solvent polarization is included in the parameters in an average sense.

In the following discussion, unless specifically stated, the $\hat{H}_{Xs}^{pol}$ term will not be explicitly included in all calculations. Consequently, Eq. [4] is reduced to only two terms: $\hat{H}_{Xs} = H_{Xs}^{el} + \hat{H}_{Xs}^{vdW}$. This assumption ignores the instantaneous polarization of the solvent molecules by a change of the solute wavefunction. The mutual solute–solvent polarization effects are particularly important for studying solvent effects on electronic spectra.

The Lennard–Jones terms in Eq. [4] account for the average dispersion interactions between the QM and MM atoms, which are neglected in the hybrid QM/MM approximation.[35,58,59] These terms contain the only adjustable parameters (σ_m and ϵ_m) in the combined QM/MM model, whose values are obtained by fitting the results for hydrogen-bonded complexes from the combined QM/MM calculation against ab initio data or experiments. This will be further discussed later. However, it should be noted here that these parameters depend only on hybridization of the QM atom and are fully transferable from one system to another. Standard combining rules are used for interactions between the QM solute and MM solvent atoms such that $\sigma_{ms} = (\sigma_m \sigma_s)^{1/2}$ and $\epsilon_{ms} = (\epsilon_m \epsilon_s)^{1/2}$, where σ_s and ϵ_s are solvent parameters.

Finally, the molecular mechanics potential energy is used to approximate the Hamiltonian for solvent–solvent interactions in the MM region (Eq. [6]).

$$\hat{H}_{\mathrm{ss}} = E_{\mathrm{ss}} \qquad [6]$$

where E_{ss} is the MM force field for interactions among solvent molecules. It generally contains bond-stretching, angle-bending, dihedral torsion, and nonbonded terms.[26–30] A number of excellent force fields are available for aqueous and organic solvents, and for biopolymers including proteins and nucleic acids.[26–30] The parameters, partial charges, and van der Waals coefficients in these force fields are derived by fitting against experimental thermodynamic data for pure liquids and solutions. Interested readers are directed to References 26–30, which contain detailed discussion concerning parameterization of empirical MM force fields.

Having defined the system effective Hamiltonian, the total energy at an instantaneous configuration sampled during a Monte Carlo or molecular dynamics simulation is determined according to Eq. [7] by the expectation value of the wavefunction, Φ:

$$E_{\mathrm{tot}} = \langle \Phi | \hat{H}_{\mathrm{eff}} | \Phi \rangle = E_{\mathrm{X}} + E_{\mathrm{Xs}} + E_{\mathrm{ss}} \qquad [7]$$

Here, the normalized wavefunction Φ minimizes the energy of the effective Hamiltonian $\hat{H}_{\mathrm{eff}}$. It should be noted that Φ corresponds to a state of the solute molecule in solution, which differs from Φ^{o} in that the latter minimizes the gas-phase energy of the Hamiltonian $\hat{H}^{\mathrm{o}}_{\mathrm{X}}$ alone. The forces needed to integrate Newton's equations of motion in molecular dynamics simulations are determined by differentiating the energy expression in Eq. [7] with respect to atomic coordinates:

$$\mathbf{F}_m = -\frac{\partial E_{\mathrm{tot}}}{\partial \mathbf{R}_m} \qquad [8]$$

where m represents either QM or MM atoms.

The effective Hamiltonian defined by Eq. [2] consists of terms that are functions of either electronic or nuclear coordinates. Only terms depending on the electronic coordinates enter the Hartree–Fock self-consistent field (HF-SCF) calculation. Besides the three terms in Eq. [3], there are two such terms in Eq. [4]. In the HF-SCF calculation, the solute–solvent interaction Hamiltonian is incorporated by modifying the Fock matrix:

$$F_{\mu\nu} = F^{o}_{\mu\nu} + H^{Xs}_{\mu\nu} \qquad [9]$$

where $F^{o}_{\mu\nu}$ is the Fock matrix element for an isolated solute in vacuum,[15] and $H^{Xs}_{\mu\nu}$ is the one-electron integral resulting from the solute–solvent interaction Hamiltonian.

The energies associated with SCF calculations in Eq. [7] are obtained from the density matrix **P** of the solute wavefunction in solution:

$$E_X = \frac{1}{2} \sum_{\mu\nu} P_{\mu\nu}(H^{o}_{\mu\nu} + F^{o}_{\mu\nu}) + E_{nuc} \qquad [10]$$

and

$$E_{Xs} = \sum_{\mu\nu} P_{\mu\nu} H^{Xs}_{\mu\nu} + \sum_{s=1}^{S} \sum_{m=1}^{M} \frac{q_s Z_m}{R_{ms}} + E^{vdW}_{Xs} \qquad [11]$$

where the indices run over all the basis set orbitals, $H^{o}_{\mu\nu}$ is an element of the gas phase one-electron Hamiltonian, and E_{nuc} is the Coulombic nuclear repulsion energy. The Hartree–Fock density matrix is defined as a sum over the occupied molecular orbitals[15]:

$$P_{\mu\nu} = 2 \sum_{i=1}^{N} C_{\mu i} C_{\nu i} \qquad [12]$$

Boundary Conditions

In all computer simulations of liquids or proteins, a finite construction has to be used to approximate a virtually infinite system.[7] Consequently, periodic boundary conditions are often used to mimic the bulk conditions.[7] In this approach, the central unit cell is repeated periodically by translation in three dimensions to remove any boundary effects. The atoms in the primary unit cell interact with the nearest images of atoms in the neighboring units. To further reduce computation time for evaluating the energy and forces, a spherical cutoff, R_{cut}, is typically used to truncate the long-range interactions. The cutoff distance is chosen to be less than half of the length of the primary unit cell. Often a smoothing function is used to allow the potential energy to vanish continuously in the cutoff region.[60,61]

An alternative approach is to employ stochastic boundary conditions where the finite molecular system employed in the simulation is not duplicated, but rather, a boundary force is applied to interact with atoms of the system.[62–67] The boundary force is chosen to mimic the bulk solvent regions that have been neglected. However, a difficulty with the use of this method is due to the ambiguity associated with the definition and parameterization of the boundary forces. Thus, there is sometimes the impression that the method is not rigorous. Recently, Beglov and Roux provided a formal definition and suggested a useful implementation of the boundary forces.[67] Their results are very promising.

Any of the methods used in classical Monte Carlo and molecular dynamics simulations may be borrowed in the combined QM/MM approach.[7] However, the use of a finite system in condensed phase simulations is always a severe approximation, even when appropriate periodic or stochastic boundary conditions are employed. A further complication is the use of potential function truncation schemes, particular in ionic aqueous solutions where the long-range Coulombic interactions are significant beyond the cutoff distance.[68,69] Thus, it is alluring to embed a continuum reaction field model in the quantum mechanical calculations in addition to the explicit solute–solvent interactions to include the dielectric effect beyond the cutoff distance.[70–72] Such an onion shell arrangement has been used in spherical systems,[33] whereas Lee and Warshel introduced an innovative local reaction field method for evaluation of long-range forces.[73]

Dividing Covalent Bonds across the QM and MM Regions

The assumption that there is no charge transfer between the QM and MM regions is quite reasonable for a dilute solution, in which the solute molecule is treated quantum mechanically and the solvent classically. The boundaries can be arranged so that there are no covalent bonds across the two regions; clearly, most organic systems fall into this category. However, a unique problem occurs for enzyme–substrate systems where the catalytic residues participate in bond formation and breaking processes. In this case, it is not clear how the reactive amino acids can be partitioned into QM and MM parts. Inevitably, one or more covalent bonds have to be divided and used to join together the QM and MM regions, as depicted in Figure 2. Particularly, when the substrate or an organic molecule is very large, such that it is impractical to treat the entire species by quantum mechanical methods, the QM and MM regions must also be divided across covalent bonds.

An early description of the representation of a covalent bond connecting QM and MM atoms was given by Warshel and Levitt, who used a single hybrid orbital on the MM atom in the QM/MM joint region.[32] In that study, the semiempirical QCFF/ALL method (quantum consistent force field) was em-

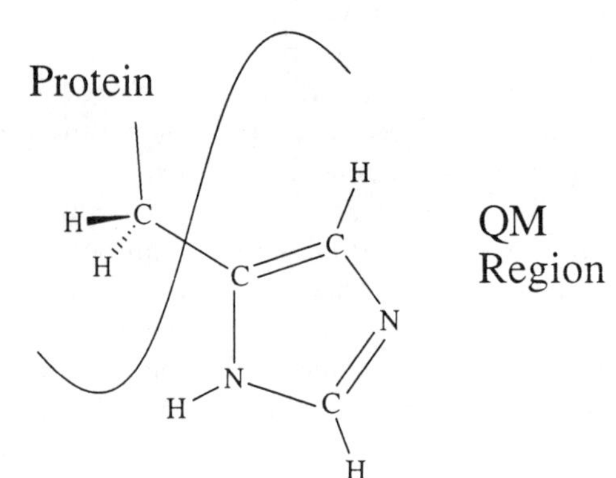

Figure 2 Illustration of the division of a covalent bond across the QM/MM regions.

ployed by including all the valence electrons with hybrid atomic orbitals. The method is an extension of a similar formulation for conjugated systems using the Pariser-Parr-Pople (PPP) model.[74,75] In the QCFF/ALL method, integral calculations were parameterized by fitting calculated and experimental results.[76–78] Warshel and Levitt reported that the semiempirical MINDO/2 method was also used in the study of the reaction of lysozyme; however, it is unclear exactly how the partitioning of covalent bonds was implemented in the latter case.[32] In recent investigations, Warshel and co-workers turned their attention to the EVB method, using experimental data to calibrate the VB Hamiltonian.[40,79–82]

Singh and Kollman, and Field, Bash, and Karplus used an alternative approximation following similar methods by others.[34,35,83] In this approach, "link" atoms are introduced to satisfy the valency of the QM atoms that are connected to any MM atoms. The link atoms are typically hydrogen atoms which are placed along the broken QM/MM bond, although other atoms or functional groups such as halogen atoms or methyl groups may certainly be used. The link atoms are treated exactly in the QM calculations, but they are not allowed to interact with (they are invisible to) the MM atoms. For computing the energy and forces resulting from the empirical *bonded* energy terms, all the internal coordinates (bond, angle, and dihedral terms) involving any one or more MM atoms are retained, whereas those including only the QM atoms are deleted from the empirical energy calculation. The latter part is evaluated by the quantum mechanical method. The van der Waals and electrostatic terms in the *nonbonded* list are treated separately: (1) the van der Waals interactions across the QM and MM regions are determined in exactly the same way as if the entire system were classical, and (2) the electrostatic interaction energy is obtained from quantum mechanical calculations of the QM molecule along with the link atoms in the presence of all MM partial charges except those replaced by the link atoms. Clearly, this method is not an ideal approach to the covalent bond division, because a perturbation will be introduced into the system, as in other truncation schemes. However, this procedure permits the partitioning of a large system into QM and MM regions separated by covalent bonds. The method should be useful particularly when the link atoms are relatively far away from atoms participating in chemical reactions.

However, as pointed out by research scientists at Hypercube Inc., the link atoms should in principle simulate the effect of the fragments that are removed from the quantum mechanical treatment.[84] These effects include electron donating and withdrawing abilities as well as hyperconjugation interactions. Clearly, standard hydrogen atoms cannot accommodate all the effects. Consequently, "phony" or pseudohalogen atoms were introduced in the program HyperChem to mimic the direct environment of the truncated system. These pseudohalogen atoms, F^s, Cl^s, Br^s, or I^s, depending on the MM atoms that the pseudohalogens represent, are specially parameterized.[84] For example, F^C, F^N, and F^O are used, respectively, to describe C, N, and O atoms, Cl^{Si} and Cl^P are used to mimic Si and P atoms, and so on.

The third approach to the separation of a covalent bond into QM and MM regions is described by Rivail and co-workers.[85–87] In a way, the method is similar to the original proposal of Warshel and Levitt.[32] The continuity between the two regions is maintained by a strictly localized bond orbital, obtained from model compounds that are assumed to be transferable. Using a local self-consistent field method, Thery et al. investigated the intramolecular proton transfer in a tetrapeptide Gly-Arg-Glu-Gly.[87] Good agreement was obtained with results from a computation for the entire system, particularly if the quantum subunit contains the side chains and the peptide lying between the proton transfer. The method was also applied to a study of the proton transfer in an Asp-Arg salt bridge in dihydrofolate reductase.[87]

Solvent Polarization Effects

One of the important features in the combined QM/MM model is that the wavefunctions for the solute both in the gas phase and in solution are determined. Consequently, all properties associated with the solvation process can in principle be computed.[54] An interesting quantity that can easily be defined, but that is difficult to determine experimentally, is the solute electronic polarization by the solvent through solute–solvent intermolecular interactions.[54] Figure 3 illustrates an energy decomposition method that has been developed in the context of Monte Carlo simulations using a combined QM/MM potential. The right vertical arrow in Figure 3 represents the experimental process of solvation with an energy of solvation ΔE_{sol}. Computationally, the solvation energy is a sum of the solute–solvent interaction energy and the solvent reorganization energy ΔE_{ss}. The latter term is the difference of the solvent energies in the presence and absence of the solute molecule. The total interaction energy between the solute and solvent is given by:

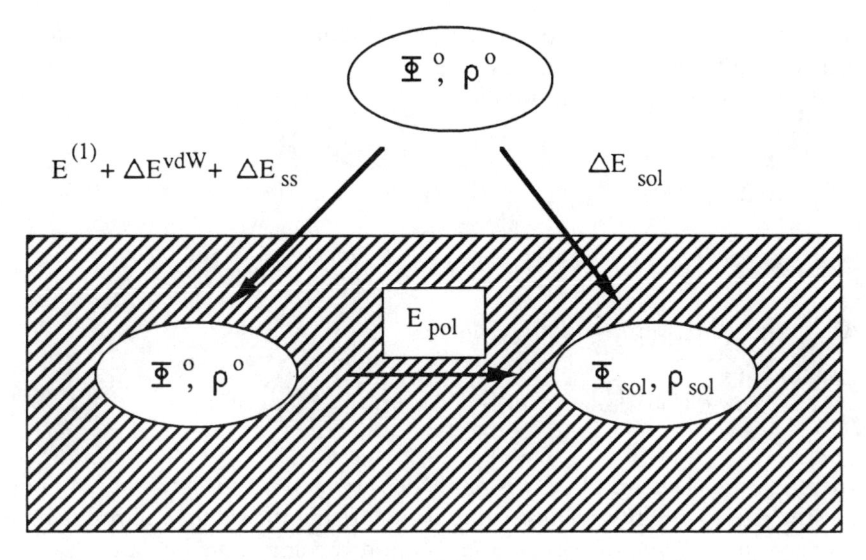

Figure 3 Schematic representation of the energy decomposition method.

$$\Delta E_{Xs} = \langle \Phi | \hat{H}^{o}_{X} + \hat{H}_{Xs} | \Phi \rangle - \langle \Phi^{o} | \hat{H}^{o}_{X} | \Phi^{o} \rangle = \Delta E^{vdW}_{Xs} + \Delta E^{el}_{Xs} \qquad [13]$$

which may be separated into a van der Waals term and an electrostatic term because of the definition of the solute–solvent interaction Hamiltonian (Eq. [4]). Accompanying the solvation is also a change of the wavefunction of the solute in the gas phase Φ^{o} to one in solution Φ.

Treating the $\hat{H}^{el}_{Xs}$ term in Eq. [13] as a perturbation to the solute wavefunction in the gas phase Φ^{o}, ΔE^{el}_{Xs} can be written as:

$$\Delta E^{el}_{Xs} = \langle \Phi | \hat{H}^{o}_{X} + \hat{H}^{el}_{Xs} | \Phi \rangle - \langle \Phi^{o} | \hat{H}^{o}_{X} | \Phi^{o} \rangle = E^{(1)} + E_{pol} \qquad [14]$$

where the first-order perturbation term, $E^{(1)} = \langle \Phi^{o} | \hat{H}^{el}_{Xs} | \Phi^{o} \rangle$, is the interaction energy between the "unpolarized" solute and the solvent, which is depicted by the vertical process on the left-hand side in Figure 3. Since all other higher-order perturbation terms consist of modifications to the gas phase wavefunction, together they contribute to the solute–solvent polarization energy, which can formally be defined by:

$$E_{pol} = E^{(2)} + E^{(3)} + \ldots \qquad [15]$$

Because both Φ^{o} and Φ_{sol} are found in the combined QM/MM Monte Carlo or molecular dynamics simulations, E_{pol} can be computed exactly taking the difference between ΔE^{el}_{Xs} and $E^{(1)}$. The E_{pol} term is further decomposed into two components[54]: a net gain in the solute–solvent interaction energy between the "polarized" (Φ_{sol}) and "unpolarized" (Φ^o) solute in solution ΔE_{stab}, and an energy penalty for distorting the solute electron distribution from its gas phase equilibrium state into that in solution ΔE_{dist}. These energy terms are expressed as follows:

$$E_{pol} = \Delta E_{dist} + \Delta E_{stab} \quad [16]$$

$$\Delta E_{stab} = \langle \Phi | \hat{H}^{el}_{Xs} | \Phi \rangle - \langle \Phi^o | \hat{H}^{el}_{Xs} | \Phi^o \rangle = \sum_{\mu\nu} (P_{\mu\nu} - P^o_{\mu\nu}) H^{Xs}_{\mu\nu} \quad [17]$$

$$\begin{aligned} \Delta E_{dist} &= \langle \Phi | \hat{H}^o_X | \Phi \rangle - \langle \Phi^o | \hat{H}^o_X | \Phi^o \rangle \\ &= \frac{1}{2} \sum_{\mu\nu} [P_{\mu\nu} (H^o_{\mu\nu} + F_{\mu\nu}) - P^o_{\mu\nu} (H^o_{\mu\nu} + F^o_{\mu\nu})] \end{aligned} \quad [18]$$

where $P^o_{\mu\nu}$ are elements of the gas phase density matrix. It is interesting to note that the energy penalty for distorting or polarizing the solute electron density to create the electron distribution in solution is compensated by a net gain in the solute–solvent interaction energy ΔE_{stab}, which is twice ΔE_{dist} according to classical linear response theory.[88–91] The numerical results from Monte Carlo simulations, which will be described later, confirm this expectation.

Simulation of Excited States and Solvatochromic Spectral Shifts

The combined QM/MM model has been recently extended to simulation of the solvent effects on excited states and electronic spectroscopy by using an MO configuration interaction (CI) formalism.[77,92,93] Using the Hartree–Fock wavefunction Φ^{HF}_{sol} generated by the procedure described above as the reference state, the CI wavefunction for the QM region can be constructed:

$$\Phi^{CI}_{sol} = a_0 \Phi^{HF}_{sol} + \sum_I a_I \Phi_I \quad [19]$$

where Φ_I values are the excited state determinants, including singly excited, doubly excited, etc., states, which are formed by replacing one or more occupied spin orbitals in Φ^{HF}_{sol} with virtual spin orbitals. However, truncated CI calculations are normally used to save computer time.[94] Standard methods can be used to diagonalize the CI matrix whose elements are defined as $\langle \Phi_I | \hat{H}_{eff} | \Phi_J \rangle$. The CI coefficients, $\{a_I\}$, are obtained as eigenvectors of the CI matrix, whereas the eigenvalues correspond to the total energies of the ground state and various excited states of the solute in solution.[15]

The total ground state (g) energy of a condensed phase system in such a CI representation is:

$$E^{CI,g}_{tot} = \langle \Phi^{CI,g}_{sol} | \hat{H}_{eff} | \Phi^{CI,g}_{sol} \rangle = E^{CI,g}_X + E^{CI,g}_{Xs} + E_{ss} \quad [20]$$

where the superscripts indicate the wavefunction and the ground electronic state. In Monte Carlo and molecular dynamics simulations, the trajectory will be carried out on the CI total potential energy surface defined by Eq. [20] and its derivatives.[93] It is of particular interest to examine the energy resulting from the Hamiltonian $\hat{H}_{Xs}$, which is given by:

$$E_{Xs}^{CI,g} = \sum_{\mu\nu} P_{\mu\nu}^{CI,g} H_{\mu\nu}^{Xs} + \sum_{s} \sum_{m} \frac{q_s Z_m}{R_{ms}} + E_{vdW} \quad [21]$$

where $P_{\mu\nu}^{CI,g}$ is the one-electron density matrix of the ground state CI wavefunction.[95] For the excited states, a similar expression can be written, with which simulations of the excited state in solution can be performed:

$$E_{tot}^{CI,e} = \langle \Phi_{sol}^{CI,e} | \hat{H}_{eff} | \Phi_{sol}^{CI,e} \rangle = E_{X}^{CI,e} + E_{Xs}^{CI,e} + E'_{ss} \quad [22]$$

where E'_{ss} is the solvent–solvent interaction energy in the presence of the excited solute.

A straightforward application of the combined QM–CI/MM model is to examine the solvent effects on absorption or emission spectra of chromophores in solution, whereas the method is of general use for studying photochemical reactions and electron transfer processes in solution and in enzymes. The solvatochromic spectral shift is defined as the difference between the excitation energy of the solute in solution $\Delta E_{tot}^{g\to e}(\Phi_{sol}^{CI})$ and that in the gas phase $\Delta E_{tot}^{g\to e}(\Phi_{gas}^{CI})$:

$$\Delta\nu = E_{tot}^{g\to e}(\Phi_{sol}^{CI}) - E_{tot}^{g\to e}(\Phi_{gas}^{CI}) \quad [23]$$

where $E_{tot}^{g\to e} = E_{tot}^{CI,e} - E_{tot}^{CI,g}$. Because the electronic transition of the solute is much faster than the time required for solvent nucleus relaxation,[96] at the instant of its formation, the excited state of the solute is surrounded by the solvent cage corresponding to the equilibrium ground state potential surface. However, the solvent electron distribution will readjust to follow the solute electron distribution of the excited state after excitation. Consequently, it is critical to consider specifically the difference in the mutual induced dipole-induced dipole interactions between the solute ground state and excited state with the same configuration of the solvent cage, and this should be treated self-consistently in the quantum mechanical calculations.[92,97–101]

However, description of the mutual solute–solvent polarization effects for both the ground and the excited states within molecular orbital theory is not straightforward, because it requires SCF calculations of the solvent-induced dipoles in each step of the HF–SCF interaction. A simplification can be made by treating the solvent molecule with fixed, effective partial charges (thus ignoring explicit solvent polarization effects). In this case, $E_{ss} = E'_{ss}$, and the calculated spectral shift may be further decomposed into two terms:

$$\Delta\nu^{o} = \Delta\Delta E_{X}^{g\to e} + \Delta E_{Xs}^{g\to e} \qquad [24]$$

where $\Delta\nu^{o}$ denotes that the calculated spectral shift does not contain the solvent polarization contribution. In Eq. [24], $\Delta\Delta E_{X}^{g\to e}$ is the change in the energy gap between the solute ground and excited states in solution and in the gas phase, and $\Delta E_{Xs}^{g\to e}$ reflects the difference in solute–solvent interaction energy between the ground and excited states upon electronic excitation of the solute. The approximation made here is a severe one because the mutual solute-solvent polarization interactions are neglected. Furthermore, the dispersion interaction between solute and solvent molecules generally favors the excited states because the polarizability of the excited state is typically larger than that of the ground state. This leads to a red spectral shift (lower transition energy).[97–99] Consequently, the dispersion red shifts for $n \to \pi^*$ transitions in nonpolar solvents will not be predicted using Eq. [24].[100,101] However, in polar, hydrogen-bonding solvents, the major contribution to the spectral shift is due to dipole–dipole and dipole–induced dipole interactions,[39] which are reasonably treated with a combined QM-CI/MM model. Thus, this model can yield reasonable results for systems involving strong solute–solvent interactions where the medium dielectric effects dominate.

To circumvent this difficulty, Luzhkov and Warshel derived a semiclassical approach in which the computed solvent spectral shift is corrected for the difference in the polarization effects in the ground and excited states:

$$\Delta\nu = \Delta\nu^{o} + \frac{1}{2}\sum_{m} \left(Q_{m}^{CI,e} V_{m}^{ind,e} - Q_{m}^{CI,g} V_{m}^{ind,g}\right) \qquad [25]$$

where m runs over all solute atoms, $Q^{CI,e}$ and $Q^{CI,g}$ are solute atomic charges in the excited and ground states, and $V^{ind,e}$ and $V^{ind,g}$ are potentials resulting from induced dipoles of the solvent in the presence of the solute excited and ground states, respectively.[92] The factor $1/2$ accounts for the energy invested to polarize the solvent-induced dipoles. In a later section, we will provide additional discussion on the calculation of solvent dispersion energies.

The QM Method

The combined QM/MM model described so far is a general procedure that can be used in various quantum mechanical schemes, although some of the energy expressions have been given in the formalism of molecular orbital theory. Because a majority of the computational results reviewed in this chapter are obtained with the use of MO theory, especially semiempirical methods, some key features of these calculations are summarized here. The combined QM/MM approach has recently been applied with the use of density functional theory in molecular dynamics simulations[50,102,103]; a summary of the latter method is also given in this section.

The Hartree–Fock MO theory is the most extensively tested and documented quantum mechanical method. The method can easily be extended to introduce electron correlation corrections via configuration interaction or many-body perturbation theory. Furthermore, there are a number of computer programs that can be conveniently used. Obviously, ab initio MO methods would be perfectly suited for the combined QM/MM model[14,15]; however, for simulation of condensed phase systems in which millions of configurations are sampled, we are forced to balance between the generality and practical applicability of the QM method. Combining the need for reasonable accuracy and computational efficiency in Monte Carlo and molecular dynamics simulations, semiempirical methods have been chosen in most calculations so far reported in the literature.[104]

Among various alternative semiempirical MO schemes (the semiempirical methods have been reviewed by Stewart and by Zerner in this series),[105–107] the Austin Model 1 (AM1) theory developed by Dewar et al. and the related MNDO and PM3 methods prove to be the best choice for applications to organic and biological problems.[108–111] The AM1 (MNDO, or PM3) model is a parameterized HF-SCF MO method that has been tested on numerous organic and organometallic compounds and reactions. The results are considered to be comparable to ab initio calculations with split valence basis sets, such as 3-21G and 4-31G, and are generally in good accord with experiments. Significantly, the computational speeds of the MNDO/AM1/PM3 methods are about 1000 times faster than their ab initio counterparts. This enables the QM energy and forces to be computed throughout Monte Carlo and molecular dynamics simulations. However, because of its empirical, parametric nature, for systems that give poor results with the semiempirical methods, there is no systematic procedure to improve the quality of the computation other than reparameterization.

The semiempirical method uses a valence electron, minimum basis set representation with fixed nuclear cores of effective charges.[104–111] The molecular orbitals are linear combinations of Slater-type atomic orbitals. Thus, hydrogen has a 1s orbital, whereas the first-row atoms retain 4 orbitals — one 2s and three 2p orbitals, and so on. The most successful semiempirical scheme adopts the Neglect of Diatomic Differential Overlap (NDDO) approximation,[112,113] which has been the basis of the MNDO, AM1, and PM3 methods. In the NDDO approximation, all integrals involving diatomic differential overlap are neglected. The remaining integrals are evaluated using empirical formulas derived by Dewar and Thiel, except the overlap integrals, $S_{\mu\nu}$, which are computed analytically and are used to determine the two-center resonance integrals. The parameters used in the MNDO, AM1, or PM3 procedures were optimized against experimental molecular geometries, heats of formation, dipole moments, and ionization potentials. Parameters for numerous elements have been derived for the AM1 and PM3 procedures. It should also be noted that geometries and heats of formation are obtained in these methods for compounds corresponding to standard conditions at 25°C and 1 atm. Thus, correlation and zero point energy are implicitly included in the parameters.[108–110]

Another empirical method that has been extensively used by Warshel and co-workers is the empirical valence bond (EVB) theory.[11,42] In this approach, it is assumed that a reaction can be described by some VB resonance structures. The analytical form of these VB functions can be approximated by appropriate molecular mechanics potentials, and the parameters of these MM potentials are calibrated to reproduce experimental or ab initio MO data in the gas phase as well as in the condensed phase. The combined EVB/MM method and its unique calibration procedure have been recently reviewed.[11,42] It should be noted that Kim and Hynes presented a similar method, yielding a nonlinear Schrödinger equation.[114] However, the solvent was treated as a dielectric continuum in the Kim–Hynes theory. Nevertheless, an interesting feature in the latter method is a consideration of nonequilibrium coupling between the solute and solvent.[114,115]

A promising alternative choice for the QM method in a hybrid QM/MM model is provided by density functional theory (DFT).[116,117] DFT techniques show considerable advantages in generality and accuracy over the semiempirical methods. Although the computational speed is comparable to ab initio Hartree–Fock calculations, DFT treatment has the advantage of including electron correlation effects.[118] The use of a hybrid DFT/MM potential in studying complex chemical reactions in solutions and enzymes might be rather difficult in the near future due to computational constraints. To this end, three groups have published results of the hydration of water or spherical ions, making use of a combined DFT/MM potential[50,102,103]; however, it appears that only one group (Stanton, Hartsough, and Merz) included quantum mechanical forces to integrate the equations of motion. The other calculations were carried out using classical potentials, followed by energy calculations on selected configurations or on all the configurations generated by the molecular dynamics simulation.

Within the local density formulation of the Kohn–Sham theory,[116,118] the ground state energy is given as a functional of the electron density $\rho(\mathbf{r})$ in the presence of an external potential $\nu(\mathbf{r})$:

$$E[\rho(\mathbf{r})] = -\frac{1}{2}\sum_i \langle\psi_i|\nabla^2|\psi_i\rangle + \int \nu(\mathbf{r})\rho(\mathbf{r})dr + \frac{1}{2}\int\int \frac{\rho(\mathbf{r})\rho(\mathbf{r}')}{\mathbf{r}-\mathbf{r}'}drdr' + E_{\mathrm{xc}}[\rho(\mathbf{r})] \qquad [26]$$

where $E_{xc}[\rho(\mathbf{r})]$ is the exchange-correlation functional and ψ is a Kohn-Sham orbital which is obtained by solving the SCF equation:

$$\left[-\frac{1}{2}\nabla^2 + V_{\mathrm{eff}}(\mathbf{r})\right]\psi_i = \epsilon_i\psi_i, \qquad i = 1, 2, \ldots, N \qquad [27]$$

in which the effective potential is defined by

$$V_{eff}(\mathbf{r}) = \nu(\mathbf{r}) + \int \frac{\rho(\mathbf{r}')}{\mathbf{r} - \mathbf{r}'} dr' + \frac{\delta E_{xc}}{\delta \rho(\mathbf{r})} \qquad [28]$$

The total electron density is obtained from Kohn–Sham orbitals:

$$\rho(\mathbf{r}) = 2 \sum_{i=1}^{N} \psi_i^* \psi_i$$

In a hybrid DFT/MM approach, interactions with the solvent molecules are introduced into the Kohn–Sham equations through the external potential $\nu(\mathrm{r})$, which is formally defined in the same way as the solute–solvent interaction Hamiltonian $\hat{H}_{\mathrm{Xs}}$ in Eq. [4]: $\nu(\mathbf{r}) = \hat{H}_{\mathrm{Xs}}$.

Implementations

The implementation of the QM/MM solute–solvent interaction Hamiltonian (Eq. [4]) within an HF MO scheme involves introduction of the one-electron integral over the operator $1/r_{is}$ into the Fock matrix (Eq. [9]). Note that three terms in Eq. [4], which do not involve positions of the solute electrons, are treated classically. The one-electron integrals have the general form:

$$H_{\mu\nu}^{\mathrm{Xs}} = -\sum_s \left(q_s \int \frac{\phi_\mu \phi_\nu}{r_{is}} dr + \frac{1}{2} \int \phi_\mu \phi_\nu \frac{\boldsymbol{\mu}_s^{\mathrm{ind}} \mathbf{r}_{is}}{r_{is}^3} dr \right) \qquad [29a]$$

where ϕ_μ and ϕ_ν are basis functions, and the negative sign specifies the interaction of a QM electron with a solvent MM partial charge, q_s, or dipole. If the induced dipolar term for the solvent molecules is not explicitly considered, as in most calculations, Eq. [29a] is simplified as follows:

$$H_{\mu\nu}^{\mathrm{Xs}} = -\sum_s q_s \int \frac{\phi_\mu \phi_\nu}{r_{is}} dr \qquad [29b]$$

With the inclusion of the $H_{\mu\nu}^{\mathrm{Xs}}$ term, the HF equations are solved exactly as in gas phase calculations. Thus, techniques developed for MO calculations can be directly used in simulation of solvation effects.

The one-electron integral of Eq. [29] can be evaluated analytically for both Gaussian and Slater-type orbitals.[119,120] However, since the one-electron integrals are nonzero only for basis functions ϕ_μ and ϕ_ν on the same atom A within the NDDO approximation, it is convenient to treat these integrals in the same way as that in semiempirical procedures:[121]

$$H_{\mu\nu}^{\mathrm{Xs}} = -\sum_s q_s (\phi_\mu \phi_\nu | S_s S_s), \qquad \mu, \nu \in A \qquad [30]$$

where S_s is an s orbital on an MM atom bearing a partial charge q_s. The s orbital is placed on MM atoms to determine the integral values approximated by Eq. [30]. It turns out that the analytical form of the one-electron integrals was found by Field, Bash, and Karplus to be inappropriate in semiempirical methods because the calculated interaction energy and forces are proportionally too large.[35]

For the AM1 or PM3 Hamiltonian, the core–core interaction between QM atom m and MM atom s (the second term in Eq. [4]) takes the form introduced by Dewar et al.[108]:

$$E_{Xs}^{core} = q_s Z_m (S_m S_m | S_s S_s)[1 + \exp(-\alpha_m R_{ms}) + \exp(-\alpha_s R_{ms})$$

$$+ \frac{q_s Z_m}{R_{ms}} \sum_{i=1}^{4} [K_i^m \exp(-L_i^m (R_{ms} - M_i^m)^2) + K_i^s \exp(-L_i^s (R_{ms} - M_i^s)^2) \quad [31]$$

The MNDO method differs from AM1 and PM3 in that the last two terms in Eq. [31] are not present in the MNDO parametrization.[109] In combining AM1 (MNDO) with classical force fields, parameters for the QM atoms are taken directly from the semiempirical method without any modification. In Eqs. [30] and [31], there are five MM-dependent parameters for each MM atom; they are α_s, ρ_s (which appears in the electron repulsion integral from the notional s orbital), K_i^s, L_i^s, and M_i^s. These are "free" parameters, which are subject to optimization in order to obtain good agreement with experimental or ab initio results. After investigating numerous hydrogen bonding complexes, Field, Bash, and Karplus suggested using values of zero for the three parameters, K_i^s, L_i^s, and M_i^s ($i = 1, \ldots, 4$), in the AM1 method, leading to the same expression for solute–solvent core–core interactions as that in MNDO.[35] They also recommended values of 5.0 and 0.0 for α_s and ρ_s to be used for all MM atoms,[35] which yield the best prediction for bimolecular interactions in comparison with ab initio results.

In addition, there are two parameters associated with the QM atoms which are treated entirely classically. They are the Lennard–Jones parameters, σ_m and ϵ_m, in the last term of the solute–solvent (QM/MM) interaction Hamiltonian. The Lennard–Jones terms are necessary in a combined QM/MM potential because some MM atoms possess no partial charges and so would be invisible to the QM particles. In cases such as chloride versus bromide ions, the only distinction between the two would be reflected in the van der Waals terms.[35] However, the existence of these parameters provides the opportunity to further optimize the performance of the combined QM/MM model, whether ab initio, DFT, or semiempirical in the QM treatment. This will also overcome some of the well-known problems in the semiempirical methods, including hydrogen bonding interactions.[111] Fortunately, the Lennard–Jones parameters depend entirely on the nature of the QM atom and sometimes its hybridization: they are fully transferable from one molecular system into another. Consequently, only a limited number of parameters are required for optimization. A set of parameters has been suggested for the AM1 Hamiltonian combined with Jorgensen's OPLS potential functions,[29] which are listed later in Table 1.[54] The performance of these parameters is summarized in a later section.

The earliest incorporation of a hybrid QM/MM potential in a computer program was made by Warshel and Levitt using the QCFF/ALL Hamiltonian,[32] although the treatment of π electrons by the PPP method on classical σ bond frameworks in trajectory calculations may also be regarded as a form of hybrid QM/MM approach.[16,78] The efforts in Warshel's laboratory have now included the EVB approach, MO methods, and DFT techniques.[33] Several groups combined ab initio molecular orbital/DFT programs with MM programs.[34,43–50,103] Subsequently, Singh and Kollman developed the QUEST program, combining Gaussian 80 and AMBER.[34] Merz and co-workers have also combined a DFT code with AMBER.[50] The semiempirical MNDO/AM1 method was first combined with an MM force field for molecular dynamics simulations within the program CHARMM by Field, Bash, and Karplus.[35] In that implementation, the public domain, QCPE program MOPAC by Stewart was used to carry out all electronic structure calculations.[122,123] Subsequently, a similar implementation was incorporated into a program developed at Buffalo.[124] The MCQUB (Monte Carlo QM/MM at the University at Buffalo) program combines the MOPAC and BOSS programs.[125] Currently, the latter program is used to perform the Metropolis Monte Carlo sampling. Recently, Thompson developed a computer program ARGUS for general purpose molecular dynamics simulations using MM and QM/MM potentials.[126] The commercial package HyperChem also contains features combining QM/MM potentials.[84] Like ARGUS, HyperChem supports a variety of semiempirical models, although the latter package currently only allows energy minimization calculations to be performed in the presence of a solvent Coulombic field.[84]

A CRITICAL EVALUATION OF THE COMBINED SEMIEMPIRICAL AM1/MM MODEL

In the following, the results obtained from Monte Carlo and molecular dynamics simulations using combined QM/MM potentials are discussed. This section describes the results that are primarily aimed at the testing and verification of the combined QM/MM potential, although these studies are of interest in their own right. Later, we provide a summary of investigations of solvent effects on chemical reactions.

Bimolecular Hydrogen Bonding Interactions

The ability to model hydrogen bonding interactions between the solute and solvent is critically important in computer simulation of condensed phase systems.[127] To verify the capability of the combined AM1/MM potential to describe intermolecular interactions, a series of bimolecular hydrogen-bonded complexes of organic compounds with water have been examined. In these investigations, the organic species are treated by the semiempirical AM1/PM3 theory, while the water molecule is represented by the empirical, three-site TIP3P model.[35,54,56,128,129] Often, the interactions between a QM water and classical organic molecules are also considered. Partial geometry optimizations are carried out to yield structural and energetic properties for these complexes, which are compared with gas phase experimental or high-level ab initio results. Similar comparisons can be made with results obtained using empirical potential functions.[127] Figure 4 illustrates two examples of the structural arrangements in these calculations.

The first systematic study of hydrogen-bonded complexes making use of a combined QM/MM approach was carried out by Field, Bash, and Karplus (FBK), who treated part of the complex quantum mechanically and the rest classically.[35] Both the MNDO and the AM1 methods were examined. They investigated a series of bimolecular complexes paired among halogen ions, ammonium ion, carboxylate, formaldehyde, formamide, hydrogen chloride, and water. In each case, one species was treated quantum mechanically, while the other was described by the standard CHARMM force field (except water, which was modeled by the TIP3P model).[27] In all cases, comparisons were made between the results of calculations using the AM1, MM, and QM/MM potentials and high-level ab initio or experimental results.[35] The combined QM/MM potential was found to reproduce geometries and energies of various complexes to the same accuracy as the AM1 and MM models. What is encouraging is that it showed that the QM/MM model yields better results for hydrogen-bonding interactions than the AM1 model alone. Importantly, their preliminary studies pointed toward a few general directions to incrementally improve the methodology. In the FBK work, none of the molecular mechanics parameters were optimized (except the α_s and ρ_s values for MM atoms).[35] Optimization of these parameters should lead to a better performance of the combined AM1/MM potential.

Optimization of the Lennard–Jones parameters for the QM atoms (Eq. [4]) was accomplished at Buffalo in 1992 (Table 1).[54] However, this was only possible with a compromise between the accuracy of the geometrical and energetic results of hydrogen-bonding interactions.[54,128,130,131] In the FBK paper, it was noted that the hydrogen bond distances for certain interactions are longer than the ab initio results, whereas the interaction energies are too weak.[35] We have noticed that the discrepancy becomes particularly severe when the QM

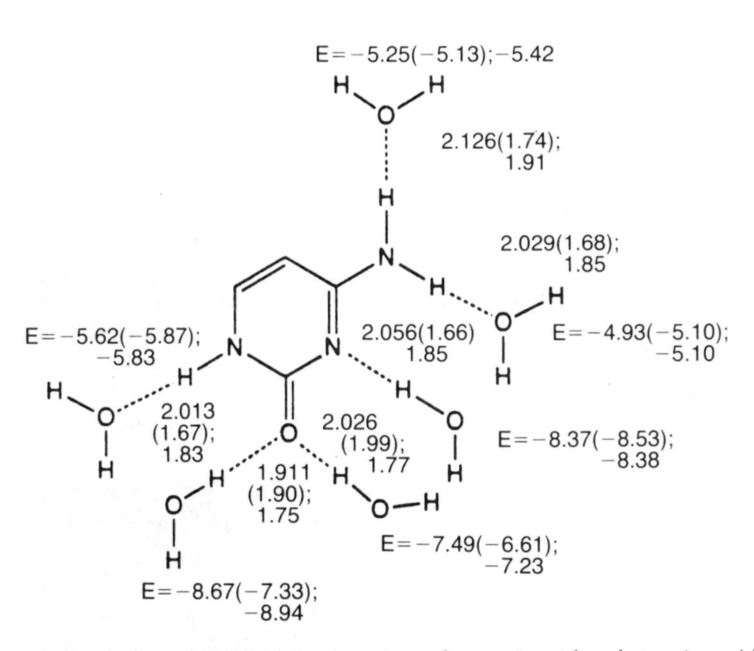

Figure 4 Illustration of bimolecular complexes for acetic acid and cytosine with water: 6-31G(d) values for the interaction energies (E) in kilocalories per mole, and distances in angstroms are followed by the AM1/TIP3P results (in parentheses), and OPLS results, respectively. Monomer geometries are fixed in the energy optimizations. The ab initio results for cytosine-water complexes are from Ref. 136.

Table 1 Lennard–Jones Parameters for the Combined AM1/OPLS Potential[a]

Atom	σ (Å)	ϵ (kcal/mol)
C	3.50	0.08
H (on C)	2.00	0.01
H (on heteroatoms)	0.80	0.10
O (sp^2)	2.95	0.20
O (sp^3)	2.10	0.15
N (no H)	2.50	0.14
N (in NH and NH_2)	2.80	0.15
N (in ions)	3.10	0.16

[a]Reference 54.

fragment acts as a hydrogen bond donor: the binding energies would be weaker by 1–2 kcal/mol compared with the corresponding ab initio results if good agreement is maintained on geometries. These systems include nitrogen and sp^3 oxygen in the QM region. Because it is critical to have the correct energy in Monte Carlo and molecular dynamics simulations, it appears reasonable to sacrifice a little of the accuracy of the predicted hydrogen bond distances for certain structures with the combined AM1/TIP3P model. Thus, the Lennard–Jones parameters for H, C, N, and O in the QM region were adjusted to obtain a best fit between the QM/MM and ab initio 6-31G(d) hydrogen bonding energies for a total of 53 complexes, covering functional groups in amino acids and nucleotide bases.[54] Overall, the agreement is excellent, with a root mean square (RMS) deviation of 0.8 kcal/mol in that study. Figure 5 includes a total of 67 such complexes, and the RMS deviation is about 0.9 kcal/mol from ab initio 6-31G(d) results. The error ranges are similar to those using empirical potential functions. The 6-31G(d) results were used because this basis set has been repeatedly demonstrated to yield excellent results for hydrogen bonding interactions in comparison with experiments.[136]

The good agreement in interaction energy is the result of a reduction in the accuracy of the geometries for some complexes. In general, hydrogen bond distances for complexes involving QM N or sp^3 O atoms are about 0.3–0.4 Å shorter than 6-31G(d) predictions, although in other cases, particularly for systems involving QM carbonyl groups, the QM/MM results are in good accord with the ab initio data.[128,130] This generalization is apparent in the data of Figure 4. However, it should be pointed out that the hydrogen bond distances predicted with empirical potentials are also about 0.1–0.2 Å too short in comparison with ab initio results.[127] In any event, this is a shortcoming with the use of semiempirical methods in combined QM/MM potentials, which tend to underestimate charge separations.[132] The situation may be improved when new semiempirical parameters are derived in the future taking into consideration the findings mentioned here, and a polarizable water model is

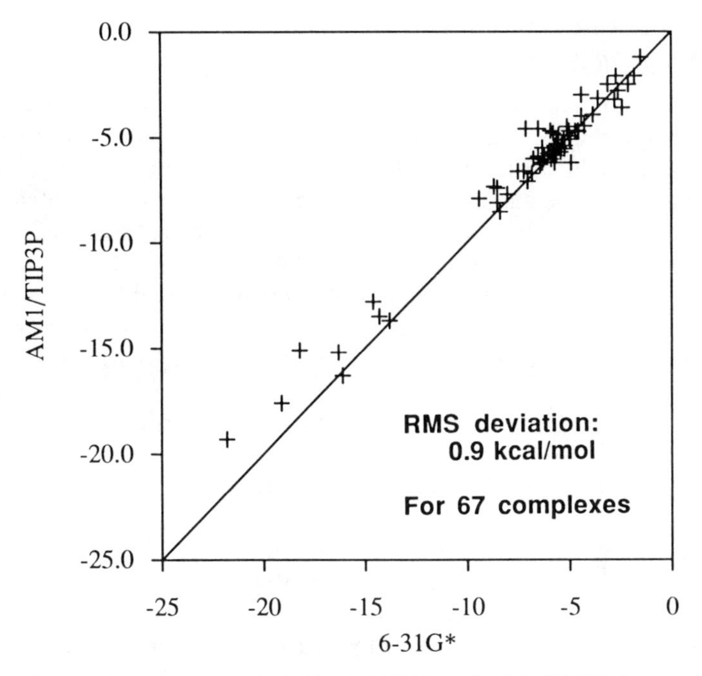

Figure 5 Comparison between ab initio 6-31G(d) and AM1/TIP3P interaction energies.

used.[133–135] Alternatively, the use of a combined ab initio QM/MM potential might overcome this difficulty.

Another thorough investigation of hydrogen-bonded systems using a combined QM/MM potential was carried out by Vasilyev, Bliznyuk, and Voityuk, who combined the MNDO, AM1, and PM3 Hamiltonian with OPLS functions.[129] Examining a great number of hydrogen-bonded complexes, these authors obtained findings similar to those noted above: the hybrid QM/MM potential yields a better agreement with experimental results than pure semiempirical methods: the average errors are 3.1–3.6 kcal/mol for 63 ion–molecule systems, and 1.0–2.4 kcal/mol for 15 neutral dimers with combined MNDO(AM1, or PM3)/OPLS potentials. Aside from their computational results, an interesting feature in their version of the implementation of the combined QM/MM model is that a scaling constant was introduced and optimized for all the $(\rho_X + \rho_s)^2$ terms in the two-electron integrals $(\mu\nu|S_sS_s)$.[129] For example,

$$(S_iS_i|S_sS_s) = e^2[R_{is}^2 + \xi(\rho_i + \rho_s)^2]^{1/2} \qquad [32]$$

where the optimized value of ξ is 0.095 for the MNDO/OPLS and AM1/OPLS combinations.

Free Energies of Solvation of Organic Compounds

Methods for calculation of solvation free energies using empirical potential functions in the contexts of Monte Carlo and molecular dynamics simulations are well established.[36,137–142] In principle, these methodologies can be directly employed when combined QM/MM potentials are used. However, in computing the relative free energies of solvation between two molecules with different numbers of atoms, one faces the difficulty of annihilating or creating nuclei and electrons in QM calculations, which is impossible within the MO schemes. Consequently, an elaborate procedure has been developed to circumvent this problem.[36] The idea is to transform the mixed QM/MM Hamiltonian into a purely empirical potential by dwindling away the QM/MM solute–solvent interaction Hamiltonian through a coupling parameter λ:

$$\hat{H}_{Xs}(\lambda) = \lambda H_{Xs}^{el} + \hat{H}_{Xs}^{vdW} \qquad [33]$$

where λ varies from 1, representing the full solute–solvent system $\hat{H}_{Xs}(1)$, to 0, a state consisting of the empirical van der Waals terms only. In essence, the method is analogous to the charge annihilation algorithm proposed by Bash et al.[143] The thermodynamic cycle used by Warshel has similar features.[144] Using the QM/MM potential, the difference in free energy of solvation between molecules **A** and **B** is determined according to Eq. [34] by consideration of the thermodynamic cycle shown in Figure 6.

$$\Delta\Delta G_{sol}(\mathbf{A} \rightarrow \mathbf{B}) = \Delta G_{sol}^{vdW}(\mathbf{A} \rightarrow \mathbf{B}) + \Delta G_{sol}^{el}(\mathbf{A}) - \Delta G_{sol}^{el}(\mathbf{B}) \qquad [34]$$

In this procedure, the electronic terms of the QM/MM interaction Hamiltonian ($\hat{H}_{Xs}^{el}$) for **A** and **B** are first annihilated, leaving only the Lennard–Jones terms, which are gradually "mutated" from structure **A** into structure **B**.[36] The latter mutation consists of empirical potentials, while standard methods are employed for the conversion.[137–142] Absolute free energies of solvation can also be determined following a procedure developed by Cieplak and Kollman,[145,146] in which A is made to vanish in solution:

$$\begin{array}{ccc}
\mathbf{A}(\hat{H}_{Xs}^{el}\ H_{Xs}^{vdW}) & \xrightarrow{\Delta\Delta G_{sol}(\mathbf{A} \rightarrow \mathbf{B})} & \mathbf{B}(\hat{H}_{Xs}^{el},\ H_{Xs}^{vdW}) \\
\Delta G_{sol}^{el}(\mathbf{A}) \downarrow & & \downarrow \Delta G_{sol}^{el}(\mathbf{B}) \\
\mathbf{A}'(0,\ H_{Xs}^{vdW}) & \xrightarrow{\Delta G_{sol}^{vdW}(\mathbf{A} \rightarrow \mathbf{B})} & \mathrm{B}'(0,\ H_{Xs}^{vdW})
\end{array}$$

Figure 6 Thermodynamic cycle used in computation of free energies of solvation between molecules A and B.

$$\Delta G_{sol}(\mathbf{A}) = \Delta G_{sol}(\mathbf{A} \rightarrow 0) - \Delta G_{sol}^{el}(\mathbf{A}) \quad [35]$$

Free energies of solvation are evaluated using a free energy perturbation method through a series of simulations to gradually interconvert the molecules by the coupling parameter λ.[147] So, the total free energy change resulting from annihilation of the QM/MM Hamiltonian is

$$\Delta G_{sol}^{el}(\mathbf{A}) = \sum_{\lambda=0}^{steps} - kT \ln \langle e^{-[\hat{H}_{eff}(\lambda+\Delta\lambda)-\hat{H}_{eff}(\lambda)]/kT} \rangle_{\lambda} \quad [36]$$

where $\Delta\lambda$ is the percent change in each step of the mutation.

A different implementation of the free energy perturbation method was introduced by Bash, Field, and Karplus,[52] who used a linear coupling between the two molecular systems:

$$\hat{H}_{eff}(\lambda) = \lambda \hat{H}_{eff}(\mathbf{B}) + (1 - \lambda)\hat{H}_{eff}(\mathbf{A}) \quad [37]$$

In this approach, the coexistence of both solute molecules in the presence of a common solvent environment is required. Applying the AM1(MNDO)/CHARMM potential, the free energy of activation for the S_N2 reaction, $Cl^- + CH_3Cl \rightarrow ClCH_3 + Cl^-$, was computed and found to be in good agreement with experiment and previous theoretical data.[2–6,52] The method has been successfully used in classical molecular dynamics simulation of proteins in aqueous solution.[148–151]

Table 2 provides a list of computed relative free energies of solvation for a series of organic compounds in water along with experimental data.[54,152–154] Table 3 compares the absolute free energies of hydration for the five nucleotide bases with results obtained by other computational methods.[155–162] The calculations were done using Monte Carlo simulations and the combined AM1/TIP3P model for systems containing 216 (for the organic solute) or 260 (for the bases) waters plus one solute.[54,155] These calculations are very demanding of computer resources because multistep simulations are required (it took more than 6 months to compute the results listed in Tables 2 and 3 on a three-processor TITAN 3030 computer). The computed free energy changes are in good agreement with experimental values, with an error range similar to that observed in calculations in which empirical force fields were used.[141] In the case of nucleotide bases, theoretical results from various methods do not appear to converge, whereas experimental data are not available for judgment.[155–162] The prediction from the combined AM1/TIP3P potential seems to be reasonable, although the computed ΔG_{sol} values for purine bases, adenine in particular, are perhaps low[155]

Using a combined DFT/SPC potential, Wei and Salahub calculated the energy and dipole moment of water in liquid water from molecular dynamics

Table 2 Computed Relative Free Energies of Hydration (kcal/mol) for Selected Organic Compounds Using the Combined AM1/TIP3P Potential

Compound	$\Delta\Delta G_h$(QM/MM)[a]	$\Delta\Delta G_h$(SCRF)[b]	$\Delta\Delta G_h$(expt)[c]
CH_3CH_3	0.0	0.0	0.0
C_6H_6	0.3	−1.7	−2.6
H_2O	−8.3	−7.5	−8.1
CH_3OH	−6.2	−7.0	−6.9
CH_3OCH_3	−3.6	−2.6	−3.7
CH_3COCH_3	−5.0	−5.3	−5.6
CH_3NH_2	−4.0	−6.4	−6.4
CH_3CN	−3.1	−5.5	−5.7
NMA[d]	8.5	−10.6	−12.0
CH_3CO_2H	−8.4	−8.9	−8.5
Imidazole	−7.2	−12.2	−6, −12.0
Cl^-	−75	−77	−77
$CH_3CO_2^-$	−80	−76	−79

[a]Reference 54.
[b]References 39 and 204, ΔG_h for ethane is 1.2 from the AM1-SM2 calculation.
[c]References 152–154.
[d]*N*-Methylacetamide.

simulations.[103] The computed dipole moment of a DFT water molecule in SPC water bath is 2.5–2.7 D with various theories, in good accord with experiment.[163] Wesolowski and Warshel also determined the free energy of solvation of water with a frozen density (FD) technique,[102,144] analogous to Yang's divide-and-conquer scheme.[164] In that study, the authors used a thermodynamic cycle to circumvent the need to integrate Newton's equations on a combined DFT/FD potential.[139] In addition, the solvation free energies for Cl^-, F^-, Na^+, and OH^- from a deMon/AMBER simulation by Stanton, Hartsough, and Merz were found in good accord with experimental results.[50]

Electronic Polarization

The polarization effect can be defined in general as the change in the molecular charge distribution induced by the environment. This change greatly influences the molecular reactivity and consequently the strength of intermolecular interactions. Fortunately, evaluation of the polarization contribution to solvation energy is rather straightforward in the framework of combined QM/MM potentials, although it is difficult to determine experimentally.[54,165,166]

Table 4 lists the computed average polarization energies along with their components using Eqs. [16]–[18].[54,165] The results are compared with those obtained from QM SCRF models. Overall, the agreement between the continuum SCRF results and the QM/MM prediction is remarkable in view of the difference in computational details. Several trends are apparent from Table 4. In all cases, there is a substantial polarization contribution to the total electrostatic interaction energy, ranging from 10% to 20%.

Table 3 Computed Free Energies of Hydration for Nucleotide Bases at 25 °C (kcal/mol)

Method	Adenine	Cytosine	Guanine	Thymine	Uracil
MC QM/MM	−5.1 ± 0.6	−16.3 ± 0.8	−13.5 ± 1.1	−8.5 ± 0.5	−9.9 ± 0.7
OPLS[a]	0.0	−8.5	−10.1	−1.5	−1.2
AMBER[b]	−12.6 ± 1.7	−12.7 ± 2.4	−19.6 ± 2.2	−7.5 ± 2.1	—
AM1-SM2[c]	−23.7 (−20.9)	−21.7 (−18.7)	−26.9 (−24.3)	−16.5 (−13.3)	−18.2 (−14.8)
SCRF(AM1)[d]	−11.3	14.4	−18.1	−8.6	—
SCRF(6–31G*)[e]	−6.5	−13.0	−16.1	−8.9	−10.0
FDPB[f]	−10.8	16.8	−19.7	−10.4	—

[a]Reference 156, relative free energies for methylated bases.
[b]Reference 143, for methylated bases.
[c]Reference 157, values for methylated bases are given in parentheses.
[d]Reference 158.
[e]Reference 159, for methylated bases.
[f]Reference 160, for methylated bases by finite difference Poisson-Boltzmann.

Table 4 Computed Polarization Energies and Components (kcal/mol)[a]

Compound	$E^{(1)}$	ΔE_{stab}	ΔE_{dist}	E_{pol}	ΔG_{pol},SCRF
H_2O	−13.7	−3.7	1.8	−1.9	−0.7
NH_3	−4.0	−0.4	0.2	−0.2	−0.2
CH_3OH	−10.3	−3.4	1.7	−1.8	−0.5
CH_3NH_2	−4.6	−0.9	0.4	−0.5	−0.3
CH_3OCH_3	−5.3	−1.5	0.8	−0.8	−0.4
CH_3CHO	−6.7	−2.3	1.1	−1.2	−1.1
CH_3COCH_3	−8.2	−3.1	1.5	−1.5	−1.2
CH_3CO_2H	−14.6	−3.9	1.9	−1.9	−1.1
NMA	−16.2	−7.9	3.9	−4.0	−2.0
Adenine	−19.5	−7.6	3.8	−3.8	−4.5
Cytosine	−34.8	−16.7	8.3	−8.5	−4.4
Guanine	−37.2	−15.6	7.8	−7.8	−5.6
Thymine	−21.2	−7.9	4.0	−4.0	−3.1
Uracil	−23.8	−7.8	3.9	−3.9	−3.8

[a]Energies are taken from simulations reported in References 54 and 165. Free energies for the SCRF(AM1) models are from References 157 (nucleotide bases) and 165.

Of course, it is well known that molecules are polarized in solution; however, quantitative contributions are not clear, particularly for proteins and DNA in aqueous solution.[90,91] This study demonstrates a significant polarization effect, which can amount to about one-half of a hydrogen bond in a single peptide unit.[54] Thus, discussion of protein allosteric mechanism, molecular recognition, and ligand–substrate binding, where change of environment is indispensable, should consider contributions specifically from the ligand polarization as well as from its surroundings.[167–170] Similar findings have been observed from classical studies.[88–90,171,172]

The qualitative and quantitative picture of polarization effects determined from a combined QM/MM potential should be valuable for deriving empirical potential functions with polarization terms.[32,90,133,134] Without these calculations, it would be rather difficult because it is not clear to what extent the polarization term contributes to the total energy. It is interesting to note that the numerical values for the penalty for creating the polarized electron distribution in solution, ΔE_{dist}, are indeed one-half of the stabilization energies of the solutes resulting from their electronic redistribution. This is in good accord with macroscopic linear response theory.[91]

Induced dipole moments, which represent the change in molecular dipole moment on transfer from the gas phase into aqueous solution, have been examined by many groups using various combined QM/MM schemes from semiempirical to ab initio methods.[33,48,52–54] Table 5 summarizes the computed dipole moments and induced dipole moments in water for various organic compounds from Monte Carlo AM1/TIP3P calculations.[54,165] Experimental dipole moments of simple organic compounds in water are not available, making direct comparison impossible; however, the agreement with predictions by Cramer and Truhlar, and by Orozco et al., among others, using the QM SCRF methods is excellent,[39,165] suggesting that the theoretical results are reasonable. Note that increases in dipole moment for the nucleotide bases are as much as 39%–75% more than their gas phase values.[155,157] However, the partial charges employed in most empirical potential functions were fitted to reproduce gas phase dipole moments.[28,136] The QM/MM simulation results indicate that it is perhaps best to include the polarization terms for these conjugated bases when solvated.

Applying the combined AM1/TIP3P potential in Monte Carlo simulations, molecular electrostatic potentials (MEPs) for five molecules were averaged in water.[166] Then, the aqueous MEPs were utilized to derive atomic partial charges by fitting the classical electrostatic potentials against the QM MEP.[173] This was the first reported calculation attempting to derive MEP-fitted charges by including average solvent effects on the solute electrostatic potential. The results were compared with similar calculations using an SCRF model at both the AM1 and ab initio 6-31G(d) levels.[166,174–176] The agreement in the MEP-fitted atomic charges based on the MEPs in water from Monte Carlo QM/MM-AM1/TIP3P simulations and those generated with the SCRF-AM1 wavefunctions is excellent, as demonstrated by Figure 7, which shows a nearly perfect fit in the linear regression analyses. Partial charges fitted in this manner should be a better approximation, at least conceptually, for use in condensed phase simulations than those fitted against ab initio MEPs in the gas

Table 5 Dipole Moment (in debyes) and Induced Dipole Moments for Selected Organic Compounds in Aqueous Solution[a]

Compound	μ(expt)	μ(gas)	μ(SCRF)[b]	$\langle\mu\rangle_{aq}$	μ_{ind}
H_2O	1.85	1.86	2.03	2.15	0.30
CH_3OH	1.70	1.60	1.85	2.01	0.41
CH_3NH_2	1.31	1.49	1.65	1.69	0.20
CH_3OCH_3	1.30	1.43	1.70	1.87	0.44
CH_3COCH_3	2.88	2.92	3.88	3.87	0.95
CH_3CO_2H	1.74	1.89	2.46	2.21	0.32
NMA	3.85	3.51	4.77	5.18	1.67
4-Nitroaniline	6.2	7.64	—	10.8	3.13
4-Pyridone	6.0	6.29	10.8	10.0	3.73
Adenine	3.16	2.17	3.1	3.81	1.64
Cytosine	7.10	6.33	9.0	9.40	3.07
Guanine	6.76	6.18	8.5	9.42	3.24
Thymine	3.58	4.24	6.2	5.89	1.65
Uracil	3.86	4.28	6.4	6.15	1.87

[a]References 54 and 155.

[b]Values for nucleotide bases are from References 157, and the rest are from Reference 165.

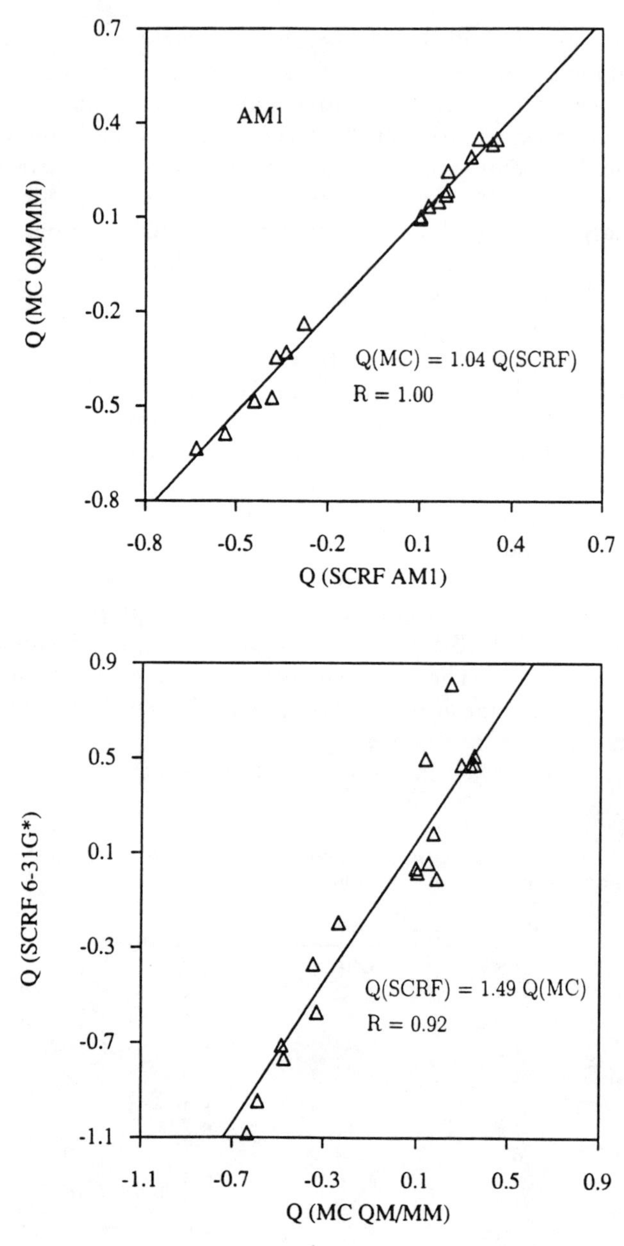

Figure 7 Linear regression analyses of the MEP-fitted atomic charges from the continuum SCRF calculations with the AM1 and 6-31G(d) wavefunction and from the Monte Carlo AM1/TIP3P simulations.

phase, because the environmental effects are averaged into the "aqueous" QM MEP.[173,177]

Marrone, Hartsough, and Merz described a similar calculation for determination of the partial charges in potential of mean force calculations of 18-crown-6 and K^+ in methanol using the MNDO Hamiltonian.[178] However, in their work, only a single solvent structure was used to calculate the MEP of the crown ether without ensemble averaging. These procedures should be valuable for deriving partial charges in empirical potential functions in future studies.

Conformational Equilibria in Aqueous and Organic Solutions

Use of the combined QM/MM model for studying conformational equilibria has been described by Warshel,[179] Jorgensen,[180] and by Field, Bash, and Karplus.[35] The first computer simulation of the solvent effects on conformational equilibrium with a combined QM/MM approach, however, was carried out in 1992 in the investigation of the relative basicity of the syn and anti lone pairs of acetate ion in water.[181] In recognizing that the basicity difference

between the two lone pairs is related to the free energy difference of the syn and anti conformation of acetic acids,[181,182] the potential of mean force (pmf) for the hydroxyl rotation in CH_3COOH in water was computed through a series of 11 free energy perturbation simulations, which gradually transform acetic acid from the syn conformation to the anti. The results show a marked solvent stabilization of the anti structure by 4.8 kcal/mol. In conjunction with the gas phase energy difference, which is 5.9 kcal/mol in favor of syn-

CH_3COOH,[183–185] the pK_a difference between the two lone pairs is reduced to less than 1 pK_a unit in water, in good accord with experimental estimates.[186–188] Recent calculations by Pranata using OPLS-type potentials yielded the same conclusion.[189] Such a small basicity difference, however, is not sufficient to explain the observation from X-ray crystal structures of proteins that the syn-oriented carboxylate side chains are predominantly preferred in enzymatic catalysis. Gandour[190] proposed that the reason for this observation is due to the much higher basicity of the syn lone pair (by about 6 kcal/mol) in the gas phase than the anti form. However, the small basicity difference in water might suggest that steric packing efficiency could operate in protein systems.[181,189]

Another system that has been studied is the cis/trans isomerization about a peptide bond. Solvent effects have a striking influence on the free energy of activation $\Delta G^{\ddagger}$. Nuclear magnetic resonance (NMR) measurements indicate that $\Delta G^{\ddagger}$ can vary as much as 4 kcal/mol for simple tertiary amides on going from the gas phase into water.[191–193] It has also been suggested that the origin of the catalytic power of peptidyl prolyl isomerases can be attributed to a change of the environment surrounding the amide group.[194] There have been numerous theoretical studies of peptide bond rotation in the gas phase[195,196]; the solvent effects were initially investigated by Duffy, Severance, and Jorgensen using OPLS potentials.[197] Then, the potential of mean force for the isomerization of dimethyl formamide (DMF) in water, $CHCl_3$, and CCl_4 was computed using the combined AM1/OPLS potential in Monte Carlo simulations.[198,199] The QM/MM results are consistent with the available experimental information and predictions by Jorgensen and co-workers.[191–193,197] An interesting finding from the QM/MM study is that the change of molecular dipole moments in solution nicely mirrors the variations in solvation free energy, and the DMF polarization is largest in the ground state where the planar geometry of the amide bond is particularly favorable for electron delocalization compared to the pyramidal transition states (TSs). Thus, the differential polarization between the ground ($\Delta\mu_{ind}$ = 1.53 D) and TSs ($\Delta\mu_{ind}$ = 0.4 and 1.0 D) is a significant contributor to the increase in barrier height of amide isomerization in aqueous solution.[198,199] This emerging concept of differential polarization effect has significant implications for chemical reactions in solution, discussed in detail later.

Thompson, Glendening, and Feller carried out a molecular dynamics study of dimethylether (DME) and 18-crown-6 (18c6) interacting with K^+ using a combined AM1/SPCE potential.[200] The results show that the most probable K^+ distance from the center of mass of 18c6 is 0.25 Å, significantly less than the minimum of 1.75 Å predicted by Dang and Kollman.[201] The QM/MM results were found to be consistent with the fact that K^+ has an optimal fit for the 18c6 cavity. Polarization energy was computed to be −14.9 kcal/mol, which is 17% of the total electrostatic interaction energy. Their results indicate the directions in which efforts should be undertaken to improve the empirical potentials for ion–macrocycle complexes in aqueous solution.

Relative Free Energies of Tautomeric Equilibria in Pyridone

The dramatic solvent effects on the tautomeric equilibria in 2- and 4-hydroxypyridine-pyridone in water and in chloroform were examined through Monte Carlo simulations using the combined AM1/OPLS potential.[202] In the case of 4-pyridone, the tautomeric equilibrium constant varies more than one millionfold on transfer from the gas phase into aqueous solution.[203] Previous theoretical studies have focused on continuum SCRF models in semiempirical, ab initio, and density functional calculations, which provided quantitative predictions and often valuable qualitative insights into the electronic properties. These results have been summarized recently and will not be repeated here.[39] The combined QM/MM model yields a relative free energy of solvation of -1.7 ± 0.1 kcal/mol for 2-pyridone relative to 2-hydroxypyridine in $CHCl_3$, and -5.7 ± 0.2 kcal/mol in water.[202] These energies may be compared with the corresponding experimental values of -1.6 and -4.6 kcal/mol.[203] For the 4-substituted system, the solvation preference is even greater for the oxo form, being -2.8 ± 0.1 and -7.4 ± 0.2 kcal/mol in $CHCl_3$ and in water, respectively. The gas phase equilibrium constant for the 4-hydroxypyridine/4-pyridone tautomerization was not determined experimentally due to detection limits, and thus only the lower bounds are available. The experimental solvent effects are <-1.5 and <-5.8 kcal/mol in $CHCl_3$ and in water, respectively. The accord with the QM/MM Monte Carlo simulation results is quite reasonable.

Analyses of the simulation results reveal that differential polarization of the tautomers in these systems contributes significantly to the overall observed solvent effects on the tautomeric equilibria, amounting to 14%–68% of the total free energies of solvation in the two solvents.[202] This is accompanied by changes in the molecular dipole moments. Of the systems considered, the most dramatic change takes place in 4-pyridone, with a computed average dipole moment of 10.3 D in water. This represents an increase of 4.0 D over its gas phase value, which may be compared with the prediction of 4.5 D reported by Cramer and Truhlar, who used a generalized Born model and AM1-SCRF method.[39,204] This change is reflected in the Mulliken population charges averaged over the Monte Carlo trajectory: the partial charge on the carbonyl O was amended from -0.34 in the gas phase to -0.54 in water. The AM1-SM2 model gives a charge of -0.52, in excellent accord with the QM/MM simulation.[204]

SOLVENT EFFECTS ON CHEMICAL REACTIONS

In this section, we provide a survey of the results obtained from computer simulations of chemical reactions in solution with the use of combined QM/MM potentials. During the last decade, methods for determining free energy profiles of chemical reactions in solution have been maturing and have been reviewed by several authors.[115,205–208] The focus here will be on the use of the combined QM/MM approach. The principal objective was determining the minimum energy reaction path (MERP) in the gas phase and then evaluating the solvent effects along this MERP.[205,209] Thus, by examining the potential of mean force, we only consider one aspect of chemical kinetics, that of equilibrium solvation. Undoubtedly, nonequilibrium solvation effects are also important to fully characterize reactions in solution, particular in cases with low intrinsic barrier heights.[206–207] Several recent articles contain discussion of these assumptions.[115,206] The combined QM/MM approach for computing the potential of mean force may be summarized by a simple two-step procedure.[210]

The initial step, which is similar to the computational procedure proposed by Jorgensen,[6,205] involves determination of the reaction path and energetics using ab initio MO calculations. The TS is first located, from which the intrinsic reaction coordinate (IRC) will be followed via standard methods available in Gaussian 92, leading to, respectively, the starting reagents and products.[211] In the empirical method, it is practical to develop potential functions for a reaction only along a one-dimensional reaction coordinate.[205,209] This would be adequate for simple reactions having "even" solvent effects on both sides of the TS.[6,212–215] However, for reactions involving heterolytic bond cleavage, such as the S_N2 Menshutkin reaction (Eq. [38]),

$$H_3N + CH_3Cl \rightarrow H_3NCH_3^+ + Cl^- \qquad [38]$$

both the position of the TS and the reaction path are significantly modified by interactions with the solvent.[216] As a result, a multidimensional surface should be employed. This can conveniently be done with the combined QM/MM potential, because intermolecular interactions are determined by QM calculations on the fly during a simulation,[210] whereas it is difficult to obtain high-quality parameters for multidimensional empirical potentials. Further, the empirical parameterization step has been described as the most critical step, which is difficult to generalize for application to a wide range of chemical reactions in solution.[115,205–209,212–215]

A particular issue should be clarified at this point: the use of a combined QM/MM potential, particularly in the context of a semiempirical QM scheme, is primarily aimed at evaluation of the effects of solvation. Although the semiempirical energies and geometries may deviate from high-level ab initio results in some cases, the intermolecular interactions between the QM and MM regions might still be adequate. However, careful examination of the performance of the combined QM/MM model for a specific problem is needed. Some of the ways to achieve this have been described in the previous section. The energetics for a chemical process in the gas phase should in general come from high-level ab initio calculations, which will then be supplemented by the solvation energy obtained from the combined QM/MM simulations.

Having defined the potential surface, the next step is to model the solvent effects on the energetics as well as the reaction path itself through statistical simulations. In the past, importance and umbrella sampling techniques have been used to obtain the solvation free energy along a reaction path.[217] This has been gradually replaced by a better-controlled procedure, the free energy perturbation (FEP) method.[205] This method effectively steps along the reaction path, thereby computing differences in free energy of solvation using the Zwanzig equation:[147]

$$\Delta G_{\text{sol}}(\mathbf{R}_j) - \Delta G_{\text{sol}}(\mathbf{R}_i) = -kT \ln\langle e^{-[E(\mathbf{R}_j)-E(\mathbf{R}_i)]/kT}\rangle_{E(\mathbf{R}_i)} \qquad [39]$$

where $\mathbf{R}$ is a structure generated from the reaction path-following calculation, and $\langle \ldots \rangle_{E(\mathbf{R}_i)}$ indicates ensemble averaging over the potential energy surface $E(\mathbf{R}_i)$. Typically, the difference between $\mathbf{R}_i$ and $\mathbf{R}_j$ is about 0.05 to 0.10 Å for distances or about 5° for bond angles. A procedure termed "double-wide" sampling allows the perturbation to run in both the forward and backward directions for "mutations" from $\mathbf{R}_i$ to $\mathbf{R}_{i-1}$, and from $\mathbf{R}_i$ to $\mathbf{R}_{i+1}$.[218] Standard errors for all computed properties are typically estimated from fluctuations of the averages of batches of 10^5 configurations in a simulation.

Finally, the potential of mean force as a function of the reaction coordinate is constructed by summing up the gas phase and solvation free energies:

$$\Delta G_{\text{tot}}(\mathbf{R}) = \Delta G_{\text{gas}}(\mathbf{R}) + \Delta G_{\text{sol}}(\mathbf{R}) \qquad [40]$$

The Claisen Rearrangement of Allyl Vinyl Ether

The aqueous solvent effects on the rates of Claisen rearrangement have been extensively investigated experimentally and theoretically.[219–229] The reaction is of particular interest because of its mechanistic and synthetic importance in organic chemistry. In addition, the Claisen rearrangement in chorismate mutase is the only known pericyclic reaction catalyzed by a naturally occurring enzyme.[224] A fundamental issue is the dipolar versus the radical nature of the transition state structure, which will determine the origin of the TS stabilization observed in water.

There have been two recent analyses of the hydration effects on the Claisen rearrangements among previous studies, one using the SCRF AM1-SM2 model by Cramer and Truhlar,[225] and the other employing Monte Carlo simulation with the OPLS potential by Severance and Jorgensen.[226] The two methods are complementary to each other, in that the former calculation includes the molecular electronic polarization whereas the latter treats solute–solvent intermolecular interactions explicitly. Cramer and Truhlar reported a rate acceleration of 21-fold, or a $\Delta\Delta G^{\ddagger}_{\text{hyd}}$ of -1.6 kcal/mol, for allyl vinyl ether

(AVE) at 25°C with the 6-31G(d) TS geometry (the rate acceleration is 3.5 using the AM1 TS), Eq. [41].[225]

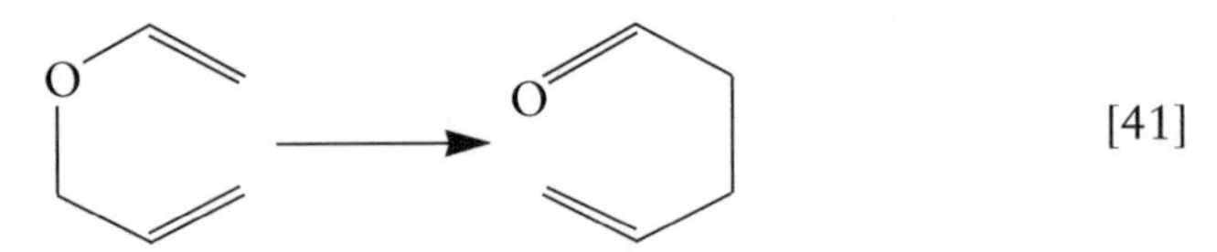

[41]

Importantly, they concluded, on the basis of electronic structure analyses, that solute polarization, in addition to the hydrophobic effects in the first solvation layer, is responsible for the rate acceleration of Claisen rearrangements in aqueous solution. On the other hand, Severance and Jorgensen predicted an aqueous rate increase by a factor of 644, or in free energy terms of -3.85 ± 0.16 kcal/mol.[226] These authors used a set of atomic charges derived from the 6-31G(d) electrostatic potentials along the reaction path and pointed out that both the number and strength of solvent hydrogen bonding to the oxallyl oxygen contribute to the rate acceleration.

Extending the initial solvation model (SM), Storer et al. introduced a specific-reaction parameterization (SRP) approach and applied the method to the Claisen reaction.[227] The new approach involves parameterization of the solvation model on the basis of a carefully selected experimental data set for the Claisen rearrangement in water and in hexadecane. In this study, Storer et al. discussed the critical role of transition state geometry on the computed solvent effects.[227] They found that a looser transition state (i.e., a larger separation of the allylic and —CH_2═CHO— units) is better solvated and consequently gives rise to greater rate acceleration. However, it is difficult to provide a definite answer, because the transition state structure for the Claisen rearrangement of AVE is very sensitive to the theoretical levels used. After analyzing the results from various theoretical treatments, these authors suggested that the "true" transition state structure may be between those from 6-31G(d) and those from MCSCF optimizations.[227] Using the AM1, HF/6-31G(d), and MCSCR/6-31G(d) transition state structures, the SM4-SRP model predicted aqueous rate accelerations of 3.4, 46, and 1400, respectively. Storer et al. pointed out that a loose transition state favors intrafragment charge polarization, leading to stronger electrostatic interaction with the solvent and better hydrogen bonding, which are responsible for the observed rate acceleration. These conclusions support the findings from a separate Monte Carlo simulation study (see below).[228]

Combined Monte Carlo AM1/TIP3P simulations were carried out to calculate the potential of mean force for the Claisen rearrangement of AVE in water along the IRC determined in the gas phase at the 6-31G(d) level.[226,228] In the reaction path calculation, a total of 143 structures were generated along the IRC using Gaussian 90, of which 69 were used in the combined QM/MM simulation.[228] This required 34 simulations using double-wide sampling, each requiring 5×10^5 to 10^6 configurations of equilibration and 1.5×10^6 configurations of averaging. The hydration effects were found to lower the gas phase activation free energy by −3.5 ± 0.1 kcal/mol in water, which translates to a predicted rate acceleration of 368. For comparison, the results are in accord with the theoretical results noted above, and an experimental estimate of about 1000 for Claisen rearrangements in water at 75 °C.[226] Another indication of the validity of the AM1/TIP3P potential is recorded in Figure 8 by the computed difference in free energy of hydration between 4-pentenal and AVE (−1.9 ± 0.2 kcal/mol). It is known experimentally that aldehydes are generally better hydrated than ethers by 1–2 kcal/mol.[152]

The origin of the aqueous solvent effects on the Claisen rearrangement was analyzed by the computation of the polarization contribution to the total solvation free energies.[228] It was found that the TS has a greater polarization free energy, ΔG_{pol}, than the reactant AVE by 1.2 kcal/mol, which amounts to 35% of the total $\Delta\Delta G^{\ddagger}_{hyd}$. Furthermore, this is nicely mirrored by the computed induced dipole moment (Figure 8), which is 1.0 ± 0.1 D for the TS in comparison with a value of 0.3 ± 0.1D for the ground state. Analyses of Mulliken population charges revealed that it is important to differentiate between solvent-induced polarization and solvent-induced heterolytic bond cleavage.[221,228] The difference in charge transfer from the allyl to the vinyl ether group between the TS and AVE increases by only 0.03 e as a result of hydration,

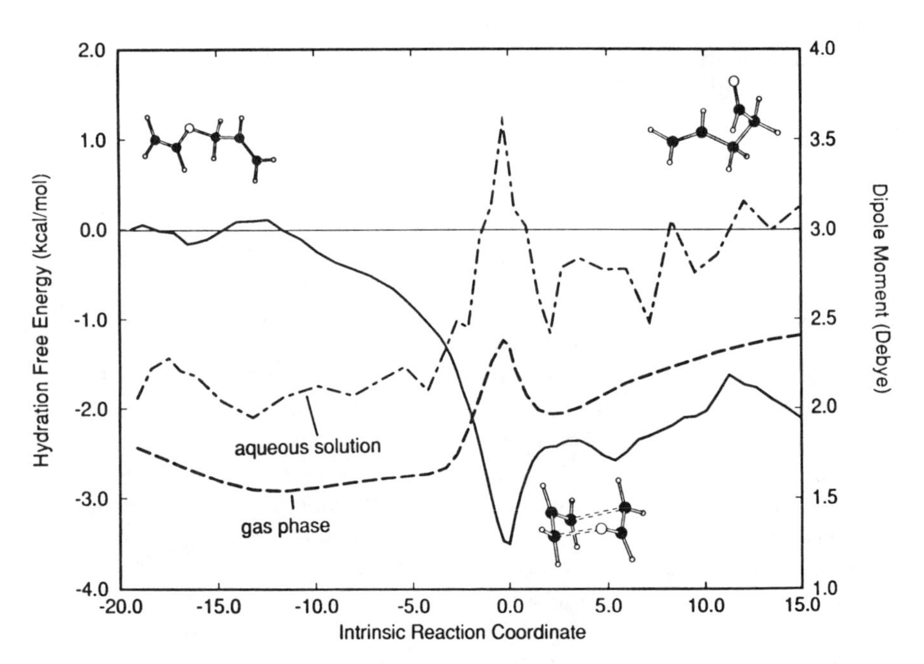

Figure 8 Computed changes in free energy of hydration (solid curve) and the AM1 dipole moment in the gas phase (dashed curve) and in aqueous solution (dash-dotted curve) along the 6-31G(d) intrinsic reaction path (units are in amu$^{1/2}$ bohr).

although the charge distribution within the $H_2C{=}CHO$— group undergoes a substantial variation. In water, the partial charge on oxygen is −0.25 and −0.42 in AVE and in the TS, respectively, an increase of 0.17 *e*, whereas the change is only 0.10 *e* in the gas phase. As a result, the enhanced charge distribution on oxygen at the TS leads to stronger hydrogen-bonding interactions with water, consistent with findings by Severance and Jorgensen.[226,227]

Simulations of Nucleophilic Substitution Reactions

In his classic work, Ingold classified nucleophilic substitution reactions into four categories according to the charge types of the nucleophile, being negative or neutral, and the substrate, being neutral or positive.[2] On the basis of this classification, the effect of solvent on nucleophilic substitution reactions can be qualitatively explained by the Hughes–Ingold rules.[230] The most common type of S_N2 reaction involves an anionic nucleophile and a neutral substrate, whereas it is less common for an anion to be the substrate.

Although one can always point to early related works, quantitative characterization of solute–solvent interactions at the atomic level for S_N2 reactions in solution began in the early 1980s.[3–6,34,52,53,212–215,231–237] These efforts were led by the seminal work by Chandrasekhar, Smith, and Jorgensen who simulated the type I process of $Cl^- + CH_3Cl \rightarrow ClCH_3 + Cl^-$ in water and in DMF using empirical potentials.[231,232] The striking solvent effects observed experimentally were demonstrated by Monte Carlo simulations with a predicted increase of 15 kcal/mol in activation free energy over the reaction in the gas phase.

Three groups have examined the same reaction with the use of combined QM/MM potentials. Singh and Kollman carried out MM energy minimizations of the reactants in a sphere of 220 water molecules, followed by single-point energy calculations using a combined ab initio/AMBER potential.[34] The first molecular dynamics simulation was performed by Bash, Field, and Karplus who computed the free energy of activation using both the combined MNDO/TIP3P and AM1/TIP3P models.[52] The AM1 calculation yields a barrier height of 27.7 kcal/mol in water, in good accord with the experimental value (26 kcal/mol). Interestingly, the effect of solvation on the charge transfer during the reaction was probed by the Mulliken populations. They found that the charge transfer lags behind that found in the gas phase, due to the stabilization of the more asymmetric charge distribution interactions with the solvent.[52]

Hwang et al. studied the energetics and dynamics of the S_N2 reaction in water by a combination of the EVB method and free energy perturbation calculation.[53] In this study, the solvent molecules were treated by the surface-constrained all-atom solvent (SCAAS),[66] while the reactants were described by the EVB. These investigators employed an elaborate mapping function to drive the reactant state (ϵ_1) to the product state (ϵ_2). The free energy of activation for the reaction along the real ground state energy surface is then recovered by an umbrella sampling-type treatment, making use of the energy gap, $\Delta\epsilon = \epsilon_2 - \epsilon_1$, as the reaction coordinate. A detailed description of the technique is beyond the scope of this review; interested readers may find References 42, 53, and 233 helpful. Effectively, the method gives the probability of reaching the transition state from the reactant state, while the energy of the transition state (defined by $\Delta\epsilon$) results from a distribution of solute geometries. Using a two-state model, Hwang et al. prescribed a recipe for deriving the force field parameters for the matrix elements and obtained an analytical function for the solute–solvent system by diagonalizing a Hamiltonian matrix. For the Cl^-/CH_3Cl system, the calculated activation barrier (26.0 kcal/mol) is similar to the experimentally observed value.[53]

The combined AM1/TIP3P potential was used in Monte Carlo simulations to examine a two-dimensional free energy surface for a type II process, the Menshutkin reaction of $H_3N + CH_3Cl \rightarrow CH_3NH_3^+ + Cl^-$ in aqueous solution (Eq. [38]).[235] This reaction differs from the type I reactions in that a charge separation occurs during the course of the reaction, and consequently the reaction is significantly enhanced by solvent effects. In fact, it is generally believed that Menshutkin reactions do not take place in the gas phase.[216] The complexity of the reaction represents a major challenge for a theoretical treatment because the geometry of the TS and the reaction path itself will be modified by interactions with the solvent. Clearly, a rigorous procedure should treat the reaction quantum mechanically, including the solute electronic polarization by the surrounding solvent, while the geometrical coordinates relating the bond formation and breaking process should be modeled independently to obtain the reaction path in solution.

QM SCRF methods would be a reasonable choice and have been used to study the Menshutkin reactions in solution;[237] however, it is of interest to investigate the contributions resulting from specific intermolecular interactions between the reagents and solvent. Thus, the effects of hydration on the TS structure are assessed by constructing a two-dimensional potential energy surface in water via a grid search method, treating the C–N and C–Cl bond variations independently.[235] A linear approach for the nucleophilic attack is assumed, which appears to be reasonable. First, a series of free energy profiles as a function of R_{C-N} were determined at a given, fixed distance of R_{C-Cl} using the free energy perturbation method. Then, the relative heights of two neighboring profiles were evaluated by another perturbation calculation with respect to R_{C-Cl} and fixed R_{C-N}. Finally, the potential surface was anchored at an energy obtained from additional, separate potential of mean force simulations. In all, a total of 87 simulations were carried out, each involving at least 5×10^5 configurations of equilibration and 1 million configurations of data collection for a system consisting of 265 water molecules plus [H_3N-CH_3-Cl].

The most striking finding from this study (Figure 9) is the change of the transition state structure on going from the gas phase (shown by an O) into

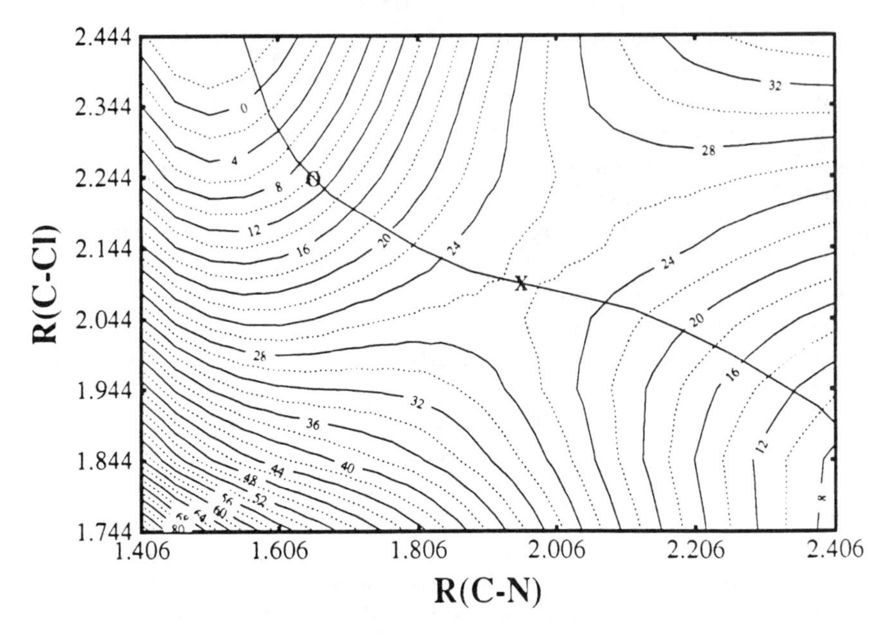

Figure 9 Computed free energy surface for the type II S_N2 reaction of NH_3 + CH_3Cl in aqueous solution. Energies are given in kilocalories per mole, and distances in angstroms.

aqueous solution (specified by an X), which features a lengthening of R_{C-N}(TS) by 0.30 Å from its gas phase value of 1.66 Å by AM1 optimization, and a decrease in R_{C-Cl}(TS) by 0.15 Å (1.94 Å in the gas phase).[235] Consequently, the TS in the Menshutkin reaction occurs much earlier in water than in the gas phase. A similar observation was made by Solá et al. using a continuum SCRF MO method.[237] These findings are consistent with the Hammond postulate because the computed free energy of activation is 26.3 ± 0.3 kcal/mol in water (the experimental estimate of $E_a^‡$ for H_3N + CH_3I in water is 23.5 kcal/mol), which is about 24 kcal/mol lower than that in the gas phase.[238,239] In addition, the solvent effects bring about a net stabilization of the products by 155 kcal/mol relative to the gas phase process, making it an exothermic reaction by −18 kcal/mol. Thus, the TS resembles the structure of the reactants more than that of the products in water. Finally, in contrast to the findings in the type I S_N2 reaction by Bash, Field, and Karplus (see above),[52] the type II Menshutkin reaction shows a solvent-induced charge separation.[235] At the TS, a charge separation of more than 65% is predicted in water, which may be compared with a charge transfer of 50% in the gas phase.

S_N1 reactions are also of interest and importance, and they have been extensively investigated.[2] Following their early work, Kim and Hynes described a theoretical model for S_N1 ionic dissociation (Eq. [42]) of t-BuCl in solution and nonequilibrium solvation using a continuum valence bond approach.[114]

$$(CH_3)_3CCl \rightarrow (CH_3)_3C^+ + Cl^- \qquad [42]$$

Combined AM1/AMBER molecular dynamics simulations were carried out by Hartsough and Merz who treated t-BuCl by the AM1 theory in a box of 483 TIP3P water molecules.[240] However, these investigators found that the barrier height was far greater than that observed experimentally.[243] The disagreement was attributed to the use of long-range cutoff (9 Å) for solute–solvent interactions. To overcome this difficulty, the solvent charges were scaled by a factor of 1.2 to include the long-range electrostatic contributions beyond the cutoff, and a reasonable potential of mean force was obtained for the S_N1 ionization process in water. An alternative approach, however, is to use a CI wavefunction to include both ionic and covalent configurations, as was used by Kim and Hynes.[114,240–242]

The Decarboxylation Reaction of 3-Carboxybenzisoxazole

The next reaction to be studied was the decarboxylation reaction of 3-carboxybenzisoxazole (Eq. [43]). The interest in this reaction stems from the elegant and thorough investigation of the molecule and derivatives in various solvents by Kemp and co-workers.[244,245] They discovered that the reaction rate can increase 10^8-fold on going from aqueous solution to hexamethylphosphoramide (HMPA).

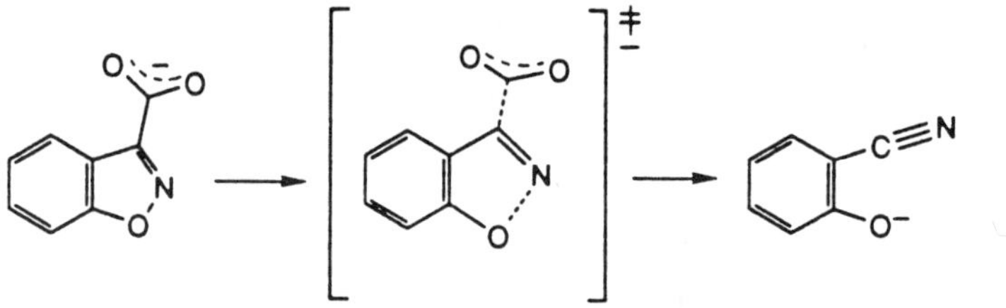

$$+ \ CO_2 \qquad [43]$$

From the kinetic data, Kemp and Paul attributed the dramatic rate change to the balancing effect of the ground state destabilization and transition state stabilization due, respectively, to hydrogen bonding and dispersion interactions. Furthermore, these authors concluded that the change in TS structure is negligible in different solvents in spite of the dramatic rate change.[245] This has been confirmed by recent measurements of the carbon kinetic isotope effects for the decarboxylation reaction in water, dioxane, and a catalytic antibody.[246] Incidentally, the rate enhancement generated by the antibody is about 17,000 times that in water but is significantly slower than that in HMPA or DMA (dimethylamide).[247,248]

The potential of mean force along the gas phase IRC obtained at the 3-21G level was determined through a series of 17 simulations in a rectangular box containing 390 water at 25 °C and 1 atm.[249] The key results are shown in Figure 10, which compares the pmf in water (solid curve) and the gas phase free energy profile (dashed curve). In aqueous solution, the reactant is better solvated than the TS by 10.6 ± 0.3 kcal/mol. In conjunction with the ab initio 3-21G results [which turned out to be in good accord with higher-level calculations at the MP2/6-31G(d) level by Zipse and Houk],[250] the calculation predicts an aqueous $\Delta G^{\ddagger}$ of 26 kcal/mol, in quantitative agreement with the experimental value (26.3 ± 1.5 kcal/mol). Notice that the position of the transition state is not affected by the large solvent effect in aqueous solution. This is somewhat surprising because ab initio 6-31G(d) optimizations with inclusion of one and two water molecules hydrogen-bonded to the leaving carboxylate group indicate that the TS has a shorter C_3–CO_2 distance in the complexes.[250] However, the gain in solvation free energy for an early TS in water is not sufficient to offset the energy required to distort the gas phase TS

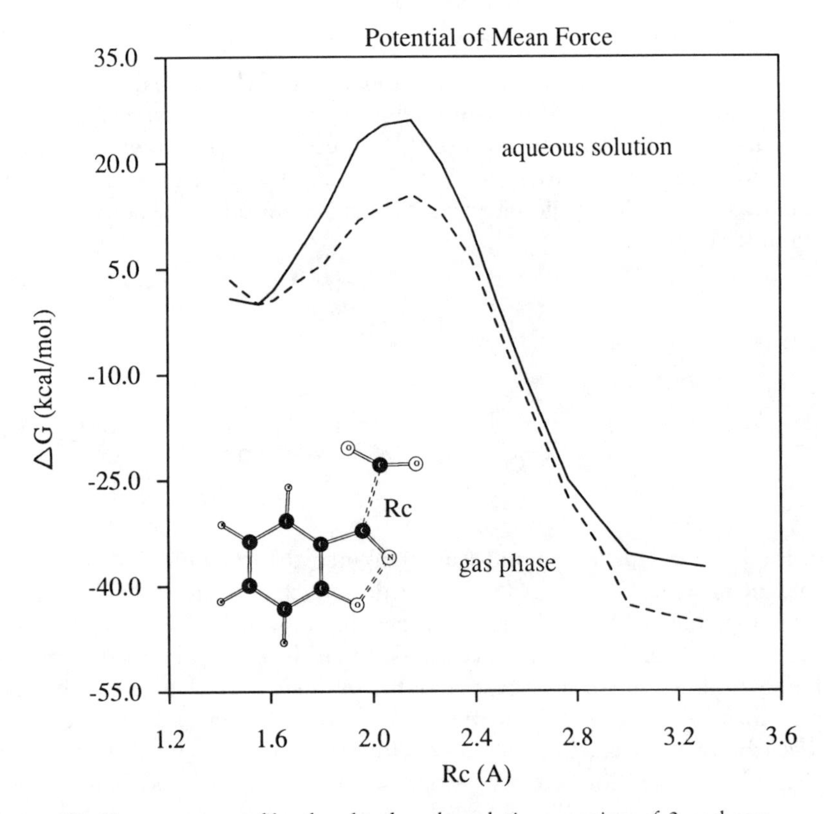

Figure 10 Free energy profiles for the decarboxylation reaction of 3-carboxybenzisoxazole in the gas phase and aqueous solution. The reaction coordinate is illustrated by the C—C distance.

geometry. This is in accord with interpretation based on experimental observations.[245,246]

The good agreement with experiments is encouraging, because it suggests that the combined AM1/TIP3P model can be used to study the solvation effects in rather complex systems. The goal of a theoretical study, however, is not limited to reproducing experimental results. Rather, agreement is a prerequisite for further analyses of the results to gain insights into specific factors that contribute to the observed solvent effects and hopefully to be able to make suggestions for new experiments.

Of particular interest are the energy components averaged in that study.[249] Remarkably, the difference in solvation between the TS and reactant comes almost entirely from the $E^{(1)}$ term, a contribution due to the "permanent" charge distributions (see Eq. [13]). This is because the charges are more delocalized in the TS than in the ground state. Polarization energies do contribute significantly to the total solute–solvent interaction energies, although the quantitative results indicate that polarization energies are the same for both the TS (−9.8 kcal/mol) and reactant (−10.1 kcal/mol).[249] The nature of the ground state and TS electronic polarization is further probed by the qualitative analyses of electron density difference plots with the AM1 wavefunction. Figure 11 illustrates the change of electron density of the solute molecule on transfer from the gas phase into aqueous solution, which is determined by subtracting the gas phase electron density from that in solution:

$$\Delta\rho(\mathbf{R}) = \rho_{sol}(\mathbf{R}) - \rho_{gas}(\mathbf{R}) = 2 \sum_{i} |\psi_i^{sol}(\mathbf{R})|^2 - |\psi_i^{gas}(\mathbf{R})|^2 \qquad [44]$$

where ψ_i^{gas} and ψ_i^{sol} are occupied molecular orbitals in the gas phase and in solution. In the ground state, electron densities are shifted to the carboxylate group, whereas the increases are moved to the isoxazole oxygen and nitrogen in the transition state. The computed Mulliken population charges mirror the same features.[249] In the ground state, the carboxylate group in water has a total charge of −0.739 *e*, which is 0.104 *e* more negative than the gas phase value; the changes on the isoxazole atoms are negligible. For the TS structure, there is virtually no change in atomic charge on the CO_2 group, whereas the gains in partial charges on the isoxazole oxygen and nitrogen are 0.113 and 0.136 *e*, respectively. This result indicates a shift of the hydrogen-bonding site from the carboxylate ion in the ground state to the isoxazole oxygen and nitrogen in the TS. Recall that the quantitative polarization energies are equal in the ground and transition state; however, the origins of the polarization effects are different for the two species.[249]

Making use of the results obtained from simulations of nonspecific interactions in water, where hydrogen bonding and polarization contributions are essentially maximized, one can propose specific catalyst design, be it host-guest

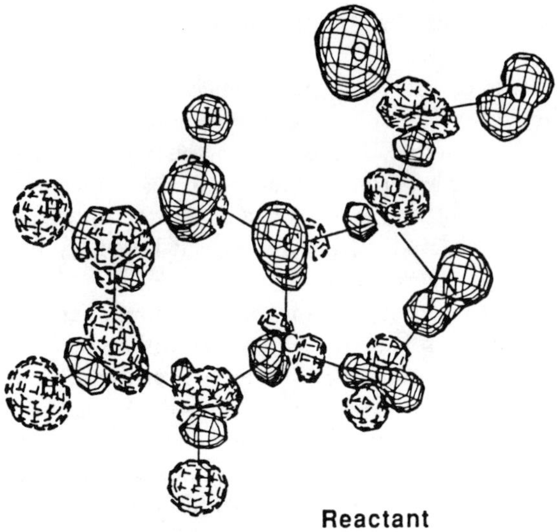

Reactant

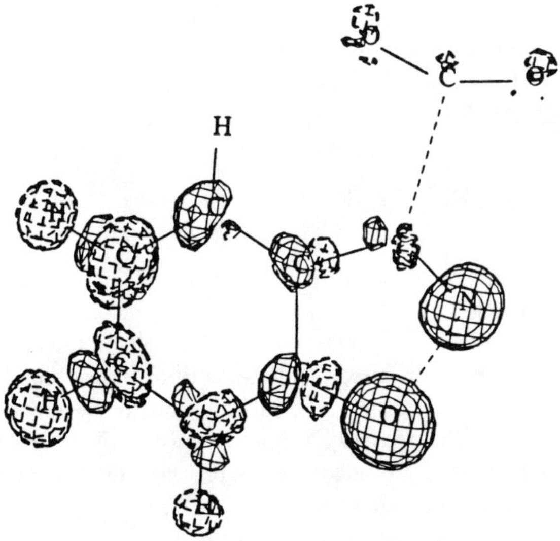

Transition State

Figure 11 Electron density difference (EDD) plots for the reactant (top) and transition state (bottom). Dotted contours represent regions where electron density is depleted, whereas solid curves indicate where there is a gain in electron density on transferring the solute from the gas phase into water. The reactant is 3-carboxybenzisoxazole.

systems or enzymes via site-directed mutagenesis. In a related work, Grate, McGill, and Hilvert suggested that efforts to improve the catalytic efficiency of antibody 21D8 should focus on reducing hydrogen-bond donation.[248] Based on the simulation results, it is suggested that the "active site" of the antibody may be tailored to introduce acidic residues capable of donating hydrogen bonds near the isoxazole oxygen/nitrogen atoms and perhaps basic residues to interact with the carboxy group.[249,251] This design ought to bring about both ground state destabilization (due to reduction in polarization of the carboxylate which is predominant in the ground state) and transition state stabilization (through enhanced polarization). Thus, introducing hydrogen bonds at specific sites is just as important as reducing such interactions at other locations for this decarboxylation process.

Potential Surface for the Proton Transfer in $[H_3N\text{-}H\text{-}NH_3]^+$

The potential surface for the proton transfer in $[H_3N\text{-}H\text{-}NH_3]^+$ in water was computed through a series of Monte Carlo simulations using the combined AM1/TIP3P potential.[252] The effort was a prelude to modeling the quantum mechanical tunneling effect in this process.[253–256] Although the quantum mechanical nuclear tunneling was not specifically considered in that study, the solvent effect on the Born–Oppenheimer potential surface for the proton transfer process is still of interest. The results indicate that aqueous solvation increases the classical barrier height of the proton transfer in $[H_3N\text{-}H\text{-}NH_3]^+$ by about 3 kcal/mol.[252] At the TS where the proton is situated in the middle of the donor and acceptor molecules, the positive charge is more delocalized than the ion–dipole complex. Consequently, the TS is not as strongly stabilized as the ion–dipole structure. The general features of the potential surface in aqueous solution are similar to those in the gas phase, which have been determined by Jaroszewski et al. at the MP2/6-31G(d) level.[257] The barrier height is sensitive to the squeezing and stretching motions of the donor-acceptor fragments. For the interested reader, a tutorial on computing the properties of hydrogen bonds has appeared in this book series.[257]

Solvatochromic Shifts of Acetone in Aqueous and Organic Solutions

Solvatochromic shifts of electronic transitions are widely used to probe solute–solvent interactions in solution and to generate quantitative solvent polarity scales.[1] Theoretical investigations of solvent effects on the observed absorption/emission spectral shift require a knowledge of the properties of a chromophore in both the ground and the excited states, so quantum mechanical methods must be used. Indeed, theoretical treatments of electronic spectra are well established; however, these methods are primarily confined to isolated molecules in the gas phase.[258–266] Study of solvent effects on electronic transitions was pioneered by McRae and Bayliss, who employed a classical reaction field method by placing the chromophore molecule in a spherical cavity embedded in a continuum medium.[100,101,264] The reaction field model has been extensively used by Zerner and co-workers,[263,267,268] among others,[41,269] in various applications,[39] with the intermediate neglect of differential overlap (INDO/S) Hamiltonian and CI calculations. Useful qualitative and quantitative insights into solvent effects on electronic transition have been obtained.

Microscopic investigations of solvent effects on spectroscopy have been described by Warshel and co-workers, who employed a combined QCFF/PI and SCAAS potential in molecular dynamics simulations of mero dyes.[92] Herman and Berne derived a Monte Carlo procedure for simulating a system in which the vibrations are treated quantum mechanically and all other degrees of freedom, including those of the solvent, are treated classically.[270] The vibrational Schrödinger equation at each solvent configuration sampled is solved by a restructured perturbation method. Using this combined QM-classical method, Herman and Berne examined the frequency shifts on vibrational transitions of Br_2 in Ar. Electronic absorption and emission line shapes and band shifts were also considered in that work.[270] Molecular dynamics simulation of solvent effects on the absorption spectrum of formaldehyde/acetone in water was carried out by Levy et al.[271–273] and by DeBolt and Kollman,[274] both employing ab initio electrostatic potential fitted charges and empirical potential functions. These calculations demonstrated the capability of microscopic treatment of the solvent effects on electronic spectra with statistical mechanical sampling.

Using molecular orbital theory, the combined AM1-CI/OPLS potential was used to simulate the solvent-induced absorption spectral shift of acetone in several solvents, including water, methanol, acetonitrile, chloroform, and carbon tetrachloride, at 25°C and 1 atm.[93] In these calculations, acetone was treated by the semiempirical AM1 theory. For each solvent configuration sampled, excitation energies were determined by CI calculations, which included a total of 100 configurations for all excitations for a 5-orbital/6-electron active space.[275,276] The Metropolis Monte Carlo sampling was carried out based on the coupled QM-CI/MM potential. The computed $n \rightarrow \pi^*$ spectral shifts accompanying solvation are listed in Table 6, along with experimental values for the $S_0 \rightarrow S_1$ transition.[93,100,101] In polar solvents such as water and methanol, the absorption bands are blue shifted by about 1700 ± 80 and 760 ± 60 cm^{-1}, respectively, which compare favorably with the experimental data (1560 and 570 cm^{-1}) and continuum models.[101,277] This suggests that the combined AM1-CI/OPLS potential is reasonable for studying solvent effects on the excited states in polar hydrogen bonding solvents.

However, discrepancy between theory and experiment becomes more apparent as the solute is moved into nonpolar solvents. In CCl_4, the absorption of acetone is predicted to be blue shifted marginally by 5 cm^{-1} from that study,[93]

Table 6 Computed Solvent Spectral Shifts and Changes in Solvation Energy for the n → π* Electronic Transition of Acetone (cm^{-1})[a]

		$S_0 \rightarrow T_1$		$S_0 \rightarrow S_1$		
Solvent	ε[b]	$\Delta E^{g \rightarrow e}_{Xs}$	$\langle \Delta \nu^o \rangle$	$\Delta E^{g \rightarrow e}_{Xs}$	$\langle \Delta \nu^o \rangle$	expt[c]
CCl_4	2.2	8	6 ± 5	6	5 ± 4	−490
$CHCl_3$	4.8	370	298 ± 22	289	277 ± 21	50
CH_3CN	35.9	769	568 ± 31	596	535 ± 29	—
CH_3OH	32.7	973	807 ± 65	789	763 ± 60	570
H_2O	78.3	1906	1765 ± 87	1625	1694 ± 84	1560

[a]The gas phase excitation energies for the 100-configuration CI calculations are 27,441 and 24,955 cm^{-1} for $S_0 \rightarrow S_1$ and $S_0 \rightarrow T_1$ transitions. Spectral shifts are relative to the gas phase values.

[b]Dielectric constant.

[c]Reference 101.

although the $S_0 \rightarrow S_1$ transition was determined experimentally to have a red shift (lower energy) of 490 cm^{-1}.[101] The red shift of electronic transitions in nonpolar solvents has been attributed to the solute–solvent dispersion interactions, which are more favorable in the excited state than the ground state.[97–101,265] Such dispersion interactions exist in all solvents but only become significant in nonpolar solvents where hydrogen bonding interactions do not exist. In the QM-CI/MM simulation, the solvent models are effective, pairwise potentials, which do not include polarization contributions due to a change in the solute charge distribution upon excitation.[93] Thus, the mutual solute-solvent induced-dipole-induced-dipole interactions are neglected. This clearly is a major limitation of a combined QM/MM model that does not take into account explicit polarization terms in the solvent potential. Improvement of the performance of the AM1-CI/OPLS model for prediction of solvent spectral shifts may be achieved by using a polarizable solvent model.[92] However, a proper QM treatment of the solute–solvent dispersion interactions, in both the ground and the excited states, requires consideration of interactions between singly excited configurations of the solute with those of the solvent molecules. Unfortunately, it is not practical to treat the entire solution system quantum mechanically in Monte Carlo simulations.

Recently, Rösch and Zerner published a paper describing a perturbation method with the use of sum rules to examine the solute–solvent dispersion contribution in solvatochromic shift within the INDO-SCRF model.[277,278] The method makes use of separate calculation of the solute and solvent in the continuum solvent and combines the results to evaluate dispersion interactions. The theory was successfully applied to acetone, benzene, naphthalene, and chrysene in cyclohexane, which all show large red absorption shifts relative to the gas phase spectra.[278] Within the same spirit, the method may be applied to correct the results predicted by combined QM-CI/MM simulations.

Enzymatic Reaction and Chemisorption on Surfaces

The combined QM/MM potentials have been used by several groups to model reactions catalyzed by enzymes. However, due to the limitation of computer speed, most calculations involved only single-point energy calculations or energy minimizations for protein systems; molecular dynamics calculations have been performed only with the EVB theory.[279] The usefulness of the QM/MM method for protein systems was first demonstrated by an examination of the electrostatic influence in the catalytic mechanism for the cleavage of hexasaccharides by lysozyme.[32] Subsequently, the active site of papain was modeled by the methanethiol-imidazole-formamide supermolecular system in ab initio calculations using the STO-3G and the extended Roos-Siegbahn basis sets.[44] The dipolar effect from a long α-helix was incorporated into the calculation by effective atomic charges.[44] The mechanisms of enzyme catalysis in trypsin and in triosephosphate isomerase (TIM) were investigated through a classical molecular dynamics/minimization study followed by energy determination using a combined quantum/AMBER potential at the STO-3G and 4-31G levels.[44,280] Using the same program, the potential energy surface for ester and amide hydrolysis by the enzyme phospholipase A_2 was probed by a series of investigations using a combined ab initio QM/MM potential at the 3-21G level.[281–283] Geometry optimizations were executed for the QM region consisting of the substrate and catalytic side chains in the presence of the protein atomic partial charges, which are fixed during the QM calculation.

The reaction path of the enzyme TIM was characterized by energy minimizations using the combined AM1/CHARMM potential.[284,285] In that study, the substrate and the catalytic side chains of His-95 and Glu-165 were treated quantum mechanically, and all other molecules in a stochastic boundary system (a 16-Å sphere) were treated by the empirical CHARMM potential. The entire system consisted of 1250 protein atoms and 100 water molecules. An energy decomposition method was used to locate residues important in catalysis. In addition to the large energy contributions from conserved residues, other amino acids make smaller contributions, but their cumulative effect is significant. Interestingly, a number of residues far from the active site also have sizable effects on the enzyme's catalytic ability.[284]

Vasilyev studied the process leading to the formation of the tetrahedral intermediate in the acylation step in acetylcholinesterase.[286] A combined PM3/OPLS potential was developed and used to determine the reaction path. In comparison with the gas phase process, the enzyme environment drastically changes the flow of the reaction, lowering the activation energy by 27 kcal/mol.

Another interesting application of the combined QM/MM potential was reported on a study of chemisorption of acetylene on the Si(100) surface.[287] The AM1 Hamiltonian was used to describe interactions between C_2H_2 and a portion of the silicon surface, whereas an empirical potential was developed to represent the rest of the surface (Figure 12). However, in this work, the QM

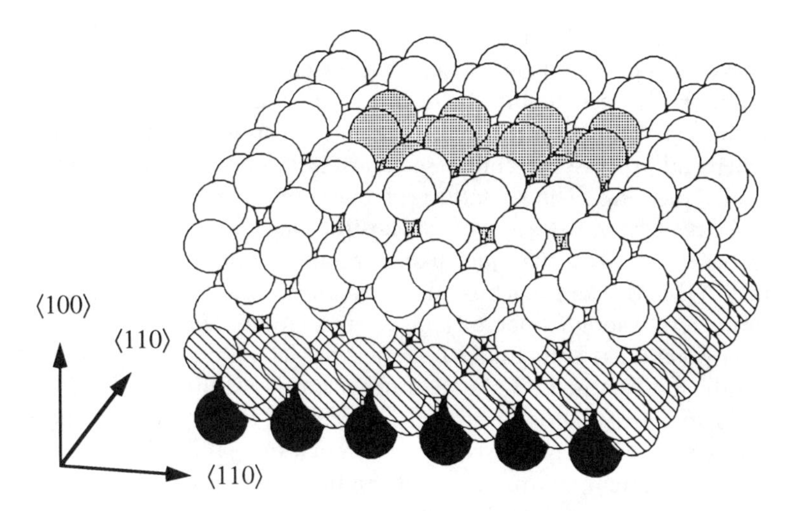

Figure 12 Model of the 360-atom Si(100) surface used in the MD simulations. Shaded circles represent quantum mechanically described atoms; white circles, empirical atoms; cross-hatched circles, empirical atoms subjected to temperature control; and black circles, empirical atoms held in fixed positions throughout the simulations. Adapted from Ref. 287.

atoms were treated as a large cluster with the terminal valences saturated by hydrogen atoms; the forces on the QM atoms due to interactions with the surrounding MM atoms and the forces at the MM sites from QM atoms are evaluated by the empirical potential. Reactions of acetylene approaching different surface sites were investigated via classical trajectory calculations. Acetylene molecules with high translational energies were found to react with a silicon dimer preferably from the same row rather than between adjacent rows. Physisorption occurs when acetylene molecules have low kinetic energy. A mechanism was proposed for chemisorption in which acetylene initially forms a single σ bond with a Si dimer site, followed by radical recombination to yield the more stable bridge structure. The procedure was recently extended to the study of reconstruction of diamond (100) surfaces with the use of the PM3 Hamiltonian.[288] These preliminary investigations indicate that the combined QM/MM potential will also find great applications in characterizing physical and chemical properties in materials science.

CONCLUSIONS

In this review, we have summarized the methodologies and applications of combined QM/MM potentials in statistical mechanical Monte Carlo and molecular dynamics simulations. The method is particularly useful for systems that cannot be appropriately described by molecular mechanics force fields and would be too large to be treated by quantum mechanical methods. The combined QM/MM potential is clearly a logical step to extend the current computational methods for studying chemical reactions in solutions and in enzymes. The method is suitable for modeling chemical reactions, electron transfer, and photochemical processes in condensed phases. Its simplicity and generality are esthetically pleasing. The versatility of the combined QM/MM potential is already evident from the array of systems that have been studied. However, it should be emphasized that just like any other computational method, the combined QM/MM approach is an approximate one, which contains empirical parameters, though hopefully more transferable and less arbitrary than empirical potentials, and should be tested for its validity in application to a given problem. The method may be further improved in several directions.

It is of particular interest and importance to incorporate the solvent quantum effects into condensed phase simulations.[289,290] Recently, Warshel and coworkers attempted to represent the quantum effect of the solvent by a frozen density functional method.[102]

Most applications have employed semiempirical quantum mechanical methods or empirical valence bond theory, because their computational speed is necessary for condensed phase simulations. However, the semiempirical methods suffer from a lack of a well-defined path for improvement. In the EVB approach, it is not clear how many valence states should be considered for complex reactions and how the empirical parameters may be calibrated when experimental data are unavailable. To improve the accuracy, it is clear that ab initio molecular orbital methods and density functional theory will be the preferred QM scheme in a combined QM/MM potential in future studies. The introduction of polarizabilities for the MM atoms will also permit the solvent charge distributions to respond to a change in the solute wavefunction and consequently will yield more accurate solute–solvent interaction energies. The computational procedure seems to be straightforward and has been described; the only difficulty is the computer time needed for self-consistent field (or inversion of large matrices) evaluation of solvent-induced dipoles along with Hartree–Fock SCF calculations. Clearly, development of parallel algorithms for these calculations will be extremely beneficial.[291–298]

Another area deserving immediate attention is the division of QM and MM regions across covalent bonds. The link atom approach is a reasonable approximation, which allows a complex system to be treated by the combined QM/MM potential. Unfortunately, any change in the QM/MM partitioning will significantly affect the results, although a clever separation of the QM and MM regions may minimize the error of the calculation. The localized bond orbital and fragment molecular orbital methods that have recently been used appear to be a promising alternative, because nonchemical link atoms are not introduced into the system. It has also been suggested that pseudopotential methods should be examined, which would permit the treatment of the entire system more rigorously. Undoubtedly, a suitable partitioning method will greatly enhance the applicability of the combined QM/MM potential for simulation of enzymatic systems in solution.[299–304]

ACKNOWLEDGMENTS

This work was supported in part by the National Institutes of Health (GM46736) and by the National Science Foundation (CHE93-19930). The author is indebted to M. J. Field and M. A. Thompson for reviewing this chapter, and to D. B. Boyd, C. J. Cramer, M. J. Field, K. B. Lipkowitz, J. W. McIver, M. Newton, D. G. Truhlar, M. A. Thompson and A. Warshel for numerous comments and suggestions. The author thanks J. J. P. Stewart and W. L. Jorgensen for making their MOPAC and BOSS programs available, M. Karplus for encouragement, and his co-workers, D. Habibollahzadeh, N. Li, J. J. Pavelites, L. Shao, and X. Xia, for their contributions to this work. In the summer of 1994, the author enjoyed the hospitality of the Center for Physics, Aspen, Colorado, during a workshop organized by S. Doniach and R. Elber, where part of this manuscript was written.

REFERENCES

1. C. Reichardt, *Solvents and Solvent Effects in Organic Chemistry,* VCH Publishers, New York, 1990.
2. C. K. Ingold, *Structure and Mechanism in Organic Chemistry,* 2nd ed., Cornell University Press, Ithaca, NY, 1969.
3. S. S. Shaik, H. B. Schlegel, and S. Wolfe, *Theoretical Aspects of Physical Organic Chemistry. The S_N2 Reaction,* John Wiley & Sons, New York, 1992.
4. J. M. Harris and S. P. McManus, Eds., *Nucleophilicity,* ACS Series 215, American Chemical Society, Washington, DC, 1987.
5. W. N. Olmstead and J. I. Brauman, *J. Am. Chem. Soc.*, **99**, 4219 (1977). Gas-Phase Nucleophilic Displacement Reactions.
6. J. Chandrasekhar, S. F. Smith, and W. L. Jorgensen, *J. Am. Chem. Soc.*, **107**, 154 (1985). Theoretical Examination of the S_N2 Reaction Involving Chloride Ion and Methyl Chloride in the Gas Phase and Aqueous Solution.
7. M. P. Allen and D. J. Tildesley, *Computer Simulations of Liquids,* Oxford University Press, London, 1987.
8. J. A. McCammon and S. C. Harvey, *Dynamics of Proteins and Nucleic Acids,* Cambridge University Press, Cambridge, 1987.
9. C. L. Brooks III, M. Karplus, and B. M. Pettitt, *Adv. Chem. Phys.*, **71**, 1 (1988). Proteins: A Theoretical Perspective of Dynamics, Structure, and Thermodynamics.
10. M. Karplus and G. A. Petsko, *Nature,* **347**, 631 (1990). Molecular Dynamics Simulations in Biology.
11. A. Warshel, *Computer Modeling of Chemical Reactions in Enzymes and Solutions,* John Wiley & Sons, New York, 1991.
12. J. M. Haile, *Molecular Dynamics Simulation,* John Wiley & Sons, New York, 1992.
13. H. F. Schaefer III, Ed., *Modern Theoretical Chemistry,* Vols. 3 and 4, Plenum Press, New York, 1977. Methods in Electronic Structure Theory.
14. W. J. Hehre, L. Radom, P. v. R. Schleyer, and J. A. Pople, *Ab Initio Molecular Orbital Theory,* John Wiley & Sons, New York, 1986.

15. A. Szabo and N. Ostlund, *Modern Quantum Chemistry,* Macmillan, New York, 1982.
16. I. S. Y. Wang and M. Karplus, *J. Am. Chem. Soc.*, **95**, 8160 (1973). Dynamics of Organic Reactions.
17. Y.-P. Liu, G. C. Lynch, T. N. Truong, D. Lu, D. G. Truhlar, and B. C. Garrett, *J. Am. Chem. Soc.*, **115**, 2408 (1993). Molecular Modeling of the Kinetic Isotope Effect for the [1,5] Sigmatropic Rearrangement of *cis*-1,3-Pentadiene.
18. K. Laasonen, M. Sprik, M. Parrinello, and R. Car, *J. Chem. Phys.*, **99**, 9080 (1993). "Ab Initio" Liquid Water.
19. E. S. Fois, M. Sprik, and M. Parrinello, *Chem. Phys. Lett.*, **223**, 411 (1994). Properties of Supercritical Water: An Ab Initio Simulation.
20. R. Car and M. Parrinello, *Phys. Rev. Lett.*, **55**, 2471 (1985). Unified Approach for Molecular Dynamics and Density-Functional Theory.
21. M. J. Field, *Chem. Phys. Lett.*, **172**, 83 (1990). Simulated Annealing, Classical Molecular Dynamics and the Hartree–Fock Method: The NDDO Approximation.
22. M. J. Field, *J. Phys. Chem.*, **95**, 5104 (1991). Constrained Optimization of Ab Initio and Semiempirical Hartree–Fock Wave Functions Using Direct Minimization or Simulated Annealing.
23. M. Field, *J. Chem. Phys.*, **96**, 4583 (1992). Time-Dependent Hartree–Fock Simulations of the Dynamics of Polyatomic Molecules.
24. P. Bala, B. Lesyng, and J. A. McCammon, *Chem. Phys. Lett.*, **219**, 259 (1994). Extended Hellmann–Feynman Theorem for Non-Stationary States and Its Application in Quantum-Classical Molecular Dynamics Simulations.
25. U. Burkert and N. Allinger, *Molecular Mechanics,* American Chemical Society, Washington, DC, 1982.
26. A. T. Hagler, E. Huler, and S. Lifson, *J. Am. Chem. Soc.*, **96**, 5319 (1974). Energy Functions for Peptides and Proteins. I. Derivation of a Consistent Force Field Including the Hydrogen Bond from Amide Crystals.
27. B. R. Brooks, R. E. Bruccoleri, B. D. Olafson, D. J. States, S. Swaminathan, and M. Karplus, *J. Comput. Chem.*, **4**, 187 (1983). CHARMM: A Program for Macromolecular Energy, Minimization, and Dynamics Calculations.
28. S. J. Weiner, P. A. Kollman, D. A. Case, U. C. Singh, C. Ghio, G. Alagona, S. Profeta, Jr., and P. Weiner, *J. Am. Chem. Soc.*, **106**, 765 (1984). A New Force Field for Molecular Mechanical Simulation of Nucleic Acids and Proteins.
29. W. L. Jorgensen and J. Tirado-Rives, *J. Am. Chem. Soc.*, **110**, 1657 (1988). The OPLS Potential Functions for Proteins. Energy Minimizations for Crystals of Cyclic Peptides and Crambin.
30. S. L. Mayo, B. D. Olafson, and W. A. Goddard III, *J. Phys. Chem.*, **94**, 8899 (1990). DREIDING: A General Force Field for Molecular Simulations.
31. P. Barnes, J. L. Finney, J. D. Nicolas, and J. E. Quinn, *Nature,* **282**, 459 (1979). Cooperative Effects in Simulated Water.
32. A. Warshel and M. Levitt, *J. Mol. Biol.*, **103**, 227 (1976). Theoretical Studies of Enzymic Reactions: Dielectric, Electrostatic and Steric Stabilization of the Carbonium Ion in the Reaction of Lysozyme.
33. V. Luzhkov and A. Warshel, *J. Comput. Chem.*, **13**, 199 (1992). Microscopic Models for Quantum Mechanical Calculations of Chemical Processes in Solutions: LD/AMPAC and SCAAS/AMPAC Calculations of Solvation Energies.
34. U. C. Singh and P. A. Kollman, *J. Comput. Chem.*, **7**, 718 (1986). A Combined Ab Initio Quantum Mechanical and Molecular Mechanical Method for Carrying Out Simulations on Complex Molecular Systems: Applications to the $CH_3Cl + Cl^-$ Exchange Reaction and Gas Phase Protonation of Polyethers.
35. M. J. Field, P. A. Bash, and M. Karplus, *J. Comput. Chem.*, **11**, 700 (1990). A Combined Quantum Mechanical and Molecular Mechanical Potential for Molecular Dynamics Simulations.

36. J. Gao, *J. Phys. Chem.*, **96**, 537 (1992). Absolute Free Energy of Solvation from Monte Carlo Simulations using Combined Quantum and Molecular Mechanical Potentials.
37. M. J. Field, in *Computer Simulation of Biomolecular Systems: Theoretical and Experimental Applications*, Vol. 2, W. F. van Gunsteren, P. K. Weiner, and A. J. Wilkinson, Eds., ESCOM, Leiden, 1993, p. 82. The Simulation of Chemical Reactions.
38. J. D. Bernal and F. D. Fowler, *J. Chem. Phys.*, **1**, 515 (1933). A Theory of Water and Ionic Solution with Particular Reference to Hydrogen and Hydroxyl Ions.
39. For a recent review of this subject, see C. J. Cramer and D. G. Truhlar, in *Reviews in Computational Chemistry*, Vol. 6, K. B. Lipkowitz and D. B. Boyd, Eds., VCH Publishers, New York, 1995, pp. 1–72. Continuum Solvation Models: Classical and Quantum Mechanical Implementations.
40. M. Karelson and M. C. Zerner, *J. Am. Chem. Soc.*, **112**, 9405 (1990). On the $n \rightarrow \pi^*$ Blue Shift Accompanying Solvation.
41. A. H. de Vries, P. Th. van Duijnen, and A. H. Juffer, *Int. J. Quantum Chem., Quantum Chem. Symp.*, **27**, 451 (1993). Success and Pitfalls of the Dielectric Continuum Model in Quantum Chemical Calculations.
42. J. Åqvist and A. Warshel, *Chem. Rev.* **93**, 2523 (1993). Simulation of Enzyme Reactions Using Valence Bond Force Fields and Other Hybrid Quantum/Classical Approaches.
43. G. Bolis, M. Ragazzi, D. Salvadenri, D. Ferro, and E. Clementi, *Gazz. Chim. Ital.*, **108**, 425 (1978). A Preliminary Attempt to Follow the Enthalpy of an Enzymic Reaction by Ab Initio Computations. The Catalytic Action of Papain.
44. P. Th. van Duijnen, B. Th. Thole, R. Broer, and W. C. Nieuwpoort, *Int. J. Quantum Chem.*, **17**, 651 (1980). Active-Site α-Helix in Papain and the Stability of the Ion Pair $RS^- \cdots ImH^+$. Ab Initio Molecular Orbital Study.
45. B. T. Thole and P. Th. van Duijnen, *Theor. Chim. Acta,* **55**, 307 (1980). On the Quantum Mechanical Treatment of Solvent Effects.
46. J. H. McCreery, R. E. Christoffersen, and G. G. Hall, *J. Am. Chem. Soc.*, **98**, 7198 (1976). On the Development of Quantum Mechanical Solvent Effect Models. Microscopic Electrostatic Contributions.
47. O. Tapia and G. Johannin, *J. Chem. Phys.*, **75**, 3624 (1981). An Inhomogeneous Self-Consistent Reaction Field Theory of Protein Core Effects. Toward a Quantum Scheme for Describing Enzyme Reactions.
48. O. Tapia, F. Colonna, and J. G. Angyan, *J. Chim. Phys.*, **87**, 875 (1990). Generalized Self-Consistent Reaction Field Theory in a Multicenter-Multipole Ab Initio LCGO Framework I. Electronic Properties of the Water Molecule in a Monte Carlo Sample of Liquid Water Molecules Studied with Standard Basis Sets.
49. N. Vaidehi, T. A. Wesolowski, and A. Warshel, *J. Chem. Phys.*, **97**, 4264 (1992). Quantum-Mechanical Calculations of Solvent Free Energies. A Combined Ab Initio Pseudopotential Free-Energy Perturbation Approach.
50. R. V. Stanton, D. S. Hartsough, and K. M. Merz Jr., *J. Phys. Chem.*, **97**, 11868 (1993). Calculation of Solvation Free Energies Using a Density Functional/Molecular Dynamics Coupled Potential.
51. S. J. Weiner, G. L. Leibel, and P. A. Kollman, *Proc. Natl. Acad. Sci. U.S.A.*, **83**, 649 (1986). The Nature of Enzyme Catalysis in Trypsin.
52. P. A. Bash, M. J. Field, and M. Karplus, *J. Am. Chem. Soc.*, **109**, 8092 (1987). Free Energy Perturbation Method for Chemical Reactions in the Condensed Phase: A Dynamical Approach Based on a Combined Quantum and Molecular Mechanics Potential.
53. J.-K. Hwang, G. King, S. Creighton, and A. Warshel, *J. Am. Chem. Soc.*, **110**, 5297 (1988). Simulation of Free Energy Relationships and Dynamics of S_N2 Reactions in Aqueous Solution.
54. J. Gao and X. Xia, *Science*, **258**, 631 (1992). A Priori Evaluation of Aqueous Polarization Effects Through Monte Carlo QM-MM Simulations.
55. G. E. Shulz and R. H. Schirmer, *Principles of Protein Structure*, Springer-Verlag, New York, 1979.

56. W. L. Jorgensen, J. Chandrasekhar, J. D. Madura, R. W. Impey, and M. L. Klein, *J. Chem. Phys.*, **79**, 926 (1983). Comparison of Simple Potential Functions for Simulating Liquid Water.
57. W. L. Jorgensen, J. M. Briggs, and M. L. Contreras, *J. Phys. Chem.*, **94**, 1683 (1990). Relative Partition Coefficients for Organic Solutes from Fluid Simulations.
58. F. Floris and J. Tomasi, *J. Comput. Chem.*, **10**, 616 (1989). Evaluation of the Dispersion Contribution to the Solvation Energy. A Simple Computational Model in the Continuum Approximation.
59. F. M. Floris, J. Tomasi, and J. L. Pascual-Ahuir, *J. Comput. Chem.*, **12**, 784 (1991). Dispersion and Repulsion Contributions to the Solvation Energy: Refinements to a Simple Computational Model in the Continuum Approximation.
60. C. L. Brooks III, B. M. Pettitt, and M. Karplus, *J. Chem. Phys.*, **83**, 5897 (1985). Structural and Energetic Effects of Truncating Long Ranged Interactions in Ionic and Polar Fluids.
61. W. C. Swope, H. C. Andersen, P. H. Berens, and K. R. Wilson, *J. Chem. Phys.*, **76**, 637 (1982). A Computer Simulation Method for the Calculation of Equilibrium Constants for the Formation of Physical Clusters of Molecules: Application to Small Water Clusters.
62. A. J. Stace and J. N. Murrell, *Mol. Phys.*, **33**, 1 (1977). Molecular Dynamics and Chemical Reactivity. A Computer Study of Iodine Atom Recombination under High Pressure Conditions.
63. M. Berkowitz and J. A. McCammon, *Chem. Phys. Lett.*, **90**, 215 (1982). Molecular Dynamics with Stochastic Boundary Conditions.
64. C. L. Brooks III and M. Karplus, *J. Chem. Phys.*, **79**, 6312 (1983). Deformable Stochastic Boundaries in Molecular Dynamics.
65. C. L. Brooks III, A. Brunger, and M. Karplus, *Biopolymers*, **24**, 843 (1985). Active Site Dynamics in Protein Molecules: A Stochastic Boundary Molecular Dynamics Approach.
66. G. King and A. Warshel, *J. Chem. Phys.*, **91**, 3647 (1989). A Surface Constrained All-Atom Solvent Model for Effective Simulations of Polar Solutions.
67. D. Beglov and B. Roux, *J. Chem. Phys.*, **100**, 9050 (1994). Finite Representation of an Infinite Bulk System: Solvent Boundary Potential for Computer Simulations.
68. J. D. Madura and B. M. Pettitt, *Chem. Phys. Lett.*, **150**, 105 (1988). Effects of Truncating Long-Range Interactions in Aqueous Ionic Solution Simulations.
69. K. Tasaki, S. McDonald, and J. W. Brady, *J. Comput. Chem.*, **14**, 278 (1993). Observations Concerning the Treatment of Long-Range Interactions in Molecular Dynamics Simulations.
70. L. Onsager, *J. Am. Chem. Soc.*, **58**, 1486 (1936). Electric Moments of Molecules in Liquids.
71. M. Born, *Z. Physik.*, **1**, 45 (1920). Volumes and Heats of Hydration of Ions.
72. J. G. Kirkwood, *J. Chem. Phys.*, **2**, 351 (1934). Solutions of Molecules Containing Widely Separated Charges—Amphoteric Ions.
73. F. S. Lee and A. Warshel, *J. Chem. Phys.*, **97**, 3100 (1993). A Local Reaction Field Method for Fast Evaluation of Long-Range Electrostatic Interactions in Molecular Simulations.
74. R. Pariser and R. G. Parr, *J. Chem. Phys.*, **21**, 466 (1953). A Semi-Empirical Theory of the Electronic Spectra and Electronic Structure of Complex Unsaturated Molecules.
75. J. A. Pople, *Trans. Faraday Soc.*, **49**, 1375 (1953). Electron Interaction in Unsaturated Hydrocarbons.
76. S. Lifson and A. Warshel, *J. Chem. Phys.*, **49**, 5116 (1968). Consistent Force Field for Calculations of Conformations, Vibrational Spectra, and Enthalpies of Cycloalkane and n-Alkane Molecules.
77. A. Warshel and M. Karplus, *J. Am. Chem. Soc.*, **94**, 5612 (1972). Calculation of Ground and Excited State Potential Surfaces of Conjugated Molecules. I. Formulation and Parametrization.
78. A. Warshel and M. Karplus, *Chem. Phys. Lett.*, **32**, 11 (1965). Semiclassical Trajectory Approach to Photoisomerization.

79. A. Warshel and R. M. Weiss, *J. Am. Chem. Soc.*, **102**, 6218 (1980). An Empirical Valence Bond Approach for Comparing Reactions in Solutions and in Enzymes.
80. A. Warshel, F. Sussman, and J.-K. Hwang, *J. Mol. Biol.*, **201**, 139 (1988). Evaluation of Catalytic Free Energies in Genetically Modified Proteins.
81. F. S. Lee, Z. T. Chu, and A. Warshel, *J. Comput. Chem.*, **14**, 161 (1993). Microscopic and Semimicroscopic Calculation of Electrostatic Energies in Proteins by the POLARIS and ENZYMIX Programs.
82. A. Yadav, R. M. Jackson, J. J. Holbrook, and A. Warshel, *J. Am. Chem. Soc.*, **113**, 4800 (1991). Role of Solvent Reorganization Energies in the Catalytic Activity of Enzymes.
83. R. E. Christoffersen and G. M. Maggiora, *Chem. Phys. Lett.*, **3**, 419 (1969). Ab Initio Calculations on Large Molecules using Molecular Fragments. Preliminary Investigations.
84. *HyperChem Users Manual, Computational Chemistry,* Hypercube, Inc., Waterloo, Ontario, Canada, 1994. Personal communication, N. Ostlund and G. Hurst.
85. D. Rinaldi, J.-L. Rivail, and N. Rguini, *J. Comput. Chem.*, **13**, 675 (1992). Fast Geometry Optimization in Self-Consistent Reaction Field Computations on Solvated Molecules.
86. G. G. Ferenczy, J.-L. Rivail, P. R. Surjan, and G. Naray-Szabo, *J. Comput. Chem.*, **13**, 830 (1992). NDDO Fragment Self-Consistent Field Approximation for Large Electronic Systems.
87. V. Thery, D. Rinaldi, J.-L. Rivail, B. Maigret, and G. G. Ferenczy, *J. Comput. Chem.*, **15**, 269 (1994). Quantum Mechanical Computations on Very Large Molecular Systems: The Local Self-Consistent Field Method.
88. A. A. Rashin, *Int. J. Quantum Chem., Quantum Biol. Symp.*, **15**, 103 (1988). Continuum Electrostatics and Molecular Electric Potential.
89. K. Sharp, A. Jean-Charles, and B. Honig, *J. Phys. Chem.*, **96**, 3822 (1992). A Local Dielectric Constant Model for Solvation Free Energies Which Accounts for Solute Polarizability.
90. D. van Belle, I. Couplet, M. Prevost, and S. J. Wodak, *J. Mol. Biol.*, **198**, 721 (1987). Calculations of Electrostatic Properties of Proteins. Analysis of Contributions from Induced Protein Dipoles.
91. C. J. F. Böttcher, *Theory of Electric Polarization,* 2nd ed., Elsevier, Amsterdam, 1973.
92. V. Luzhkov and A. Warshel, *J. Am. Chem. Soc.*, **113**, 4491 (1991). Microscopic Calculations of Solvent Effects on Absorption Spectra of Conjugated Molecules.
93. J. Gao, *J. Am. Chem. Soc.*, **116**, 9324 (1994). Monte Carlo Quantum Mechanical-Configuration Interaction and Molecular Mechanics Simulation of Solvent Effects on the $n \rightarrow \pi^*$ Blue Shift of Acetone.
94. J. B. Foresman, M. Head-Gordon, and J. A. Pople, *J. Phys. Chem.*, **96**, 135 (1992). Toward a Systematic Molecular Orbital Theory for Excited States.
95. R. McWeeny, *Proc. R. Soc. (London) A*, **253**, 242 (1959). The Density Matrix in Many-Electron Quantum Mechanics I. Generalized Product Functions. Factorization and Physical Interpretation of the Density Matrices.
96. N. J. Turro, *Modern Molecular Photochemistry,* Benjamin/Cummings, Menlo Park, CA, 1978.
97. W. Liptay, in *Modern Quantum Chemistry,* Part II, O. Sinanoglu, Ed., Academic Press, New York, 1966, pp. 173–198. The Solvent Dependence of the Wavenumber of Optical Absorption and Emission.
98. A. T. Amos and B. L. Burrows, *Adv. Quantum Chem.*, **7**, 289 (1973). Solvent-Shift Effects on Electronic Spectra and Excited-State Dipole Moments and Polarizabilities.
99. P. Suppan, *J. Photochem. Photobiol. A*, **50**, 293 (1990). Solvatochromic Shifts—The Influence of the Medium on the Energy of Electronic States.
100. N. S. Bayliss and E. G. McRae, *J. Phys. Chem.*, **58**, 1002 (1954). Solvent Effects in Organic Spectra: Dipole Forces and the Franck–Condon Principle.

101. N. S. Bayliss and E. G. McRae, *J. Phys. Chem.*, **58**, 1006 (1954). Solvent Effects in Organic Spectra of Acetone, Crotonaldehyde, Nitromethane and Nitrobenzene.

102. T. A. Wesolowski and A. Warshel, *J. Phys. Chem.*, **97**, 8050 (1993). Frozen Density Functional Approach for Ab Initio Calculations of Solvated Molecules.

103. D. Wei and D. R. Salahub, *Chem. Phys. Lett.*, **224**, 291 (1994). A Combined Density Functional and Molecular Dynamics Simulation of a Quantum Water Molecule in Aqueous Solution.

104. J. A. Pople and D. L. Beveridge, *Approximate Molecular Orbital Theory*, McGraw-Hill, New York, 1970.

105. W. Thiel, *Tetrahedron*, **44**, 7393 (1988). Semiempirical Methods: Current Status and Perspectives.

106. J. J. P. Stewart, in *Reviews in Computational Chemistry*, Vol. 1, K. B. Lipkowitz and D. B. Boyd, Eds., VCH Publishers, New York, 1990, pp. 45–81. Semiempirical Molecular Orbital Methods.

107. M. C. Zerner, in *Reviews in Computational Chemistry*, Vol. 2, K. B. Lipkowitz and D. B. Boyd, Eds., VCH Publishers, New York, 1991, pp. 313–365. Semiempirical Molecular Orbital Methods.

108. M. J. S. Dewar, E. G. Zoebisch, E. F. Healy, and J. J. P. Stewart, *J. Am. Chem. Soc.*, **107**, 3902 (1985). AM1: A New General Purpose Quantum Mechanical Molecular Model.

109. M. J. S. Dewar and W. Thiel, *J. Am. Chem. Soc.*, **99**, 4899 (1977). Ground States of Molecules. 38. The MNDO Method. Approximations and Parameters.

110. J. J. P. Stewart, *J. Comput. Chem.*, **10**, 209 (1989). Optimization of Parameters for Semiempirical Methods. I. Method.

111. J. J. P. Stewart, *J. Comput. Chem.*, **10**, 221 (1989). Optimization of Parameters for Semiempirical Methods. II. Applications.

112. J. A. Pople, D. P. Santry, and G. A. Segal, *J. Chem. Phys.*, **43**, S129 (1965). Approximate Self-Consistent Molecular Orbital Theory. I. Invariant Procedures.

113. J. A. Pople and G. A. Segal, *J. Chem. Phys.*, **43**, S136 (1965). Approximate Self-Consistent Molecular Orbital Theory. I. Calculations with Complete Neglect of Differential Overlap.

114. H. J. Kim and J. T. Hynes, *J. Am. Chem. Soc.*, **114**, 10508 (1992). A Theoretical Model for S_N1 Ionic Dissociation in Solution. 1. Activation Free Energetics and Transition-State Structure.

115. D. G. Truhlar, G. K. Schenter, and B. C. Garrett, *J. Chem. Phys.*, **98**, 5756 (1993). Inclusion of Nonequilibrium Continuum Solvation Effects in Variation Transition State Theory.

116. R. G. Parr and W. Yang, *Density Functional Theory of Atoms and Molecules*, Oxford University Press, London, 1989. J. K. Labanowski and J. W. Andzelm, Eds., *Density Functional Methods in Chemistry*, Springer-Verlag, New York, 1991.

117. L. J. Bartolotti and K. Flurchick, this volume. A. St-Amant, this volume.

118. W. Kohn and L. J. Sham, *Phys. Rev.*, **A140**, 1133 (1965). Self-Consistent Equations Including Exchange and Correlation Effects.

119. S. F. Boys, *Proc. R. Soc. (London) Ser. A*, **200**, 542 (1950). Electronic Wave Functions. II. A Calculation for the Ground State of the Beryllium Atom.

120. C. C. J. Roothaan, *J. Chem. Phys.*, **19**, 1445 (1951). A Study of Two-Center Integrals Useful in Calculations on Molecular Structure. I.

121. M. J. S. Dewar and W. Thiel, *Theoret. Chim. Acta*, **46**, 89 (1977). A Semiempirical Model for the Two-Center Repulsion Integrals in the NDDO Approximation.

122. J. J. P. Stewart, *J. Comput.-Aided Mol. Design*, **4**, 1 (1990). MOPAC: A Semiempirical Molecular Orbital Program.

123. J. J. P. Stewart, *QCPE Bull.*, **3**, 43 (1983). QCPE Program 455, MOPAC: A General Molecular Orbital Package. *QCPE Bull.*, **10**, 86 (1990). MOPAC Version 6.0.

124. J. Gao, *MCQUB*, State University of New York at Buffalo, Buffalo, NY, 1992.

125. W. L. Jorgensen, *BOSS*, Version 3.0, Yale University, New Haven, CT, 1990.

126. M. A. Thompson, *ARGUS*, Pacific Northwest Laboratory, Richland, WA, 1994.

127. W. L. Jorgensen, *Chemtracts-Org. Chem.*, **4**, 91 (1991). Computational Insights on Intermolecular Interactions and Binding in Solution.

128. J. Gao, in *Modeling the Hydrogen Bond*, D. A. Smith, Ed., American Chemical Society Symp. Ser. 569, 1994, pp. 8–20. Computation of Intermolecular Interactions with a Combined Quantum Mechanical and Classical Approach.

129. V. V. Vasilyev, A. A. Bliznyuk, and A. A. Voityuk, *Int. J. Quantum Chem.*, **44**, 897 (1992). A Combined Quantum Chemical/Molecular Mechanical Study of Hydrogen-Bonded Systems.

130. J. Gao, *J. Phys. Chem.*, **96**, 6432 (1992). Comparison of the Hybrid AM1/TIP3P and the OPLS Functions through Monte Carlo Simulation of Acetic Acid in Water.

131. J. Gao, L. W. Chou, and A. Auerbach, *Biophys. J.*, **65**, 43 (1993). The Nature of Cation-π Binding: Interactions between Tetramethylammonium Ion and Benzene in Aqueous Solution.

132. J. W. Storer, D. J. Giesen, C. J. Cramer, and D. G. Truhlar, *J. Comput.-Aided Mol. Design*, **9**, 87 (1995). Class IV Charge Models: A New Semiempirical Approach in Quantum Chemistry.

133. L. X. Dang, J. E. Rice, J. W. Caldwell, and P. A. Kollman, *J. Am. Chem. Soc.*, **113**, 2481 (1991). Ion Solvation in Polarizable Water: Molecular Dynamics Simulations.

134. A. Wallqvist and B. J. Berne, *J. Phys. Chem.*, **97**, 13841 (1993). Effective Potentials for Liquid Water Using Polarizable and Nonpolarizable Models.

135. D. N. Bernardo, Y. Ding, K. Krogh-Jespersen, and R. M. Levy, *J. Phys. Chem.*, **98**, 4180 (1994). An Anisotropic Polarizable Water Model: Incorporation of All-Atom Polarizabilities into Molecular Mechanics Force Fields.

136. J. Pranata, S. G. Wierschke, and W. L. Jorgensen, *J. Am. Chem. Soc.*, **113**, 2810 (1991). OPLS Potential Functions for Nucleotide Bases. Relative Association Constants of Hydrogen-Bonded Base Pairs in Chloroform.

137. D. L. Beveridge and F. M. DiCapua, *Annu. Rev. Biophys. Chem.*, **18**, 431 (1989). Free Energy via Molecular Simulation—Applications to Chemical and Biomolecular Systems.

138. W. L. Jorgensen, *Acc. Chem. Res.*, **22**, 184 (1989). Free Energy Calculations: A Breakthrough for Modeling Organic Chemistry in Solution.

139. W. F. van Gunsteren and H. J. C. Berensen, *Angew, Chem., Int. Ed. Engl.*, **29**, 992 (1990). Computer Simulation of Molecular Dynamics: Methodology, Applications, and Perspectives in Chemistry.

140. T. P. Straastma and J. A. McCammon, *Annu. Rev. Phys. Chem.*, **43**, 407 (1992). Computational Alchemy.

141. P. Kollman, *Chem. Rev.*, **93**, 2395 (1993). Free Energy Calculations: Applications to Chemical and Biochemical Phenomena.

142. C. A. Reynolds, P. M. King, and W. G. Richards, *Mol. Phys.*, **76**, 251 (1992). Free Energy Calculations in Molecular Biophysics.

143. P. A. Bash, U. C. Singh, R. Langridge, and P. A. Kollman, *Science*, **236**, 564 (1987). Free Energy Calculations by Computer Simulation.

144. T. Wesolowski and A. Warshel, *J. Phys. Chem.*, **98**, 5183 (1994). Ab Initio Free Energy Perturbation Calculations of Solvation Free Energy Using the Frozen Density Functional Approach.

145. P. Cieplak and P. A. Kollman, *J. Am. Chem. Soc.*, **110**, 3734 (1988). Calculation of the Free Energy of Association of Nucleic Acid Bases in Vacuo and Water Solution.

146. W. L. Jorgensen, J. F. Blake, and J. K. Buckner, *Chem. Phys.*, **129**, 193 (1989). Free Energy of TIP4P Water and the Free Energies of Hydration of CH_4 and Cl^- from Statistical Perturbation Theory.

147. R. W. Zwanzig, *J. Chem. Phys.*, **22**, 1420 (1954). High Temperature Equation of State by a Perturbation Method. I. Nonpolar Gases.

148. J. Gao, K. Kuczera, B. Tidor, and M. Karplus, *Science,* **244,** 1069 (1989). Hidden Thermodynamics of Mutant Proteins: A Molecular Dynamics Analysis.

149. K. Kuczera, J. Gao, B. Tidor, and M. Karplus, *Proc. Natl. Acad. Sci. U.S.A.,* **87,** 8481 (1990). Free Energy of Sickling: A Simulation Analysis.

150. B. Tidor and M. Karplus, *Biochemistry,* **30,** 3217 (1991). Simulation Analysis of the Stability of Mutant R96H of T4 Lysozyme.

151. D. J. Tobias and C. L. Brooks III, *Biochemistry,* **30,** 6059 (1991). Thermodynamics and Mechanism of Alpha Helix Initiation in Alanine and Valine Peptides.

152. A. Ben-Naim and Y. Marcus, *J. Chem. Phys.,* **81,** 2016 (1984). Solvation Thermodynamics of Nonionic Solutes.

153. R. Wolfenden, L. Andersson, P. M. Cullis, and C. C. B. Southgate, *Biochemistry,* **20,** 849 (1981). Affinities of Amino Acid Side Chains for Solvent Water.

154. R. G. Pearson, *J. Am. Chem. Soc.,* **108,** 6109 (1986). Ionization Potentials and Electron Affinities in Aqueous Solution.

155. J. Gao, *Biophys. Chem.,* **51,** 253 (1994). The Hydration and Solvent Polarization Effects of Nucleotide Bases.

156. A. H. Elcock and W. G. Richards, *J. Am. Chem. Soc.,* **115,** 7930 (1993). Relative Hydration Free Energies of Nucleic Acid Bases.

157. C. J. Cramer and D. G. Truhlar, *Chem. Phys. Lett.,* **198,** 74 (1992); **202,** 567 (1993). Polarization of the Nucleic Acid Bases in Aqueous Solution.

158. M. Orozco and F. J. Luque, *Biopolymers,* **33,** 1851 (1993). Self-Consistent Reaction Field Computation of the Reactive Characteristics of DNA Bases in Water.

159. P. E. Young and I. H. Hillier, *Chem. Phys. Lett.,* **215,** 405 (1993). Hydration Free Energies of Nucleic Acid Bases Using an Ab Initio Continuum Model.

160. V. Mohan, M. E. Davis, J. A. McCammon, and B. M. Pettitt, *J. Phys. Chem.,* **96,** 6428 (1992). Continuum Model Calculations of Solvation Free Energies: Accurate Evaluation of Electrostatic Contributions.

161. A. R. Katritzky and M. Karelson, *J. Am. Chem. Soc.,* **113,** 1561 (1991). AM1 Calculations of Reaction Field Effects on the Tautomeric Equilibria of Nucleic Acid Pyrimidine and Purine Bases and Their 1-Methyl Analogues.

162. D. M. Ferguson, D. A. Pearlman, W. C. Swope, and P. A. Kollman, *J. Comput. Chem.,* **13,** 362 (1992). Free Energy Perturbation Calculations Involving Potential Function Changes.

163. D. Eisenberg and W. Kauzmann, *The Structure and Properties of Water,* Oxford University Press, Oxford, 1969, pp. 105–110, 191–193.

164. W. Yang, *Phys. Rev. Lett.,* **66,** 1438 (1991). Direct Calculation of Electron Density in Density-Functional Theory.

165. M. Orozco, F. J. Luque, D. Habibollahzadeh, and J. Gao, *J. Chem. Phys.,* **102,** 6145 (1995). The Polarization Contribution to the Free Energy of Hydration.

166. J. Gao, F. J. Luque, and M. Orozco, *J. Chem. Phys.,* **98,** 2975 (1993). Induced Dipole Moment and Atomic Charges Based on Average Electrostatic Potentials in Aqueous Solution.

167. M. Perutz, *Mechanisms of Cooperativity and Allosteric Regulation in Proteins,* Cambridge University Press, Cambridge, U.K., 1990.

168. J.-M. Lehn, *Angew. Chem., Int. Ed. Engl.,* **27,** 89 (1988). Supramolecular Chemistry—Scope and Perspectives. Molecules, Supermolecules, and Molecular Devices.

169. D. J. Cram, *Angew. Chem., Int. Ed. Engl.,* **27,** 1009 (1988). The Design of Molecular Hosts, Guests, and Their Complexes.

170. J. W. Caldwell, D. A. Agard, and P. A. Kollman, *Proteins,* **10,** 140 (1991). Free Energy Calculations on Binding and Catalysis by Alpha-Lytic Protease: The Role of Substrate Size in the P1 Pocket.

171. A. A. Rashin, *J. Phys. Chem.,* **94,** 1725 (1990). Hydration Phenomena, Classical Electrostatics, and Boundary Element Method.

172. M. E. Davis and J. A. McCammon, *Chem. Rev.,* **90,** 509 (1990). Electrostatics in Biomolecular Structure and Dynamics.

173. B. H. Besler, K. M. Merz Jr., and P. A. Kollman, *J. Comput. Chem.,* **11,** 431 (1990). Atomic Charges Derived from Semiempirical Methods.

174. S. Mietus, E. Scrocco, and J. Tomasi, *Chem. Phys.,* **55,** 117 (1981). Electrostatic Interaction of a Solute with a Continuum. A Direct Utilization of Ab Initio Molecular Potentials for the Provision of Solvent Effects.

175. J. L. Pascual-Ahuir, E. Silla, J. Tomasi, and R. Bonaccorsi, *J. Comput. Chem.,* **8,** 778 (1987). Electrostatic Interaction of a Solute with a Continuum. Improved Description of the Cavity and of the Surface Cavity Bound Charge Distribution.

176. M. J. Negre, M. Orozco, and F. J. Luque, *Chem. Phys. Lett.,* **196,** 27 (1992). A New Strategy for the Representation of Environment Effects in Semiempirical Calculations Based on Dewar's Hamiltonians.

177. C. I. Bayly, P. Cieplak, W. D. Cornell, and P. A. Kollman, *J. Phys. Chem.,* **97,** 10269 (1993). A Well-Behaved Electrostatic Potential Based Method Using Charge Restraints for Deriving Atomic Charges—The Resp Model.

178. T. J. Marrone, D. S. Hartsough, and K. M. Merz Jr., *J. Phys. Chem.,* **98,** 1341 (1994). Determination of Atomic Charges Including Solvation and Conformational Effects.

179. A. Warshel, *J. Phys. Chem.,* **83,** 1840 (1979). Calculations of Chemical Processes in Solutions.

180. W. L. Jorgensen, *J. Phys. Chem.,* **87,** 5304 (1983). Theoretical Studies of Medium Effects on Conformational Equilibria.

181. J. Gao and J. J. Pavelites, *J. Am. Chem. Soc.,* **114,** 1912 (1992). Aqueous Basicity of the Carboxylate Lone Pairs and the C-O Barrier in Acetic Acid: A Combined Quantum and Statistical Mechanical Study.

182. F. H. Allen and A. J. Kirby, *J. Am. Chem. Soc.,* **113,** 8829 (1991). Stereoelectronic Effects at Carboxylate Oxygen: Similar Basicity of the E and Z Lone Pairs in Solution.

183. K. B. Wiberg and K. E. Laidig, *J. Am. Chem. Soc.,* **109,** 5935 (1987). Barriers to Rotation Adjacent to Double Bonds. 3. The Carbon-Oxygen Barrier in Formic Acid, Methyl Formate, Acetic Acid, and Methyl Acetate. The Origin of Ester and Amide Resonance.

184. K. B. Wiberg and K. E. Laidig, *J. Am. Chem. Soc.,* **110,** 1872 (1988). Acidity of (Z)- and (E)-Methyl Acetates: Relationship to Meldrum's Acid.

185. Y. Li and K. N. Houk, *J. Am. Chem. Soc.,* **111,** 4505 (1989). Theoretical Assessments of the Basicity and Nucleophilicity of Carboxylate Syn and Anti Lone Pairs.

186. B. M. Tadayoni, K. Parris, and J. Rebek, Jr., *J. Am. Chem. Soc.,* **111,** 4503 (1989). Intramolecular Catalysis and Enolization: A Probe for Stereoelectronic Effects at Carboxyl Oxygen.

187. B. M. Tadayoni and J. Rebek, Jr., *Bioorg. Med. Chem. Lett.,* **1,** 13 (1991). Intramolecular Nucleophilic Displacement Reactions at Carboxyl Oxygen.

188. K. D. Cramer and S. C. Zimmerman, *J. Am. Chem. Soc.,* **112,** 3680 (1990). Kinetic Effect of a syn-Oriented Carboxylate on a Proximate Imidazole in Catalysis: A Model for the Histidine-Aspartate Couple in Enzymes.

189. J. Pranata, *J. Comput. Chem.,* **14,** 685 (1993). Relative Basicities of Carboxylate Lone Pairs in Aqueous Solution.

190. R. D. Gandour, *Bioorg. Chem.,* **10,** 169 (1981). On the Importance of Orientation in General Base Catalysis by Carboxylate.

191. T. Drakenberg, K.-I. Dahlqvist, and S. Forsen, *J. Phys. Chem.,* **76,** 2178 (1972). Barrier to Internal Rotation in Amides. IV. *N,N*-Dimethylamides. Substituents and Solvent Effects.

192. M. Feigel, *J. Phys. Chem.,* **87,** 3054 (1983). Rotational Barriers and Amides in the Gas Phase.

193. B. D. Ross and N. S. True, *J. Am. Chem. Soc.,* **106,** 2451 (1984). Gas-Phase ^{13}C NMR Spectra and Exchange Kinetics of *N,N*-Dimethylformamide.

194. M. W. Albers, C. T. Walsh, and S. L. Schreiber, *J. Org. Chem.*, **55**, 4984 (1990). Substrate Specificity for the Human Rotamase FKBP: A View of FK506 and Rapamycin as Leucine-(Twisted Amide)-Proline Mimics.

195. S. Tsuzuki and K. Tanabe, *J. Chem. Soc., Perkin Trans. 2*, 1255 (1991). Basis Set and Electron Correlation Effects on the Internal Rotational Barrier Heights of Formamide and Acetamide.

196. K. B. Wiberg and C. M. Breneman, *J. Am. Chem. Soc.*, **114**, 831 (1992). Resonance Interactions in Acyclic Systems. 3. Formamide Internal Rotation Revisited. Charge and Energy Redistribution along the C-N Bond Rotational Pathway.

197. E. M. Duffy, D. L. Severance, and W. L. Jorgensen, *J. Am. Chem. Soc.*, **114**, 7535 (1992). Solvent Effects on the Barrier to Isomerization for a Tertiary Amide from Ab Initio and Monte Carlo Calculations.

198. J. Gao, *J. Am. Chem. Soc.*, **115**, 2930 (1993). Potential of Mean Force for the Isomerization of DMF in Aqueous Solution: A Monte Carlo QM/MM Simulation Study.

199. J. Gao, *Proc. Ind. Acad. Sci., Chem. Sci.*, **106**, 507 (1994). Origin of the Solvent Effects on the Barrier to Amide Isomerization from the Combined QM/MM Monte Carlo Simulations.

200. M. A. Thompson, E. D. Glendening, and D. Feller, *J. Phys. Chem.*, **98**, 10465 (1994). The Nature of K^+/Crown-Ether Interactions: A Hybrid Quantum Mechanical–Molecular Mechanical Study.

201. L. X. Dang and P. A. Kollman, *J. Am. Chem. Soc.*, **112**, 5716 (1990). Free Energy of Association of the 18-Crown-6:K^+ Complex in Water: A Molecular Dynamics Simulation.

202. J. Gao and L. Shao, *J. Phys. Chem.*, **98**, 13772 (1994). Polarization Effects on the Tautomeric Equilibria of 2- and 4-Hydroxypyridine in Aqueous and Organic Solution.

203. P. Beak, *Acc. Chem. Res.*, **10**, 186 (1977). Energies and Alkylations of Tautomeric Heterocyclic Compounds—Old Problems New Solutions.

204. C. J. Cramer and D. G. Truhlar, *J. Comput.-Aided Mol. Design*, **6**, 629 (1992). AM1-SM2 and PM3-SM3 Parameterized SCF Solvation Models for Free Energies in Aqueous Solution.

205. W. L. Jorgensen, *Adv. Chem. Phys.*, **70**, 469 (1988). Energy Profiles for Organic Reactions in Solution.

206. For a review, see R. M. Whitnell and K. R. Wilson, in *Reviews in Computational Chemistry*, Vol. 4, K. B. Lipkowitz and D. B. Boyd, Eds., VCH Publishers, New York, 1993, pp. 67–148. Computational Molecular Dynamics of Chemical Reactions in Solution.

207. J. T. Hynes, in *Theory of Chemical Reaction Dynamics*, Vol. 4, M. Baer, Ed., CRC Press, Boca Raton, FL, 1985, pp. 171–234. The Theory of Reactions in Solution.

208. D. G. Truhlar, W. L. Hase, and J. T. Hynes, *J. Phys. Chem.*, **87**, 2664 (1983). Current Status of Transition-State Theory.

209. Y. Zheng and K. M. Merz Jr., *J. Am. Chem. Soc.*, **114**, 2733 (1992). The Gas-Phase and Solution-Phase Free Energy Surfaces for Carbon Dioxide Reaction with Hydroxide.

210. J. Gao and X. Xia, in *Structure and Reactivity in Aqueous Solution: Characterization of Chemical and Biological Systems*, C. J. Cramer and D. G. Truhlar, Eds., ACS Symp. Series 568, 1994, pp. 212–228. Simulating Solvent Effects on Reactivity and Interactions in Ambient and Supercritical Water.

211. M. J. Frisch, M. Head-Gordon, G. W. Trucks, J. B. Foresman, H. B. Schlegel, K. Raghavachari, M. Robb, J. S. Binkley, C. Gonzales, D. J. DeFrees, D. J. Fox, R. A. Whiteside, R. Seeger, C. F. Melius, J. Baker, R. L. Martin, L. R. Kahn, J. J. P. Stewart, S. Topiol, and J. A. Pople, *Gaussian 90*, Gaussian Inc., Pittsburgh, PA, 1990.

212. S. E. Huston, P. J. Rossky and D. A. Zichi, *J. Am. Chem. Soc.*, **111**, 5680 (1980). Hydration Effects on S_N2 Reactions: An Integral Equation Study of Free Energy Surfaces and Corrections to Transition-State Theory.

213. B. J. Gertner, R. M. Whitnell, K. R. Wilson, and J. T. Hynes, *J. Am. Chem. Soc.*, **113**, 74 (1991). Activation to the Transition State: Reactant and Solvent Energy Flow for a Model S_N2 Reaction in Water.

214. S. C. Tucker and D. G. Truhlar, *Chem. Phys. Lett.*, **157**, 164 (1989). Generalized Born Fragment Charge Model for Solvation Effects as a Function of Reaction Coordinate.

215. S. C. Tucker and D. G. Truhlar, *J. Am. Chem. Soc.*, **112**, 3347 (1990). The Effect of Nonequilibrium Solvation on Chemical Reaction Rates. Variational Transition State Theory Studies of the Microsolvated Reaction $Cl^-(H_2O)_n + CH_3Cl$.

216. J. M. Abboud, R. Notario, J. Bertran, and M. Sola, *Prog. Phys. Org. Chem.*, **19**, 1 (1993). One Century of Physical Organic Chemistry: The Menshutkin Reaction.

217. J. P. Valleau and G. M. Torrie, in *Statistical Mechanics Part A: Equilibrium Techniques*, B. J. Berne, Ed., Plenum, New York, 1977, p. 137. A Guide to Monte Carlo for Statistical Mechanics: 1. Highways.

218. W. J. Jorgensen and C. Ravimohan, *J. Chem. Phys.*, **83**, 3050 (1985). Monte Carlo Simulation of Differences in Free Energies of Hydration.

219. R. M. Coates, B. Rogers, S. J. Hobbs, D. R. Peck, and D. P. Curran, *J. Am. Chem. Soc.*, **109**, 1160 (1987). Synthesis and Claisen Rearrangement of Alkoxyallyl Enol Ethers. Evidence for a Dipolar Transition State.

220. J. J. Gajewski, J. Jurayj, D. R. Kimbrough, M. E. Gande, B. Ganem, and B. K. Carpenter, *J. Am. Chem. Soc.*, **109**, 1170 (1987). On the Mechanism of Rearrangement of Chorismic Acid and Related Compounds.

221. J. J. Gajewski, *J. Am. Chem. Soc.*, **116**, 3165 (1995). Secondary Deuterium Kinetic Isotope Effects in the Aqueous Claisen Rearrangement: Evidence Against an Ionic Transition State.

222. M. J. S. Dewar and C. Jie, *J. Am. Chem. Soc.*, **111**, 511 (1989). Mechanism of the Claisen Rearrangement of Allyl Vinyl Ethers.

223. R. L. Vance, N. G. Rondan, K. N. Houk, F. Jensen, W. T. Borden, A. Komornicki, and E. Wimmer, *J. Am. Chem. Soc.*, **110**, 2314 (1988). Transition Structure for the Claisen Rearrangement.

224. W. J. Guilford, S. D. Copley, and J. R. Knowles, *J. Am. Chem. Soc.*, **109**, 5013 (1987). The Mechanism of the Chorismate Mutase Reaction.

225. C. J. Cramer and D. G. Truhlar, *J. Am. Chem. Soc.*, **114**, 8794 (1992). What Causes Aqueous Acceleration of the Claisen Reaction?

226. D. L. Severance and W. L. Jorgensen, *J. Am. Chem. Soc.*, **114**, 10966 (1992). Effect of Hydration on the Claisen Rearrangement of Allyl Vinyl Ether from Computer Simulations.

227. J. W. Storer, D. J. Giesen, G. D. Hawkins, G. C. Lynch, C. J. Cramer, D. G. Truhlar, and D. A. Liotard, in *Structure and Reactivity in Aqueous Solution*, C. J. Cramer and D. G. Truhlar, Eds., American Chemical Society Symposium Series 568, Washington DC, 1994, pp. 24–49. Solvation Modeling in Aqueous and Nonaqueous Solvents.

228. J. Gao, *J. Am. Chem. Soc.*, **116**, 1563 (1994). Combined QM/MM Simulation Study of the Claisen Rearrangement of Allyl Vinyl Ether in Aqueous Solution.

229. J. J. Gajewski, *J. Org. Chem.*, **57**, 5500 (1992). A Semitheoretical Multiparameter Approach to Correlate Solvent Effects on Reactions and Equilibria.

230. E. D. Hughes and C. K. Ingold, *J. Chem. Soc.*, 244 (1935). Mechanism of Substitution at a Saturated Carbon Atom. Part IV. A Discussion of Constitutional and Solvent Effects on the Mechanism. Kinetics, Velocity, and Orientation of Substitution.

231. J. Chandrasekhar, S. F. Smith, and W. L. Jorgensen, *J. Am. Chem. Soc.*, **106**, 3049 (1984). S_N2 Reaction Profiles in the Gas Phase and Aqueous Solution.

232. J. Chandrasekhar and W. L. Jorgensen, *J. Am. Chem. Soc.*, **107**, 2975 (1985). Energy Profile for a Nonconcerted S_N2 Reaction in Solution.

233. J.-K. Hwang and A. Warshel, *J. Am. Chem. Soc.*, **109**, 715 (1987). Microscopic Examination of Free-Energy Relationships for Electron Transfer in Polar Solvents.

234. J. R. Mathis, R. Bianco, and J. T. Hynes, *J. Mol. Liquids*, **61**, 81 (1994). On the Activation Free Energy of the $Cl^- + CH_3Cl$ S_N2 Reaction in Solution.

235. J. Gao and X. Xia, *J. Am. Chem. Soc.*, **115**, 9667 (1993). A Two-Dimensional Energy Surface for a Type II S_N2 Reaction in Aqueous Solution.

236. J. Gao, *J. Am. Chem. Soc.*, **113**, 7796 (1991). A Priori Computation of a Solvent-Enhanced S_N2 Reaction Profile in Water: The Menshutkin Reaction.

237. M. Sola, A. Lledos, M. Duran, J. Bertran, and J. M. Abboud, *J. Am. Chem. Soc.*, **113**, 2873 (1991). Analysis of Solvent Effects on the Menshutkin Reaction.

238. N. S. Isaacs, *Physical Organic Chemistry*, Wiley, New York, 1987.

239. K. Okamoto, S. Fukui, I. Nitta, and H. Shingu, *Bull. Chem. Soc. Jpn.*, **40**, 2354 (1967). Kinetic Studies of Bimolecular Nucleophilic Substitution. VIII. The Effect of Hydroxylic Solvents on the Nucleophilicity of Aliphatic Amines in the Menschutkin Reaction.

240. D. S. Hartsough and K. M. Merz Jr., *J. Phys. Chem.*, **99**, 384 (1995). Potential of Mean Force Calculations on the S_N1 Fragmentation of tert-Butyl Chloride.

241. W. L. Jorgensen, J. K. Buckner, S. E. Huston, and P. J. Rossky, *J. Am. Chem. Soc.*, **109**, 1891 (1987). Hydration and Energetics for $(CH_3)_3CCl$ Ion Pair in Aqueous Solution.

242. N. Li and J. Gao, manuscript to be submitted. A Combined QM-CI/MM Potential for the S_N1 Ionic Dissociation of t-BuCl in Water.

243. M. H. Abraham, *Pure Appl. Chem.*, **57**, 1055 (1985). Solvent Effects on Reaction Rates.

244. D. S. Kemp and K. G. Paul, *J. Am. Chem. Soc.*, **97**, 7305 (1975). The Physical Organic Chemistry of Benzisoxazoles. III. The Mechanism and the Effects of Solvents on Rates of Decarboxylation of Benzisoxazole-3-Carboxylic Acids.

245. D. S. Kemp, D. D. Cox, and K. G. Paul, *J. Am. Chem. Soc.*, **97**, 7305 (1975). The Physical Organic Chemistry of Benzisoxazoles. IV. The Origins and Catalytic Nature of the Solvent Rate Acceleration for the Decarboxylation of 3-Carboxybenzisoxazoles.

246. C. Lewis, P. Paneth, M. H. O'Leary, and D. Hilvert, *J. Am. Chem. Soc.*, **115**, 1410 (1993). Carbon Kinetic Isotope Effects on the Spontaneous and Antibody-Catalyzed Decarboxylation of 5-Nitro-3-Carboxybenzisoxazole.

247. C. T. Lewis, T. Kramer, S. Robinson, and D. Hilvert, *Science*, **253**, 1019 (1991). Medium Effects in Antibody-Catalyzed Reactions.

248. J. W. Grate, R. A. McGill, and D. Hilvert, *J. Am. Chem. Soc.*, **115**, 8577 (1993). Analysis of Solvent Effects on the Decarboxylation of Benzisoxazole-3-Carboxylate Ions Using Linear Solvation Energy Relationships: Relevance to Catalysis in an Antibody Binding Site.

249. J. Gao, *J. Am. Chem. Soc.*, **117**, submitted (1995). An Automated Procedure for Simulating Chemical Reactions in Solution. Application to the Decarboxylation of 3-Carboxybenzisoxazole in Water.

250. H. Zipse and K. N. Houk, *J. Am. Chem. Soc.*, **117**, in press (1995). A Quantum Mechanical and Statistical Mechanical Exploration of the Thermal Decarboxylation of Kemp's Other Acid (Benzisoxazole-3-Carboxylic Acid). The Influence of Solvation on the Transition State Geometries and Kinetic Isotope Effects of a Reaction with a Gigantic Solvent Effect.

251. D. C. Ferris and R. S. Drago, *J. Am. Chem. Soc.*, **116**, 7509 (1994). Rate of Decarboxylation of Benzisoxazole-3-Carboxylate Ions as a Probe of Solvation in Biological and Other Media.

252. J. Gao, *Int. J. Quantum Chem., Quantum Chem. Symp.*, **27**, 491 (1993). Solvent Effect on the Potential Surface of the Proton Transfer in $[H_3N\text{-}H\text{-}NH_3]^+$.

253. J.-K. Hwang, Z. T. Chu, A. Yadav, and A. Warshel, *J. Phys. Chem.*, **95**, 8445 (1991). Simulations of Quantum Mechanical Corrections for Rate Constants of Hydride-Transfer Reactions in Enzymes and Solutions.

254. G. A. Voth, D. Chandler, and W. W. Miller, *J. Chem. Phys.*, **91**, 7749 (1989). Rigorous Formulation of Quantum Transition State Theory and Its Dynamical Corrections.

255. M. M. Kreevoy, D. Ostovic, D. G. Truhlar, and B. C. Garrett, *J. Phys. Chem.*, **90**, 3766 (1986). Phenomenological Manifestations of Large Curvature Tunneling in Hydride Transfer Reactions.

256. D. G. Truhlar, Y.-P. Liu, G. K. Schenter, and B. C. Garrett, *J. Phys. Chem.*, **98**, 8396 (1994). Tunneling in the Presence of a Bath: A Generalized Transition State Theory Approach.

257. L. Jaroszewski, B. Lesyng, and J. A. McCammon, *J. Mol. Struct. (THEOCHEM),* **102,** 57 (1990). Ab Initio Potential-Energy Functions for Proton-Transfer in $(H_3N\text{-}H\text{-}NH_3)^+$ and $(H_3N\text{-}H\text{-}OH_2)^+$. See also, S. Scheiner, in *Reviews in Computational Chemistry,* Vol. 2, K. B. Lipkowitz and D. B. Boyd, Eds., VCH Publishers, New York, 1991, pp. 165–218. Calculating the Properties of Hydrogen Bonds by Ab Initio Methods.

258. H. H. Jaffe and M. Orchin, *Theory and Application of UV Spectroscopy.* John Wiley & Sons, New York, 1962.

259. R. J. Bartlett and J. F. Stanton, in *Reviews in Computational Chemistry,* Vol. 5, K. B. Lipkowitz and D. B. Boyd, Eds., VCH Publishers, New York, 1994, pp. 65–169. Applications of Post-Hartree–Fock Methods: A Tutorial.

260. B. O. Roos, *Adv. Chem. Phys.,* **69,** 399 (1987). The Complete Active Space Self-Consistent Field Method and Its Application in Electronic Structure Calculations.

261. J. M. Bofill and P. Pulay, *J. Chem. Phys.,* **90,** 3637 (1989). The Unrestricted Natural Orbital-Complete Active Space (UNO-CAS) Method: An Inexpensive Alternative to the Complete Active Space-Self-Consistent-Field (CAS-SCF) Method.

262. H. Nakatsuji, *J. Chem. Phys.,* **94,** 6716 (1991). Exponentially Generated Configuration Interaction Theory. Descriptions of Excited, Ionized, and Electron Attached States.

263. J. Ridley and M. C. Zerner, *Theor. Chim. Acta,* **32,** 111 (1973). An Intermediate Neglect of Differential Overlap Technique for Spectroscopy: Pyrrole and the Azines.

264. N. S. Bayliss, *J. Chem. Phys.,* **18,** 292 (1950). The Effect of the Electrostatic Polarization of the Solvent on Electronic Absorption Spectra in Solution.

265. S. Basu, *Adv. Quantum Chem.,* **1,** 145 (1964). Theory of Solvent Effects on Molecular Electronic Spectra.

266. Y. Dimitrova and S. D. Peyerimhoff, *J. Phys. Chem.,* **97,** 12731 (1993). Theoretical Study of Hydrogen-Bonded Formaldehyde–Water Complexes.

267. M. M. Karelson and M. C. Zerner, *J. Phys. Chem.,* **96,** 6949 (1992). Theoretical Treatment of Solvent Effects on Electronic Spectroscopy.

268. M. Szafran, M. M. Karelson, A. R. Katrizky, J. Koput, and M. C. Zerner, *J. Comput. Chem.,* **14,** 371 (1993). Reconsideration of Solvent Effects Calculated by Semiempirical Quantum Chemical Methods.

269. T. Fox and N. Rösch, *J. Mol. Struct. (THEOCHEM),* **276,** 279 (1992). On the Cavity Model for Solvent Shifts of Excited States—A Critical Appraisal.

270. M. F. Herman and B. J. Berne, *J. Chem. Phys.,* **78,** 4103 (1983). Monte Carlo Simulation of Solvent Effects on Vibrational and Electronic Spectra.

271. R. M. Levy, J. D. Westbrook, D. B. Kitchen, and K. Krogh-Jespersen, *J. Phys. Chem.,* **95,** 6756 (1991). Solvent Effects on the Adiabatic Free Energy Difference between the Ground and Excited States of Methylindole in Water.

272. J. T. Blair, K. Krogh-Jespersen, and R. M. Levy, *J. Am. Chem. Soc.,* **111,** 6948 (1989). Solvent Effects on Optical Absorption Spectra: The $^1A_1 \rightarrow {}^1A_2$ Transition of Formaldehyde in Water.

273. R. M. Levy, D. B. Kitchen, J. T. Blair, and K. Krogh-Jespersen, *J. Phys. Chem.,* **94,** 4470 (1990). Molecular Dynamics Simulation of Time-Resolved Fluorescence and Nonequilibrium Solvation of Formaldehyde in Water.

274. S. E. DeBolt and P. A. Kollman, *J. Am. Chem. Soc.,* **112,** 7515 (1990). A Theoretical Examination of Solvatochromism and Solute–Solvent Structuring in Simple Alkyl Carbonyl Compounds. Simulations Using Statistical Mechanical Free Energy Perturbation Methods.

275. D. R. Armstrong, R. Fortune, P. G. Perkins, and J. J. P. Stewart, *J. Chem. Soc. Faraday Trans. 2,* **68,** 1839 (1972). Molecular Orbital Theory for the Excited States of Transition Metal Complexes.

276. M. McCourt and J. W. McIver, Jr., *J. Comput. Chem.,* **8,** 454 (1987). On the SCF Calculation of Excited States: Singlet States in the Two-Electron Problem.

277. T. Fox and N. Rösch, *Chem. Phys. Lett.*, **191**, 33 (1992). The Calculation of Solvatochromic Shifts: The n → π* Transition of Acetone.
278. N. Rösch and M. C. Zerner, *J. Phys. Chem.*, **98**, 5817 (1994). On the Calculation of Dispersion Energy Shifts in Molecular Electronic Spectra.
279. J. Åqvist, M. Fothergill, and A. Warshel, *J. Am. Chem. Soc.*, **115**, 631 (1993). Computer Simulation of the CO_2/HCO_3^- Interconversion Step in Human Carbonic Anhydrase I.
280. G. Alagona, P. Desmeules, C. Ghio, and P. A. Kollman, *J. Am. Chem. Soc.*, **106**, 3623 (1984). Quantum Mechanical and Molecular Mechanical Studies on a Model for the Dihydroxyacetone Phosphate-Glyceraldehyde Phosphate Isomerization Catalyzed by Triosephosphate Isomerase (TIM).
281. B. Waszkowycz, I. H. Hillier, N. Gensmantel, and D. W. Payling, *J. Chem. Soc., Perkin Trans. 2*, 225 (1991). A Combined Quantum Mechanical/Molecular Mechanical Model of the Potential Energy Surface of Ester Hydrolysis by the Enzyme Phospholipase A_2.
282. B. Waszkowycz, I. H. Hillier, N. Gensmantel, and D. W. Payling, *J. Chem. Soc., Perkin Trans. 2*, 1819 (1991). A Combined Quantum Mechanical/Molecular Mechanical Model of Inhibition of the Enzyme Phospholipase A_2.
283. B. Waszkowycz, I. H. Hillier, N. Gensmantel, and D. W. Payling, *J. Chem. Soc., Perkin Trans. 2*, 2025 (1991). Combined Quantum Mechanical–Molecular Mechanical Study of Catalysis by the Enzyme Phospholipase A_2: An Investigation of the Potential Energy Surface for Amide Hydrolysis.
284. P. A. Bash, M. J. Field, R. C. Davenport, G. A. Petsko, D. Ringe, and M. Karplus, *Biochemistry*, **30**, 5826 (1991). Computer Simulation and Analysis of the Reaction Pathway of Triosephosphate Isomerase.
285. M. Karplus, J. D. Evenseck, D. Joseph, P. A. Bash, and M. J. Field, *Faraday Discussions*, **93**, 239 (1992). Simulation Analysis of Triose Phosphate Isomerase—Conformational Transition and Catalysis.
286. V. V. Vasilyev, *J. Mol. Struct. (THEOCHEM)*, **304**, 129 (1994). Tetrahedral Intermediate Formation in the Acylation Step of Acetylcholinesterase. A Combined Quantum Chemical and Molecular Mechanical Model.
287. C. S. Carmer, B. Weiner, and M. Frenklach, *J. Chem. Phys.*, **99**, 1356 (1993). Molecular Dynamics with Combined Quantum and Empirical Potentials: C_2H_2 Adsorption on Si(100).
288. S. Skokov, C. S. Carmer, B. Weiner, and M. Frenklach, *Phys. Rev. B*, **49**, 5662 (1994). Reconstruction of (100) Diamond Surfaces Using Molecular Dynamics with Combined Quantum and Empirical Forces.
289. D. Ceperley and B. Alder, *Science*, **231**, 555 (1986). Quantum Monte Carlo.
290. J. Doll and D. L. Freeman, *IEEE Comput. Sci. Eng.*, **1**, 22 (1994). Monte Carlo Methods in Chemistry.
291. J. E. Mertz, D. J. Tobias, C. L. Brooks III, and U. C. Singh, *J. Comput. Chem.*, **12**, 1270 (1991). Vector and Parallel Algorithms for the Molecular Dynamics Simulation of Macromolecules on Shared-Memory Computers.
292. S. L. Lin, J. Mellor-Crummey, B. M. Pettitt, and G. N. Phillips, Jr., *J. Comput. Chem.*, **13**, 1022 (1992). Molecular Dynamics on a Distributed-Memory Multiprocessor.
293. S. E. DeBolt, D. A. Pearlman, and P. A. Kollman, *J. Comput. Chem.*, **15**, 351 (1994). Free Energy Perturbation Calculations on Parallel Computers: Demonstrations of Scalable Linear Speedup.
294. W. Eggers, R. Rico, J. W. McIver, Jr., and P. J. Eberlein, *Proc. 4th Conference on Hypercube Multiprocessors*, Monterey, CA, 1999. MNDO on the iPSC/2.
295. M. W. Feyereisen, R. A. Kendall, J. Nichols, D. Dame, and J. T. Golab, *J. Comput. Chem.*, **14**, 818 (1993). Implementation of the Direct SCF and RPA Methods on Loosely Coupled Networks of Workstations.

296. M. E. Colvin, C. L. Janssen, R. A. Whiteside, and C. H. Tong, *Theor. Chim. Acta,* **84,** 301 (1993). Parallel Direct SCF for Large Scale Calculations.

297. T. R. Furlani and H. F. King, *J. Comput. Chem.,* **16,** 91 (1995). Implementation of a Parallel Direct SCF Algorithm on Distributed Memory Computers.

298. R. A. Kendall, R. A. Harrison, R. J. Littlefield, and M. F. Guest, in *Reviews in Computational Chemistry,* Vol. 6, K. B. Lipkowitz and D. B. Boyd, Eds., VCH Publishers, New York, 1995, pp. 209–316. High Performance Computing in Computational Chemistry: Methods and Machines.

299. A number of publications have appeared making use of combined QM/MM potentials after the writing of this chapter was completed. Thompson described a molecular dynamics (MD) study of 18-crown-6 in water and in carbon tetrachloride.[300] The AM1 method was used to describe the crown ether with a modification by including a molecular mechanics torsional term to better describe the energetics of OCCO and COCO dihedrals. The use of MM terms to modify specific properties that are poorly described by semiempirical methods is of interest and should be further investigated. Thompson and Schenter extended their QM/MM program to include MM polarizations, and the method was presented in a study of the excited states of the bacteriochlorophyll b dimer of *R. Viridis.*[301] Liu and Shi employed a combined AM1/GROMOS MD simulation method to study the nucleophilic addition reaction between hydroxide ion and formaldehyde in water.[302] These authors found that the reaction path in solution is different from the gas phase minimum energy path, thereby demonstrating that the phase reaction path may not always be utilized to model condensed phase processes.[302] Liu, Muller-Plathe, and van Gunsteren reported an MD study of the solvation effects on the conformational equilibrium of dimethoxyethane in water.[303] In this study the organic molecule was modeled by the MNDO method along with the SPC/E potential for water. The computed gauche stabilization of −3.7 kcal/mol relative to trans is in accord with experimental estimates. An interesting observation was made that, although the MNDO-type semiempirical methods may not yield accurate energetics for a system in the gas phase, the interaction with its environment may still be appropriately described by the QM/MM potential,[303] a point that has been emphasized in this chapter. Another application of the combined QM/MM potential was described by Elcock et al. who studied the DNA cross-linking by nitrous acid.[304] The CHARMM program was used, combining the AM1 Hamiltonian and the CHARMM 23 force field.

300. M. A. Thompson, *J. Phys. Chem.,* **99,** 4794 (1995). Hybrid Quantum Mechanical-Molecular Mechanical Force Field Development for Large Flexible Molecules: A Molecular Dynamics Study of 18-Crown-6.

301. M. A. Thompson and G. K. Schenter, *J. Phys. Chem.,* **99,** 6374 (1995). Excited States of the Bacteriochlorophyll b Dimer of *Rhodopseudomonas viridis:* A QM/MM Study of the Photosynthetic Reaction Center that Includes MM Polarization.

302. H. Liu and Y. Shi, *J. Comput. Chem.,* **15,** 1311 (1994). Combined Molecular Mechanical and Quantum Mechanical Potential Study of a Nucleophilic Addition Reaction in Solution.

303. H. Liu, F. Muller-Plathe, and W. F. van Gunsteren, *J. Chem. Phys.,* **102,** 1702 (1995). A Molecular Dynamics Simulation Study with a Combined Quantum Mechanical and Molecular Mechanical Potential Energy Function: Solvation Effects of the Conformational Equilibrium of Dimethoxyethane.

304. A. H. Elcock, P. D. Lyne, A. J. Mulholland, A. Nandra, and W. G. Richards, *J. Am. Chem. Soc.,* **117,** 4706 (1995). Combined Quantum and Molecular Mechanical Study of DNA Cross-Linking by Nitrous Acid.

CHAPTER 4

An Introduction to Density Functional Theory

Libero J. Bartolotti and Ken Flurchick

Research Institute, North Carolina Supercomputing Center, MCNC, Research Triangle Park, North Carolina 27709

INTRODUCTION

For many years, wavefunction functional theory (Schrödinger theory) has been the method of choice when performing electronic structure calculations on chemical systems. However, within the last few years, density functional theory (DFT)[1–3] has experienced a rise in its popularity for calculating structures and properties of chemical systems.[4] The need to include electron-correlation in calculations on large chemical systems and the discovery of new and more accurate approximations to the exchange-correlation energy density functional has contributed to the increased popularity of DFT. Not only does DFT provide computational advantages over wavefunction functional theory (WFT),[4] but quantities of interest to chemists (such as electronegativity, hardness, softness, reactivity indices, etc.) are readily defined.[1]

Density functional theory is equivalent to solving Schrödinger's equation and is, therefore, an exact theory for describing the electronic structure and properties of matter. An important difference between DFT and WFT is the principal quantity of interest. In DFT this quantity is the electron density which, unlike the wavefunction, is an observable. As a result, quantities not readily accessible through WFT are conveniently defined. The computational

Reviews in Computational Chemistry, Volume 7
Kenny B. Lipkowitz and Donald B. Boyd, Editors
VCH Publishers, Inc. New York, © 1996

advantage of DFT originates with the fact that the electron density has three spatial coordinates, regardless of the number of electrons in the chemical system. Thus, DFT allows the calculation of structures and properties of molecules with a couple of hundred atoms, a feat not generally possible with high-level WFT methods.

Considerable advances in the formal development of DFT as well as in its practical application have been made these last few years. In this chapter, we will not attempt to provide a comprehensive review of the subject, but rather provide an introduction to DFT, covering many of the important attributes that pertain to chemical problems. In the next section, we provide some of the basic calculus associated with functionals. Then some early models used in DFT are presented, followed by a formal presentation of DFT and the formalism used in practical applications. Finally, in the last section, we discuss the ability of DFT to compete with WFT in electronic structure calculations. A more detailed and comprehensive discussion of DFT can be found in the 1989 book by Parr and Yang.[1] In what follows, we assume the Born–Oppenheimer nonrelativistic approximation and use atomic units throughout. Since the nuclei are assumed to be fixed in space, we will not explicitly show this dependence in equations and variables.

SOME BASIC CALCULUS

Simply speaking, a *functional* maps a function into a number. An example is the area under a curve, which is a functional of the function that defines the curve between two points. An arbitrary functional, which we call Ω, can be written as

$$\Omega = \int f(\mathbf{r}) d\tau \qquad [1]$$

where $f(\mathbf{r})$ is the integrand of the integral, $\mathbf{r}$ represents the three spatial coordinates, and $d\tau$ represents the volume element of integration. In general, the integrand $f(\mathbf{r})$ has a complex dependence upon one or more well-defined functions. For instance, suppose it is known that

$$f(\mathbf{r}) = \rho(\mathbf{r}) \qquad [2]$$

where $\rho(\mathbf{r})$ is a function of interest (the electron density in our case). Thus, Ω would be a functional of $\rho(\mathbf{r})$, i.e.,

$$\Omega[\rho] = \int \rho(\mathbf{r}) d\tau \qquad [3]$$

where the square brackets denote the functional dependence. Now consider

$$f(\mathbf{r}) = \frac{\nabla\rho(\mathbf{r})\cdot\nabla\rho(\mathbf{r})}{\rho(\mathbf{r})} \qquad [4]$$

where ∇ is the gradient operator and "$\cdot$" represents a vector dot product. Ω is still a functional of $\rho(\mathbf{r})$, i.e.,

$$\Omega[\rho] = \int \frac{\nabla\rho(\mathbf{r})\cdot\nabla\rho(\mathbf{r})}{\rho(\mathbf{r})}\, d\tau \qquad [5]$$

Another example is given by

$$f(\mathbf{r}) = \rho(\mathbf{r})\nu(\mathbf{r}) \qquad [6]$$

where ρ and ν are different functions. In this case, Ω is considered to be a functional of both functions, denoted as $\Omega[\rho,\nu]$. For simplicity, the above discussion is restricted to 3D functions. However, without any loss of generality, the integrand could be multidimensional, as is the case in WFT, which has three coordinates for each electron.

The concepts of local and nonlocal functionals represent some important terminology associated with functionals of the electron density. Functionals whose integrand contains gradients (or higher derivatives) of ρ, such as in Eq. [4], are called *nonlocal functionals* of ρ, whereas those functionals that have integrands which are only simple functions of ρ, such as in Eq. [2] above, are called *local functionals* of ρ. In DFT, the terms "gradient corrected functionals" and "nonlocal functionals" are interchangeable. The significance of these two kinds of functionals, local and nonlocal, will become apparent later. Other definitions of "local" and "nonlocal" can also apply. For example, in the context of some property of the system, the term "local" implies that the property has a spatial dependence. In WFT the term "nonlocal" is used to describe the Hartree–Fock exchange energy operator, but it does not have the same meaning as does "nonlocal" in DFT. Which definition of local or nonlocal applies will be apparent from the context in which it is used.

As in the theory of functions, a calculus exists for functionals.[5] This calculus provides the tools necessary to develop and implement density functional theory. We begin with the discussion of expansions of functionals, which plays an important role in developing models within DFT and in deriving perturbation expansions. Analogous to the Taylor Series expansion for a function, a functional can be expanded about a reference function. This expansion, called a Volterra expansion,[5,6] exists provided the functional has functional derivatives to any order and provided the last term in the infinite expansion has limit zero. Assuming these conditions, the Volterra expansion of $\Omega[\rho]$ about a reference function, ρ_0, is given by

$$\Omega[\rho] = \Omega[\rho_0] + \int \left(\frac{\delta\Omega[\rho]}{\delta\rho(\mathbf{r})} \right)_{\rho=\rho_0} \Delta\rho(\mathbf{r})\, d\tau$$

$$+ \frac{1}{2} \int\int \left(\frac{\delta^2\Omega[\rho]}{\delta\rho(\mathbf{r}_1)\delta\rho(\mathbf{r}_2)} \right)_{\rho=\rho_0} \Delta\rho(\mathbf{r}_1)\Delta\rho(\mathbf{r}_2) d\tau_1 d\tau_2 + \cdots$$

$$= \sum_{n=0}^{\infty} D^n_{\underbrace{\Delta\rho(\mathbf{r}_1) \cdots \Delta\rho(\mathbf{r}_n)}_{n}} \Omega[\rho] \quad [7]$$

where $\Delta\rho(\mathbf{r})$ measures the deviation of the reference function ρ_0 from ρ, i.e.,

$$\rho = \rho_0 + \Delta\rho \quad [8]$$

and

$$D^n_{\underbrace{\Delta\rho(\mathbf{r}_1) \cdots \Delta\rho(\mathbf{r}_n)}_{n}} \Omega[\rho]$$

$$= \frac{1}{n!} \underbrace{\int \cdots \int}_{n} \left(\frac{\delta^n\Omega[\rho]}{\underbrace{\delta\rho(\mathbf{r}_1) \cdots \delta\rho(\mathbf{r}_n)}_{n}} \right)_{\rho=\rho_0} \underbrace{\Delta\rho(\mathbf{r}_1) \cdots \Delta\rho(\mathbf{r}_n)}_{n} \underbrace{d\rho_1 \cdots d\tau_n}_{n} \quad [9]$$

The definition of the operator $D^n_{\underbrace{\Delta\rho(\mathbf{r}_1) \cdots \Delta\rho(\mathbf{r}_n)}_{n}}$ implies that for $n = 0$ the following relationship holds

$$D^0\Omega[\rho] = \Omega[\rho_0] \quad [10]$$

Here the notation $\left(\frac{\delta^n\Omega[\rho]}{\underbrace{\delta\rho(\mathbf{r}_1) \cdots \delta\rho(\mathbf{r}_n)}_{n}} \right)_{\rho=\rho_0}$ represents the nth functional derivative of the functional $\Omega[\rho]$ evaluated with the reference density. A perturbation expansion of Ω is obtained upon expanding $\Delta\rho$ as a power series in the perturbation parameter(s) λ, namely,

$$\Delta\rho = \sum_{i=1}^{\infty} \lambda^i \rho_i \quad [11]$$

where ρ_i is the "ith" correction to ρ. Substituting this expression into Eq. [7] leads to the desired perturbation expansion

$$\Omega[\rho] = \Omega[\rho_0] + \sum_{n=1}^{\infty} \underbrace{\sum_{i=1}^{\infty} \cdots \sum_{j=1}^{\infty}}_{n} \lambda^{\overbrace{i+ \cdots +j}^{n}} D^n_{\underbrace{\rho_i(\mathbf{r}_1) \cdots \rho_j(\mathbf{r}_n)}_{n}} \Omega[\rho] \quad [12]$$

Isolating like powers of λ gives the order-by-order corrections to $\Omega[\rho]$. When applying the above perturbation expansion, we assume that to any order of the perturbation expansion all lower order solutions are exactly known. This greatly simplifies the resulting order-by-order corrections to $\Omega[\rho]$. The above perturbation expansion is completely general and can also apply to wavefunction functional theory.

Functional derivatives are easily performed using the following expression:

$$\frac{\delta\Omega[\rho]}{\delta\rho(\mathbf{r})} = \frac{\partial f(\mathbf{r})}{\partial\rho(\mathbf{r})} - \nabla\cdot\left(\frac{\partial f(\mathbf{r})}{\partial\nabla\rho(\mathbf{r})}\right) + \nabla^2\left(\frac{\partial f(\mathbf{r})}{\partial\nabla^2\rho(\mathbf{r})}\right)\cdots \qquad [13]$$

Because this operational expression for the functional derivative relies on the integrand of Eq. [1], care must be exercised to properly define the volume element $d\tau$. Because the first functional derivative is not in the form of an integral, an interesting question surfaces. How is Eq. [13] used to calculate the higher functional derivatives associated with the expansion in Eq. [7]? We proceed by writing an arbitrary function ($\delta\Omega/\delta\rho$ in this case) as an integral using the delta function, namely,

$$\frac{\partial\Omega[\rho]}{\delta\rho(\mathbf{r}_1)} = \int \frac{\delta\Omega[\rho]}{\delta\rho(\mathbf{r}_2)}\,\delta(\mathbf{r}_1 - \mathbf{r}_2)d\tau_2 \qquad [14]$$

We now apply Eq. [13] to the right-hand side of Eq. [14] and thereby calculate the second functional derivative. This process is repeated again and again to calculate the functional derivative to any order. When applying this process to calculate functional derivatives, care must be taken to keep the delta function as part of the integrand. This ensures that the spatial derivatives in Eq. [13] also operate on the delta function. We see that the first functional derivative is a scalar function, but all higher derivatives are, in general, operators. Since the higher derivatives usually appear in an integral, the delta functions are removed by integration by parts. Functional derivatives are widely utilized in DFT (and also in WFT), and they are instrumental in deriving many of the expressions associated with chemical concepts.

A set of rules for derivatives of functionals exists, analogous to derivatives of functions. For an arbitrary small change $\delta\Omega$, we have

$$\delta\Omega = \int \frac{\delta\Omega}{\delta\rho(\mathbf{r})}\,\delta\rho(\mathbf{r})d\tau \qquad [15]$$

If $\rho(\mathbf{r}_1)$ at each point in space is a functional of $y(\mathbf{r}_2)$, we can write Eq. [15] as

$$\delta\Omega = \iint \frac{\delta\Omega}{\delta\rho(\mathbf{r}_1)}\frac{\delta\rho(\mathbf{r}_1)}{\delta y(\mathbf{r}_2)}\,\delta y(\mathbf{r}_2)d\tau_2 d\tau_1 \qquad [16]$$

which leads to the chain rule for functional derivatives

$$\frac{\delta \Omega}{\delta y(\mathbf{r}_2)} = \int \frac{\delta \Omega}{\delta \rho(\mathbf{r}_1)} \frac{\delta \rho(\mathbf{r}_1)}{\delta y(\mathbf{r}_2)} \, d\tau_1 \qquad [17]$$

Care must be exercised when looking at the inverse of a functional derivative. One cannot simply invert the functional derivative, but must ensure that the inverse satisfies the identity

$$\int \frac{\delta y(\mathbf{r}_1)}{\delta \rho(\mathbf{r}_2)} \frac{\delta \rho(\mathbf{r}_2)}{\delta y(\mathbf{r}_3)} \, d\tau_2 = \delta(\mathbf{r}_1 - \mathbf{r}_3) \qquad [18]$$

where

$$\left[\frac{\delta y(\mathbf{r}_3)}{\delta \rho(\mathbf{r}_2)} \right]^{-1} \equiv \frac{\delta \rho(\mathbf{r}_2)}{\delta y(\mathbf{r}_3)} \qquad [19]$$

is the inverse functional derivative (if it exists) of $\frac{\delta y(\mathbf{r}_1)}{\delta \rho(\mathbf{r}_2)}$. Inverses of functionals become important in the derivation of Maxwell relations between functionals and are useful in the formal development of DFT. Although we do not explicitly use them in this chapter, they have been used to derive some of the concepts discussed later, for instance, in the discussion of local hardness and local softness.

Scaling rules (when they exist) involving functionals play important roles in DFT. Functionals are considered homogeneous of degree n if the following condition holds:

$$\Omega[\lambda \rho(\mathbf{r})] = \lambda^n \, \Omega[\rho(\mathbf{r})] \qquad [20]$$

or equivalently

$$\Omega[\rho] = n \int \rho(\mathbf{r}) \frac{\delta \Omega[\rho]}{\delta \rho(\mathbf{r})} \, d\tau \qquad [21]$$

where λ is an arbitrary constant. Taking $\rho(\mathbf{r})$ to be the electron density and scaling the spatial coordinates of the electrons and nuclei by λ rather than scaling the function, we obtain a different kind of scaling rule. Any well-behaved functional of the electron density that obeys

$$\Omega[\rho_\lambda] = \lambda^n \, \Omega[\rho(\mathbf{r})] \qquad [22]$$

where

$$\rho_\lambda = \lambda^3 \, \rho(\lambda \mathbf{r}) \qquad [23]$$

is said to scale as λ^n. The λ^3 in Eq. [23] was introduced to preserve the normalization of ρ to the number of electrons.[1]

Likewise, we could have scaled the spatial coordinates of the electrons and nuclei in the wavefunction.[7,8] For example, the kinetic energy as a functional of the wavefunction is given by

$$T[\Psi] = -\frac{1}{2}\sum_{i=1}^{N}\int \Psi^*(\mathbf{r})\nabla^2\Psi(\mathbf{r})\, d\tau \qquad [24]$$

where $\mathbf{r}$ and $d\tau$ now refer to the $3N$ spatial coordinates of the wavefunction for the N electron system. Now the scaled $3N$-dimensional wavefunction, with normalization preserved,

$$\Psi_\lambda = \lambda^{3N/2}\Psi(\lambda\mathbf{r}) \qquad [25]$$

leads to

$$T[\Psi_\lambda] = \lambda^2 T[\Psi(\mathbf{r})] \qquad [26]$$

That is, the kinetic energy scales as λ^2. In a similar manner, it can be shown that the electron–electron repulsion energy, V_{ee}, as a functional of the wavefunction scales as λ, as does the nuclear–electron attraction energy, V_{ne}. In general, kinetic energy contributions scale as λ^2, whereas potential energy components scale as λ. Coordinate scaling can be used to prove the virial theorem in wavefuncton functional theory by minimizing the energy with respect to the scaling parameter.[7] Note that for T and V_{ee} it is only necessary to scale the coordinates of the electrons, because the associated operators do not explicitly depend upon the nuclear coordinates. However, the nuclear–electron attraction energy must have both the electronic and nuclear coordinates scaled in order for it to scale linearly in λ.

Finally, we discuss locating the extremals of functionals. Locating extremals is an important ingredient in both DFT and WFT. The associated equations provide a working set of expressions for discussing the respective theories, and the resulting solutions provide us with chemical and physical knowledge about the system. A necessary condition for finding an extremal is

$$\delta\Omega = 0 = \int \frac{\delta\Omega}{\delta\rho(\mathbf{r})}\,\delta\rho(\mathbf{r})\, d\tau + \text{“surface terms”} \qquad [27]$$

Both terms on the right-hand side of Eq. [27] identically vanish. The “surface terms” are to be evaluated on the surface of the system (at the boundary). An expression for the “surface terms” is not given, because their exact form varies from functional to functional. The vanishing of the first term in Eq. [27] gives the equation

$$\frac{\delta\Omega}{\delta\rho(\mathbf{r})} = 0 \qquad [28]$$

which provides the $\rho(\mathbf{r})$ that makes $\Omega[\rho]$ an extremal. The vanishing of the "surface terms" provides the natural boundary conditions that the solution to Eq. [28] must satisfy. Later, explicit examples will be given to elucidate the meaning of the "surface terms" (the natural boundary conditions). To classify whether the extremal is a minimum or a maximum, we need the second variation of the functional,

$$\delta^2\Omega = \frac{1}{2}\int\int \frac{\delta^2\Omega}{\delta\rho(\mathbf{r}_1)\delta\rho(\mathbf{r}_2)}\, d\tau_2 d\tau_1 \qquad [29]$$

As with extremals of functions, a necessary condition for a minimum is that Eq. [29] be positive, and for a maximum, Eq. [29] must be negative.

In many instances, it is necessary to find an extremal of a functional, subject to some constraint. These constraints can be introduced via the method of Lagrange multipliers, which can be global or have a local (pointwise) dependence. For instance, in WFT, one minimizes the energy of a molecular system, $\langle\Psi|\hat{H}|\Psi\rangle$, keeping the wavefunction, Ψ, normalized to unity, namely,

$$\delta\{\langle\Psi|\hat{H}|\Psi\rangle - E(\langle\Psi|\Psi\rangle - 1)\} = 0 \qquad [30]$$

where $\hat{H}$ is the Hamiltonian operator[7] and the Lagrange multiplier that ensures normalization of Ψ is the energy of the system. Assuming that Ψ vanishes at the boundary (the "surface terms" are neglected), Eqs. [27], [28], and [30] give the Schrödinger equation

$$\hat{H}\,\Psi = E\,\Psi \qquad [31]$$

As another example, suppose we wish to find the extremal of $T_s[\{\phi_k(\mathbf{r})\}]$, a functional of a set of N functions $\phi_k(\mathbf{r})$, subject to the constraints

$$\rho(\mathbf{r}) = \sum_{k=1}^{N} \phi_k(\mathbf{r})^2 \qquad [32]$$

and

$$1 = \int \phi_k^2\, d\tau \qquad \text{for all } k \qquad [33]$$

where N is a known constant and $\rho(\mathbf{r})$ is a known function. The $\phi_k(\mathbf{r})$ that minimize $T_s[\{\phi_k\}]$ are found from the condition set forth in Eq. [27], namely,

$$\delta\{T_s[\{\phi_k\}] - \int [\beta(\mathbf{r})(\rho(\mathbf{r}) - \sum_{i=1}^{N} \phi_k(\mathbf{r})^2]d\tau - \sum_{i=1} \epsilon_k(1 - \int \phi_k^2 d\tau)\} = 0 \quad [34]$$

Here $\beta(\mathbf{r})$ is a local Lagrange multiplier, and the ϵ_i values are global Lagrange multipliers that ensure that Eqs. [32] and [33], respectively, are satisfied at the solution point. The signs in front of the Lagrange multipliers in Eq. [34] are arbitrary and have been set negative for convenience. Assuming that the functions $\phi_k(\mathbf{r})$ vanish at the boundary, then the $\{\phi_k(\mathbf{r})\}$, which satisfy the constraints and make $T_s[\{\phi_k(\mathbf{r})\}]$ a minimum, are the solutions of the equation

$$\frac{\delta T_s[\{\phi_k\}]}{\delta\phi_k(\mathbf{r})} - 2\beta(\mathbf{r})\phi_k(\mathbf{r}) - 2\epsilon_k\phi_k(\mathbf{r}) = 0 \quad [35]$$

This equation, as well as Eq. [31], is called an Euler–Lagrange equation. Equation [28] for unconstrained variations is called an Euler equation.

EARLY MODELS

Models describing the energy as a functional of the electron density have a long history, dating back to the early days of quantum mechanics.[9–13] It was as appealing then as it is now to have a set of equations based upon the electron density as opposed to the wavefunction. These early developments proceeded in spite of the lack of a rigorous justification that such a set of equations existed. However, using the properties of a uniform electron gas, models were developed. These early models were not able to provide accurate electron densities for chemical systems, which is not surprising, since a uniform electron gas is a poor approximation to an atom or a molecule. However, these models laid the foundation for a theory based upon the electron density, and they were useful in predicting certain trends and behaviors in chemical systems.[1]

An early, well-known model was developed by Thomas and Fermi.[9–12] In the Thomas–Fermi (TF) model, the ground state energy of an atom is given by

$$E_{\mathrm{TF}}[\rho] = c_{\mathrm{TF}}\int \rho^{5/3}(\mathbf{r})\, d\tau + \frac{1}{2}\int\int \frac{\rho(\mathbf{r}_1)\rho(\mathbf{r}_2)}{r_{12}}\, d\tau_2 d\tau_1 + \int \rho v\, d\tau$$

$$= T_{\mathrm{TF}}[\rho] + J[\rho] + V_{\mathrm{ne}}[\rho] \quad [36]$$

where r_{12} is the distance between two particles, v is the nuclear–electron attraction potential, $v = -Z/r$ (Z being the charge on the nucleus), and

$c_{TF} = \frac{3}{10}(3\pi^2)^{2/3}$. The first term in the right-hand side of Eq. [36] approximates the kinetic energy of the system $T[\rho]$; the second term is the Coulomb energy $J[\rho]$ for the system, which is a component of, and used to approximate, the electron–electron repulsion energy $V_{ee}[\rho]$; and the last term is the nuclear–electron attraction energy $V_{ne}[\rho]$. In general, the energy and V_{ne} are also functionals of the potential ν. Although we do not explicitly show this dependence, it is important to realize that it exists.

The density that makes Eq. [36] a minimum must conserve the number of particles, namely,

$$N = \int \rho \, d\tau \qquad [37]$$

where N is the number of electrons in the atom. Thus, the minimization principle that determines the TF density is

$$\delta\left\{E_{TF}[\rho] - \mu\left(\int \rho d\tau - N\right)\right\} = 0 \qquad [38]$$

where μ is the Lagrange multiplier introduced to preserve the number of electrons. In the literature, one may see Eq. [38] in a slightly different, but equivalent, form. If we define a new functional

$$L[\rho] = E_{TF}[\phi] - \mu \int \rho \, d\tau \qquad [39]$$

then it is apparent that $\delta L[\rho] = 0$ leads to the same Euler–Lagrange equation as found from Eq. [38], namely

$$\frac{\delta E_{TF}[\rho]}{\delta \rho} = \mu \qquad [40]$$

The Lagrange multiplier μ associated with particle conservation, here and elsewhere, has a physical meaning: it is the *chemical potential* of the system. After performing the functional derivative, Eq. [40] becomes

$$\frac{5}{3} c_{TF} \, \rho^{2/3}(\mathbf{r}) + \nu^* = \mu \qquad [41]$$

where

$$\nu^* = \frac{\delta J[\rho]}{\delta \rho} + \frac{\delta V_{ne}[\rho]}{\delta \rho} = \int \frac{\rho(\mathbf{r}_2)}{r_{12}} \, d\tau_2 - \frac{Z}{r} \qquad [42]$$

is the classical *electrostatic potential* for the system. The natural boundary condition satisfied on the surface of the atom is

$$c_{TF}\,\rho^{5/3}(R) + \rho(R)\nu^{*}(R) = 0 \qquad [43]$$

where R denotes the boundary (size) of the atom. Here, R is the radius of the sphere that encompasses the TF atom, the value of which will be determined later. Equation [41] is an integral equation

$$\rho(\mathbf{r}) = \left[\frac{3}{5c_{TF}}(\mu - \nu^{*}(\mathbf{r}))\right]^{3/2} \qquad [44]$$

whose solution is the density of the atom. An analytic solution of this equation cannot be determined, and rather than seek approximate solutions of the integral equation, Eq. [41] is usually transformed into a differential equation through the *Poisson equation*

$$\nabla^2\nu^{*}(\mathbf{r}) = 4\pi\rho(\mathbf{r}) \qquad [45]$$

This expression can be conveniently solved using numerical methods.[11,12]

Without actually solving the Euler–Lagrange equation, some characteristics of a TF atom can be determined. First, we multiply Eq. [41] by $\rho(\mathbf{r})$ and then evaluate the resulting expression on the boundary of the atom

$$\frac{5}{3}c_{TF}\,\rho^{5/3}(R) + \rho(R)\nu^{*}(R) = \rho(R)\mu \qquad [46]$$

Subtracting this equation from Eq. [43] gives the value of the density on the surface of the atom, namely,

$$\frac{2}{3}c_{TF}\,\rho^{5/3}(R) = 0 \qquad [47]$$

which shows that the TF density vanishes at the boundary of the atom. For a neutral system there is no net charge on the surface of the atom, which means that $\nu^{*}(R) = 0$. Thus from Eq. [41] we find that the chemical potential for a neutral atom is zero,

$$\mu = \frac{5}{3}c_{TF}\,\rho^{2/3}(R) = 0 \qquad [48]$$

Normalization of the density gives the size of the TF atom, which is of infinite extent ($R = \infty$). For a nonneutral atom, it can be shown that the chemical potential is $-(Z - N)/R$, and the extent of the atom is finite.[11,12] Thus, some important properties of the TF atom have been determined without explicitly solving for the electron density.

This is a very simple model, and therefore it does have some shortcomings when used to describe chemical systems. The TF density has the undesirable characteristic that it does not go to zero exponentially in the limit $r \to \infty$. Furthermore, as $r \to 0$, the TF density becomes infinite. Not only are the calculated energy components poor approximations to their true counterparts, but the nonbinding theorem proves that stable molecules (relative to their individual atoms) are not possible within the TF model.[14–17]

The scaling properties of this simple model are easily determined. The TF kinetic energy is homogeneous of degree 5/3 in the density, whereas its coordinates scale as λ^2. The Coulomb energy density functional is homogeneous of degree 2, whereas the nuclear–electron attraction energy is homogeneous of degree 1. For both, the coordinates scale linearly in λ. Interesting energy relationships can be obtained from this information. Using the viral theorem and the expression that results from multiplying Eq. [41] by $\rho(\mathbf{r})$ and integrating yields the *Fraga relations*.[1,18] One such relationship for atoms is

$$V_{ee}[\rho] = -\frac{1}{3} E_{TF}[\rho] + N\mu \qquad [49]$$

In the TF model, μ vanishes and Eq. [49] is simplified. Other relationships involving ratios of the energy components can also be derived. Such expressions provide a convenient test of the functional form of the TF model. When the TF values for the energy components are replaced by their accurate counterparts, the Fraga ratios are found to be satisfied[1] surprisingly well. The accuracy of such relationships, based on the TF model, originates in the fact that the TF model becomes exact in the limit of an infinite number of electrons,[16,17,19] a result only proved in the last 20 years. Thus there is considerable merit in studying this model in the context of molecular systems.

Dirac[13] made an improvement to the TF model by introducing a formula for the exchange energy of a uniform electron gas. The Dirac exchange formula is

$$K_D[\rho] = c_D \int \rho^{4/3}(\mathbf{r})\, d\tau \qquad [50]$$

where $c_D = -\frac{3}{4}(3/\pi)^{1/3}$. Adding this term to the TF model yields the Thomas–Fermi–Dirac (TFD) model,

$$\begin{aligned} E_{TFD}[\rho] &= E_{TF}[\rho] + K_D[\rho] \\ &= c_{TF} \int \rho^{5/3}(\mathbf{r})\, d\tau + c_D \int \rho^{4/3} d\tau + \frac{1}{2} \iint \frac{\rho(\mathbf{r}_1)\rho(\mathbf{r}_2)}{r_{12}}\, d\tau_2 d\tau_1 \\ &\quad + \int \rho(\mathbf{r})v(\mathbf{r})\, d\tau \end{aligned} \qquad [51]$$

Again, for simplicity, we have written the energy expression as applied to atoms. Finding the electron density that makes Eq. [51] a minimum and conserves the number of electrons, Eq. [37] leads to the Euler–Lagrange equation

$$\frac{5}{3} c_{\mathrm{TF}} \, \rho^{2/3}(\mathbf{r}) + \frac{4}{3} c_{\mathrm{D}} \, \rho^{1/3}(\mathbf{r}) + \nu^{*}(\mathbf{r}) = \mu \qquad [52]$$

with natural boundary conditions

$$c_{\mathrm{TF}} \, \rho^{5/3}(R) + c_{\mathrm{D}} \, \rho^{4/3}(R) + \rho(R)\nu^{*}(R) = \rho(R)\mu \qquad [53]$$

As with the TF model, the integral equation given in Eq. [52] is transformed into a differential equation through the Poisson equation. Properties of the TFD electron density can also be determined without actually solving Eq. [52], as was done with the TF model. Multiplying Eq. [52] by $\rho(\mathbf{r})$, evaluating the resulting expression on the boundary, and then subtracting from Eq. [53] yields

$$\rho(R) = \begin{cases} 0 \\ \left(\dfrac{-c_{\mathrm{D}}}{2\, c_{\mathrm{TF}}}\right)^{3} \end{cases} \qquad [54]$$

This shows that the density has two solutions on the surface of a TFD atom, which implies that the density is discontinuous at the boundary, taking the values $\rho(R) = (-c_{\mathrm{D}}/2c_{\mathrm{TF}})^3$ when $r = R$ and $\rho(\mathbf{r}) = 0$ for $r > R$. For neutral atoms, the TFD model predicts atoms of finite size with a chemical potential equal to $-c_{\mathrm{D}}^2/4c_{\mathrm{TF}}$. This value of the chemical potential is obtained by evaluating Eq. [52] at the boundary of the TFD atom.

Unfortunately, adding the Dirac exchange formula to the TF model does not improve the quality of the calculated electron density. The TFD density suffers from the same undesirable characteristics as does the TF density. A major enhancement of these two overly simplified models was made through addition of an inhomogeneity of the electron density correction to the kinetic energy density functional. This was first investigated by von Weizsacker,[20] who derived a correction that depends upon the gradient of the density, namely,

$$T_{\mathrm{W}}[\rho] = \frac{1}{2} \int \frac{\nabla\rho(\mathbf{r})\cdot\nabla\rho(\mathbf{r})}{\rho(\mathbf{r})} \, d\tau \qquad [55]$$

The addition of this term to the TF and TFD energy density functionals improves the behavior of the calculated electron density.[1] The density is no longer infinite at the nucleus, and the asymptotic behavior far from the nucleus is greatly improved. Another benefit from using this gradient correction is that molecules are now stable entities with respect to the separated atoms; the nonbinding theorems (whether or not atoms will bind together to form a molecule) that plagued the TF and TFD models are no longer applicable. An important point is hidden in Eq. [55] because of our selection of atomic units. Equation [55] explicitly depends upon Planck's constant, h, formally introducing some form of a quantum effect. Unfortunately, the Weizsacker correction did not provide a set of improved models that could compete with the Schrödinger equation. The calculated densities still did not demonstrate the periodicity found in the elements of the periodic table, nor were the calculated properties accurate. To generate improved kinetic energy density functionals, additional gradient corrections with terms up through sixth order have been derived.[1,21–24] However, as an approximation to the exact kinetic energy, these corrections fare poorly.

As with the kinetic energy, gradient corrections have been derived for the exchange energy density functional. Although formally instructive, these gradient expansions also did not prove useful in actual calculations. There is an inherent flaw in these gradient expansions (of both the kinetic and exchange energies) in that to some order they become divergent, as do their functional derivatives.

FORMAL DEVELOPMENT OF DENSITY FUNCTIONAL THEORY

In 1964 the perception of DFT was forever changed. That year, Hohenberg and Kohn[25] provided the long sought-after proof that DFT is in fact an exact theory for describing the electronic behavior of matter. This was accomplished by proving the following remarkable theorem:

> *There exists a variational principle in terms of the electron density which determines the ground state energy and electron density. Further, the ground state electron density determines the external potential, within an additive constant.*

This theorem means that the ground state electron density, as obtained from the Hohenberg–Kohn variational principle, uniquely determines the ground state properties of the system of interest. The electron density is obtained from the variational principle

$$\delta\left\{E[\rho,\nu] - \mu\left(\int \rho d\tau - N\right)\right\} = 0 \qquad [56]$$

Here $E[\rho,\nu]$ is the ground state energy, as a functional of both the electron density ρ and the external potential ν (the nuclear–electron attraction potential in many cases); μ (the chemical potential) is a Lagrange multiplier introduced to preserve the number of particles; and N is the number of electrons. For emphasis, we now explicitly denote the functional dependence of the energy on the external potential.

The solution of the resulting Euler–Lagrange equation

$$\left(\frac{\delta E[\rho,\nu]}{\delta \rho}\right)_{\nu} = \mu \qquad [57]$$

determines the ground state density and, therefore, all the ground state properties of the system. The subscript ν on the functional derivative implies that ν is kept fixed when performing the functional derivative with respect to ρ. Equation [57] is the DFT counterpart to the Schrödinger equation. The functional for the ground state energy can be partitioned

$$E[\rho,\nu] = F[\rho] + \int \rho\nu \, d\tau \qquad [58]$$

where $F[\rho]$ is a universal functional of ρ. This universal functional contains kinetic and potential energy contributions,

$$\begin{aligned} F[\rho] &= T[\rho] + V_{ee}[\rho] \\ &= T[\rho] + V_{xc}[\rho] + J[\rho] \\ &= G[\rho] + J[\rho] \end{aligned} \qquad [59]$$

where $T[\rho]$ is a universal functional for the kinetic energy, $V_{ee}[\rho]$ is a universal functional for the electron–electron repulsion energy, and $V_{xc}[\rho]$ is a universal functional of the exchange-correlation energy. Unfortunately, the exact forms of $T[\rho]$ and $V_{xc}[\rho]$ are not known. For this reason, Eqs. [56] and [57] are not used in practical applications of DFT. However, this formal development of DFT does provide important relationships, which lead to useful expressions for properties of chemical interest.

The chemical potential is actually the negative of the *electronegativity*[1,26,27]

$$-\chi = \mu = \left(\frac{\partial E}{\partial N}\right)_{\nu} \qquad [60]$$

The finite difference approximation for the electronegativity is given in terms of the ionization potential (I) and electron affinity (A), namely,

$$\chi = \frac{I + A}{2} \qquad [61]$$

a formula also derived by Mulliken[28] using very different arguments. The second derivative of the energy defines the global hardness[1,29,30]

$$\eta = \frac{1}{2}\left(\frac{\partial^2 E}{\partial N^2}\right)_\nu = \frac{1}{2}\left(\frac{\partial \mu}{\partial N}\right)_\nu \qquad [62]$$

Like χ, the hardness can also be approximated from a knowledge of I and A, namely,

$$\eta = (I - A)/2 \qquad [63]$$

Global softness is just the inverse of η. The quantities χ and η prove to be useful concepts for the discussion of chemical bonding. Together χ and η provide an expression for charge transfer.[1,29,30] For instance, consider the equilibration of two systems A and B. The fractional number of electrons transferred can be written as

$$\Delta N = \frac{\chi_A - \chi_B}{2(\eta_A + \eta_B)} \qquad [64]$$

Other such expressions are also derivable.[1,30,31]

The Gibbs–Duhem equation relating the change in chemical potential to the number of electrons is given by the formula[1,32]

$$N d\mu = \int \rho d\nu^* d\tau + \int dP\, d\tau + dX[\rho] \qquad [65]$$

where

$$X[\rho] = \int \rho \frac{\delta T[\rho]}{\delta \rho} d\tau - \frac{5}{3} T[\rho] - \int \rho \frac{\delta V_{xc}[\rho]}{\delta \rho} d\tau + \frac{4}{3} V_{xc}[\rho] \qquad [66]$$

and P is the pressure defined in terms of $\delta G[\rho]/\delta\rho$. In the limit of infinitely large N, the functional $X[\rho]$ vanishes, and we recover the familiar form of the Gibbs–Duhem equation. This suggests that the quantum effects are included in the mysterious functional $X[\rho]$.

An equivalent expression for changes in the chemical potential can be written[1]

$$d\mu = 2\eta dN + \int f(\mathbf{r}) d\nu(\mathbf{r})\, d\tau \qquad [67]$$

The quantity[1,33]

$$f(\mathbf{r}) = \left(\frac{\partial \rho}{\partial N}\right)_\nu = \left(\frac{\delta \mu}{\delta \nu(\mathbf{r})}\right)_N \qquad [68]$$

is called the *Fukui function*. This quantity measures the sensitivity of the system's chemical potential to an external perturbation. This is a local property of the system, that is, it is a function of the electron's position and generally differs from point to point in coordinate space. This quantity represents a chemical reactivity index in the sense of Fukui's frontier orbital theory.[34–37] For integer N, the derivative $(\partial\rho/\partial N)_v$ is not continuous. This means that the derivative taken from the left, $N - \delta N$, may differ from the derivative taken from the right, $N + \delta N$, as $\delta N \rightarrow 0$. This discontinuity has important implications. The left-hand derivative, $f_-(\mathbf{r})$, provides information about those sites in the molecule most susceptible to electrophilic attack. The right-hand derivative, $f_+(\mathbf{r})$, provides information about those sites most susceptible to nucleophilic attack, and their average provides information about homolytic reactions.

The concepts of local softness and hardness are also definable[1,30,38–40] in DFT. Local softness is defined

$$s(\mathbf{r}) = \left(\frac{\partial\rho(\mathbf{r})}{\partial\mu}\right)_v \qquad [69]$$

The local softness is related to the global softness

$$S = \int s(\mathbf{r})d\tau \qquad [70]$$

and to the Fukui function

$$s(\mathbf{r}) = f(\mathbf{r})S \qquad [71]$$

Local hardness can be defined as

$$2\eta(\mathbf{r}) = \frac{1}{N}\int\int \frac{\delta^2 F[\rho]}{\delta\rho(\mathbf{r})\delta\rho(\mathbf{r}')}\,\rho(\mathbf{r}')d\tau' \qquad [72]$$

and determines the global hardness

$$\eta = \int f(\mathbf{r})\eta(\mathbf{r})d\tau \qquad [73]$$

Inverse relationships relate softness to hardness, i.e.,

$$\eta S = 1 \qquad [74]$$

and

$$2\int \eta(\mathbf{r})s(\mathbf{r})d\tau = 1 \qquad [75]$$

The reactivity indices identify sites in a molecule where chemical bonds are likely to form. For reacting species, soft prefer soft and hard prefer hard; thus the above ideas can be used to formulate a theory of chemical reactivity. A noteworthy approach has been developed by Nalewajski.[41] Using these concepts, he formulated a systematic charge sensitivity analysis of molecular systems, which he has successfully applied to various problems in the theory of chemical reactivity.

This formal description of DFT provides formulas that are useful in understanding chemical phenomena, but it does not, as yet, provide a convenient method for determining the electron density of chemical systems. The problem results from our lack of knowledge of the exact form of both $T[\rho]$ and $V_{xc}[\rho]$, with (as it turns out) the uncertainty in $T[\rho]$ being most critical. Systematic improvements through gradient expansions have been tried but have not proved useful in finding approximate solutions to the density.[1] The failure of the gradient expansions is a consequence of their being based upon model systems that are far removed from actual atoms and molecules.

Finally, we briefly mention the important work of Levy.[42] One problem with the proof of Hohenberg and Kohn was that of ν-representability. Their proof applies only to an electron density that is ν-representable, that is, only to that density associated with the antisymmetric ground state wavefunction obtained from a Hamiltonian containing some external potential $\nu(\mathbf{r})$. Not all densities are ν-representable, but this does not present any problems. Levy has presented a formalism of DFT that eliminates the ν-representable constraint imposed on the density by the proof of Hohenberg–Kohn. He proposed a constrained-search approach based upon the bounding property of Schrödinger theory. Levy's prescription for obtaining the electron density has the further benefit of eliminating the requirement that only nondegenerate ground states can be considered.

ORBITAL-DENSITY FORMALISM

A practical description of DFT was given by Kohn and Sham,[43] just one year after the momentous Hohenberg–Kohn paper. The Kohn–Sham (KS) implementation is based upon an orbital-density description of DFT that removes the necessity of knowing the exact form of $T[\rho]$. They proposed focusing on the kinetic energy of a noninteracting system of electrons as a functional of a set of single-particle orbitals that give the exact density. Levy[42,44] has since presented a constrained search formulation of KS theory in which the kinetic energy of a noninteracting system of electrons is minimized with respect to a set of single-particle orbitals, subject to the constraint that the orbitals are orthonormal and that the sum of the squares of the orbitals gives the exact ground state density.

Adopting the constrained search procedure, a time-dependent formulation of KS theory is also possible.[1,45,46] Here we outline the hydrodynamic formulation of time-dependent KS theory.[47–49] For an N-electron system with a periodic time-dependent potential, we define

$$L[\{\phi_k(\mathbf{r},t),S_k(\mathbf{r},t)\}] = \sum_{k=1}^{N} \left\langle \phi_k(\mathbf{r},t) \middle| -\frac{1}{2}\nabla^2 + \frac{1}{2}(\nabla S_k(\mathbf{r},t)\cdot\nabla S_k(\mathbf{r},t)) + \partial S_k(\mathbf{r},t)/\partial t \middle| \phi_k(\mathbf{r},t) \right\rangle + J[\rho(\mathbf{r},t)] + E_{xc}[\rho(\mathbf{r},t)] + \int \rho(\mathbf{r},t)v(\mathbf{r})\, d\tau + \int \rho(\mathbf{r},t)\, H^{(1)}(\mathbf{r},t)\, d\tau \qquad [76]$$

where $v(\mathbf{r})$ is the time-independent contribution to the external potential, $E_{xc}[\rho(\mathbf{r},t)]$ is the unknown exchange-correlation energy density functional, $H^{(1)}(\mathbf{r},t)$ is the as yet undefined external time-dependent single particle potential, and

$$\rho(\mathbf{r},t) = \sum_{k=1}^{N} \phi_k(\mathbf{r},t)^2 \qquad [77]$$

is the time-dependent electron density. In the above formulation of DFT, we have restricted the discussion to integral occupancy of the orbitals ϕ_k.

The minimization of Eq. [76] is equivalent to performing the constrained search minimization,

$$T_s[\{\phi_k(\mathbf{r},t),S_k(\mathbf{r},t)\}] = \min_{\substack{\langle\phi_k|\phi_k\rangle=0 \\ \rho(\mathbf{r},t)=\sum_{k=1}^{N}\phi_k(\mathbf{r},t)^2 \\ \frac{\partial\phi_k^2}{\partial t}+\nabla\cdot(\phi_k^2\nabla S_k)=0}} \left\langle \phi_k(\mathbf{r},t) \middle| -\frac{1}{2}\nabla^2 + \frac{1}{2}(\nabla S_k(\mathbf{r},t)\cdot\nabla S_k(\mathbf{r},t)) \middle| \phi_k(\mathbf{r},t) \right\rangle \qquad [78]$$

where T_s is the kinetic energy density functional for a noninteracting system of electrons. The notation in the right-hand side of Eq. [78] means that we are to find the minimum of the expectation value with respect to independent variations of $\{\phi_k\}$ and $\{S_k\}$ subject to the three constraints that appear under min. The first constraint is a normalization condition imposed on the $\{\phi_k\}$; the second constraint ensures that the $\{\phi_k\}$ give the exact density; and the last constraint is an equation of continuity. The minimum expectation value is then assigned to T_s. This quantity is not quite equal to the total kinetic energy. The difference $T[\rho(\mathbf{r},t)] - T_s[\rho(\mathbf{r},t)]$ has been put into the functional $E_{xc}[\rho(\mathbf{r},t)]$. Thus, the exchange-correlation energy density functional contains both kinetic and potential contributions. This is an important point to remember.

The set of time-dependent KS amplitudes $\{\phi_k(\mathbf{r},t)\}$ and phases $\{S_k(\mathbf{r},t)\}$ are real single-particle functions of time and space. They are the solutions of the coupled nonlinear differential equations

$$\left\{-\frac{1}{2}\nabla^2 + \frac{1}{2}(\nabla S_k \cdot \nabla S_k) + \frac{\delta\{J[\rho] + E_{xc}[\rho]\}}{\delta\rho} + v + H^{(1)}\right\}\phi_k = -\frac{\partial S_k}{\partial t}\phi_k = \epsilon_k\,\phi_k \qquad [79]$$

and

$$\frac{\partial\phi_k^2}{\partial t} + \nabla\cdot(\phi_k^2\nabla S_k) = 0 \qquad [80]$$

The KS Lagrange multipliers $\epsilon_k(\mathbf{r},t) = -\partial S_k(\mathbf{r},t)/\partial t$ are functions of both time and spatial coordinates and not constants as in time-independent KS theory. These Euler–Lagrange equations are obtained from the stationary principle

$$\delta L[\{\phi_k, S_k\}] = 0 \qquad [81]$$

where the ϕ_k and the S_k are independently varied. This set of coupled nonlinear differential equations is consistent with the time-independent KS equation. In the static limit ($t \to -\infty$), ∇S_k vanishes and Eqs. [76]–[80] reduce to their (correct) time-independent counterparts,[43]

$$L\Big[\{\phi_k(\mathbf{r})\}\Big] = \sum_{k=1}^{N}\left\langle\phi_k(\mathbf{r})\left| -\frac{1}{2}\nabla^2 - \epsilon_k\right|\phi_k(\mathbf{r})\right\rangle + J[\rho(\mathbf{r})] + E_{xc}[\rho(\mathbf{r})] + \int\rho(\mathbf{r})v(\mathbf{r})\,d\tau \qquad [82]$$

and

$$\left\{-\frac{1}{2}\nabla^2 + \frac{\delta\{J[\rho] + E_{xc}[\rho]\}}{\delta\rho} + v\right\}\phi_k = \epsilon_k\phi_k \qquad [83]$$

where the time-independent component of $H^{(1)}$ has been put into v. The ϵ_k values are now constants and not functions, as in the time-dependent case.

Actually, the time-independent KS equations were derived earlier in Eqs. [34] and [35]. The above equations, which must be solved self-consistently, are reminiscent of Hartree and Hartree–Fock theory, but they are fundamentally different because they yield (and are defined in terms of) the exact ground state electron density. Payne[50] has used the constrained search procedure to develop a DFT for the Hartree–Fock model.

Since time-independent KS theory is commonly used in calculations of molecular structures and properties, we will focus on this form of the theory. We will briefly return to the time-dependent formalism later. The KS Lagrange multipliers in Eqs. [82] and [83] reflect the response of the total electronic energy to changes in occupation number (n_k),[51] i.e.,

$$\frac{\partial E}{\partial n_k} = \epsilon_k \qquad [84]$$

This implies that the ϵ_k values provide information about the electronegativity of the system. Indeed, χ can be obtained from ϵ_{HOMO} and ϵ_{LUMO}, where HOMO refers to the highest occupied KS orbital and LUMO refers to the lowest unoccupied KS orbital. Excitation energies are also obtained from the ϵ_k values. For instance, the first ionization energy is given by[1]

$$-\mathrm{I} = \mathrm{E}_N - \mathrm{E}_{N-1} = \int_0^1 \epsilon_{\mathrm{HOMO}}(n)\, dn \qquad [85]$$

and the first electron affinity by

$$-A = E_{N+1} - E_N = \int_0^1 \epsilon_{\mathrm{LUMO}}(n)\, dn \qquad [86]$$

Slater[52] has derived a fruitful numerical approximation to these expressions, namely,

$$-I \approx \epsilon_{\mathrm{HOMO}}\left(n = \frac{1}{2}\right) \qquad [87]$$

and

$$-A \approx \epsilon_{\mathrm{LUMO}}\left(n = \frac{1}{2}\right) \qquad [88]$$

That is, the first ionization energy is found by removing ½ an electron from the HOMO, and the electron affinity is found by placing ½ an electron in the LUMO. Alternatively, the electron affinity can be obtained directly from a knowledge of the ionization energy and electronegativity,[1,53]

$$A = 2\chi - I \qquad [89]$$

Of course, both I and A can be calculated from ground state energy differences, as given by Eqs. [85] and [86], respectively.

Because the KS orbitals give the total ground state electron density of the system, they also provide a practical method for calculating the local chemical reactivity properties discussed above. The Fukui reactivity indices, discussed earlier, may be obtained from the following equations[1,54]:

$$f_-(\mathbf{r}) = \phi_{\mathrm{HOMO}}(\mathbf{r})^2 + \sum_{i=1}^{N-1} \frac{\partial}{\partial N} \phi_k(\mathbf{r})^2 \qquad [90]$$

$$f_+(\mathbf{r}) = \phi_{\mathrm{LUMO}}(\mathbf{r})^2 + \sum_{i=1}^{N} \frac{\partial}{\partial N} \phi_k(\mathbf{r})^2 \qquad [91]$$

In many instances, the relaxation terms, $\frac{\partial}{\partial N} \phi_k^2$, can be ignored, and the frontier molecular orbital theory as first put forth by Fukui[34–37] is recovered. For particular molecules, the identification of reactive sites using $f(\mathbf{r})$ has been successful.[55,56] A convenient way to visualize the reactive sites in a molecule using the above reactivity index is to first display an isosurface of the electron density that just encloses the van der Waals volumes of the individual atoms in the molecule. Typically, the value of this isosurface is between 0.002 and 0.005.[57,58] Next, the values of the reactivity index are mapped upon this surface and color coded from blue (zero) to red (most positive).

In particular, suppose one is interested in those sites of a molecule most susceptible to electrophilic attack. The meaningful quantity is $f_-(\mathbf{r})$, which is calculated from Eq. [68] using finite differences,

$$f_-(\mathbf{r}) = \rho_N(\mathbf{r}) - \rho_{N-1}(\mathbf{r}) \qquad [92]$$

Here ρ_N is the electron density of the N electron system and ρ_{N-1} is the electron density of the same system with one electron removed. Alternatively, $f_-(\mathbf{r})$ may be approximated by ϕ_{HOMO}^2 as in Eq. [90] (provided the relaxation term can be neglected). The calculated $f_-(\mathbf{r})$ is mapped upon the isosurface of the density and color coded from blue to red. The portions of the surface colored red indicate those sites in the molecule most susceptible to attack by a soft electrophile.[58]

DISCUSSION AND PRACTICE

Although the KS formalism is, in principle, exact, its application is made inexact because the exchange-correlation energy as a functional of the electron density, $E_{xc}[\rho]$, is not known. Thus, the ability of the KS formalism to yield quantitative results for calculated structures and properties of molecular systems is directly connected to the accuracy of the approximation to $E_{xc}[\rho]$ used in Eqs. [82] and [83]. Fortunately, reasonable approximations to $E_{xc}[\rho]$ are known. The simplest approximation, due to Dirac[13] and later parameterized by Slater,[52] works surprisingly well. More elaborate approximations to $E_{xc}[\rho]$ have been proposed and have proven useful in the calculation of molecular properties. For many applications, local approximations to $E_{xc}[\rho]$ suffice. However, for those situations in which elongated bonds (transition states, hydrogen bonds, etc.) are involved, nonlocal (gradient) corrections must be utilized. The consequence of using an approximation to $E_{xc}[\rho]$ makes KS theory nonvariational with respect to the exact ground state energy. However, KS theory is variational with respect to the model system described by the approximate $E_{xc}[\rho]$.

The search for the exact exchange-correlation energy as a functional of the electron density is one of the most important endeavors in DFT. Although the exact form of $E_{xc}[\rho]$ is not known, Levy[59–62] employed coordinate scaling arguments to derive certain properties which the exact $E_{xc}[\rho]$ must satisfy. These scaling properties have been used to improve existing approximations to $E_{xc}[\rho]$[63,64] and should serve as a guide in developing new approximations to $E_{xc}[\rho]$. Unfortunately, an approximation to $E_{xc}[\rho]$ that can be systematically improved has not been formulated. Thus, knowledge about the uniform and inhomogeneous electron gas, as well as chemical insight, has directed the development of the functional form of many of the currently known approximations to $E_{xc}[\rho]$.

As in self-consistent wavefunction functional methods, the linear variation of constants method is commonly employed when solving the KS equations.[4] That is, the ϕ_k values are approximated by

$$\tilde{\phi}_k(\mathbf{r}) = \sum_{l=1}^{M} C_{k,l}\chi_{k,l}(\mathbf{r}) \qquad [93]$$

Employing this trial function in Eq. [82], the constants $C_{k,l}$ are determined from the stationary condition

$$\frac{\partial L[\{\tilde{\phi}_k\}]}{\partial C_{k,l}} = 0 \qquad [94]$$

That is, a variational solution is obtained. If a complete set of functions is used to represent the basis functions, $\chi_{k,l}$, then, in principle, the exact solution of the problem is found. However, to keep the problem tractable, a finite set of functions is employed in the expansion in Eq. [93], analogous to that found in WFT.

The basis functions have been represented as Slater-type orbitals,[65,66] plane waves,[67,68] Gaussian functions,[69,70] and numerical functions.[71,72] The advantages of Gaussian functions lie in their computational tractability and in the wealth of experience gained from years of usage in WFT. Numerical basis sets, on the other hand, take advantage of the fact that for a given $E_{xc}[\rho]$, the atomic problem can be solved exactly (at least to some given number of significant figures). These numerical solutions of the atomic KS orbitals are tables of numbers over a grid of points. Employing numerical basis sets when solving the KS equations has certain advantages. The effects of basis set superposition should be minimal, and the molecule will dissociate correctly to its atoms.[73,74] Also, it is relatively simple to generate new numerical basis functions.

The use of basis sets is not necessary when solving the molecular KS equations. Becke[75,76] has developed a basis-set-free numerical algorithm and has applied this method to structure calculations of many molecules. That is, he numerically solves the differential equation in Eq. [83], without having to rely on the variational method, and obtains the ϕ_k values as a table of numbers over a three-dimensional grid. This work has allowed the investigation of the error introduced when using various types of basis sets. Becke found that in general, basis set methods (using double zeta quality basis sets) possess a strong tendency to yield longer bond lengths than basis-set-free calculations. Improvement of the basis set (to triple zeta quality) often leads to shorter bond lengths that more closely agree with the basis-set-free calculations.

Detailed and systematic investigations that compare approximate KS calculations of molecular structures and properties to WFT and experiment have been performed.[77–84] These studies have employed various approximations to $E_{xc}[\rho]$ (both local and nonlocal), and although we cite only a few, they illustrate the suitability of approximate KS methods for describing molecular systems. We especially note the work of Pople and co-workers,[83] who did a comparison study of KS theory versus WFT for a subset of his G2 molecule data set.[85] It has been shown that approximate KS methods generally yield molecular structures in which the bond lengths are accurate to about 0.01 Å, the angles are accurate to within a couple of degrees, and the harmonic frequencies are accurate to within a few tens of cm^{-1}. In general, it has been observed that approximate KS methods, as described above, provide results that are comparable to or better than those obtained from a high-level WFT second-order Møller-Plesset (MP2) calculation.

There are advantages and disadvantages associated with both KS theory and WFT. We mention only a few and refer the reader to the works discussed above[73–84] and to reviews of DFT for a more comprehensive comparison.[1–4] In WFT we have an equation that we cannot (in general) solve, whereas in the KS formalism we can in principle solve the equation but we do not know its complete form. In going from WFT to KS theory, we have traded one problem for another, that is, replacing the difficulty of finding accurate solutions of the multidimensional Schrödinger equation with the search for accurate approximations to $E_{xc}[\rho]$. We believe that the latter holds the most promise.

Although not knowing the exact form of $E_{xc}[\rho]$ is a disadvantage in KS theory, we do not think that it is as problematic as the difficulty experienced in accurately solving Schrödinger's equation. In applications of approximate KS methods to problems of chemical interest, approximate forms of $E_{xc}[\rho]$ work rather well. As mentioned above, approximate KS calculations are comparable to the high-level MP2 method found in WFT. One area that needs more study and one in which the approximate implementation of KS theory may have difficulties is with predicting structures and properties of chemical systems that have bond lengths greater than normal, i.e., transition state structures and hydrogen bonds.

A major advantage of KS theory originates from the three-dimensional nature of the equations in KS theory. Thus, larger molecular systems can be investigated in the KS formalism than is currently possible in WFT (given the same computer resources). This follows from the scaling properties associated with KS theory and WFT. Scaling provides information about how the computer time required to perform a calculation increases with basis set size. Hartree–Fock methods scale formally as n^4, where n is the number of basis functions employed. Further, WFT methods that include electron correlation (such as MP2, MP4, and CI) scale as n^l, where $l \geq 5$. Hence, as one goes to higher levels of theory within WFT, the computational effort (cost) increases dramatically. In contrast, the KS formalism of DFT scales formally as n^3. Another advantage of KS theory lies in its ability to provide formal expressions for quantities of chemical interest, most of which are not easily defined in WFT. An example is the reactivity index mentioned earlier.

Over the last several years, several computer programs for performing approximate KS calculations have become available. In general, they offer the researcher several different approximations to $E_{xc}[\rho]$, both local and nonlocal. As mentioned earlier, the selection of a local approximation to $E_{xc}[\rho]$ should suffice for most calculations. However, for systems that contain unusually long interatomic interactions, such as transition states[86–88] and hydrogen bonds,[89,90] nonlocal corrections are necessary to attain useful results. Even then, however, one may find that KS theory does not provide useful information. Also, because the nonlocal approximations to $E_{xc}[\rho]$ contain derivatives to the electron density, the computational effort needed to do a calculation is increased relative to a calculation that uses only a local approximation to $E_{xc}[\rho]$.

Using a KS computer program is usually not much different from using a popular WFT computer program except for selecting the desired approximation to $E_{xc}[\rho]$. That is, a basis set must be selected, convergence criteria for the self-consistent field (SCF) and geometry optimizations must be selected, and the properties to be calculated must be selected. The available basis sets, as discussed earlier, are usually provided in the form of Slater-type functions, Gaussian functions, or numbers distributed over a grid. Generally, only one type of basis set is available in a given KS computer program. When available, we prefer using numerical basis sets, because these make it easy to generate an atomic basis set that is optimal for the selected $E_{xc}[\rho]$. Compared to WFT, smaller basis sets can be employed when doing KS calculations. A basis set of double zeta (with polarization) or triple zeta (with polarization) quality is generally sufficient.

Because there is not a long history behind most KS computer programs, unlike most WFT computer programs, KS computer programs may not be as robust as many of the WFT programs. A known problem area is with SCF convergence failures. This can often be corrected by changing the starting geometry, selecting an option (if available) that performs charge smearing at the Fermi level, invoking level shifting (if available), or using a different starting solution for the SCF. We find that the user must interact more frequently with KS computer programs than with WFT codes. This should change as the KS computer codes evolve.

Applications of time-dependent DFT to the calculation of polarizabilities have proved promising.[45,91,92] DFT has also been successful in deriving non-bonded interaction parameters for molecular force fields.[93] The attractive term in the Lennard–Jones or Lennard–Jones-like potential was obtained from the calculated long-range dipole–dipole dispersion term (R^{-6}; R is the internuclear separation). The repulsive term can be obtained by fitting the potential to experimental information about internuclear separations or some other size-dependent parameter such as the van der Waals radius.[93]

We have presented an introduction to DFT and have discussed some of its more important concepts. Density functional theory is an exciting field of study that has only recently become competitive with wavefunction functional methods. Many new methodological advances have been made in recent years. Of particular interest is the divide-and-conquer density functional approach developed by Yang.[94,95] He has developed a technique that directly calculates the electron density and that allows him to divide a molecular system into subsystems, the smallest possible subsystem being a single atom. This greatly simplifies the computational complexity of the problem. This, and similar theories[96,97] that scale linearly with the number of basis functions, have the potential of allowing quantum mechanical calculations on molecules containing thousands of atoms.

Present-day implementations of DFT have allowed applications not only to many inorganic systems,[1–4,38,71,74,78,81,87] but also to small biomolecules.[98] With methodological advances and requisite benchmarking and validations, even larger molecular systems can be modeled quantum mechanically in the future.

REFERENCES

1. R. G. Parr and W. Yang, *Density-Functional Theory of Atoms and Molecules*, Oxford University Press, New York, 1989.
2. R. M. Dreizler and E. K. U. Gross, *Density Functional Theory*, Springer, Berlin, 1990.
3. R. O. Jones and O. Gunnarsson, *Rev. Mod. Phys.*, **61**, 689 (1989). The Density Functional Formalism, Its Application and Prospects.

4. J. K. Labanowski and J. W. Andzelm, Eds., *Density Functional Methods in Chemistry,* Springer-Verlag, New York, 1991.
5. I. M. Gelfand and S. V. Fomin, *Calculus of Variations,* Prentice Hall, Englewood Cliffs, NJ, 1963.
6. V. Volterra, *Theory of Functionals,* Dover, New York, 1959.
7. I. Levine, *Quantum Chemistry,* Simon and Schuster, Englewood Cliffs, NJ, 1991.
8. L. Szasz, I. Berrios-Pagan, and G. McGinn, *Z. Naturforsch.,* **30a,** 1516 (1975). Density-Functional Formalism.
9. L. H. Thomas, *Proc. Cambridge Philos. Soc.,* **23,** 542 (1927). The Calculation of Atomic Fields.
10. E. Fermi, *Z. Physik.,* **48,** 73 (1928). A Statistical Method for the Determination of Some Atomic Properties and the Application of This Method to the Theory of the Periodic System of Elements.
11. P. Gombas, *Die statistichen Theorie des Atomes und ihre Anwendungen,* Springer-Verlag, Wein, 1949.
12. N. March, *Self-Consistent Fields in Atoms,* Pergamon, Oxford, 1975.
13. P. A. M. Dirac, *Proc. Cambridge Philos. Soc.,* **26,** 376 (1930). Note on Exchange Phenomena in the Thomas Atom.
14. E. Teller, *Rev. Mod. Phys.,* **34,** 627 (1962). On the Stability of Molecules in the Thomas-Fermi Theory.
15. N. L. Balazs, *Phys. Rev.* **156,** 42 (1967). Formation of Stable Molecules Within the Statistical Theory of Atoms.
16. E. H. Lieb and B. Simon, *Phys. Rev. Lett.,* **31,** 681 (1973). Thomas-Fermi Theory Revisited.
17. E. H. Lieb and B. Simon, *Adv. Math.,* **23,** 22 (1977). The Thomas-Fermi Theory of Atoms, Molecules and Solids.
18. S. Fraga, *Theoret. Chim. Acta,* **2,** 406 (1964). Non-Relativistic Self-Consistent Field Theory. II.
19. E. H. Lieb, *Rev. Mod. Phys.,* **48,** 553 (1976). Stability of Matter.
20. C. F. von Weizsacker, *Z. Physik,* **96,** 431 (1935). Zur Theorie der Kernmassen.
21. D. A. Kirzhnits, *Sov. Phys. JETP* **5,** 64 (1957). Quantum Corrections to the Thomas-Fermi Equation.
22. D. A. Kirzhnits, *Field Theoretical Methods in Many Body Systems,* Pergamon, Oxford, 1967.
23. C. H. Hodges, *Can. J. Phys.,* **51,** 1428 (1973). Quantum Corrections to the Thomas-Fermi Approximation—The Kirzhnits Method.
24. D. R. Murphy, *Phys. Rev. A,* **24,** 1682 (1981). The Sixth-Order Term of the Gradient Expansion of the Kinetic Energy Density Functional.
25. P. Hohenberg and W. Kohn, *Phys. Rev.,* **136,** B864 (1964). Inhomogeneous Electron Gas.
26. R. G. Parr, R. A. Donnelly, M. Levy, and W. E. Palke, *J. Chem. Phys.,* **68,** 3801 (1978). Electronegativity: The Density Functional Viewpoint.
27. K. D. Sen, Ed., *Structure and Bonding: Electronegativity,* Vol. 66, Springer-Verlag, New York, 1987.
28. R. S. Mulliken, *J. Chem. Phys.,* **2,** 782 (1934). A New Electroaffinity Scale: Together with Data on Valence States and Ionization Potentials and Electron Affinities.
29. R. G. Parr and R. G. Pearson, *J. Am. Chem. Soc.,* **105,** 7512 (1983). Absolute Hardness: Companion Parameter to Absolute Electronegativity.
30. K. D. Sen, Ed., *Structure and Bonding: Chemical Hardness,* Vol. 80, Springer-Verlag, New York, 1992.
31. A. Tachibana and R. G. Parr, *Int. J. Quantum Chem.,* **41,** 527 (1992). On the Redistribution of Electrons for Chemical Reaction Systems.
32. L. J. Bartolotti and R. G. Parr, *J. Chem. Phys.,* **72,** 1593 (1980). The Concept of Pressure in Density Functional Theory.

33. R. G. Parr and W. Yang, *J. Am. Chem. Soc.*, **106,** 4049 (1984). Density Functional Approach to the Frontier-Electron Theory of Chemical Reactivity.
34. K. Fukui, T. Yonezawa, and H. Shingu, *J. Chem. Phys.*, **20,** 722 (1952). A Molecular Orbital Theory of Reactivity in Aromatic Hydrocarbons.
35. K. Fukui, T. Yonezawa, C. Nagata, and H. Shingu, *J. Chem. Phys.*, **22,** 1433 (1954). Molecular Orbital Theory of Orientation in Aromatic Heteroaromatic, and Other Conjugated Molecules.
36. K. Fukui, *Theory of Orientation and Stereoselection,* Springer-Verlag, Berlin, 1975.
37. K. Fukui, *Science,* **218,** 747 (1987). Role of Frontier Orbitals in Chemical Reactions.
38. W. Yang and R. G. Parr, *Proc. Natl. Acad. Sci. U.S.A.*, **82,** 6723 (1985). Hardness, Softness, and the Fukui Functions in the Electronic Theory of Metals and Catalysis.
39. M. Berkowitz and R. G. Parr, *J. Chem. Phys.*, **88,** 2554 (1988). Molecular Hardness and Softness, Local Hardness and Softness, Hardness and Softness Kernels, and Relations among These Quantities.
40. M. K. Harbola, P. K. Chattaraj, and R. G. Parr, *Israel J. Chem.*, **31,** 395 (1991). Aspects of the Softness and Hardness Concepts of Density-Functional Theory.
41. R. Nalewajski, in *Structure and Bonding: Chemical Hardness,* Vol. 80, K. D. Sen, Ed., Springer-Verlag, New York, 1993. The Hardness Based Molecular Charge Sensitivities and Their Use in the Theory of Chemical Reactivity.
42. M. Levy, *Proc. Natl. Acad. Sci. U.S.A.*, **76,** 6062 (1979). Universal Variational Functionals of Electron Densities, First-Order Density Matrices, and Natural Spin-Orbitals and Solution of the v-Representability Problem.
43. W. Kohn and L. J. Sham, *Phys. Rev.*, **140,** A1133 (1965). Self-Consistent Equations Including Exchange and Correlation Effects.
44. M. Levy, *Phys. Rev. A,* **26,** 1200 (1982). Electron Densities in Search of Hamiltonians.
45. G. D. Mahan, *Local Density Theory of Polarizability,* Plenum Press, New York, 1990.
46. E. Runge and E. K. U. Gross, *Phys. Rev. Lett.*, **52,** 997 (1984). Density-Functional Theory for Time-Dependent Systems.
47. L. J. Bartolotti, *Phys. Rev. A,* **24,** 1661 (1981). Time-Dependent Extension of the Hohenberg–Kohn–Levy Energy-Density Functional.
48. L. J. Bartolotti, *Phys. Rev. A,* **26,** 2248 (1982). Erratum: *Phys. Rev. A,* **27,** 2248 (1983). Time-Dependent Kohn–Sham Density-Functional Theory.
49. B. M. Deb and S. K. Ghosh, *J. Chem. Phys.*, **77,** 342 (1982). Schrödinger Fluid Dynamics of Many-Electron Systems in a Time-Dependent-Density-Functional Framework.
50. P. W. Payne, *J. Chem. Phys.*, **71,** 490 (1979). Density Functionals in Unrestricted Hartree–Fock Theory.
51. J. F. Janak, *Phys. Rev. B,* **18,** 7165 (1978). Proof that $\partial E/\partial n_i = \epsilon_i$ in Density Functional Theory.
52. J. C. Slater, *The Self-Consistent Field for Molecules and Solids,* McGraw-Hill, New York, 1974.
53. L. J. Bartolotti, S. R. Gadre, and R. G. Parr, *J. Am. Chem. Soc.*, **102,** 2945 (1980). Electronegativities of the Elements from Simple $X\alpha$ Theory.
54. W. Yang, R. G. Parr, and R. Pucci, *J. Chem. Phys.*, **81,** 2862 (1984). Electron Density, Kohn–Sham Frontier Orbitals, and Fukui Functions.
55. C. Lee, W. Yang, and R. G. Parr, *THEOCHEM,* **163,** 305 (1987). Local Softness and Chemical Reactivity in the Molecules CO, SCN^- and H_2CO.
56. F. Méndez and M. Galván, in *Density Functional Methods in Chemistry,* J. K. Labanowski and J. W. Andzelm, Eds., Springer-Verlag, New York, 1991, pp. 387–400. Nucleophilic Attacks on Maleic Anhydride: A Density Functional Theory Approach.
57. J. Kendrick and M. Fox, *J. Mol. Graphics,* **9,** 182 (1991). Calculation and Display of Electrostatic Potentials.

58. K. Flurchick and L. Bartolotti, *J. Mol. Graphics,* **13,** 10 (1995). Visualizing Properties of Atomic and Molecular Systems.

59. M. Levy, in *Density Functional Methods in Chemistry,* J. K. Labanowski and J. W. Andzelm, Eds., Springer-Verlag, New York, 1991, pp. 175–193. Nonlocal Correlation Energy Functionals and Coupling Constant Integration.

60. M. Levy, *Phys. Rev. A,* **43,** 4637 (1991). Density-Functional Exchange Correlation through Coordinate Scaling in Adiabatic Connection and Correlation Hole.

61. H. Oui-Yang and M. Levy, *Phys. Rev. A,* **44,** 54 (1991). Theorem for Functional Derivatives in Density-Functional Theory.

62. A. Görling and M. Levy, *Phys. Rev. A,* **45,** 1509 (1992). Requirements for Correlation Energy Density Functionals from Coordinate Transformations.

63. A. Görling, M. Levy, and J. Perdew, *Phys. Rev. B (Condensed Matter),* **47,** 1167 (1993). Expectation Values in Density-Functional Theory, and Kinetic Contribution to the Exchange-Correlation Energy.

64. Q. Zhao and L. J. Bartolotti, *Phys. Rev. A,* **48,** 3983 (1993). Further Tests of a Scaled Local Exchange-Correlation Functional.

65. E. J. Baerends, D. E. Ellis, and P. Ross, *Chem. Phys.,* **2,** 41 (1973). Self-Consistent Molecular Hartree–Fock–Slater Calculations I. The Computational Procedure.

66. A. Rosen and D. E. Ellis, *J. Chem. Phys.,* **65,** 3629 (1976). Calculation of Molecular Ionization Energies Using a Self-Consistent-Charge Hartree–Fock–Slater Method.

67. R. Car and M. Parrinello, *Phys. Rev. Lett.,* **55,** 2471 (1985). Unified Approaches for Molecular Dynamics and Density-Functional Theory.

68. M. P. Teter, M. C. Payne, and D. C. Allan, *Phys. Rev. B,* **40,** 12255 (1989). Solution of Schrödinger's Equation for Large Systems.

69. H. Sambe and R. H. Felton, *J. Chem. Phys.,* **62,** 1122 (1975). A New Computational Approach to Slater's SCF-Xα Equation.

70. B. I. Dunlap, J. W. D. Connolly, and J. R. Sabin, *J. Chem. Phys.,* **71,** 3396 (1979). On Some Approximations in Applications of Xα Theory.

71. F. W. Averill and D. E. Ellis, *J. Chem. Phys.,* **59,** 6412 (1973). An Efficient Numerical Multicenter Basis Set for Molecular Orbital Calculations: Applications to $FeCl_4$.

72. B. Delley, *J. Chem. Phys.,* **92,** 508 (1990). An All-Electron Numerical Method for Solving the Local Density Functional for Polyatomic Molecules.

73. J. Andzelm and E. Wimmer, *J. Chem. Phys.,* **96,** 1280 (1992). Density Functional Gaussian-Type-Orbital Approach to Molecular Geometries, Vibrations, and Reaction Energies.

74. C. Sosa, J. Andzelm, B. C. Elkin, and E. Wimmer, *J. Phys. Chem.,* **96,** 6630 (1992). A Local Density Functional Study of the Structure and Vibrational Frequencies of Molecular Transition-Metal Compounds.

75. A. D. Becke, *Int. J. Quantum Chem., Quantum Chem. Symp.,* **23,** 599 (1989). Basis-Set-Free Density-Functional Quantum Chemistry.

76. R. M. Dickson and A. D. Becke, *J. Chem. Phys.,* **99,** 3898 (1993). Basis-Set-Free Local Density-Functional Calculations of Geometries of Polyatomic Molecules.

77. J. Andzelm and E. Wimmer, *J. Chem. Phys.,* **96,** 1280 (1992). Density Functional Gaussian-Type-Orbital Approach to Molecular Geometries, Vibrations, and Reaction Energies.

78. C. Sosa, J. Andzelm, B. C. Elkin, and E. Wimmer, *J. Phys. Chem.,* **96,** 6630 (1992). A Local Density Functional Study of the Structure and Vibrational Frequencies of Molecular Transition-Metal Compounds.

79. J. Guan, P. Duffy, J. T. Carter, D. P. Chong, K. C. Casida, M. E. Casida, and M. Wrinn, *J. Chem. Phys.,* **98,** 4753 (1993). Comparison of Local-Density and Hartree-Fock Calculations of Molecular Polarizabilities and Hyperpolarizabilities.

80. R. Fournier and A. E. DePristo, *J. Chem. Phys.,* **96,** 1183 (1992). Predicted Bond Energies in Peroxides and Disulfides by Density Functional Methods.

81. R. Fournier, S. B. Sinnott, and A. E. DePristo, *J. Chem. Phys.*, **97**, 4149 (1992). Density Functional Study of the Bonding in Small Silicon Clusters.
82. P. M. W. Gill, B. G. Johnson, and J. A. Pople, *Chem. Phys. Lett.*, **197**, 499 (1992). The Performance of the Becke-Lee-Yang-Parr (B-LYP) Density Functional Theory with Various Basis Sets.
83. B. G. Johnson, P. M. W. Gill, and J. A. Pople, *J. Chem. Phys.*, **98**, 5612 (1993). The Performance of a Family of Density Functional Methods.
84. G. J. Laming, N. C. Handy, and R. D. Amos, *Mol. Phys.*, **80**, 1121 (1993). Kohn–Sham Calculations and Open-Shell Diatomic Molecules.
85. L. A. Curtiss, K. Raghavachari, G. W. Trucks, and J. A. Pople, *J. Chem. Phys.*, **94**, 7221 (1991). Gaussian-2 Theory for Molecular Energies of First- and Second-Row Compounds.
86. L. Fan and T. Ziegler, *J. Chem. Phys.*, **92**, 3645 (1990). The Application of Density Functional Theory to the Optimization of Transition State Structures. I. Organic Migration Reactions.
87. C. Sosa and C. Lee, *J. Chem. Phys.*, **98**, 8004 (1993). Density Functional Description of Transition Structures Using Nonlocal Corrections. Silylene Insertion Reactions into the Hydrogen Molecule.
88. R. V. Stanton and K. M. Merz Jr., *J. Chem. Phys.*, **100**, 434 (1994). Density Functional Transition States of Organic and Organometallic Reactions.
89. F. Sim, A. St-Amant, I. Papai, and D. R. Salahub, *J. Am. Chem. Soc.*, **114**, 4391 (1992). Gaussian Density Functional Calculations on Hydrogen-Bonded Systems.
90. K. Laasonen, M. Parrinello, R. Car, C. Lee, and D. Vanderbilt, *Chem. Phys. Lett.*, **207**, 208 (1993). Structures of Small Water Clusters Using Gradient-Corrected Density Functional Theory.
91. L. J. Bartolotti and Q. Xie, *Theor. Chim. Acta*, **77**, 239 (1990). Dipole Cauchy Moments of the Atoms H through Ar: An Application of Kohn–Sham Theory with the Atomic Gradient Expansion of the Exchange-Correlation Energy Density Functional.
92. L. J. Bartolotti, Q. Xie, and L. Ortiz, *Int. J. Quantum Chem.*, **49**, 449 (1994). Quadrupole and Octupole Cauchy Moments of the Atoms H through Ar.
93. L. J. Bartolotti, L. G. Pedersen, and P. S. Charifson, *J. Comput. Chem.*, **12**, 1125 (1994). Long Range Nonbonded Attractive Constants for Some Charged Atoms.
94. W. Yang, *Phys. Rev. Lett.*, **66**, 1438 (1991). Direct Calculation of Electron Density in Density-Functional Theory.
95. W. Yang, *Phys. Rev. A*, **44**, 7823 (1991). Direct Calculation of Electron Density in Density-Functional Theory: Implementation for Benzene and a Tetrapeptide.
96. G. Galli and M. Parrinello, *Phys. Rev. Lett.*, **69**, 3547 (1992). Large Scale Electronic Structure Calculations.
97. S. Baroni and P. Giannozzi, *Europhys. Lett.*, **17**, 547 (1992). Towards Very Large-Scale Electronic-Structure Calculations.
98. A. St-Amant, this volume. Density Functional Methods in Biomolecular Modeling.

CHAPTER 5

Density Functional Methods in Biomolecular Modeling

Alain St-Amant

Department of Chemistry, University of Ottawa, 10 Marie Curie, Ottawa, Ontario, Canada K1N 6N5

INTRODUCTION

Density functional theory (DFT)[1,2] is based on the Hohenberg–Kohn theorem,[3] which states that the total energy of a system in its ground state is a functional of that system's electronic density, $\rho(\mathbf{r})$, and that any density, $\rho'(\mathbf{r})$, other than the true density will necessarily lead to a higher energy. The Hohenberg–Kohn theorem therefore introduces an alternative approach to performing exact, variational, ab initio electronic structure calculations. In conventional ab initio methodology, Schrödinger's equation

$$H\Psi = E\Psi \quad [1]$$

must be solved. Meanwhile, DFT requires only that we minimize the energy functional, $E[\rho(\mathbf{r})]$. The conceptual simplification thus offered cannot be overstated. Rather than working with a complex $3N$-dimensional wavefunction, Ψ, describing the behavior of each electron in an N-electron system, DFT allows us to work with a simple three-dimensional function, the total electronic density. Unfortunately, the exact nature of the energy functional is not known, and

Reviews in Computational Chemistry, Volume 7
Kenny B. Lipkowitz and Donald B. Boyd, Editors
VCH Publishers, Inc. New York, © 1996

we cannot simply output the total energy of a system when we are given a trial density, $\rho'(\mathbf{r})$, as input. As will be seen further on, we must therefore turn to approximate DFT methods, and though we will not have a wavefunction, we will have to make use of one-electron Kohn–Sham[4] molecular orbitals which rather closely resemble molecular orbitals from the well-known Hartree–Fock (HF) method.[5,6]

Early applications of DFT tended to be within the physics community, concentrating on systems where the HF approximation is a particularly poor starting point.[7] Therefore, the vast majority of the applications were on metallic systems, because a single determinantal approach is notoriously bad in such cases. Since DFT works with the density, and not the wavefunction, systems that would require a great number of electronic configurations to be well described by conventional ab initio approaches are in principle no harder, and no costlier, for DFT than systems that are well described by a single configuration. Correlation effects, absent within the HF approximation, are also built into the approximate energy functionals used in modern DFT applications. Therefore, DFT methods are in principle able to treat the entire periodic table with unvarying ease and accuracy.

Nevertheless, DFT applications on simple main group organic and inorganic molecules have only recently become popular.[2,8] Though DFT could yield results on systems with metallic centers that were as good as, or even better than, very expensive correlated post-HF techniques,[9] the state of DFT methods 10 years ago was such that it was not clear whether or not DFT methods offered any real advantage over even the familiar HF method for simple models of biomolecular systems. The approximate exchange and correlation (XC) energy functionals of the time would indeed yield results that were essentially between HF and MP2 in terms of quality[2,10–12] for simple organic and inorganic systems. However, for systems with hydrogen bonds, DFT results were invariably poor and could yield even qualitatively incorrect descriptions of such systems.[13] Such a glaring deficiency would necessarily have to be rectified before DFT could provide any meaningful insights into a large number of biochemical processes.

It must also be taken into account that DFT programs of the time did not have analytic energy gradients, making the search for stationary points on the potential energy surfaces of polyatomic molecules a daunting task.[14] It is therefore not surprising that biomolecular modelers of the time chose to stick with the conventional ab initio approaches and the "tried and true" wavefunction-based software packages such as the Gaussian series of programs.[6] However, with the advent of more accurate XC energy functionals,[15–19] the descriptions of weakly bound systems, including the all-important hydrogen-bonded systems, have now been significantly improved.[13,20,21] Also, the past decade has seen the development of several DFT software packages featuring both analytic gradients and second derivatives. It is therefore not surprising to see that biomolecular modelers are now turning to these DFT programs to study systems that previously may have been too large to treat by conventional ab initio techniques at the level of sophistication demanded of these types of calculations.

This chapter concentrates on the practical implementation of density functional methods in chemistry, leaving the review of the formal aspects of DFT to others.[1] The solution of the Kohn–Sham equations is our starting point. Along the way, the subtle variations between the various DFT codes currently available will be pointed out, as will the computational simplifications that make DFT calculations considerably less computationally demanding than any correlated post-HF method. When we consider actual applications, we will see that they have to date been primarily limited to simple systems that are of interest only as models of much more complex biomolecular systems. These applications will be reviewed, discussing both the strengths and the weaknesses of DFT. The review of the applications will be by no means comprehensive, but it is hoped that it will give the reader a better appreciation of what modern DFT can provide in terms of quality. These methods hold great promise for biomolecular applications, but as will be seen, there are instances where significant improvements can still be made to the fundamental theory.

DENSITY FUNCTIONAL THEORY

The Kohn–Sham Approach

Though the Hohenberg–Kohn theorem clearly established that one could, in principle, work directly with the density in ab initio calculations, it was the subsequent work of Kohn and Sham (KS)[4] that offered a practical approach to performing DFT calculations. In the KS approach, the unknown Hohenberg–Kohn energy functional, $E[\rho(\mathbf{r})]$, is partitioned in the following manner:

$$E[\rho(\mathbf{r})] = U[\rho(\mathbf{r})] + T[\rho(\mathbf{r})] + E_{xc}[\rho(\mathbf{r})] \qquad [2]$$

In this partitioning scheme, $U[\rho(\mathbf{r})]$ is simply the classical electrostatic energy, the sum of the electron–nucleus attractions and the electron–electron repulsions

$$U[\rho(r)] = \left(\sum_A \int \frac{-Z_A\rho(\mathbf{r})}{|\mathbf{r} - \mathbf{R}_A|}\, d\mathbf{r}\right) + \frac{1}{2}\iint \frac{\rho(\mathbf{r})\rho(\mathbf{r}')}{|\mathbf{r} - \mathbf{r}'|}\, d\mathbf{r}d\mathbf{r}' \qquad [3]$$

The next term, $T[\rho(\mathbf{r})]$, is defined as the kinetic energy of a system of noninteracting electrons with the same density, $\rho(\mathbf{r})$, as the real system of interacting electrons being studied. This may seem to be introducing a severe error. However, this is not the case, because the final term, $E_{xc}[\rho(\mathbf{r})]$, is made to contain, in addition to the exchange and correlation (XC) contributions to the energy, the difference between $T[\rho(\mathbf{r})]$ and the true electronic kinetic energy of the system.

Following Kohn and Sham, $\rho(\mathbf{r})$ of an N-electron system (with N^α spin up electrons and N^β spin down electrons) is expressed as the sum of the square moduli of singly occupied, orthonormal Kohn–Sham (KS) molecular orbitals,

$$\rho(\mathbf{r}) = \rho^\alpha(\mathbf{r}) + \rho^\beta(\mathbf{r}) = \sum_i^{N^\alpha} |\psi_i^\alpha(\mathbf{r})|^2 + \sum_i^{N^\beta} |\psi_i^\beta(\mathbf{r})|^2 \qquad [4]$$

Having done this, $T[\rho(\mathbf{r})]$ can now be defined as

$$T[\rho(\mathbf{r})] = \sum_{\sigma=\alpha,\beta} \sum_i^{N^\sigma} \int \psi_i^\sigma(\mathbf{r}) \frac{-\nabla^2}{2} \psi_i^\sigma(\mathbf{r})\, d\mathbf{r} \qquad [5]$$

One should note that $T[\rho(\mathbf{r})]$ is not a true density functional, because the KS orbitals are required. Alternate forms of $T[\rho(\mathbf{r})]$ that depend only on the electronic density and do not require KS orbitals have been proposed.[1] However, they are too imprecise to be of any practical use in chemistry.

Finally, recalling the fact that the energy functional is minimized by the true ground state density, $\rho(\mathbf{r})$, the energy functional $E[\rho(\mathbf{r})]$ must be stationary with respect to any arbitrary variation in either of the spin densities, i.e.,

$$\frac{\delta E[\rho(\mathbf{r})]}{\delta \rho^\alpha(\mathbf{r})} = \frac{\delta E[\rho(\mathbf{r})]}{\delta \rho^\beta(\mathbf{r})} = 0 \qquad [6]$$

This condition yields the one-electron KS equations,

$$\left\{ \frac{-\nabla^2}{2} - \left(\sum_A \frac{Z_A}{|\mathbf{r} - \mathbf{R_A}|} \right) + \int \frac{\rho(\mathbf{r}')}{|\mathbf{r} - \mathbf{r}'|}\, d\mathbf{r}' + \frac{\delta E_{xc}[\rho(\mathbf{r})]}{\delta \rho^\sigma(\mathbf{r})} \right\} \psi_i^\sigma(\mathbf{r}) = \epsilon_i \psi_i^\sigma(\mathbf{r}),$$

$$\sigma = \alpha,\beta \qquad [7]$$

A scheme for performing practical DFT calculations thus emerges. With an initial guess at the total spin densities, $\rho^\alpha(\mathbf{r})$ and $\rho^\beta(\mathbf{r})$, the KS equations are constructed and solved, and the resulting set of KS spin-orbitals, $\{\psi_i^\sigma(\mathbf{r})\}$, are then used to generate new guesses at $\rho^\alpha(\mathbf{r})$ and $\rho^\beta(\mathbf{r})$. This procedure is repeated until self-consistency is achieved so the same densities and KS orbitals are regenerated.

In the preceding discussion, we avoided dealing with the precise nature of the XC energy functional, $E_{xc}[\rho(\mathbf{r})]$, and the XC potentials, which are the functional derivatives of $E_{xc}[\rho(\mathbf{r})]$ with respect to $\rho^{\alpha}(\mathbf{r})$ and $\rho^{\beta}(\mathbf{r})$; $v_{xc}^{\alpha}(\mathbf{r})$ and $v_{xc}^{\beta}(\mathbf{r})$ are formally given by

$$v_{xc}^{\sigma}(\mathbf{r}) = \frac{\delta E_{xc}[\rho(\mathbf{r})]}{\delta \rho^{\sigma}(\mathbf{r})} \quad [8]$$

If the true XC energy functional, $E_{xc}[\rho(\mathbf{r})]$, were known, this scheme would yield the true ground state density, and in turn, exact values for all ground state properties. Unfortunately, the precise nature of $E_{xc}[\rho(\mathbf{r})]$ is not known, and at first glance, it may seem that we are no further along to performing practical DFT calculations then when we had only the Hohenberg–Kohn theorem and an unknown total energy functional, $E[\rho(\mathbf{r})]$. However, as we shall see, very simple approximations to $E_{xc}[\rho(\mathbf{r})]$ can, perhaps surprisingly to some, yield fairly accurate results. This issue is addressed in the following section. The KS approach is therefore of great practical importance and has become the cornerstone of all modern DFT applications.

Exchange and Correlation Energy Functionals

Since the quality of any DFT calculation is limited by the quality of its approximation to the true XC energy functional, $E_{xc}[\rho(\mathbf{r})]$, it is not surprising that a great deal of effort has been invested in the development of ever more sophisticated XC functionals. Fortunately, one of the simplest and most popular approximations to $E_{xc}[\rho(\mathbf{r})]$, the Local Spin Density Approximation (LSDA),[22] is capable of yielding results that compete favorably with those from HF calculations.[2,8–11] In the LSDA, $E_{xc}[\rho(\mathbf{r})]$ is approximated by

$$E_{xc}[\rho(\mathbf{r})] = \int \rho(\mathbf{r})\, \epsilon_{xc}(\rho^{\alpha}(\mathbf{r}), \rho^{\beta}(\mathbf{r}))\, d\mathbf{r} \quad [9]$$

where $\epsilon_{xc}(\rho^{\alpha}(\mathbf{r}), \rho^{\beta}(\mathbf{r}))$ is the XC energy density at a point $\mathbf{r}$ in space. Within the LSDA, ϵ_{xc} is but a function of $\rho^{\alpha}(\mathbf{r})$ and $\rho^{\beta}(\mathbf{r})$ at that specific point $\mathbf{r}$ in space. When $E_{xc}[\rho(\mathbf{r})]$ is expressed in such a fashion, the XC potentials, $v_{xc}^{\sigma}(\mathbf{r})$, from Eq. [8] are given by

$$v_{xc}^{\sigma}(\mathbf{r}) = \rho(\mathbf{r}) \frac{d\epsilon_{xc}(\rho^{\alpha}(\mathbf{r}), \rho^{\beta}(\mathbf{r}))}{d\rho^{\sigma}(\mathbf{r})} + \epsilon_{xc}(\rho^{\alpha}(\mathbf{r}), \rho^{\beta}(\mathbf{r})), \; \sigma = \alpha, \beta \quad [10]$$

The heart of the LSDA is the approximation that a particular point in space of an inhomogeneous distribution of electrons (such as an atom or molecule) with densities $\rho^{\alpha}(\mathbf{r})$ and $\rho^{\beta}(\mathbf{r})$ has the same values of ϵ_{xc}, v_{xc}^{α}, and v_{xc}^{β} as any point in a homogeneous distribution of electrons of the exact same densities ρ^α and ρ^β. The value of $\epsilon_{xc}(\rho^\alpha(\mathbf{r}), \rho^\beta(\mathbf{r}))$ has been determined for a large number of homogeneous gases of interacting electrons of varying total density, $\rho = \rho^\alpha + \rho^\beta$, and net spin density, $\rho^s = \rho^\alpha - \rho^\beta$, by means of quantum Monte Carlo methods.[23,24] The results have since been parameterized by Vosko, Wilk, and Nusair (VWN)[22] and by Perdew and Zunger (PZ),[25] in such a way that the XC potentials can be easily evaluated via Eq. [10]. Since the quantum Monte Carlo calculations on the homogeneous electron gases are "exact," the PZ and VWN parameterizations have essentially brought us to the limit of the LSDA.

Though this may seem to be a crude approximation, theoretical considerations[26] do justify the LSDA's ability to provide quantitatively accurate geometries, charge distributions, and vibrational spectra on a wide variety of systems.[2,8–11] It should not be surprising to see the LSDA results surpass HF and even challenge correlated post-HF methods. One must remember that correlation effects are included within the LSDA XC functional. Systems that prove to be extremely problematic in HF theory, where correlation effects are by definition absent, are therefore not necessarily systems that offer any great challenge to the LSDA. A simple molecule like ozone is one such classic example.[27]

Experience has shown that the LSDA exchange energy is roughly 10% too small,[17] whereas the correlation energy is almost 100% too large.[16] These errors do not seem to cause serious problems when constructing potential energy surfaces, because these errors are consistent and therefore do not greatly affect the shape of the surface. By far, the most troubling aspect of the LSDA is its systematic overestimation of binding energies.[11,28,29] For general chemistry, this means that the LSDA is not particularly well suited for thermochemical applications such as computing atomization energies, heats of reactions, or activation energies. For weakly bound systems, such as hydrogen-bonded systems[13] or van der Waals complexes,[20] the overestimation of the strength of certain interactions can disrupt what is often a delicate balance of forces, driving a system to a geometry that is in very poor agreement with experiment. Needless to say, for a great deal of biomolecular applications, this flaw is fatal.

Malonaldehyde (Figure 1) is a system with an intramolecular hydrogen bond where the LSDA fails to provide even a qualitatively correct description.[13] Experiment finds that this system forms a six-membered ring with a short covalent O—H bond and a long O·····H hydrogen bond. As seen in Table 1, HF theory does not do particularly well, as the hydrogen bond is 0.2 Å too long. However, the MP2 calculation cleans up the geometry very nicely, with the hydrogen bond now in error by less than 0.02 Å. Within the LSDA, malonaldehyde is very poorly described. An overestimation of the strength of the hydrogen bond essentially drives the system towards a C_{2v} structure, with the O·····H hydrogen bond almost 0.5 Å shorter than experiment and only 0.02 Å longer than the covalent O—H bond, which is more than 0.2 Å too long. Clearly, the LSDA's propensity to overestimate binding energies can become intolerable when one deals with systems where relatively weak intra- and intermolecular forces are involved.

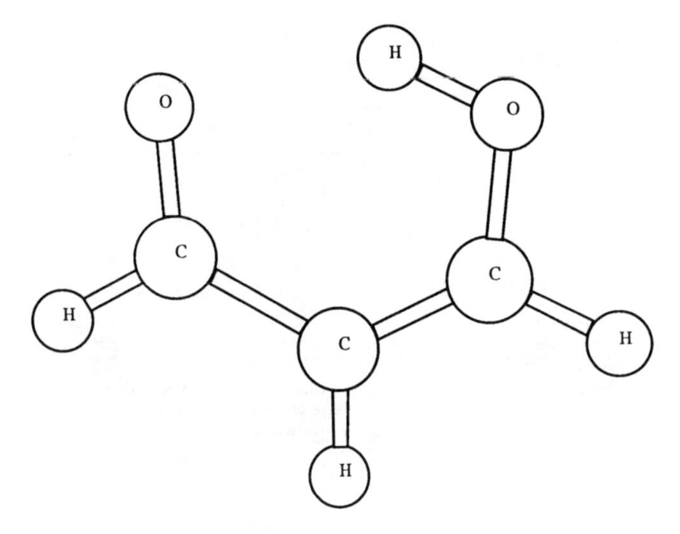

Figure 1 Malonaldehyde.

For such applications, one must therefore go beyond the LSDA and adopt gradient-corrected XC energy functionals. The XC energy functional still has the general form of Equation [9], but as the name suggests, the gradients of $\rho^{\alpha}(\mathbf{r})$ and $\rho^{\beta}(\mathbf{r})$ appear so that $E_{xc}[\rho(\mathbf{r})]$ is now given by

$$E_{xc}[\rho(\mathbf{r})] = \int \rho(\mathbf{r})\, \epsilon_{xc}(\rho^{\alpha}(\mathbf{r}), \rho^{\beta}(\mathbf{r}), \nabla\rho^{\alpha}(\mathbf{r}), \nabla\rho^{\beta}(\mathbf{r}))\, d\mathbf{r} \qquad [11]$$

Unfortunately, once we go to gradient-corrected functionals, the conceptual simplicity of the LSDA is lost and various models may be adopted. Consequently, many gradient-corrected XC energy functionals have been proposed.[15–19,30] It is important to note that though these functionals have been referred to in the past as "nonlocal" functionals, the gradient-corrected functionals are indeed mathematically local functionals in that the values of $\epsilon_{xc}(\mathbf{r})$, $v_{xc}^{\alpha}(\mathbf{r})$, and $v_{xc}^{\beta}(\mathbf{r})$ at any point in space depend only the values and derivatives of $\rho^{\alpha}(\mathbf{r})$ and $\rho^{\beta}(\mathbf{r})$ at that precise point in space.

Returning to malonaldehyde in Table 1, we see that gradient-corrected XC functionals lead to a much improved description of the hydrogen bond. Malonaldehyde's structure is now at least qualitatively correct, and the description of the hydrogen bond is better than that within the HF approximation. Considerable room for improvement remains because the O·····H hydrogen bond is still too short by 0.11 Å. The fact that it is underestimated suggests that the strength of weak bonds will still be overestimated with gradient-corrected XC functionals, but to a much lesser extent than within the LSDA.

Even though gradient-corrected XC functionals are local functionals, they do impose a significantly greater computational burden than do the LSDA

Table 1 HF[a], MP2[a], LSDA[b], Gradient-Corrected DFT[b], and Experimental Bond Lengths (in Å) of Malonaldehyde

	O—H	O····H
HF	0.956	1.880
MP2	0.994	1.694
LSDA[c]	1.204	1.220
DFT/PP (self-consistent)[d]	1.042	1.568
DFT/PP (perturbative)[d]	1.055	1.543
experiment[e]	0.969	1.680

[a]As cited in Reference 13, 6-31G** basis set.
[b]From Reference 13, triple-ζ + polarization basis set for heavy atoms, double-ζ + polarization basis set for hydrogen atoms.
[c]Vosko–Wilk–Nusair LSDA XC functional.[22]
[d]Gradient-corrected Perdew–Wang exchange[15] and Perdew correlation[16] functionals.
[e]As cited in Reference 13.

functionals. Derivatives of $\rho^{\alpha}(\mathbf{r})$ and $\rho^{\beta}(\mathbf{r})$ must now be evaluated before $\epsilon_{xc}(\mathbf{r})$, $v_{xc}^{\alpha}(\mathbf{r})$, and $v_{xc}^{\beta}(\mathbf{r})$ can be synthesized. Fortunately, experience has shown that it is often not essential to include gradient corrections in the course of the self-consistent procedure.[31] In other words, the KS equations can be iteratively solved using LSDA XC potentials. Once self-consistency is achieved, and it is time to calculate the total energy and the forces acting on the atoms, the derivatives of the LSDA SCF density are evaluated and plugged into the gradient-corrected forms of $\epsilon_{xc}(\mathbf{r})$, $v_{xc}^{\alpha}(\mathbf{r})$, and $v_{xc}^{\beta}(\mathbf{r})$. Gradient corrections are thus simply treated in a perturbative fashion. This simplification essentially renders gradient-corrected DFT calculations as inexpensive as their LSDA counterparts.

Tests of the perturbative approach to gradient-corrected DFT calculations have concentrated on systems where the LSDA is adequate.[31] Malonaldehyde provides an interesting, and demanding, test case. Returning to Table 1, it is clear that even though the LSDA is deficient, the perturbative gradient-corrected calculation employing the LSDA density recovers the improvements afforded by the fully self-consistent gradient-corrected calculations almost entirely. More than 90% of the correction to each of the O—H and O····H bonds is recovered in the perturbative scheme. The success of the perturbative approach indicates that corrections brought to the electronic structure by gradient corrections are not nearly as important as the corrections they may bring to geometries and energies. Consequently, past work on the electronic structure of molecular systems within the LSDA should still be viewed with confidence.

It has been suggested that the LSDA's propensity for generating hydrogen bonds that are too short and therefore too strong arises from the fact that the LSDA electronic densities decay too abruptly in the outer regions.[21] However, the perturbative approach to incorporating gradient corrections does not alter the LSDA electronic structure, yet repairs the poor LSDA hydrogen bond lengths and strengths. The key would therefore seem to be the gradient-corrected XC functionals' evaluation of the tail regions' contribution to the XC energy, and not corrections to the tail regions' electronic structure itself, as suggested.[21]

Several gradient-corrected exchange and correlation energy functionals exist. Typically, they are formulated so as to address either the exchange or the correlation component of $E_{xc}[\rho(\mathbf{r})]$. Popular gradient-corrected exchange functionals are those of Perdew and Wang[15] and Becke.[17] Perdew's gradient-corrected correlation functional[16] has been extensively used, as has that of Miehlich, Savin, Stoll, and Preuss,[18] which is a reformulation of the Lee, Yang, and Parr correlation functional.[32] Since each functional addresses a specific component of $E_{xc}[\rho(\mathbf{r})]$, any of the two exchange functionals can be coupled with any two of the correlation functionals. Perdew and Wang have recently introduced a coupled exchange and correlation gradient-corrected functional.[15]

Despite their often radically different forms, no particular gradient-corrected XC functional has emerged as being clearly superior. This may be due in part to the fact that the LSDA is very often a very good starting point: gradient corrections have little to correct. This is also an indication that it will be very difficult to derive new gradient-corrected functionals that will dramatically improve DFT results. Future XC functionals will most likely have to take on new forms. One promising example is that of Becke,[33,34] where (1) the exact exchange energy, obtained in the same fashion as an HF calculation but with the KS orbitals replacing the HF orbitals, (2) the LSDA XC energy, and (3) gradient corrections to the LSDA XC energy are all mixed together to create a new XC energy functional. Though preliminary results are promising, it is not clear whether these improvements will warrant the extra computational burden imposed by the calculation of the exact exchange energy and potentials. Approaches that tack on correlation functionals to HF calculations have not been successful, because their starting point, the HF approximation, is simply too poor.[34]

Computational Strategies for Solving the Kohn–Sham Equations

Once an approximate XC functional has been selected, the need to solve the KS equations remains. Several strategies have emerged, which differ mainly in two ways: the type of basis in which the KS orbitals are expanded and the manner in which the electrostatic component of the KS equations is handled. The large number of feasible DFT approaches is in sharp contrast to the more familiar HF formalism where the linear combination of Gaussian-type orbitals (LCGTO) approach clearly dominates.[5,6] The wide variation in DFT schemes is a direct result of the fact that exchange is no longer a nonlocal phenomenon as in HF theory. Therefore, as will be seen, any linear combination of atomic orbitals (LCAO) approach can be made to avoid the need to calculate the burdensome four-centered integrals that become the bottleneck in an LCAO-HF calculation.

It should be first pointed out that the KS equations can be solved numerically in a basis-set-free approach. Such a scheme is used in NUMOL.[35] Assuming the grids employed are sufficiently accurate, this approach has the distinct advantage of being limited only by the quality of the XC energy functional employed and is, therefore, ideally suited for benchmarking their performance on various systems. However, in the vast majority of DFT programs, the KS orbitals, $\{\psi_i\}$, are expressed as a linear combination of atom-centered basis functions,

$$\psi_i(\mathbf{r}) = \sum_{\mu}^{N_b} C_{\mu i}\chi_{\mu}(\mathbf{r}) \qquad [12]$$

where $\{\chi_\mu\}$ forms an LCAO basis consisting of N_b functions and $\{C_{\mu i}\}$ are the expansion coefficients for the ith KS orbital. In programs such as DeFT,[36] deMon,[12,37,38] DGauss,[10] and Gaussian 92/DFT,[11] Gaussians are used as basis functions. One of the distinct advantages of the LCGTO-DFT variant is the possibility of using the vast wealth of experience in basis sets and rapid integral evaluation gained by those working within the LCGTO-HF approach. In other LCAO approaches, Slater-type orbitals (STOs), as in the Amsterdam DFT code,[39,40] or numerical basis functions, as in DMol,[41] may be used. One promising non-LCAO approach is the use of plane waves.[42] Though such approaches have been extensively used in the study of crystalline materials, only recently have they been used for the study of systems containing first-row elements.[21,43–45]

One may now ask why such a large number of LCAO approaches are feasible. Expanding the KS orbitals in an LCAO basis as in Eq. [12], the KS equations can be recast in matrix form,

$$\mathbf{H}^{\sigma}\mathbf{C}^{\sigma} = \mathbf{S}\mathbf{C}^{\sigma}\boldsymbol{\epsilon}^{\sigma} \qquad [13]$$

where the KS and overlap matrix elements, $H^{\sigma}_{\mu\nu}$ and $S_{\mu\nu}$, are given by

$$H^{\sigma}_{\mu\nu} = \int \chi_\mu(\mathbf{r}) \left\{ -\frac{\nabla^2}{2} - \sum_{A}^{\nu} \frac{Z_A}{|\mathbf{r} - \mathbf{R}_A|} + v_{\text{el}}(\mathbf{r}) + v^{\sigma}_{\text{xc}}(\mathbf{r}) \right\} \chi_\nu(\mathbf{r})\, d\mathbf{r} \qquad [14]$$

and

$$S_{\mu\nu} = \int \chi_\mu(\mathbf{r})\chi_\nu(\mathbf{r})\, d\mathbf{r} \qquad [15]$$

where $v_{\mathrm{el}}(\mathbf{r})$ is given by

$$v_{\mathrm{el}}(\mathbf{r}) = \int \frac{\rho(\mathbf{r}')}{|\mathbf{r} - \mathbf{r}'|} d\mathbf{r}' \qquad [16]$$

Expanding the Coulomb repulsion component of $H^{\sigma}_{\mu\nu}$, we obtain

$$\int \chi_{\mu}(\mathbf{r}) v_{\mathrm{el}}(\mathbf{r}) \chi_{\nu}(\mathbf{r})\, d\mathbf{r} = \int\int \frac{\chi_{\mu}(\mathbf{r})\chi_{\nu}(\mathbf{r})\rho(\mathbf{r}')}{|\mathbf{r} - \mathbf{r}'|} d\mathbf{r} d\mathbf{r}'$$

$$= \sum_{\theta,\phi}^{N_b} P_{\theta\phi} \int\int \frac{\chi_{\mu}(\mathbf{r})\chi_{\nu}(\mathbf{r})\chi_{\theta}(\mathbf{r}')\chi_{\phi}(\mathbf{r}')}{|\mathbf{r} - \mathbf{r}'|} d\mathbf{r} d\mathbf{r}' \qquad [17]$$

where the density, $\rho(\mathbf{r})$, can be expressed within an LCAO basis as

$$\rho(\mathbf{r}) = \sum_{\sigma=\alpha,\beta} \sum_{i}^{N^{\sigma}} |\psi_i^{\sigma}(\mathbf{r})|^2 = \sum_{\sigma=\alpha,\beta} \sum_{i}^{N^{\sigma}} \left| \sum_{\mu} C^{\sigma}_{\mu i} \chi_{\mu}(\mathbf{r}) \right|^2 = \sum_{\mu,\nu}^{N_b} P_{\mu\nu} \chi_{\mu}(\mathbf{r}) \chi_{\nu}(\mathbf{r}) \qquad [18]$$

and the density matrix element, $P_{\mu\nu}$, is given by

$$P_{\mu\nu} = \sum_{\sigma=\alpha,\beta} \sum_{i}^{N^{\sigma}} C^{\sigma}_{\mu i} C^{\sigma}_{\nu i} \qquad [19]$$

as with the conventional LCAO HF approaches.

At this point, we are confronted with the same four-centered integrals

$$\int\int \frac{\chi_{\mu}(\mathbf{r})\chi_{\nu}(\mathbf{r})\chi_{\theta}(\mathbf{r}')\chi_{\phi}(\mathbf{r}')}{|\mathbf{r} - \mathbf{r}'|} d\mathbf{r} d\mathbf{r}' \qquad [20]$$

that are the bane of LCAO-HF calculations. One would think that the KS orbitals, like their HF counterparts, should be expanded in Gaussians to make the evaluation of the four-centered integrals as efficient as possible. When GTOs are used, these four-centered integrals can be efficiently evaluated, as is the case in Gaussian 92/DFT.[11] However, DFT applications can be made less computationally demanding by eliminating these costly four-centered integrals in one of two ways.

The first option is to fit the charge density. The fitted density, $\tilde{\rho}(\mathbf{r})$, given by

$$\rho(\mathbf{r}) \approx \tilde{\rho}(\mathbf{r}) = \sum_{k} c_k \chi'_k(\mathbf{r}) \qquad [21]$$

is expanded in an auxiliary basis, $\{\chi'_k\}$, that is typically composed of the same type of functions (GTOs or STOs) as the orbital basis. The values of the coefficients, $\{c_k\}$, are determined by a least-squares fitting procedure.[39,46,47] The computational demands are thus greatly reduced as the four-centered two-electron integrals are replaced by three-centered two-electron integrals,

$$\int\int \frac{\chi_\mu(\mathbf{r})\chi_\nu(\mathbf{r})\chi'_k(\mathbf{r}')}{|\mathbf{r}-\mathbf{r}'|}\, d\mathbf{r}d\mathbf{r}' \quad [22]$$

Such a fitting approach is used in the Amsterdam code[39,40] (STOs) as well as in DeFT,[36] deMon,[12,37,38] and DGauss[10] (GTOs).

A second option is to use Poisson's equation,

$$\nabla^2 v_{\mathrm{el}}(\mathbf{r}) = -4\pi\rho(\mathbf{r}) \quad [23]$$

Efficient schemes have been proposed for solving for $v_{\mathrm{el}}(\mathbf{r})$ on a grid.[41,48] After Poisson's equation is solved, Eq. [17] can be evaluated by numerical integration

$$\int \chi_\mu(\mathbf{r})\ v_{\mathrm{el}}(\mathbf{r})\chi_\nu(\mathbf{r})\ d\mathbf{r} \approx \sum_I^M \chi_\mu(\mathbf{R}_I)\ v_{\mathrm{el}}(\mathbf{R}_I)\ \chi_\nu(\mathbf{R}_I)\ W_I \quad [24]$$

over the molecular grid composed of M points positioned at $\{\mathbf{R}_I\}$ with quadrature weights $\{W_I\}$. This approach is used in DMol.[41]

It is important to remember that the above two approaches to simplifying the four-centered integrals are only made worthwhile by the fact that in DFT applications, the XC potentials are local functions. In conventional HF, the above four-centered integrals are still required to evaluate the exchange contribution,[5,6]

$$\sum_{\theta,\phi} P^\sigma_{\theta\phi} \int\int \frac{\chi_\mu(\mathbf{r})\chi_\theta(\mathbf{r})\chi_\nu(\mathbf{r}')\chi_\phi(\mathbf{r}')}{|\mathbf{r}-\mathbf{r}'|}\, d\mathbf{r}d\mathbf{r}' \quad [25]$$

to the Fock matrix element, $F^\sigma_{\mu\nu}$. In this case, exchange is a nonlocal functional that cannot be readily simplified. Therefore, these four-centered integrals must be calculated. Because the electrostatic repulsion contribution requires the same primitive four-centered two-electron integrals, there is no real advantage to finding a computational shortcut, such as fitting the density or solving Poisson's equation, within an LCGTO-HF scheme.

However, in DFT, the XC contribution to $H^\sigma_{\mu\nu}$ is simply expressed as

$$\int \chi_\mu(\mathbf{r})\ v^\sigma_{\mathrm{xc}}(\mathbf{r})\ \chi_\nu(\mathbf{r})\ d\mathbf{r} \quad [26]$$

The complexity of the XC functionals that generate $v^\alpha_{\mathrm{xc}}(\mathbf{r})$ and $v^\beta_{\mathrm{xc}}(\mathbf{r})$ from $\rho^\alpha(\mathbf{r})$ and $\rho^\beta(\mathbf{r})$ (and their derivatives, in the case of gradient-corrected functionals) does not permit the analytical, exact evaluation above. Invariably, $\rho^\alpha(\mathbf{r})$ and

$\rho^\beta(\mathbf{r})$ and their derivatives must be synthesized on a grid similar to those used to solve Poisson's equation. The XC potentials, $v^\alpha_{xc}(\mathbf{r})$ and $v^\beta_{xc}(\mathbf{r})$, are then evaluated at each of the grid points and the XC contribution to $H^\sigma_{\mu\nu}$ is numerically integrated,

$$\int \chi_\mu(\mathbf{r}) v^\sigma_{xc}(\mathbf{r}) \chi_\nu(\mathbf{r})\, d\mathbf{r} \approx \sum_I^M \chi_\mu(\mathbf{R}_I)\, v^\sigma_{xc}(\mathbf{R}_I)\, \chi_\nu(\mathbf{R}_I)\, W_I \qquad [27]$$

This is done in the Amsterdam code,[39,40] DMol,[41] and Gaussian 92/DFT.[11]

Another possibility is to synthesize $v^\sigma_{xc}(\mathbf{r})$ again on a grid, but rather than using numerical integration, a least-squares fit of $v^\sigma_{xc}(\mathbf{r})$ to yet another auxiliary basis, $\{\chi''_l\}$, is performed,[46,47]

$$v^\sigma_{xc}(\mathbf{r}) \approx \tilde{v}^\sigma_{xc}(\mathbf{r}) = \sum_l a^\sigma_l \chi''_l(\mathbf{r}) \qquad [28]$$

This fit is used to evaluate analytically an approximate XC contribution to $H^\sigma_{\mu\nu}$,

$$\int \chi_\mu(\mathbf{r})\, v^\sigma_{xc}(\mathbf{r})\, \chi_\nu(\mathbf{r})\, d\mathbf{r} \approx \int \chi_\mu(\mathbf{r})\, \tilde{v}^\sigma_{xc}(\mathbf{r})\, \chi_\nu(\mathbf{r})\, d\mathbf{r} = \sum_l a^\sigma_l \int \chi_\mu(\mathbf{r}) \chi''_l(\mathbf{r}) \chi_\nu(\mathbf{r})\, d\mathbf{r} \qquad [29]$$

This scheme is used in DeFT,[36] deMon,[12,37,38] and DGauss.[10]

As can be seen, the most computationally demanding parts in any DFT application (save Gaussian 92/DFT[11] where the four-centered two-electron integrals are evaluated) scale formally as N^2M, where N is the number of functions in the LCAO basis and M is either the number of grid points or the number of functions in either of the auxiliary bases. Because the size of M usually depends on the size of the LCAO basis, it is often said that DFT applications scale as N^3, a dramatic improvement over the N^4 scaling of conventional HF calculations. However, with the use of efficient cutoff schemes, today's LCGTO-HF programs scale better than N^3. The computational efficiency of DFT codes is not that dramatic when one compares them with HF codes. However, we must keep in mind that the DFT XC functionals contain the effects of electron correlation. When we compare DFT against even the simplest correlated post-HF approaches, which formally scale as at least N^5, the computational efficiency of DFT techniques becomes an important factor in favor of their use.

Analytic Gradients and Hessians

The development of gradient-corrected XC functionals has definitely been one of the keys to the recent surge in the popularity of DFT. Nevertheless, such functionals could not have had as much impact had it not been for the concurrent development of analytic DFT energy gradients, the first derivatives of the DFT total energy with respect to any of the nuclear coordinates. Without energy gradients, complete geometry optimizations by DFT methods were limited to small polyatomics with perhaps only four or five independent degrees of freedom. With analytic energy gradients, complete geometry optimizations of large polyatomics have become routine.[2,8,10–12]

Analytic energy gradients are rather trivially implemented in both Becke's fully numerical code, NUMOL,[35] and the plane wave codes.[42] The gradients are simply the Hellman–Feynman forces,[49] so neither type of code deals with the so-called Pulay forces[49] that arise from the incompleteness of an LCAO basis. The former code is free of these, because it is a fully numerical approach and basis sets are not an issue. In the latter codes, plane waves are delocalized basis functions that are not centered on any particular atom. In the LCAO codes, gradients were first introduced by Versluis and Ziegler for STOs.[40] Subsequently, an energy gradient expression was derived by Fournier, Andzelm, and Salahub for GTOs when auxiliary fitting functions are used.[50] Johnson, Gill, and Pople did the same for GTOs when auxiliary fitting functions are not used.[11] Finally, Delley has developed analytic energy derivatives when numerical basis functions are used.[51]

It should be noted that the grid techniques used for the XC terms make it very difficult to have an exact expression for the energy gradient. Issues such as the changes in the grid points' quadrature weights can be addressed to deliver more accurate energy gradients.[11] However, even when these issues are not addressed, errors in geometries of less than a few thousandths of an angstrom in bond lengths arise from the approximate energy gradients. Because this is well under the intrinsic errors of today's DFT methods and these errors do not seem to lead to spurious behavior in the course of geometry optimizations, these errors in the gradient can be tolerated easily.[10]

In addition to making full geometry optimizations commonplace, analytic energy gradients also make vibrational analyses a feasible option. The required Hessian, the matrix of the second derivatives of the total energy with respect to the nuclear coordinates, must be constructed by finite differentiation of the gradients. This approach has been used successfully to perform vibrational analyses on polyatomics.[9,52] However, this procedure can be costly. For an N-atom system, this type of procedure would require the evaluation of the gradient at, the very least, $3N$–6 geometries. Consequently, a great deal of effort has been invested in the development of analytic second derivatives. However, to date, only DGauss[53] and Gaussian 92/DFT[54] feature this option.

Car–Parrinello Methods

Molecular dynamics studies have yielded a great deal of insight into biomolecular systems.[55] Traditionally, the large number of time steps involved, each requiring the evaluation of the forces acting on the system's nuclei, has limited molecular dynamics simulations to simple empirical force fields. For noncovalent processes in biomolecular systems, this may not be a severe limitation, as molecular dynamics has enjoyed great success. Nevertheless, the process by which all the interatomic forces are parameterized is time-consuming. A fundamental weakness of empirical force fields, however, is their inability to describe covalent processes, such as the bond-making and bond-breaking steps in an enzymatic reaction.

The past decade has seen the arrival of the Car–Parrinello (CP) approach[56] to molecular dynamics where interatomic forces are calculated by ab initio means. By so doing, the need for empirical force fields, and hence the parameterization process, is eliminated. Depending on the sophistication of the quantum mechanical ab initio approach employed, the computed forces can be significantly more accurate than their empirical counterparts. However, the key feature of ab initio quantum mechanical forces is their ability to accurately model reactive processes. CP approaches can therefore, in principle, simulate the key steps in an enzyme's mechanism.

Though CP dynamics is by no means limited to DFT approaches,[57,58] the vast majority of CP applications have been within DFT, and more specifically, within plane wave DFT approaches.[21,43,44,56] CP dynamics involves the simultaneous evolution of both the electronic structure and the geometry by means of a fictitious Lagrangian,

$$L = \frac{1}{2}\sum_{\mu}^{N}\sum_{i}^{M} m_{\mu i}\left(\frac{\partial C_{\mu i}}{\partial t}\right)^2 + \frac{1}{2}\sum_{L}^{\text{nuclei}} M_I\left(\frac{\partial \mathbf{R}_\mathrm{I}}{\partial t}\right)^2 - E[\{C_{\mu i}\},\{\mathbf{R}_\mathrm{I}\}] \qquad [30]$$

where $\{C_{\mu i}\}$ is the set of coefficients for the M occupied KS orbitals, $\{\psi_i\}$, within a basis of N functions (in an HF application, they would be the expansion coefficients of the HF orbitals), and $\{\mathbf{R}_\mathrm{I}\}$ is the set of nuclear coordinates. Associated with these two sets of parameters is a set of arbitrary masses, $\{m_{\mu i}\}$ and $\{M_I\}$. The potential energy component of the Lagrangian, $E[\{C_{\mu i}\}, \{\mathbf{R}_\mathrm{I}\}]$, is the total energy of the system as would be calculated by a DFT calculation at the geometry defined by $\{\mathbf{R}_\mathrm{I}\}$ and the density generated by the set of KS orbital expansion coefficients $\{C_{\mu i}\}$. The two remaining terms of Eq. [30] comprise the kinetic energy component of the Lagrangian. Clearly, a physical significance can be assigned to the kinetic energy arising from the motion of the nuclei,

$$\frac{1}{2}\sum_{L}^{\text{nuclei}} M_I\left(\frac{\partial \mathbf{R}_\mathrm{I}}{\partial t}\right)^2 \qquad [31]$$

provided the masses assigned are the true masses of the nuclei. However, it is also clear that the kinetic energy component of the Lagrangian arising from the motion of the expansion coefficients,

$$\frac{1}{2}\sum_{\mu}^{N}\sum_{i}^{M} m_{\mu i}\left(\frac{\partial C_{\mu i}}{\partial t}\right)^2 \qquad [32]$$

is a purely fictitious term. The real kinetic energy of the electrons is included in the potential energy component, $E[\{C_{\mu i}\}, \{\mathbf{R}_I\}]$, of the CP Lagrangian.

The CP Lagrangian in turn generates the following equations of motion for the nuclei:

$$M_I \frac{\partial^2 \mathbf{R}_I}{\partial t^2} = -\nabla_{\mathbf{R}_I} E[\{C_{\mu i}\}, \{\mathbf{R}_I\}] \tag{33}$$

The equations of motion for the expansion coefficients are not as simple. One must first remember that they are subject to the constraint that the KS orbitals in the set are orthonormal,

$$\int \psi_i(\mathbf{r})\psi_j(\mathbf{r})\, d\mathbf{r} = \delta_{ij} \tag{34}$$

Taking these constraints into consideration, the equations of motion for the expansion coefficients are given by

$$m_{\mu i} \frac{\partial^2 C_{\mu i}}{\partial t^2} = \frac{-\partial E[\{C_{\mu i}\}, \{\mathbf{R}_I\}]}{\partial C_{\mu i}} + \frac{\partial \Sigma_j^M \Sigma_k^M \int \psi_i(\mathbf{r})\psi_j(\mathbf{r})\, d\mathbf{r}}{\partial C_{\mu i}} \tag{35}$$

The trajectories of the nuclear coordinates and expansion coefficients are controlled by these equations of motion. The constraints on the $\{C_{\mu i}\}$ are rigorously enforced at each time step with the aid of the SHAKE[59] algorithm, in the same manner as SHAKE is used to enforce geometrical constraints in traditional molecular dynamics studies. One must, however, not forget that a component of the CP Lagrangian is completely fictitious, and as such, these equations of motion determine the dynamics of a fictitious system.

Despite this conceptual difficulty, CP dynamics can be made to describe the temporal behavior of a real system. The key is to choose both an appropriate set of expansion coefficient mass parameters, $\{m_{\mu i}\}$, and an appropriate timestep, including the possibility of a separate, shorter time step for the $\{C_{\mu i}\}$. If these choices are taken, the velocities of the expansion coefficients, $\left\{\frac{\partial C_{\mu i}}{\partial t}\right\}$, can be made, for all intents and purposes, negligible. At this point, the fictitious component of the CP Lagrangian essentially vanishes,

$$\frac{1}{2}\sum_{\mu}^{N}\sum_{i}^{M} m_{\mu i}\left(\frac{\partial C_{\mu i}}{\partial t}\right)^2 \approx 0 \tag{36}$$

and a real system is now described. If the same is done for the accelerations of the expansion coefficients, $\left\{\frac{\partial^2 C_{\mu i}}{\partial t^2}\right\}$, then Eq. [35] is reduced to

$$-\frac{\partial E[\{C_{\mu i}\}, \{\mathbf{R}_\mathrm{I}\}]}{\partial C_{\mu i}} + \frac{\partial \Sigma_j^M \Sigma_k^M \int \psi_i(\mathbf{r})\psi_j(\mathbf{r})\, d\mathbf{r}}{\partial C_{\mu i}} \approx 0 \qquad [37]$$

The above says that $E[\{C_{\mu i}\}, \{\mathbf{R}_\mathrm{I}\}]$ is stationary with respect to any arbitrary change in any of the expansion coefficients, provided that this change obeys the orthonormality constraints. This is just a restatement of that which allowed the formulation of the KS equations. $E[\{C_{\mu i}\}, \{\mathbf{R}_\mathrm{I}\}]$ is therefore, at this point, the total energy that would be calculated by a conventional, self-consistent DFT calculation at the geometry defined by $\{\mathbf{R}_\mathrm{I}\}$. In other words, if the expansion coefficients' velocities and accelerations are negligible at each point in the trajectory of the nuclei, then the electronic structure is "instantaneously" adjusting itself to each new geometry. One can therefore say that one is performing ab initio molecular dynamics along a system's Born–Oppenheimer surface.

The CP method is ideally suited to plane wave approaches, because the basis functions are delocalized and their number is great. As mentioned in the preceding section, the fact that plane waves are delocalized means that the evaluation of the energy gradients, $\nabla_{R_I} E[\{C_{\mu i}\}, \{\mathbf{R}_\mathrm{I}\}]$, is greatly simplified. This is important because a molecular dynamics simulation consists of a large number of time steps, each demanding an evaluation of the forces acting on the nuclei. To appreciate the second factor favoring CP methods with plane waves, one should note that a diagonalization procedure is not required to obtain the KS orbitals within the CP approach. Unlike conventional DFT methods, one avoids the cost of matrix diagonalization which scales as N^3. Because plane wave basis sets are often quite large, CP dynamics is an attractive alternative to determining a system's electronic structure.

To date, CP dynamics has been limited primarily to semiconductors and alkali metal systems.[60] However, recent improvements in the plane wave methodology[42] have now made CP dynamics feasible for systems containing first-row elements.[21,43,44] Consequently, the first CP applications on systems of any biochemical interest have now appeared.[21,43,44] These are discussed in greater detail in the following section on applications of DFT.

APPLICATIONS OF DENSITY FUNCTIONAL THEORY IN BIOMOLECULAR MODELING

Molecular Geometries and Vibrational Frequencies

Molecular geometries and vibrational frequencies obtained by various DFT approaches have been the subject of many papers in the recent literature.[2,9–13,40,52] Though gradient-corrected XC functionals have led to dramatic improvements in some areas, they have done very little to improve these two sets of properties, because the LSDA usually predicts them well. Nevertheless, there is room for improvement in the LSDA results, and as we shall see, the lack of any systematic improvement in molecular geometries and vibrational frequencies is a sobering reminder that it is very difficult to formulate new gradient-corrected XC functionals that do anything other than fix the LSDA's glaring deficiencies.

Two of the earliest systematic studies of polyatomic molecular structures were that of Versluis and Ziegler[40] and that of Andzelm and Wimmer.[10] Both of these studies were performed within the LSDA, though the former was limited to an Xα, exchange-only approach. Andzelm and Wimmer's results with the VWN functional,[22] in general, tend to indicate that the LSDA underestimates single bonds between heavy atoms by roughly 0.01 to 0.02 Å. However, for multiple bonds, bond lengths can be overestimated by the same amount. Any bonds involving an H atom display a similar overestimation.

Upon going to gradient-corrected XC functionals, these trends change, but the overall accuracy is essentially unaltered. With a set of test molecules similar to that of Andzelm and Wimmer, Johnson, Gill, and Pople[11] found similar results with the VWN XC potentials. The mean and the mean absolute errors in bond lengths in the Johnson et al. data set were respectively +0.014 and 0.021 Å. Gradient-corrected DFT calculations with the Becke exchange[17] and Lee–Yang–Parr correlation[32] functionals (known as the B-LYP approach) only brought the mean absolute error down by 0.001 Å to 0.020 Å. The most striking feature of the B-LYP results is that the mean error, +0.020 Å, is of the same magnitude as the mean absolute error. All the single bonds between heavy atoms, underestimated within the LSDA, are now overestimated in a fashion similar to the multiple bonds and bonds involving H atoms. The B-LYP functional also does little to improve bond angles. In fact, the mean absolute error rises from 1.93° to 2.33° upon introduction of the gradient corrections. For the sake of comparison, their HF and MP2 mean absolute errors are 0.020 and 0.014 Å for bond lengths and 1.99° and 1.78° for bond angles.

Incorporating the gradient-corrected Becke exchange[17] and Perdew correlation[16] functionals in a perturbative fashion, St-Amant et al.[12] found results that were similar to those of Johnson et al.,[11] but on a set of significantly larger organic molecules. Though some bond lengths were underestimated, the vast majority were overestimated. One disconcerting set of results was that for bonds involving S atoms. For example, the S—C and S—S bonds in dimethyl disulfide were overestimated by 0.038 Å and 0.064 Å, respectively, and the S—H bond in thiomethanol was overestimated by 0.028 Å. No such bonds involving sulfur were studied by Johnson et al.[11] Though not as bad as sulfur-containing bonds, C—F bonds and C—Cl bonds were found to be as much as 0.03 and 0.05 Å too long, respectively, within the perturbative gradient-corrected DFT scheme.

Overall, the average absolute error within the perturbative approach[12] was 0.015 Å for bond lengths and 0.6° for bond angles; these errors are no worse than the fully self-consistent results of Johnson et al.[11] The corresponding MP2 numbers are significantly better, at 0.007 Å and 0.5°. Results on a set of 11 torsions gave average absolute errors of 3.4°, 3.6°, and 3.8° for perturbative gradient-corrected DFT, MP2, and HF calculations, respectively. It should be noted, however, that the experimental data for the torsions come with appreciably larger uncertainties than do the bond lengths and bond angles.

A systematic evaluation of the LSDA, gradient-corrected DFT, HF, and MP2 vibrational frequencies has also been carried out by Johnson, Gill, and Pople.[11] Previous studies[9,10,52] had employed only the LSDA. As has been well documented, the HF and MP2 methods systematically overestimate harmonic frequencies; the mean absolute errors were respectively 168 and 99 cm^{-1}, and their mean errors were +165 and +69 cm^{-1}. The LSDA harmonic frequencies were already better than their MP2 counterparts, as their mean and mean absolute errors were −51 and 75 cm^{-1}, respectively. Unfortunately, the B-LYP gradient-corrected functional was arguably worse than the LSDA at estimating harmonic frequencies, the mean and mean absolute errors being −63 and 73 cm^{-1}. Though it is unfair to compare frequencies calculated within the harmonic approximation with experimentally observed frequencies, it should be noted that the fortuitous systematic underestimation of harmonic frequencies by both the LSDA and gradient-corrected DFT caused the calculated and observed frequencies to agree even better. Within the LSDA, the mean and mean absolute errors fell to +28 and 46 cm^{-1}. With the B-LYP functional, they fell to +13 and 45 cm^{-1}.

Though gradient-corrected XC functionals have gone a long way towards improving energetics within DFT, they have done little to improve equilibrium geometries and harmonic frequencies of main group compounds. Consequently, modern DFT finds itself in the same state it found itself in 10 years ago when we were limited to the LSDA:

- For geometries of small organic and main group compounds, DFT is as good as, if not better than, HF, but not better than MP2, and
- For the vibrational analyses of these same systems, DFT can perform as well as MP2.

However, when dealing with weakly bound systems, such as hydrogen-bonded systems, gradient corrections are essential to obtain meaningful geometries. The case of malonaldehyde has already been addressed, and further on, we will see the dramatic effects of gradient corrections on hydrogen-bonding distances in the water dimer and in the DNA base pairs.

Modeling Reactions

The ability of modern DFT to accurately model reactions has been the most important advance in DFT applications over the past decade and is the result of more accurate gradient-corrected XC functionals. As described above, these functionals offered little in the way of improvements in geometries and vibrational analyses. However, the impact of gradient corrections on the calculation of heats of reactions and reaction barriers has been dramatic. Previously, the LSDA was notorious for its systematic overestimation of binding. Atomization energies and bond energies were therefore invariably overestimated.

The most systematic tests performed on DFT's ability to reproduce heats of reactions in small main group compounds are those of Becke on the G1 database.[28,29,33,34] Using the gradient-corrected Becke exchange[17] and Perdew–Wang correlation[19] functionals, the mean absolute error within the G1 database was 5.7 kcal/mol.[29] Compared to the corresponding value for the LSDA, 35.7 kcal/mol, with almost all the atomization energies overestimated, this result is a truly dramatic improvement. It is also important to note that once gradient corrections were added, there remained a slight tendency to overestimate atomization energies, but atomization energies of some molecules were underestimated.

Andzelm and Wimmer also provided systematic studies of heats of reactions of more complex processes.[10] Using double-ζ plus polarization basis sets and the gradient-corrected Becke exchange[17] and Perdew correlation[16] functionals, they found that the LSDA and gradient-corrected DFT had average errors of 11.8 and 3.8 kcal/mol, respectively, for the C—H bond energies in CH_4, CH_3, CH_2, and CH. For the sake of comparison, the HF/6-31G**, MP2/6-31G**, and MP4/6-31G** mean errors were 22.0, 5.5, and 3.5 kcal/mol. In the DFT calculations, the strength of the C—H bond was always overestimated. For the carbon–carbon bonds in ethane, ethylene, and acetylene, the errors in the LSDA bond strengths were, respectively, +18, +25, and +32 kcal/mol. By incorporating gradient corrections, the errors fell to −2, −1, and −1 kcal/mol. Over a small set of molecules, the average error in bond energies between the first row elements C, N, O, and F, were found to be 47 kcal/mol for HF, 28 kcal/mol for the LSDA, 5 kcal/mol for gradient-corrected DFT, and 2 kcal/mol for MP2. For this small database, the average MP4 error, 6 kcal/mol, was actually higher than that of MP2.

A final application of Andzelm and Wimmer was the study of hydrogenation reactions.[10] It should be noted that the number of bonds is unchanged upon going from reactants to products. Therefore, the systematic overestimation of bond strengths within the LSDA should not be as critical a factor. For a set of 17 hydrogenation reactions, the LSDA error was 10.7 kcal/mol. Gradient corrections provided a systematic improvement, the average error being 5.4 kcal/mol. Though the gradient-corrected errors were somewhat worse when double and triple bonds were involved, the LSDA results got dramatically worse. Upon going from systems with only single bonds to systems with double bonds, and to systems with triple bonds, the average LSDA errors in the hydrogenation energies rose from 3.1 kcal/mol, to 10.0 kcal/mol, to 30.0 kcal/mol. Consistent with these errors in the carbon–carbon single, double, and triple bonds, the signs of the errors in the hydrogenation energies indicated a systematic underestimation of the strengths of double and triple bonds relative to the corresponding number of single bonds. However, we must remember that the strengths of the multiple bonds estimated within the LSDA were still too large relative to experiment. Surprisingly, and perhaps disturbingly, the average error rose from 4.0 kcal/mol for HF, to 4.4 kcal/mol for MP2, and to 6.2 kcal/mol for MP4. Considered together, Andzelm and Wimmer's results strongly suggest that gradient-corrected DFT is a viable alternative to considerably more expensive MP2 and MP4 calculations for heats of reactions in organic systems.

Density functional studies of transition states have been performed by Fan and Ziegler.[61,62] Their original study[61] was within the LSDA, but subsequent work[62] included the effects of gradient corrections. Among the reactions studied at the double-ζ plus polarization level were the following:

$$\cdot CH_3 + CH_4 \rightarrow CH_4 + \cdot CH_3$$

$$H_2CO \rightarrow H_2 + CO$$

$$CH_3CN \rightarrow CNCH_3$$

For the abstraction of a hydrogen atom from methane by a methyl radical, the LSDA did a particularly poor job, as it predicted a barrier height of 1.9 kcal/mol. The experimental value is 14.1 kcal/mol. Upon including gradient corrections, Becke's exchange[17] and Perdew's correlation[16] functionals, DFT predicted a barrier height of 11.7 kcal/mol. This value compares well with barriers of 29.7 and 19.7 kcal/mol, respectively, from modest unrestricted HF/6-31G and subsequent configuration interaction (CI) calculations (assuming the same corrections as DFT for zero-point energies, these barriers would fall to 28.8 and 18.8 kcal/mol).

For the dissociation of formaldehyde, the true barrier height is believed to be roughly 80.6 kcal/mol. For this system, the LSDA and gradient-corrected DFT calculations provided almost identical barrier heights of 74.0 and 75.0 kcal/mol, respectively. The quoted double-ζ plus polarization HF, singles and doubles CI, and multiple configuration SCF plus CI calculations gave 100.2, 92.8, and 84.4 kcal/mol, respectively.

Finally, for the isomerization of CH_3CN to $CNCH_3$, the LSDA and gradient-corrected results were again very similar for the reaction barrier. The LSDA value was 42.1 kcal/mol, whereas the gradient-corrected value was 39.5 kcal/mol. Nevertheless, it should be noted that the gradient corrections did offer an improvement, as the experimental value is 38.4 kcal/mol. Double-ζ plus polarization HF calculations predicted a barrier height of 48.1 kcal/mol. Even after performing a large CI calculation, the barrier height still remained too high, at 45.3 kcal/mol. These DFT results are very encouraging and clearly suggest that modern DFT is quite capable of outperforming sophisticated post-HF methods.

Though the LSDA was quite accurate in two of the three reactions just discussed, it must be remembered that its performance in the first case was abysmal.

Stanton and Merz examined more complex organic and organometallic reactions.[63] Of particular interest is their work on the Diels–Alder reaction, the [1,5] sigmatropic shift of 1,3–pentadiene, and the reaction of $ZnOH^-$ and CO_2 to produce Zn^{2+} and HCO_3^-. These studies used the gradient-corrected Becke exchange[17] and Perdew correlation[16] functionals. All calculations employed double-ζ plus polarization level basis sets.

For the Diels-Alder reaction, the formation of cyclohexene from ethylene and butadiene, the LSDA and gradient-corrected DFT calculations agreed on a symmetric transition state structure,[63] in accord with the wavefunction-based ab initio approaches. The cited experimental values ranged from 24.2 to 27.5 kcal/mol. Though the LSDA and gradient-corrected transition state structures were similar, the reaction barrier within the LSDA was severely underestimated, by about 20 kcal/mol, at 4.5 kcal/mol. The gradient-corrected DFT result was in much better agreement, at 18.7 kcal/mol. This result was better than the HF and MP2 values, which were 45.9 and 15.9 kcal/mol, respectively.

The cited experimental value for the [1,5] sigmatropic shift of 1,3 pentadiene is 35.4 kcal/mol.[63] Again, the LSDA and gradient-corrected DFT transition state structures were very similar, with the bond lengths agreeing within 0.02 Å. However, in this case, the LSDA estimation of the height of the reaction barrier was also quite good at 26.0 kcal/mol. It was significantly better than the HF value of 59.8 kcal/mol. The MP2 activation energy was somewhat high, at 38.0 kcal/mol, whereas the gradient-corrected DFT value was somewhat low, at 32.9 kcal/mol.

The final system studied by Stanton and Merz[63] is the reaction

$$ZnOH^+ + CO_2 \rightarrow Zn^{2+} + HCO_3^-$$

It is of particular interest because it is a simple model of the zinc metalloenzyme carbonic anhydrase. Though Zn^{2+} is not a particularly challenging transition metal ion, this is still a good test of DFT methods. The performance of DFT methods relative to wavefunction-based ab initio approaches should only get better when enzymatic reactions involving true transition metals are studied. No experimental number is available to gauge the accuracy of each method. The transition state energies within the HF, MP2, LSDA, and gradient-corrected DFT approaches all agreed well, spanning the range of 24.8 kcal/mol (LSDA) to 31.8 kcal/mol (gradient-corrected DFT). The only major difference was the LSDA's estimate of the energy of the $CO_2/ZnOH^+$ association process. This is consistent with the LSDA's tendency to overestimate binding energies.

Overall, these results show that gradient-corrected DFT presents us with exciting possibilities for modeling reactions accurately. Heats of reactions and activation energies obtained within this approach clearly rival, and arguably surpass, correlated post-HF methods that require considerably more computing power. Though equilibrium and transition state structures are generally very well reproduced within the LSDA, one should incorporate gradient corrections before the energetics can be really trusted.

Conformational Energies

The overwhelming majority of DFT energy studies have focused on properties such as atomization energies and bond energies. In these types of applications, one may be satisfied if DFT agrees within 2–4 kcal/mol of the experimental results. One cannot be satisfied with this level of accuracy when trying to calculate the relative energies of low-lying conformers of organic molecules. Because relative conformational energies often differ by only a few tenths of a kilocalorie per mole, such calculations are a demanding test of DFT methods. The calculation of relative conformational energies is potentially useful because experimental data can be scarce, and such numbers play a pivotal role in the parameterization of empirical molecular mechanical force fields used in molecular dynamics simulations. Two DFT studies that have already appeared in the literature and that are of particular interest to biomolecular modelers are Oie, Topol, and Burt's study of the conformers of ethylene glycol[64] and Liang, Ewig, Stouch, and Hagler's conformational analysis of inositols.[65]

Table 2 contains HF, MP2, LSDA, and gradient-corrected DFT results for a subset of 35 molecules taken from a larger study conducted by St-Amant, Cornell, Halgren, and Kollman[12] on the ability of DFT methods to reproduce accurate relative conformational energies in small organic molecules spanning a wide range of functional groups. These 35 compounds were chosen because they were ones for which experimental data were available. The experimental data come with appreciable error bars. Moreover, they often correspond to enthalpies, free energies, or energy differences between ground vibrational states rather than differences between the bottoms of wells in the potential energy surface, which is the quantity actually calculated. Furthermore, some of the experimental data are obtained in solution. Nevertheless, the experimental data selected allow us to evaluate the merits of the various theoretical approaches.

Over the set of 35 molecules, the RMS errors in the relative conformational energies are, in kcal/mol: 0.64 for HF, 0.43 for MP2, 0.74 for the LSDA, and 0.47 for gradient-corrected DFT. The LSDA would therefore seem to be significantly worse than the simple HF approach for such applications. Gradient corrections, however, clearly improve DFT beyond the HF approximation, but not quite up to the level of MP2 theory. In three cases the gradient corrections have a profound effect on the energetics. These systems are glyoxylic acid, methyl vinyl ether, and 1-butene, for which gradient corrections bring the DFT results closer to the experimental values by 1.13, 1.13, and 0.95 kcal/mol, respectively.

Considering the relative costs of MP2 and DFT calculations and the marginal superiority of MP2, gradient-corrected DFT could very well be the

Table 2 HF[a], MP2[a], LSDA[b], Gradient-Corrected DFT[b], and Experimental Relative Conformational Energies for Small Organic Molecules (Energies in kcal/mol, Taken from Reference 12)

		ΔE				
Molecule	Conformation[c]	HF	MP2	VWN[d]	VWN/BP[e]	Expt[f]
Amides						
N-Methylformamide	trans	—	—	—	—	—
	cis	1.01	1.21	2.11	1.58	1.45
N-Methylacetamide	trans	—	—	—	—	—
	cis	2.62	2.00	2.10	2.22	2.3
Carboxylic acids						
Formic acid	O═C—O—H c	—	—	—	—	—
	O═C—O—H t	5.70	4.94	4.91	4.61	3.90
Glyoxylic acid	O═C—C═O t, O═C—O—H t	0.08	—	—	—	—
	O═C—C═O t, O═C—O—H c	—	0.54	2.55	1.42	1.2
Esters						
Methyl formate	O═C—O—C c	—	—	—	—	—
	O═C—O—C t	6.13	5.74	5.15	4.77	3.85, 4.75
Methyl acetate	O═C—O—C c	—	—	—	—	—
	O═C—O—C t	9.42	8.34	7.04	7.20	8.5
Ethyl formate	O═C—O—C c, C—O—C—C a	—	—	—	—	—
	O═C—O—C c, C—O—C—C g	0.67	0.22	0.01	0.38	0.19
Aldehydes and ketones						
Propionaldehyde	C—C—C═O c	—	—	—	—	—
	C—C—C═O s	0.74	0.84	1.43	0.82	0.95
Butanone	C—C—C═O c	—	—	—	—	—
	C—C—C═O s	1.20	0.85	2.14	1.78	2.02, 1.15

Conjugated systems						
1,3-Butadiene	C=C—C=C t	—	—	—	—	—
	C=C—C=C ≈ 30°	3.09	2.47	3.67	3.80	2.49
2-Methyl-1,3-butadiene	C=C—C=C t	—	—	—	—	—
	C=C—C=C ≈ 30°	2.43	2.33	3.90	3.34	2.65
Acrolein	C=C—C=O t	—	—	—	—	—
	C=C—C=O c	2.07	2.12	2.23	2.23	2.0
Alcohols						
Ethanol	C—C—O—H a	—	0.02	0.38	0.34	—
	C—C—O—H g	0.30	—	—	—	0.4, 0.12
Isopropanol	H—O—C—H g	—	—	0.10	0.07	—
	H—O—C—H a	0.38	0.23	—	—	0.28
Cyclohexanol	Equatorial C_s	0.34	0.21	—	—	—
	Equatorial C_1	—	—	0.01	0.06	—
	Axial C_1	0.69	0.26	0.71	0.96	0.52
Ethers						
Methyl ethyl ether	C—C—O—C a	—	—	—	—	—
	C—C—O—C g	1.75	1.45	1.07	1.25	1.5
Methyl vinyl ether	C=C—O—C c	—	—	—	—	—
	C=C—O—C s	1.75	2.68	3.60	2.47	1.7
Diethyl ether	C—C—O—C a, C—O—C—C a	—	—	—	—	—
	C—C—O—C a, C—O—C—C g	1.76	1.49	0.91	1.13	1.1
Methoxycyclohexane	Equatorial C_1	—	0.21	0.13	—	—
	Axial C_1	0.57	—	—	0.36	0.45

(continued)

Table 2 *(continued)*

Molecule	Conformation[c]	ΔE				
		HF	MP2	VWN[d]	VWN/BP[e]	Expt[f]
Acetals and hemiacetals						
2-Methoxytetrahydropyran	Axial	—	—	—	—	—
	Equatorial	1.06	1.49	1.16	1.11	1.05
2,5-Dimethyl-1,3-dioxane	2 Equatorial, 5 equatorial	—	—	—	—	—
	2 Equatorial, 5 axial	1.07	0.47	1.00	1.16	0.9
Amines						
Isopropylamine	l.p.—N—C—H a	—	—	—	—	—
	l.p.—N—C—H g	0.47	0.54	0.49	0.34	0.45
Cyclohexylamine	Equatorial	—	—	—	—	—
	Axial	1.15	0.56	1.15	1.46	1.1–1.8
Piperidine	Equatorial	—	—	—	—	—
	Axial	0.95	0.87	0.88	0.64	0.4
N-Methylpiperidine	Equatorial	—	—	—	—	—
	Axial	3.93	3.65	3.19	3.15	3.15
Alkanes						
Butane	C—C—C—C a	—	—	—	—	—
	C—C—C—C g	0.96	0.61	0.31	0.67	0.75
Cyclohexane	Chair	—	—	—	—	—
	Twist-boat	6.75	6.22	6.61	6.15	5.5

Methylcyclohexane	Equatorial	—	—	—	—	—
	Axial	2.34	1.64	2.15	2.41	1.75
2,3-Dimethylbutane	H—C_2—C_3—H a	0.11	—	—	—	—
	H—C_2—C_3—H g	—	0.08	0.47	0.36	0.17
Alkenes						
1-Butene	C=C—C—C s	—	—	0.83	—	—
	C=C—C—C c	0.74	0.37	—	0.12	0.53
2-Butene	trans	—	—	—	—	—
	cis	1.54	1.31	0.41	0.93	1.0
Alkyl halides						
1,2-Difluoroethane	F—C—C—F g	0.13	—	—	—	—
	F—C—C—F a	—	0.68	1.86	1.14	0.8
1,2-Dichloroethane	Cl—C—C—Cl a	—	—	—	—	—
	Cl—C—C—Cl g	1.92	1.29	0.98	1.54	1.20
Fluoropropane	F—C—C—C g	—	—	—	—	—
	F—C—C—C a	0.04	0.05	0.28	0.05	0.35
Chloropropane	Cl—C—C—C g	0.37	—	—	0.29	—
	Cl—C—C—C a	—	0.01	0.23	—	0.36, 0.05

[a] 6-31+G** basis sets, calculated at geometries optimized using 6-31G* basis sets.
[b] Triple-ζ + polarization on heavy atoms, double-ζ + polarization on hydrogen atoms.
[c] a indicates anti; c cis; g gauche; s skew; t trans.
[d] Vosko–Wilk–Nusair LSDA XC functional.[22]
[e] Gradient-corrected Becke exchange[17] and Perdew correlation[16] functionals used perturbatively with self-consistent Vosko–Wilk–Nusair LSDA densities.
[f] As cited in Reference 12.

method of choice for such calculations. In light of the scarcity of accurate experimental data in this area, such calculations will undoubtedly continue to play an important role in future force field parameterizations. The DFT study of the glycine and alanine dipeptide analogs, discussed in an upcoming section, is one such example.

Electrostatics

A molecule's electronic distribution is perhaps the most fundamental property that must be reproduced by an electronic structure method. An overview of how to compute molecular electrostatic potentials and their influence on chemical reactivity has been presented by Politzer and Murray.[66] Without an accurate electronic structure, no other property is likely to be accurately described by a quantum mechanical model. It is sometimes unclear just how to determine the accuracy of a theoretical method, but one way is to compare calculated and experimental dipole moments. From good electronic distributions, one can extract atom-centered point charges for use in molecular mechanical applications.

For 21 organic molecules spanning a wide variety of functional groups, St-Amant et al.[12] calculated dipole moments at the LSDA and gradient-corrected DFT levels (employing Becke's exchange[17] and Perdew's correlation[16] functionals). With triple-ζ plus polarization basis sets, the LSDA and gradient-corrected DFT calculations systematically overestimated experimental dipole moments by 11.7% and 9.1%, respectively. Consistent with the argument that perturbative incorporation of gradient corrections is sufficient, the LSDA and the fully self-consistent gradient-corrected average errors are fairly similar. By scaling down the dipole moments by roughly 10%, the average errors fall to about 6%. By adding diffuse functions and doubling polarization functions, the average unscaled errors fall to 6.6% for the LSDA and 5.5% for gradient-corrected DFT. With these basis sets, scaling is no longer as effective because there no longer exists a systematic overestimation of dipole moments.

Despite these fairly accurate dipole moments, DFT molecular electrostatic potential fitted charges are most likely not appropriate for simple additive molecular mechanical force fields. For these types of force fields to be accurate in aqueous environments, a priori polarization of molecules is required. The DFT charges are too gas phase-like, and initial tests[67] indicate that the charges are too small and their electrostatic interactions with prepolarized water molecules (such as the SPC, TIP3P, and TIP4P water models[68]) are too weak. On a brighter note, as nonadditive, polarizable, force fields[69] become more popular, the ability to accurately reproduce gas phase dipole moments will be extremely desirable.

Finally, a very interesting set of applications was that of Hagler and coworkers.[70,71] Employing the LSDA, they studied the changes in the electronic structure of folate, dihydrofolate, and NADPH upon binding to the enzyme dihydrofolate reductase. In these calculations, the enzyme was represented by simple point charges, while a substrate's electronic structure was solved for at the LSDA level within the reaction field generated by the enzyme's point charges. By studying the polarization of the substrate's electronic structure by the enzyme, Bajorath et al.[70,71] were able to rationalize the role of critical charged residues and to better understand aspects of the enzyme's catalytic activity. Though LSDA calculations were, at the time, really the only feasible, accurate, ab initio calculations for such a large system, the qualitative nature of the results was such that perhaps less sophisticated semiempirical quantum mechanical approaches[72] would have been adequate.

Water Dimer, Water Clusters, and Liquid Water

The presence of water in every living system has made the study of water–water interactions of great importance. The simplest relevant system one can consider is the water dimer. It has in fact been the subject of calculations at various levels of conventional ab initio theory[73] and is almost surely the most studied and best understood hydrogen-bonded system. Experiment and theory conclude the equilibrium structure to be the C_s trans-linear structure depicted in Figure 2.

The results of HF, MP2, LSDA, and gradient-corrected DFT calculations are presented in Table 3. The DFT calculations were performed within a plane wave approach by Laasonen, Csajka, and Parrinello.[21] The PZ parameterization[25] was used within the LSDA calculations, whereas the gradient-corrected DFT calculations used Becke's exchange functional[17] and Perdew's correlation[16] functional. As seen in Table 3, the LSDA does a very poor job on the water dimer. The O····O distance is 0.26 Å shorter than the experimental value of 2.98 Å, while the LSDA binding energy is 8.3 kcal/mol, much higher than

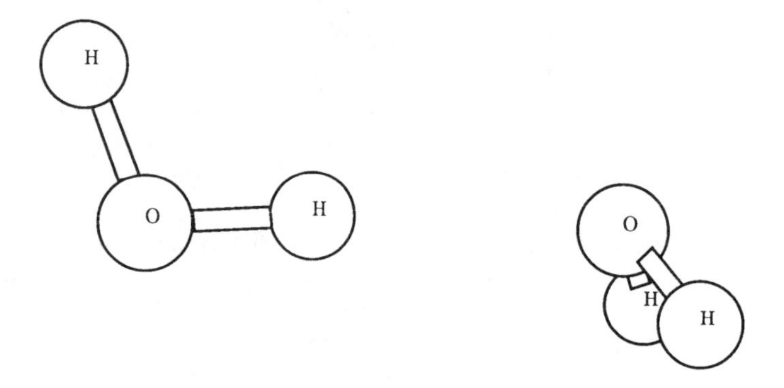

Figure 2 Water dimer, C_s trans-linear structure.

Table 3 HF, MP2, LSDA, Gradient-Corrected DFT, and Experimental Structures and Dimerization Energies for the Water Dimer

	HF[a]	MP2[a]	DFT/PZ[b]	DFT/BP[c]	Experiment[d]
d O····O (Å)	3.04	2.91	2.72	2.93	2.98 ± 0.01
∠ O····H_b—O (deg)	176	174	177	176	174 ± 10
Tilt ∠, θ (deg)	131	123	134	116	123 ± 10
Dimerization energy (kcal/mol)	3.85	5.30	8.30	4.11	4.8 ± 0.2

[a]From Reference 73, with 6-311++G(2d,2p) basis set.
[b]From Reference 21, plane wave basis, Perdew–Zunger LSDA parameterization.[25]
[c]From Reference 21, plane wave basis, gradient-corrected Becke exchange[17] and Perdew correlation[16] functionals.
[d]As cited in Reference 21.

the experimental value of 4.8 kcal/mol. When gradient corrections are implemented, DFT does indeed yield an accurate description of the water dimer. The O····O distance is now only 0.05 Å shorter than experiment, and the dimerization energy is only 0.7 kcal/mol smaller than experiment. Results from LCGTO-DFT calculations with a different XC functional provide similar results.[13] The angles listed in Table 3 are associated with extremely flat potentials, and the significant variations in their values should not be alarming.

After the success of their plane wave gradient-corrected DFT calculations, Laasonen et al. proceeded to the study of water clusters.[43] No pertinent experimental data exist for any of these. The key findings are that the water trimer and tetramer form flat ring structures. For the hexamer, a chairlike ring structure is preferred, but other structures have energies within a few tenths of a kilocalorie per mole. Referring back to the water dimer where experimental data are available, such energy differences are smaller than the errors in the gradient-corrected DFT calculations. Therefore one cannot be sure that the DFT calculations on the hexamer have found the truly preferred equilibrium structure. For the octamer, the largest cluster studied, a cubic structure is found to be the lowest in energy, but again, other structures are only slightly higher in energy. In all these water clusters, the O····O distances and the hydrogen bond strengths are respectively shorter and stronger than in the dimer. This is likely the result of increased dipole moments in the water molecules. For the octamer, O····O bonds can be as short as 2.6 Å, whereas the average strength of the hydrogen bonds can be as high as 8 kcal/mol.

With these benchmark calculations in hand, Laasonen et al. went on to perform ab initio molecular dynamics simulations on liquid water using the CP approach.[44] The simulation had 32 D_2O molecules at 300 K in a cubic cell of length 9.6 Å with periodic boundary conditions. Deuterated water was used instead of ordinary water to slow down nuclear motion. This decrease makes it easier for the electronic parameters to adjust to nuclear motion and have the simulation remain on the Born–Oppenheimer surface. Using Becke's gradient-corrected exchange functional[17] and Perdew's gradient-corrected correlation functional,[16] a 1.5 ps simulation resulted in the first peak of the O····O radial distribution function being at 2.69 Å. This is almost 0.2 Å shorter than the experimental value of 2.87 Å. The low prediction is consistent with the short O····O distance in the water dimer. When the Perdew correlation functional was replaced by LSDA correlation, the first peak in the O····O radial distribution function agreed almost perfectly with experiment. However, this may be fortuitous, and it is unlikely that falling back upon any aspect of the LSDA is an adequate general option within modern DFT.

The utility of CP dynamics is that it allows one to follow the electronic structure throughout the course of the simulation. This is not the case when empirical force fields are used. One simple property that can be evaluated throughout the CP liquid water simulation is the average dipole moment of a water molecule. The value extracted from these simulations, 2.66 D, is in good agreement with their cited experimental value of 2.6 D.[44] Within CP dynamics, polarization of the water molecules (their gas phase dipole moment is 1.85 D) is automatically achieved, bypassing the parameterization process that use of a polarizable, nonadditive force field[69] would entail. The ab initio molecular dynamics simulation also allowed them to follow the nature of the lowest unoccupied molecular orbital (LUMO) which plays a role in the conduction of an excess electron.

Though this ab initio molecular dynamics simulation required roughly a month of workstation CPU time to complete 1 ps of dynamics, the promise of future applications makes the Laasonen et al.[44] study of liquid water one of great importance. With the advent of more efficient algorithms and more powerful computers, CP dynamics could be used to study reactions in solution. Such processes are in principle no more complicated to simulate than the intermolecular interactions considered in the above-mentioned CP dynamics simulation.

Glycine and Alanine Dipeptide Analogs

The glycine and alanine dipeptide analogs (GDA and ADA) have recently garnered theoretical attention, because they are the two simplest models of a polypeptide.[74] Molecular mechanical force fields designed to model proteins are often parameterized to reproduce, as well as possible, the geometries and relative energies of the various GDA and ADA conformers. Since few experimental data are available, one must rely on theoretical studies. To date, DFT calculations on peptides have been limited to GDA and ADA.[12]

The two geometrical parameters of greatest interest are the C—N—C_α—C and N—C_α—C—N torsions, known respectively as ϕ and ψ. The series of ϕ and ψ values for each residue in a protein ultimately describes that protein's secondary and tertiary structures. Table 4 contains the ϕ and ψ values for the

Table 4 HF[a], MP2[a], DFT[b] ϕ, ψ Values (in Degrees) for the Glycine and Alanine Dipeptide Analogs (from Reference 12)

Conformation	Dihedral angle	HF	MP2	DFT/BP[c]
Glycine dipeptide analog				
C_7	ϕ	−85.2	−82.7	−78.6
	ψ	67.4	74.0	61.8
C_5	ϕ	180.0	−178.2	−177.8
	ψ	180.0	−179.7	177.1
Alanine dipeptide analog				
C_{7eq}	ϕ	−85.8	−82.9	−80.0
	ψ	78.1	77.5	65.2
C_5	ϕ	−155.6	−159.3	−154.4
	ψ	160.2	166.8	164.9
C_{7ax}	ϕ	75.1	73.9	69.3
	ψ	−54.1	−65.6	−57.9
β_2	ϕ	−110.4	−137.0	−121.2
	ψ	12.0	22.6	21.8
α_L	ϕ	69.5	62.3	89.8
	ψ	24.9	37.2	−0.9
α'	ϕ	−165.6	−168.0	−164.2
	ψ	−40.7	−36.2	−41.0

[a]6-31G* basis sets.
[b]Triple-ζ + polarization on heavy atoms, double-ζ + polarization on hydrogen atoms.
[c]Gradient-corrected Becke exchange[17] and Perdew correlation[16] functionals used perturbatively with self-consistent Vosko–Wilk–Nusair LSDA[22] densities.

local minima on the potential energy surfaces of GDA and ADA. Results at the HF and MP2 levels of theory are included with those from DFT calculations employing gradient-corrected XC functionals. The first system, GDA, has only two local minima. The greatest discrepancy between the MP2 and DFT values for GDA is that for ψ in the C_7 conformer (Figure 3). The 12° difference may be disconcerting at first, but one must remember that these torsional potentials are soft. For the C_5, or extended, conformation (Figure 4), DFT and MP2 both display a slight pyramidalization at the N atoms. Within the HF approximation, the peptide linkages remain perfectly planar.

ADA has a slightly more complicated potential energy surface than GDA. Six local minima exist. Discrepancies of roughly 10° are found between the MP2 and DFT ϕ and ψ values for five of the conformations: C_5, C_{7eq}, C_{7ax}, β_2, and α'. Again, it is noted that these are all soft torsions and that no experimental results are available to determine which set of theoretical results are more accurate. However, for the α_L conformer, glaring differences of ca. 30° exist between the MP2 and DFT results. These discrepancies may be attributable in large part to DFT's tendency, even with gradient-corrected XC functionals, to overestimate the strength of hydrogen bonds. The ψ value of about 0° brings the terminal amide H as near to the other peptide linkage's N atom as possible. In turn, the ϕ value of almost 90° reduces steric hindrance as the hydrogen bond is

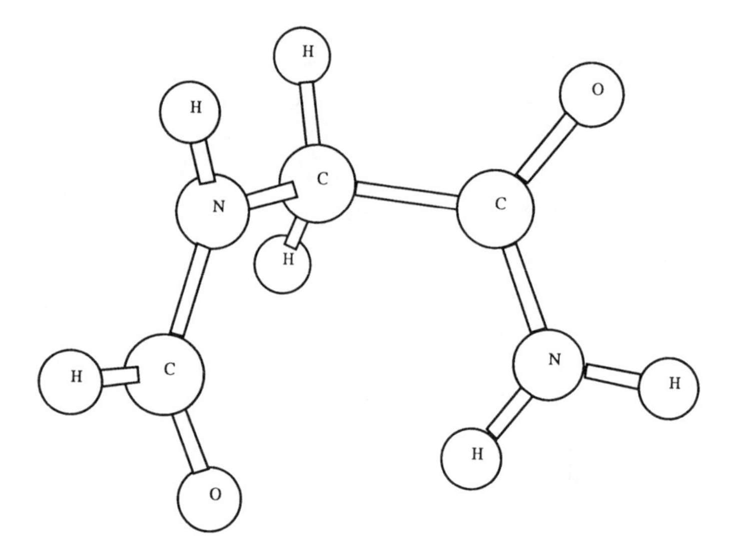

Figure 3 Glycine dipeptide analog, C_7 conformer.

formed. This situation represents an example of how a slight overestimation of the hydrogen bond strength coupled with a soft potential combine to give unrealistic geometrical values.

Table 5 lists the relative conformational energies of the GDA and ADA conformers. For GDA, the MP2 and DFT calculations indicate the seven-membered hydrogen-bonded ring to be lower in energy than the five-membered ring. Assuming this is correct, the HF calculations would be significantly in

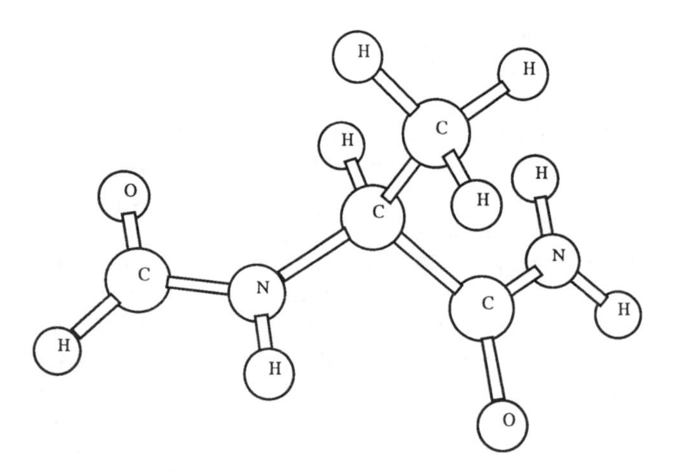

Figure 4 Alanine dipeptide analog, C_5 conformer.

Table 5 HF[a], MP2[a], and DFT[b] Relative Conformational Energies (kcal/mol) for the Glycine and Alanine Dipeptide Analogs (from Reference 12)

Conformation	HF	MP2	DFT/VWN[c]	DFT/BP[d]
Glycine dipeptide analog				
C_7	0.62	—	—	—
C_5	—	1.22	1.16	0.45
Alanine dipeptide analog				
C_{7eq}	—	—	—	—
C_5	0.13	1.18	2.61	1.32
C_{7ax}	2.52	2.17	2.22	2.07
β_2	2.36	2.96	4.39	2.77
α_L	4.57	4.25	4.78	3.15
α'	5.71	5.14	8.67	6.53

[a]6-31+G** basis sets, calculated at geometries optimized using 6-31G* basis sets.
[b]Triple-ζ + polarization on heavy atoms, double-ζ + polarization on hydrogen atoms.
[c]Vosko–Wilk–Nusair LSDA XC functional.[22]
[d]Gradient-corrected Becke exchange[17] and Perdew correlation[16] functionals used perturbatively with self-consistent Vosko–Wilk–Nusair LSDA densities.

error, preferring C_5 to C_7 by about 0.6 kcal/mol. The LSDA and MP2 results are in nearly perfect agreement, finding the C_7 conformer lower in energy by 1.2 kcal/mol. However, once gradient corrections are introduced, the DFT and MP2 energies differ by nearly 0.8 kcal/mol. The true value most likely lies between the MP2 and gradient-corrected DFT values, because MP4 calculations carried out at the MP2-optimized geometries find the C_5 conformer to lie roughly 0.9 kcal/mol above the C_7 conformer.[12]

Results for the relative conformational energies of ADA clearly indicate that the excellent agreement between the LSDA and MP2 results for the C_7/C_5 split in GDA is completely fortuitous. The only results that agree to within 0.5 kcal/mol are those for the C_{7eq}/C_{7ax} split. This agreement may not be very surprising, because this energy difference arises from steric considerations, not hydrogen bonding which is one of the LSDA's weaknesses. The discrepancies between the LSDA and MP2 results can be as high as 3.5 kcal/mol in the case of the C_{7eq} and α' conformers. Turning to the gradient-corrected DFT calculations, the agreement with the MP2 results is outstanding, within 0.2 kcal/mol, for the four lowest-lying conformers. The HF result that seems to be incorrect is the relative conformational energy of the C_5 conformer. And again, as in GDA, the energy of the C_5 conformer would appear to be far too low.

Returning to the MP2 and gradient-corrected DFT results for ADA (Table 5), significant discrepancies arise within the two highest-lying conformers, α_L and α'. In light of the radically different ϕ and ψ angles for the α_L conformer, it should not be surprising that the gradient-corrected DFT C_{7eq}/α_L split is 1.1 kcal/mol lower than its MP2 counterpart. The fact that it is lower is consonant with the premise that the DFT ϕ and ψ torsions are wrong because of an overestimation of the strength of the hydrogen bond in this conformer. Meanwhile, the α' conformer, which has the longest hydrogen bond, is found to be 1.4 kcal/mol higher in energy within the gradient-corrected DFT calculations than within the MP2 calculations. This seemingly poor result for the α' conformer from DFT may be more a consequence of overestimated hydrogen bond strengths in the other five conformers than an actual problem in the α' conformer itself.

Nucleic Acid Base Pairs

The hydrogen bonds formed between nucleic acid base pairs are crucial to the structure and function of DNA and have therefore been the subject of considerable theoretical work.[75,76] Two systems of particular interest are the Watson–Crick guanine–cytosine (GC) and adenine–thymine (AT) base pairs depicted in Figures 5 and 6. Hydrogen bond distances and strengths from high-level ab initio calculations on the GC and AT base pairs could be used to parameterize molecular mechanical force fields that then could be used to study the structure and dynamics of extended double-stranded DNA helices.

The first DFT calculations on GC and AT base pairs recently appeared.[75] Calculations were carried out within the LSDA using the VWN parameterization.[22] Calculations beyond the LSDA were also performed using Becke's gradient-corrected exchange functional[17] and Perdew's gradient-corrected correlation functional.[16] The results from these calculations are listed in Tables 6 and 7, along with those from HF and MP2 calculations.[76]

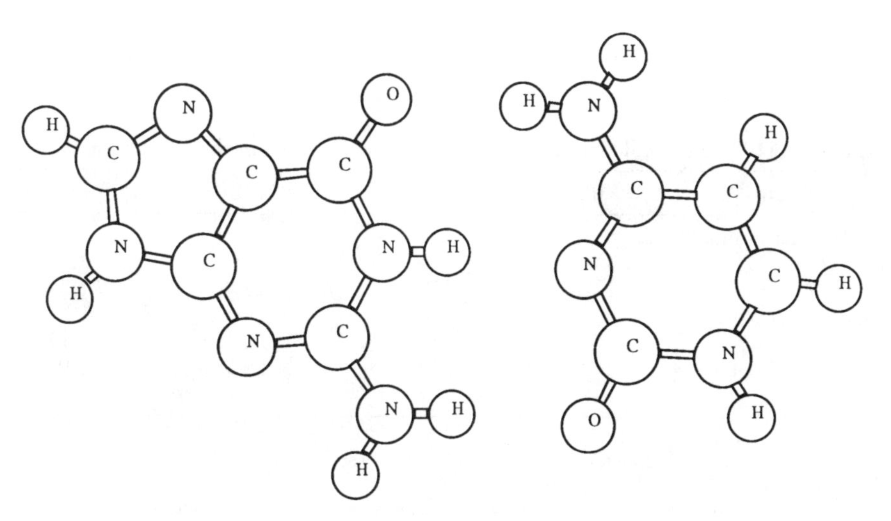

Figure 5 Watson–Crick guanine–cytosine base pair.

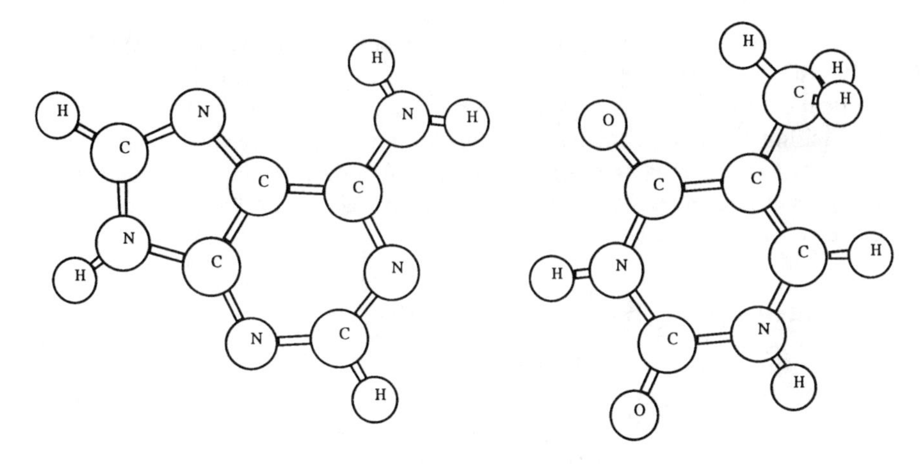

Figure 6 Watson–Crick adenine–thymine base pair.

Table 6 lists the LSDA, gradient-corrected DFT, MP2, and experimental AT and GC base pair hydrogen bond distances. It should be pointed out that the MP2 numbers are obtained by first performing full HF optimizations of the individual base pairs and then using single-point MP2 calculations to adjust the hydrogen bond distances, keeping each base of the pair frozen. The MP2 calculations are also carried out on slightly modified systems in which each of the DNA bases is methylated at the N atom that would otherwise be bound to the sugar unit in the nucleoside (it is felt that this is a better model with which to parameterize force fields[76]).

From the results in Table 6, it is clear that the gradient-corrected DFT results agree best with experiment. Both gradient-corrected DFT hydrogen bond distances in the AT base pair are within 0.05 Å of experiment. Mean-

Table 6 MP2[a] and DFT[b] Hydrogen-Bonding Distances (Å) in the DNA Base Pairs

Distance	MP2	DFT/VWN[c]	DFT/BP[d]	Experiment[e]
Adenine/thymine				
N6····O4	3.08	2.78	2.95	2.95
N1····N3	3.01	2.70	2.87	2.82
Guanine/cytosine				
O6····N4	2.93	2.65	2.82	2.91
N1····N3	3.05	2.81	2.93	2.95
N2····O1	3.01	2.78	2.93	2.86

[a]From Reference 76, 6-31G* basis set.
[b]From Reference 75, double-ζ plus polarization basis set.
[c]Vosko–Wilk–Nusair[22] LSDA functional.
[d]Gradient-corrected Becke exchange[17] and Perdew correlation[16] functionals.
[e]As cited in Reference 76.

Table 7 HF[a], MP2[a], and DFT[b] Hydrogen Bond Strengths (kcal/mol) in the DNA Base Pairs

Base pair	HF	MP2	DFT/VWN[c]	DFT/BP[d]	Experiment[e]
Adenine/thymine	7.8	11.9	22.9	12.3	—
Guanine/cytosine	19.7	25.4	39.6	26.2	21.0

[a]From Reference 76, 6-31G* basis set.
[b]From Reference 75, double-ζ plus polarization basis set.
[c]Vosko–Wilk–Nusair[22] LSDA functional.
[d]Gradient-corrected Becke exchange[17] and Perdew correlation[16] functionals.
[e]As cited in Reference 76.

while, the MP2 distances are 0.13 Å and 0.19 Å too long. As expected, the LSDA hydrogen bonds are too short by 0.17 Å and 0.12 Å. Compared to the MP2 errors, the LSDA is doing relatively well. Turning to the GC base pair, the LSDA and MP2 errors for its three hydrogen bond distances are similar to those errors seen in the AT base pair. Unfortunately, the gradient-corrected DFT results for the GC base pair are not in as good agreement with experiment as their AT base pair counterparts. Though the middle N····H hydrogen bond distance is predicted well (the distance between the two heavy atoms involved in the hydrogen bond is only 0.02 Å shorter than the experimental value of 2.95 Å), the hydrogen bond between the carbonyl oxygen of guanine and the amide hydrogen of cytosine is too short by roughly 0.1 Å, and the hydrogen bond between the carbonyl oxygen of cytosine and the amide hydrogen of guanine is too long by roughly 0.1 Å. It seems that gradient-corrected DFT makes a slight error in the relative strengths of these last two hydrogen bonds, and the base pair undergoes a hinge movement, making one hydrogen bond too short and the other too long, relative to experiment (assuming the experimental arrangement is not a by-product of crystal packing). Nonetheless, gradient-corrected DFT yields the best picture of the GC base pair geometry.

Table 7 lists the strengths of the hydrogen bonds in the DNA base pairs at the LSDA, gradient-corrected DFT, HF, and MP2 levels of theory. To obtain the enthalpy of binding at 298 K, ΔH°_{298}, LSDA vibrational analyses were used to correct both the LSDA and the gradient-corrected DFT calculations for zero point energies. Results from an HF/6-31G* vibrational analysis were used to do the same for the HF and MP2 calculations. Because the original DFT paper[75] did not tabulate the vibrational frequencies for either of the DNA base pairs, we have chosen to use the same HF/6-31G* frequencies to correct all levels of DFT and conventional ab initio theory for thermal effects.

Table 7 makes it clear that the LSDA is severely deficient. The gradient-corrected DFT and MP2 hydrogen bond strengths agree to within 1.0 kcal/mol for both the AT and GC base pairs. In contrast, the HF results indicate significantly weaker hydrogen bonding. Consistent with all results obtained to date, the LSDA grossly overestimates the strength of hydrogen bonding, by almost

100% in the AT base pair and 60% in the GC base pair compared to gradient-corrected DFT. The excellent agreement between the gradient-corrected DFT and MP2 results for the GC base pair lends credence to the claims of Gould and Kollman[76] that the experimental data for the GC base pair should be reevaluated. Unfortunately, no data are available for the Watson–Crick AT base pair.

Gradient-corrected DFT provides a very accurate description of hydrogen bonding in the DNA base pairs. It yields better hydrogen bond distances than MP2, but recall that the MP2 results did not come from a full MP2 geometry optimization. If such full MP optimizations were to be carried out, they might very well provide better results. However, they have not been performed to date because they would require an inordinate amount of CPU time. The important point here is that the gradient-corrected DFT optimizations have been affordable enough to undertake, and once completed, have yielded accurate results on this very important pair of systems.

CONCLUSIONS: THE OUTLOOK

The use of density functional methods in biomolecular modeling is still in its infancy. Systematic benchmarks have clearly shown that results from modern DFT, with gradient-corrected XC functionals, easily surpass those from HF theory. At the same time, today's DFT packages allow these calculations to be performed more rapidly than their HF counterparts. Little doubt remains that DFT will allow modelers to study larger biomolecular systems with greater accuracy than HF theory ever did before. However, today's state-of-the-art XC functionals have brought DFT only to the point where it is more or less as accurate as MP2 theory for organic systems. Clearly, DFT is dramatically less expensive than MP2. Modelers wishing to use any kind of correlated post-HF calculations must therefore restrict themselves to smaller model systems. Consequently, the DFT applications can be expected to generate more useful results, because their computational efficiency allows larger, more realistic, model systems to be tackled.

The greatest drawback of DFT today can be summarized by posing the following simple question about the XC functionals: Where do we go from here? The Vosko–Wilk–Nusair[22] and Perdew–Zunger[25] XC functionals have brought us to the limit of the local spin density approximation. The past decade has seen the introduction of several gradient-corrected XC functionals. Once the major problems of the initial functionals[30] were addressed, subsequent functionals[15–19] failed to systematically improve the DFT results. Meanwhile, the great strength of conventional ab initio theory is its well-established hierarchy of formalisms that allows one to proceed systematically toward more accurate and dependable results. For example, if one has the desire and the computational resources to go beyond MP2, MP4 may be undertaken. No such option presently exists within DFT.

Unless new and better XC functionals are introduced and DFT results are systematically improved, conventional wavefunction-based approaches should have a renaissance and remain competitive. Ultimately, when computing power evolves to the point where the cost of correlated post-HF calculations on large biomolecular systems is no longer prohibitive, DFT in its present form will have outlived its usefulness. All is not gloom and doom, however. A great deal of ongoing research is concentrated on the development of new XC functionals that hopefully will push DFT beyond MP2 and other post-HF methods. The form of these XC functionals and the increased computational effort they will introduce are not yet known.

Beyond improving the quality of DFT results, another avenue of vigorous research is to widen the gap between computational demands of DFT and conventional ab initio methodology. We now see the emergence of methods that scale linearly with the size of the system under study.[77–81] Within an LCAO context, there has been Yang's divide-and-conquer approach to DFT.[81] In it, the locality of the XC functionals is exploited and a large system is divided into smaller subsystems. Subsystem electronic densities arising from subsystem calculations are then pieced together to build the global system's electronic density. The first important feature of this method is that the computational cost scales linearly with the number of subsystems and hence only linearly with the size of the global system. The second important feature of the divide-and-conquer philosophy is its inherent parallelism. It is therefore ideally suited for massively parallel machines, the type of architectures that will soon deliver teraflop computer performance. The divide-and-conquer approach should therefore soon allow DFT calculations to be performed on large systems clearly beyond the abilities of conventional DFT and wavefunction-based approaches.

Hopefully, the systematic improvements in XC functionals will indeed come about, and efficient codes will be developed to exploit such exciting new linear-scaling approaches for performing DFT calculations on large biomolecular systems. Ultimately, we could use these DFT methods to accurately and realistically model reactions in condensed media. DFT studies of enzymatic reactions will undoubtedly yield valuable insights into often poorly understood mechanisms. Such applications may be overly ambitious and may not even become remotely feasible until well past the turn of the century. However, it is safe to say that at the present time, such goals are far more realistic, and far closer, within DFT than within any wavefunction-based ab initio approach.

ACKNOWLEDGMENTS

Financial support from the National Science and Engineering Research Council of Canada and the University of Ottawa is gratefully acknowledged. I would like to thank Wendy Cornell of the University of California, San Francisco for her comments on the preliminary versions of this chapter.

REFERENCES

1. R. G. Parr and W. Yang, *Density-Functional Theory of Atoms and Molecules,* Oxford University Press, New York, 1989. See also, L. J. Bartolotti and K. Flurchick, this volume. An Introduction to Density Functional Theory.
2. T. Ziegler, *Chem. Rev.,* **91,** 651 (1991). Approximate Density Functional Theory as a Practical Tool in Molecular Energetics and Dynamics.
3. P. Hohenberg and W. Kohn, *Phys. Rev. B,* **136,** 864 (1964). Inhomogeneous Electron Gas.
4. W. Kohn and L. J. Sham, *Phys. Rev. A,* **140,** 1133 (1965). Self-Consistent Equations Including Exchange and Correlation Effects.
5. A. Szabo and N. S. Ostlund, *Modern Quantum Chemistry: Introduction to Advanced Electronic Structure Theory,* 1st ed. Revised, McGraw-Hill, New York, 1989.
6. W. J. Hehre, L. Radom, P. v. R. Schleyer, and J. A. Pople, *Ab Initio Molecular Orbital Theory,* John Wiley & Sons, New York, 1986.
7. J. P. Dahl and J. Avery, Eds., *Local Density Approximations in Quantum Chemistry and Physics,* Plenum, New York, 1984.
8. J. K. Labanowski and J. W. Andzelm, Eds., *Density Functional Methods in Chemistry,* Springer-Verlag, New York, 1991.
9. I. Papai, A. St-Amant, J. Ushio, and D. Salahub, *Int. J. Quantum Chem., Quantum Chem. Symp.* **24,** 29 (1990). Calculation of Equilibrium Geometries and Harmonic Frequencies by the LCGTO-MCP-Local Spin Density Method. See also R. J. Bartlett and J. F. Stanton, in *Reviews in Computational Chemistry,* Vol. 5, K. B. Lipkowitz and D. B. Boyd, Eds., VCH Publishers, New York, 1994, pp. 65–169. Applications of Post–Hartree–Fock Methods: A Tutorial.
10. J. Andzelm and E. Wimmer, *J. Chem. Phys.,* **96,** 1280 (1992). Density Functional Gaussian-Type-Orbital Approach to Molecular Geometries, Vibrations, and Reaction Energies.
11. B. G. Johnson, P. M. W. Gill, and J. A. Pople, *J. Chem. Phys.,* **98,** 5612 (1993). The Performance of a Family of Density Functional Methods.
12. A. St-Amant, W. D. Cornell, T. A. Halgren, and P. A. Kollman, *J. Comput. Chem.,* in press (1995). Calculation of Molecular Geometries, Relative Conformational Energies, Dipole Moments, and Molecular Electrostatic Potential Fitted Charges of Small Organic Molecules of Biochemical Interest by Density Functional Theory.
13. F. Sim, A. St-Amant, I. Papai, and D. R. Salahub, *J. Am. Chem. Soc.,* **114,** 4391 (1992). Gaussian Density Functional Calculations on Hydrogen-Bonded Systems.
14. H. B. Schlegel, *Adv. Chem. Phys.,* **67,** 249 (1987). Optimization of Equilibrium Geometries and Transition Structures.
15. J. P. Perdew and Y. Wang, *Phys. Rev. B,* **33,** 8800 (1986). Accurate and Simple Density Functional for the Electronic Exchange Energy: Generalized Gradient Approximation.
16. J. P. Perdew, *Phys. Rev. B,* **33,** 8822 (1986). Density Functional Approximation for the Correlation Energy of the Inhomogeneous Electron Gas.
17. A. D. Becke, *Phys. Rev. A,* **38,** 3098 (1988). Density Functional Exchange Energy Approximation with Correct Asymptotic Behavior.
18. B. Miehlich, A. Savin, H. Stoll, and H. Preuss, *Chem. Phys. Lett.,* **157,** 200 (1989). Results Obtained with the Correlation Energy Density Functionals of Becke and Lee, Yang and Parr.
19. J. P. Perdew, J. A. Chevary, S. H. Vosko, K. A. Jackson, M. R. Pederson, D. J. Singh, and C. Fiolhais, *Phys. Rev. A,* **46,** 6671 (1992). Atoms, Molecules, Solids, and Surfaces: Applications of the Generalized Gradient Approximation for Exchange and Correlation.
20. P. Mlynarski and D. R. Salahub, *Phys. Rev. B,* **43,** 1399 (1991). Self-Consistent Implementation of Nonlocal Exchange and Correlation in a Gaussian Density-Functional Method.
21. K. Laasonen, F. Csajka, and M. Parrinello, *Chem. Phys. Lett.,* **194,** 172 (1992). Water Dimer Properties in the Gradient-Corrected Density Functional Theory.

22. S. H. Vosko, L. Wilk, and M. Nusair, *Can. J. Phys.*, **58,** 1200 (1980). Accurate Spin-Dependent Electron Liquid Correlation Energies for Local Spin Density Calculations: A Critical Analysis.

23. D. M. Ceperley, *Phys. Rev. B,* **18,** 3126 (1978). Ground State of the Fermion One-Component Plasma: A Monte Carlo Study in Two and Three Dimensions.

24. D. M. Ceperley and B. J. Alder, *Phys. Rev. Lett.*, **45,** 566 (1980). Ground State of the Electron Gas by a Stochastic Method.

25. J. P. Perdew and A. Zunger, *Phys. Rev. B,* **23,** 5048 (1981). Self-Interaction Correction to Density Functional Approximations for Many-Electron Systems.

26. A. D. Becke, in *The Challenge of d and f Electrons,* D. R. Salahub and M. C. Zerner, Eds., *ACS Symp. Ser.* 394, American Chemical Society, Washington, DC, 1989, p. 165. Density-Functional Theories in Quantum Chemistry: Beyond the Local Density Approximation.

27. M. Morin, A. E. Foti, and D. R. Salahub, *Can. J. Chem.*, **63,** 1982 (1985). Molecular and Electronic Structure of Ozone and Thiozone from LCAO Local Density Calculations.

28. A. D. Becke, *J. Chem. Phys.*, **96,** 2155 (1992). Density-Functional Thermochemistry. 1. The Effect of Exchange-Only Gradient Correction.

29. A. D. Becke, *J. Chem. Phys.*, **97,** 9173 (1992). Density-Functional Thermochemistry. 2. The Effect of the Perdew–Wang Generalized-Gradient Correlation Correction.

30. D. C. Langreth and M. J. Mehl, *Phys. Rev. B,* **28,** 1809 (1983). Beyond the Local-Density Approximation in Calculations of Ground State Electronic Properties.

31. L. Y. Fan and T. Ziegler, *J. Chem. Phys.*, **94,** 6057 (1991). The Influence of Self-Consistency on Nonlocal Density Functional Calculations.

32. C. Lee, W. Yang, and R. G. Parr, *Phys. Rev. B,* **37,** 785 (1988). Development of the Colle–Salvetti Correlation-Energy Formula into a Functional of the Electron Density. (The reformulation of the Lee, Yang, and Parr correlation energy functional eliminates a term involving the Laplacian of the electronic density, leaving an energy functional involving only the gradient of the density, making subsequent calculations computationally more efficient.)

33. A. D. Becke, *J. Chem. Phys.*, **98,** 1372 (1993). A New Mixing of Hartree–Fock and Local Density-Functional Theories.

34. A. D. Becke, *J. Chem. Phys.*, **98,** 5648 (1993). Density-Functional Thermochemistry. III. The Role of Exact Exchange.

35. A. D. Becke, *Int. J. Quantum Chem., Quantum Chem. Symp.*, **23,** 599 (1989). Basis-Set-Free Density Functional Quantum Chemistry.

36. DeFT and related documentation can be obtained by sending inquiries to the author at the University of Ottawa. E-mail address: st–amant@theory.chem.uottawa.ca.

37. A. St-Amant, Ph. D. Thesis, Université de Montréal (1992). deMon, un programme LCGTO-MCP-DF, et une étude théorique de l'adsorption de l'hydrogène sur les monomères et dimères de Ni, Rh et Pd.

38. D. R. Salahub, R. Fournier, P. Mlynarski, I. Papai, A. St-Amant, and J. Ushio, in *Density Functional Methods in Chemistry,* J. K. Labanowski and J. W. Andzelm, Eds., Springer-Verlag, New York, 1991, pp. 77–100. Gaussian-Based Density Functional Methodology, Software, and Applications.

39. E. J. Baerends, D. E. Ellis, and P. Ros, *Chem. Phys.*, **2,** 41 (1973). Self-Consistent Molecular Hartree–Fock–Slater Calculations. I. The Computational Procedure.

40. L. Versluis and T. Ziegler, *J. Chem. Phys.*, **88,** 322 (1988). The Determination of Molecular Structures by Density Functional Theory. The Evaluation of Analytical Energy Gradients by Numerical Integration.

41. B. Delley, *J. Chem. Phys.*, **92,** 508 (1990). An All-Electron Numerical Method for Solving the Local Density Functional for Polyatomic Molecules.

42. K. Laasonen, R. Car, C. Lee, and D. Vanderbilt, *Phys. Rev. B,* **43,** 6796 (1991). Implementation of Ultrasoft Pseudopotentials in *Ab Initio* Molecular Dynamics.

43. K. Laasonen, M. Parrinello, R. Car, C. Lee, and D. Vanderbilt, *Chem. Phys. Lett.*, **207**, 208 (1993). Structures of Small Water Clusters Using Gradient-Corrected Density Functional Theory.

44. K. Laasonen, M. Sprik, M. Parrinello, and R. Car, *J. Chem. Phys.*, **99**, 9080 (1993). Ab Initio Liquid Water.

45. A. M. Rappe, J. D. Joannopoulos, and P. A. Bash, *J. Am. Chem. Soc.*, **114**, 6466 (1992). A Test of the Utility of Plane Waves for the Study of Molecules from First Principles.

46. H. Sambe and R. H. Felton, *J. Chem. Phys.*, **62**, 1122 (1975). A New Computational Approach to Slater's-Xα Equation.

47. B. I. Dunlap, J. W. D. Connolly, and J. R. Sabin, *J. Chem. Phys.*, **71**, 3396 (1979). On Some Approximations in Applications of Xα Theory.

48. A. D. Becke and R. M. Dickson, *J. Chem. Phys.*, **89**, 2993 (1988). Numerical Solution of Poisson's Equation in Polyatomic Molecules.

49. P. Pulay, in *Applications of Electronic Structure Theory*, H. F. Schaefer III, Ed., Plenum, New York, 1977. Direct Use of the Gradient for Investigating Molecular Energy Surfaces.

50. R. Fournier, J. Andzelm, and D. R. Salahub, *J. Chem. Phys.*, **90**, 6371 (1989). Analytical Gradient of the Linear Combination of Gaussian-Type—Local Spin Density Energy.

51. B. Delley, *J. Chem. Phys.*, **94**, 7245 (1991). Analytic Energy Derivatives in the Numerical Local Density-Functional Approach.

52. L. Fan, L. Versluis, T. Ziegler, E. J. Baerends, and W. Ravenek, *Int. J. Quantum Chem., Quantum Chem. Symp.*, **22**, 173 (1988). Calculation of Harmonic Frequencies and Harmonic Force Fields by the Hartree–Fock–Slater Method.

53. A. Komornicki and G. Fitzgerald, *J. Chem. Phys.*, **98**, 1398 (1993). Molecular Gradients and Hessians Implemented in Density Functional Theory.

54. B. G. Johnson and M. J. Frisch, *Chem. Phys. Lett.*, **216**, 133 (1993). Analytic Second Derivatives of the Gradient-Corrected Density Functional Energy—Effect of Quadrature Weight Derivatives.

55. C. L. Brooks III, M. Karplus, and B. M. Pettitt, *Proteins: A Theoretical Perspective of Dynamics, Structure, and Thermodynamics*, Wiley-Interscience, New York, 1988.

56. R. Car and M. Parrinello, *Phys. Rev. Lett.*, **55**, 2471 (1985). Unified Approach for Molecular Dynamics and Density-Functional Theory.

57. M. J. Field, *J. Chem. Phys.*, **96**, 4583 (1992). Time-Dependent Hartree–Fock Simulations of the Dynamics of Polyatomic Molecules.

58. B. Hartke and E. A. Carter, *Chem. Phys. Lett.*, **216**, 324 (1993). Ab Initio Molecular Dynamics Simulated Annealing at the Generalized Valence Bond Level—Application to a Small Nickel Cluster.

59. J. P. Ryckaert, G. Ciccotti, and H. J. C. Berendsen, *J. Comput. Phys.*, **23**, 327 (1977). Numerical Integration of the Cartesian Equations of Motion of a System with Constraints: Molecular Dynamics of n-Alkanes.

60. M. C. Payne, M. P. Teter, D. C. Allan, T. A. Arias, and J. D. Joannopoulos, *Rev. Mod. Phys.*, **64**, 1045 (1992). Iterative Minimization Techniques for *Ab Initio* Total Energy Calculations —Molecular Dynamics and Conjugate Gradients.

61. L. Fan and T. Ziegler, *J. Chem. Phys.*, **92**, 3645 (1990). The Application of Density Functional Theory to the Optimization of Transition State Structures. I. Organic Migration Reactions.

62. L. Fan and T. Ziegler, *J. Am. Chem. Soc.*, **114**, 10890 (1992). Nonlocal Density Functional Theory as a Practical Tool in Calculations on Transition States and Activation Energies. Applications to Elementary Reaction Steps in Organic Chemistry.

63. R. V. Stanton and K. M. Merz Jr., *J. Chem. Phys.*, **100**, 434 (1994). Density Functional Transition States of Organic and Organometallic Reactions.

64. T. Oie, I. A. Topol, and S. K. Burt, *J. Phys. Chem.*, **98**, 1121 (1994). Ab Initio and Density Functional Calculations on Ethylene Glycol.

65. C. Liang, C. S. Ewig, T. R. Stouch, and A. T. Hagler, *J. Am. Chem. Soc.*, **116**, 3904 (1993). Ab Initio Studies of Lipid Model Species. 2. Conformational Analysis of Inositols.

66. P. Politzer and J. S. Murray, in *Reviews in Computational Chemistry*, Vol. 2, K. B. Lipkowitz and D. B. Boyd, Eds., VCH Publishers, New York, 1991, pp. 273–312. Molecular Electrostatic Potentials and Chemical Reactivity.

67. P. Cieplak and P. Kollman, private communication.

68. W. L. Jorgensen, J. Chandrasekhar, J. D. Madura, R. W. Impey, and M. L. Klein, *J. Chem. Phys.*, **79**, 926 (1983). Comparison of Simple Potential Functions for Simulating Liquid Water.

69. L. X. Dang, *Chem. Phys. Lett.*, **213**, 541 (1993). Solvation of Ammonium Ion—A Molecular Dynamics Simulation with Nonadditive Potentials.

70. J. Bajorath, J. Kraut, Z. Li, D. H. Kitson, and A. T. Hagler, *Proc. Natl. Acad. Sci. USA*, **88**, 6423 (1991). Theoretical Studies of the Dihydrofolate Reductase Mechanism: Electronic Polarization of Bound Substrates.

71. J. Bajorath, D. H. Kitson, G. Fitzgerald, J. Andzelm, J. Kraut, and A. T. Hagler, *Proteins*, **9**, 217 (1991). Electron Redistribution on Binding of a Substrate to an Enzyme: Folate and Dihydrofolate Reductase. J. Bajorath, Z. Li, G. Fitzgerald, D. H. Kitson, M. Farnum, R. M. Fine, J. Kraut, and A. T. Hagler, *Proteins*, **11**, 263 (1991). Changes in the Electron Density of the Cofactor NADPH on Binding to *E. coli* Dihydrofolate Reductase.

72. J. Gao, this volume. Methods and Applications of Combined Quantum Mechanical and Molecular Mechanical Potentials.

73. M. J. Frisch, J. E. Del Bene, J. S. Binkley, and H. F. Schaefer III, *J. Chem. Phys.*, **84**, 2279 (1986). Extensive Theoretical Studies of the Hydrogen-Bonded Complexes $(H_2O)_2$, $(H_2O)_2H^+$, $(HF)_2$, $(HF)_2H^+$, F_2H^-, and $(NH_3)_2$. See also S. Scheiner, in *Reviews in Computational Chemistry*, Vol. 2, K. B. Lipkowitz and D. B. Boyd, Eds., VCH Publishers, New York, 1991, pp. 165–218. Calculating the Properties of Hydrogen Bonds by Ab Initio Methods.

74. C. L. Brooks III and D. A. Case, *Chem. Rev.*, **93**, 2487 (1993). Simulations of Peptide Conformational Dynamics and Thermodynamics.

75. R. Santamaria and A. Vazquez, *J. Comput. Chem.*, **15**, 981 (1994). Structural and Electronic Property Changes of the Nucleic Acid Bases upon Base Pair Formation.

76. I. R. Gould and P. A. Kollman, *J. Am. Chem. Soc.*, **116**, 2493 (1994). Theoretical Investigation of the Hydrogen Bond Strengths in Guanine-Cytosine and Adenine-Thymine Base Pairs.

77. S. Baroni and P. Giannozzi, *Europhys. Lett.*, **17**, 547 (1992). Towards Very Large Scale Electronic Structure Calculations.

78. F. Mauri, G. Galli, and R. Car, *Phys. Rev. B*, **47**, 9973 (1993). Orbital Formulation for Electronic Structure Calculations with Linear System-Size Scaling.

79. P. Ordejon, D. A. Drabold, M. P. Grumback, and R. M. Martin, *Phys. Rev. B*, **48**, 14646 (1993). Unconstrained Minimization Approach for Electronic Computations Which Scales Linearly with System Size.

80. X. P. Li, R. W. Nunes, and D. Vanderbilt, *Phys. Rev. B*, **47**, 10891 (1993). Density Matrix Electronic Structure Method with Linear System-Size Scaling.

81. W. Yang, *J. Mol. Struct. (THEOCHEM)*, **225**, 461 (1992). Electron Density as the Basic Variable: A Divide-and-Conquer Approach to the *Ab Initio* Computation of Large Molecules.

CHAPTER 6

The A Priori Calculation of Vibrational Circular Dichroism Intensities

Danya Yang and Arvi Rauk

Department of Chemistry, The University of Calgary, Calgary, Alberta, Canada T2N 1N4

INTRODUCTION

A study of the structures of molecules requires techniques that are sensitive to the three-dimensional shapes, and ideally the property of handedness. Handed molecules that have been separated, at least in part, from their antipodes, are said to be *optically active* in that they behave differently toward the right and left components of circularly polarized light (CPL). All biological molecules of any complexity have this characteristic. The differential interaction with CPL may manifest itself as a differential scattering, which is measured as a non-zero *rotation* of the plane of polarization of plane polarized light, measurable with a polarimeter at specific wave lengths, or as a function of wave length, λ, as *optical rotatory dispersion* (ORD).

Optical activity also manifests itself as a differential *absorption* of left and right CPL, $\Delta A = A_L - A_R$ or $\Delta\epsilon = \epsilon_L - \epsilon_R$, the wavelength dependence of which is called circular dichroism (CD), and is measured on a CD spectrometer. The area under the CD absorption band of a given transition, j, is the *rotatory strength*, $[R]_j$,

Reviews in Computational Chemistry, Volume 7
Kenny B. Lipkowitz and Donald B. Boyd, Editors
VCH Publishers, Inc. New York, © 1996

$$[R]_j = 0.229 \times 10^{-38} \int_0^\infty \Delta\epsilon_j(\lambda) \frac{d\lambda}{\lambda} \qquad [1]$$

in units of esu^2cm^2, where $\Delta\epsilon$ is in units of molar extinction coefficient. An analogous measure of the area under the normal absorption band determines the *oscillator strength* of the transition. A substance that consists of a single antipode is said to be *optically pure*. The *absolute configuration* of a chiral substance distinguishes between the two antipodes. The antipodes (*enantiomers*) of a substance have rotatory strengths of equal magnitude and opposite in sign for every transition. Accordingly, their CD spectra are mirror images of each other. It is sufficient to predict correctly the sign of the rotatory strength of any single transition in order to assign the absolute configuration of a substance, provided the relative spatial dispositions of the nuclei, i.e., the conformation, is known. Different conformations of the same optically pure substance may have very different CD spectra, with rotatory strengths differing in magnitude and possibly in sign.

In general, the principal objective of chiroptical techniques is not to determine absolute configurations, because for the most part these are known, but rather to determine conformations in solution and in the gas phase. Long established in the UV/visible region of the electromagnetic spectrum, CD spectroscopy has been applied to the infrared (IR) region in the last two decades, and significant advances in the instrumental, theoretical, and computational aspects have taken place, both in Raman[1–3] and conventional IR spectroscopy. We will focus here on *vibrational circular dichroism* (VCD), the circular dichroism analog of conventional IR spectroscopy.

HISTORICAL OVERVIEW

The first definitive measurements of VCD spectra were carried out in the early 1970s in Holtzwarth's group at the University of Chicago, first in the solid state,[4] and then in isotropic solution,[5] using optics sensitive to the hydrogenic stretching region of the spectrum. A second instrument built by L. A. Nafie, T. A. Keiderling, and P. J. Stephens at the University of Southern California was used to measure the 2500–4000 cm^{-1} VCD spectra of a large variety of chiral molecules, and essentially established the tractability and usefulness of such measurements.[6,7] From these seminal groups, improved instruments have been assembled[8–11] and applied to a wide range of studies of absolute configurations and conformations of small molecules (see below), and to proteins, nucleic acids,[12–15] and synthetic polymers.[16]

Theoretical developments have kept pace with or surpassed developments in the experimental side of VCD. Since the earliest theory developed by

Deutsche and Moscowitz,[17,18] a number of semiempirical models have been proposed to explain features in experimental spectra. Although none has enjoyed any but the most modest successes, experience is gradually accumulating on the limits of applicability. Some of these are reviewed below. It is readily shown that within the Born–Oppenheimer (BO) approximation,[19] the electronic contribution to the magnetic dipole transition moment associated with a vibrational transition of a molecule in its ground electronic state vanishes. A non-BO theory of VCD intensities was independently developed by several groups.[20–24]

Ab initio computer codes have been available in the last five or six years to calculate VCD intensities by several variations of the theory. Agreement of simulated theoretical spectra with experimentally measured spectra for small molecules is often excellent and sufficiently accurate to establish VCD as the first spectroscopic method for the determination of absolute configurations of small molecules in the gas or liquid phase. Most of the available computational work is reviewed in the sections below.

EXPERIMENTAL ASPECTS

The overwhelming fact that determines the ability to measure VCD spectra experimentally is the intrinsic weakness of the signal compared to normal IR. Typically, $\Delta\epsilon/\epsilon \approx 10^{-5}$. Current experimental aspects of VCD spectroscopy have been reviewed in detail.[9,25,26] As of this writing, there are no commercial CD spectrometers that operate in the usual IR region of the spectrum (400–4000 cm^{-1}). Existing instruments are laboratory modifications of conventional IR spectrometers using additional commercially available components and fall into two categories, namely scanning dispersive (most commonly for the H and C=O stretching region, 2800–3500 cm^{-1} and 1500–1800 cm^{-1}, respectively)[27–30] and Fourier transform CD instruments (almost exclusively for the mid-IR, 800–1500 cm^{-1}).[9,26,31,32] The spectral range that can presently be accessed is evidently limited. In general, neither of the two kinds of spectrometers is capable of covering the entire fundamental vibrational spectrum and neither can reach below 800 cm^{-1}, although measurements down to 650 cm^{-1} have been reported on two occasions.[28,33]

The light source is either a xenon arc lamp, a silicon carbide rod, or a Nernst glower, depending on the spectral region and intensity that is thought to be necessary. Whereas there are several methods for producing circularly polarized radiation, the polarizing element that is presently most commonly used is a photoelastic modulator (PEM, commercially available from Hinds Instruments) for reasons of technical feasibility. It is largely the PEM crystal that further restricts the accessible spectral range. Of the two optical materials for the PEM that are most frequently used, namely CaF_2 and ZnSe, the former material offers greater radiation throughout its range from 900 cm^{-1} up into the H stretching region, while the latter transmits in principle down to 600 cm^{-1} but is less transparent than the former from 900 cm^{-1} up.

One further consideration is the frequency with which the normally isotropic crystal is piezoelectrically compressed to produce the polarized radiation. Typically, a CaF_2 crystal is driven at 50 kHz, and a ZnSe crystal at 37 kHz. Light, after traversing the grating monochromator in the case of a dispersive instrument or after a beamsplitter in the case of a Fourier transform (FT) interferometer, is passed through a polarizer of suitable and matching optical material oriented at 45° with respect to the stress axis of the PEM. It is then focused through the PEM onto the sample and refocused usually through a lens of the appropriate optical material to a liquid nitrogen-cooled semiconductor detector. Above 2000 cm^{-1}, an InSb detector provides greater sensitivity, but below 2000 cm^{-1}, a HgCdTe (MCT) detector must be used. With an FT interferometer, filtering of the incoming light within a certain wave number range is normally necessary to avoid detector saturation. The signal is processed electronically slightly differently for the two types of spectrometers requiring demodulation through a lock–in amplifier in both cases. To produce adequate signal to noise, upwards of 10 scans are typically accumulated with a resolution of 10 cm^{-1} with a dispersive spectrometer, and from 5000 to 20,000 scans with a resolution of 4 cm^{-1} with an FT interferometer.

Absorption and baseline artifacts, especially with the latter instrument, invariably pose severe problems requiring caution when interpreting the observed spectra. Such artifacts, which can arise from any reflective surfaces in the optical train or from inherent strain in the detector windows or lenses, are usually removed by subtracting spectra of the racemic compounds recorded with the same conditions, or the CD spectra of the opposite enantiomer. In a recent report, especially high signal to noise and essentially complete removal of artifacts were claimed with an FT interferometer in the range 900–1800 cm^{-1}, making it possible to record high-quality CD spectra and requiring only between 100 and 5000 scans.[26]

Standard IR cells with windows appropriate for the materials under investigation are used. To avoid erroneous VCD signals, absorption intensities should be kept in the range 0.2–0.8, and it is usually good practice to check instrument performance regularly with a compound for which the VCD spectrum is well documented, e.g., α-pinene. With an FT interferometer, VCD intensities must be calibrated,[8–10] a procedure in which the sample is replaced by a second polarizer. Samples are usually in solution in a suitable solvent, in a matrix of solid inert gas, or neat liquids. The spectra of a few selected gaseous materials have been reported.[34,35] Solid samples of chiral compounds are not suitable for CD measurements because they also exhibit linear dichroism that generally dominates and obscures the circular dichroism.

THEORY IN BRIEF

Rosenfeld[36] showed that the rotatory strength of a transition, *j,* assumed to be between states 0 and *n,* is the imaginary part of the scalar product of the electric dipole and magnetic dipole transition moments for the transition:

$$[R]_j = [R_{0n}] = \mathrm{Im}(\langle\Psi_0|\vec{\boldsymbol{\mu}}|\Psi_n\rangle\cdot\langle\Psi_n|\vec{\mathbf{m}}|\Psi_0\rangle) \qquad [2]$$

where the electric and magnetic dipole moment operators, $\vec{\boldsymbol{\mu}}$ and $\vec{\mathbf{m}}$, respectively, are the sums of electron and nuclear operators,

$$\vec{\boldsymbol{\mu}} = \vec{\boldsymbol{\mu}}^e + \vec{\boldsymbol{\mu}}^n = -\sum_i e\vec{\mathbf{r}}_i + \sum_I Z_I e\vec{\mathbf{R}}_I \qquad [3]$$

and

$$\vec{\mathbf{m}} = \vec{\mathbf{m}}^e + \vec{\mathbf{m}}^n = -\sum_i \frac{e}{2mc}\vec{\mathbf{r}}_i \times \vec{\mathbf{p}}_i + \sum_I \frac{Z_I e}{2M_I c}\vec{\mathbf{R}}_I \times \vec{\mathbf{P}}_I \qquad [4]$$

The quantities, e, m, $\vec{\mathbf{r}}_i$, and $\vec{\mathbf{p}}_i$, are the charge, mass, position, and momentum of the *i*th electron, respectively; $Z_I e$, M_I, $\vec{\mathbf{R}}_I$ and $\vec{\mathbf{P}}_I$, are the charge, mass, position, and momentum of the *I*th nucleus, respectively; and c is the speed of light. The transition may be between electronic states, typically the ground state and an electronically excited state, or it may be between different vibrational levels of the ground state. The theoretical description of electronic CD and VCD[37,38] therefore reduces to the task of evaluating the transition moments by Eq. [2] in a computationally viable manner. Because of the inherent complexities involved, reasonably reliable evaluation of transition moments has only been available relatively recently for small molecules.

For vibrational transitions it is inappropriate to use the BO approximation, because the ground and excited states in the context of vibrational transitions have the same electronic wavefunction and differ only in the nuclear wavefunctions, a consequence of which is that the electronic contribution to the magnetic dipole transition matrix element vanishes in the BO approximation, and this is physically unacceptable. To include the important electronic contribution to magnetic dipole transition moments, one must choose either to make further approximations to the magnetic dipole operator yielding effective nonvanishing magnetic transition moments or to go beyond the BO approximation. Various approximate models and exact a priori methods have been investigated in the last 25 years.

Approximate models include the coupled oscillator (CO),[39] the fixed partial charge (FPC),[17,18,40] the charge flow (CF),[41] the ring current,[42] the dynamic polarizability (DP),[43] the bond dipole (BD),[44] the localized molecular orbital (LMO),[45] the nonlocalized molecular orbital (NMO),[46] the atomic polar tensor (APT),[47] and the very recent locally distributed origins gauge (LDO) model.[48] In the simplest CO and FPC models, the BO approximation is bypassed completely. In the CO model, VCD is considered to result from the interaction of two achiral dipole oscillators chirally disposed relative to each other in space. In the FPC model, the electronic effects are introduced by considering the electrons as serving to screen the nucleus, resulting in an effective nuclear charge that rigidly moves with the nucleus during vibrations. Charge redistribution during vibrations is totally missing in the FPC model, and later efforts were made to include this effect. In the CF model, the charge is allowed to flow along bonds from one atom to another resulting in changes to the screened nuclear charges. The ring current model accounts specifically for possible charge flow around a closed ring system. The charge redistribution is viewed as inducing the electric dipole moments in the DP model. The BD model is closely related to the CF model and is different from the CF model in that the equivalent of the "fixed charge" contribution comes from the bond dipole contributions.

A number of approximate VCD models are based on electronic structure calculations. The LMO model[45] is a more accurate molecular orbital (MO) approach in evaluation of VCD intensities and has been implemented at the ab initio level. The nuclear and the electronic contributions to the dipole transition moments are treated completely separately in this model. The expressions of nuclear contributions are derived with full nuclear charges. For the expression of the electronic contributions, the BO approximation is invoked after the electronic part of the magnetic dipole operator is modified in such a way that the BO difficulty is avoided. Nonvanishing electronic contributions to the magnetic transition moments are thus obtained, which come from the displacements of the centroids of the localized MOs during vibrations.

The NMO model,[46] which uses nonlocalized MOs and is based on atomic contributions to the dipole transition moments, is the MO analog of the empirical CF model. An additional term, which describes the contribution due to quantum mechanical rehybridization effects on atoms, appears in the NMO model. The APT model is an approximate form of the NMO model. In the APT model,[47] the magnetic dipole transition moment is evaluated through the infrared atomic polar tensors (APT), which are derivatives of the full molecular electric dipole moment with respect to nuclear Cartesian coordinates. The LDO model[48] is derived from the exact a priori vibronic coupling theory (VCT) formalism by ignoring terms involving local atomic magnetic dipole moment operators and converting the remaining term involving velocity-form atomic APT contributions to position-form in a well-defined manner.

Exact a priori formulations for VCD intensities are obtained by adding high-order corrections to the BO wavefunctions. Nafie and Freedman,[22] and to a certain extent also Craig and Thirunamachandran,[20] have developed the vibronic coupling theory of VCD, in which a general molecular wavefunction is derived as a linear combination of BO wavefunctions. The mixing of electronic and nuclear contributions is accomplished as a vibronic coupling effect through a perturbation approach using a part of the nuclear kinetic energy as the perturbation operator. Alternatively, it was shown that at the level of linear couplings, the same result for the magnetic dipole transition moment may be derived from a magnetic field interaction as a perturbation operator, and this theory is widely referred to as the magnetic field perturbation (MFP)[23] theory of VCD. Another method based on a floating basis set formalism was proposed by Freedman and Nafie.[49] In this model, the atomic orbitals are treated to float not only with the nuclear positions but also with the nuclear velocities. This method has not yet been implemented at the ab initio level. More recently, Nafie developed a nuclear velocity perturbation (NVP) velocity-gauge formalism for VCD and IR intensities.[50] This approach involves the exact incorporation of all or part of the dependence of the electronic wavefunction on an electron-velocity perturbation, such as the vector potential or the velocities of the nuclei, into the atomic orbital basis functions as a gauge transformation. This method has not been implemented yet either.

The following is a brief description of the VCT and the MFP approaches intended to lay a foundation for the discussion of the practical aspects of VCD calculations. A fuller description was provided earlier[38] and is available from the original literature cited.

Within the BO approximation, the total molecular Hamiltonian has the following form

$$H\Psi_e(\vec{r},Q)\Phi_{ev}(Q) = [H_E(Q) + T_N]\Psi_e(\vec{r},Q)\Phi_{ev}(Q) \quad [5a]$$

$$\approx \Psi_e(\vec{r},Q)[E_e(Q) + T_N]\Phi_{ev}(Q) \quad [5b]$$

$$= E_{ev}\Psi_e(\vec{r},Q)\Phi_{ev}(Q) \quad [5c]$$

where $H_E(Q)$ is the electronic Hamiltonian, T_N is the nuclear kinetic energy operator, $E_e(Q)$ and $\Psi_e(Q)$ are, respectively, the electronic energy (but including nuclear–nuclear repulsion) and the electronic wavefunction, $\Phi_{ev}(Q)$ is the nuclear wavefunction, and E_{ev} is the total energy. The variables $\vec{r}$ and Q refer collectively to electron positions and mass-weighted normal nuclear coordinates, respectively.

The approximation inherent in Equation [5b] means that the nuclear kinetic operator, $(-\hbar^2/2)(\partial^2/\partial Q^2)$, has no influence on the electronic wavefunction, namely, the terms, $-\hbar^2(\partial\Psi_e/\partial Q)_{el}(\partial/\partial Q)_n$ and $(-\hbar^2/2)(\partial^2\Psi_e\partial Q^2)$, which operate on the electronic part of the wavefunction, are omitted. This implies that infinitesimal changes in the nuclear configuration do not affect the electronic wavefunction, Ψ_e. This approximation is, however, not good enough in evaluations of the magnetic dipole transition moments. The disturbance to the electronic state caused by changes in the nuclear configuration has to be taken into account. The neglected terms can be brought back by using them as perturbation operators to mix BO wavefunctions, namely, an arbitrary electronic-vibrational state, (e',v'), can be expanded in terms of adiabatic BO wavefunctions, $\{\Psi_e\Phi_{ev}\}$, as

$$\Psi(\vec{r},Q)_{e',v'} \approx \Psi(\vec{r},Q)_{e'}\Phi(Q)_{e',v'} - \sum_{e(\neq e')} a(Q)_{e'v',ev}\Psi(\vec{r},Q)_e\Phi(Q)_{e,v} \quad [6]$$

where the coefficients are defined, on the basis of Rayleigh–Schrödinger perturbation theory, as

$$a(Q)_{e'v',ev} = \frac{\langle\Psi(\vec{r},Q)_e\Phi(Q)_{e,v}|\hat{T}'_N|\Psi(\vec{r},Q)_{e'}\Phi(Q)_{e',v'}\rangle}{E(Q)_{e,v} - E(Q)_{e',v'}} \quad [7]$$

and $\hat{T}'_N$ refers to the two nuclear kinetic energy perturbation operators that are omitted in the BO approximation.

The unperturbed electronic wavefunctions, Ψ_0 and Ψ_e, which depend on the nuclear coordinates, Q, can be expressed as a Taylor series about the equilibrium geometry of the ground electronic state, where $Q = 0$, that is,

$$\Psi(\vec{r},Q)_{e'} = (\Psi_{e'})_0 + \left(\frac{\partial\Psi_{e'}}{\partial Q}\right)_0 Q + \cdots \quad [8]$$

Equation [8] is then substituted into [7] and [6]. Only first-order terms are kept in the wavefunction and in the derived matrix elements, and an average energy approximation is invoked to permit closure of the sums over vibrational states. For the electronic parts of the dipole transition moments, one obtains, for the jth normal mode

$$\langle\psi_{0,0}|\vec{\mu}^E|\psi_{0,1}\rangle^j = 2\left(\frac{\hbar}{2\omega_j}\right)^{1/2}\left\langle\Psi_0|\vec{\mu}^E|\frac{\partial\Psi_0}{\partial Q_j}\right\rangle_0 \quad [9]$$

and

$$\langle\psi_{0,0}|\vec{m}^E|\psi_{0,1}\rangle^j = 2\hbar^2\left(\frac{\omega_j}{2\hbar}\right)^{1/2}\sum_{e\neq 0}\frac{\langle\Psi_0|\vec{m}^E|\Psi_e\rangle_0\left\langle\Psi_e|\frac{\partial\Psi_0}{\partial Q_j}\right\rangle_0}{(E_e - E_0)_0} \quad [10]$$

Thus, a nonvanishing electronic contribution to the magnetic dipole transition moment is obtained. Assuming nuclear wavefunctions to be harmonic oscillator wavefunctions, the final expressions for the individual electric and magnetic dipole transition moments are obtained respectively as,

$$(\vec{\boldsymbol{\mu}})^{VC}_{(00,01)j} = e\left(\frac{\hbar}{2\omega_j}\right)^{1/2}\left[\sum_{I}^{atoms} Z_I\vec{\mathbf{S}}^j_I - 2\sum_{I}^{atoms}\sum_{i}^{electrons}\left\langle\Psi_0|\vec{\mathbf{r}}_i\frac{\partial}{\partial\vec{\mathbf{R}}_I}|\Psi_0\right\rangle\vec{\mathbf{S}}^j_I\right] \qquad [11]$$

$$(\vec{\mathbf{m}})^{VC}_{(01,00)j} = \frac{e}{2c}\left(\frac{\hbar\omega_j}{2}\right)^{1/2}\left[i\sum_{I}^{atoms} Z_I(\vec{\mathbf{R}}^0_j \times \vec{\mathbf{S}}^j_I) + \frac{2\hbar}{m_e}\sum_{I}^{atoms}\sum_{i}^{electrons}\sum_{e\neq 0}^{excited}\langle\Psi_0|\vec{\mathbf{r}}_i \times \vec{\mathbf{p}}_i|\Psi_e\rangle\left\langle\Psi_e\left|\frac{\partial}{\partial\vec{\mathbf{R}}_I}\right|\Psi_0\right\rangle\vec{\mathbf{S}}^j_I\frac{1}{E_e - E_0}\right] \qquad [12]$$

where Z_I is the bare nuclear charge of atom I with positional vector $\vec{\mathbf{R}}_I$, $\vec{\mathbf{S}}^j_I = (\partial\vec{\mathbf{R}}_I/\partial\mathbf{Q}_j)$ is the displacement vector during a normal vibrational mode $\mathbf{Q}_j$, $\vec{\mathbf{r}}_i$ and $\vec{\mathbf{p}}_i$ are the position and momentum of the ith electron, ω_j is the frequency of the jth normal mode, and $E_e - E_0$ is the vertical electronic excitation energy. The electronic wavefunctions are denoted as Ψ_0 and Ψ_e for ground and excited states, respectively. The superscript *VC* indicates that these expressions describe a vibronic coupling mechanism for IR and VCD intensities. The subscripts (00,01) and (01,00) signify that the transition is between the 0 and 1 vibrational levels of the ground electronic state of the molecule.

In the MFP approach, the magnetic field dependence of the electronic ground state wavefunction may be elicited by expansion in Taylor and perturbation series. Thus, Taylor expansion, truncating after the linear term, yields

$$\Psi_0(\vec{\mathbf{R}}^0,\vec{\mathbf{B}}) \approx (\Psi_0(\vec{\mathbf{R}}^0,\vec{\mathbf{B}}))_{\vec{\mathbf{B}}=0} + \left(\frac{\partial\Psi^0(\vec{\mathbf{R}}^0,\vec{\mathbf{B}})}{\partial\vec{\mathbf{B}}}\right)_{\vec{\mathbf{B}}=0}\cdot\vec{\mathbf{B}} \qquad [13]$$

where $\vec{\mathbf{B}}$ is the external magnetic field. Alternatively, expansion as a Rayleigh–Schrödinger perturbation series, with

$$\mathrm{H}'(\vec{\mathbf{B}}) = -\vec{\mathbf{m}}^E\cdot\vec{\mathbf{B}} \qquad [14]$$

yields, to first order,

$$\Psi_0(\vec{\mathbf{R}}^0,\vec{\mathbf{B}}) \approx (\Psi_0(\vec{\mathbf{R}}^0,\vec{\mathbf{B}}))_{\vec{\mathbf{B}}=0} + \sum_{e(\neq 0)}\frac{\langle\Psi_e|\vec{\mathbf{m}}^E|\Psi_0\rangle}{E_e - E_0}\cdot\vec{\mathbf{B}}\Psi_e \qquad [15]$$

From comparison of Eqs. [13] and [15], one may identify the relationship

$$\left(\frac{\partial\Psi_0(\vec{\mathbf{R}}^0,\vec{\mathbf{B}})}{\partial\vec{\mathbf{B}}}\right)_{\vec{\mathbf{B}}=0} \approx \sum_{e(\neq 0)}\frac{\langle\Psi_e|\vec{\mathbf{m}}^E|\Psi_0\rangle}{E_e - E_0}\Psi_e \qquad [16]$$

Insertion of Eq. [16] into [10] yields an expression in which the sum over states does not appear. The interesting result is that the electronic contribution to the magnetic dipole transition moment is proportional to the overlap of two derivatives of the ground state electronic wavefunction:

$$\langle \psi_{0,0} | \vec{m}^E | \psi_{0,1} \rangle = 2\hbar^2 \left(\frac{\omega_i}{2\hbar} \right)^{1/2} \left\langle \left(\frac{\partial \Psi_0(\vec{R}^0, \vec{B})}{\partial \vec{B}} \right)_{\vec{B}=0} \middle| \left(\frac{\partial \Psi_0(\vec{R})}{\partial Q} \right) \right\rangle_{\vec{R}=0} \qquad [17]$$

where the subscripts, $\vec{B} = 0$ and $\vec{R} = 0$, indicate that the derivatives are evaluated at zero magnetic field and at the equilibrium geometry, respectively.

COMPUTATIONAL CONSIDERATIONS

We give here a description of the practicalities to consider if one undertakes the computation of vibrational circular dichroism intensities. Some of these will be general for any of the theoretical approaches, and some will be specific for the VCT method which we have implemented at the ab initio level.

In practice, the electric and magnetic dipole transition moments are usually expressed as summations of atomic properties, namely the atomic polar tensor (APT), $\mathbf{P}^n_{a\beta}$, and atomic axial tensor (AAT), $\mathbf{M}^n_{a\beta}$,[37] respectively. In the VCT approach, these tensors can be extracted from Eqs. [11] and [12] as

$$\mathbf{P}^n_{a\beta} = eZ^n \delta_{a\beta} - 2e \sum_i^{electrons} \left\langle \Psi_0 \middle| \vec{r}^{\,i}_\beta \frac{\partial}{\partial \vec{R}^n_a} \middle| \Psi_0 \right\rangle \qquad [18]$$

and

$$\mathbf{M}^n_{a\beta} = \frac{e}{2c} Z^n \vec{R}^n_{0,a} \epsilon_{a\beta\gamma}$$

$$- \frac{i\hbar e}{m_e c} \sum_i^{electrons} \sum_{e \neq 0}^{excited} \frac{\langle \Psi_0 | (\vec{r}^{\,i} \times \vec{p}^{\,i})_\beta | \Psi_e \rangle}{E_e - E_0} \left\langle \Psi_e \middle| \frac{d}{\partial \vec{R}^n_a} \middle| \Psi_0 \right\rangle \qquad [19]$$

The definition of $\mathbf{M}^n_{a\beta}$ here differs by a factor of $i/2\hbar$ from the one introduced by Stephens.[51] The rotatory strengths of the fundamental vibrational transitions are thus determined by the APT, AAT, and the molecular force field from which the descriptions of the vibrational normal modes are derived. The evaluation of the force field and APTs, which determine the IR intensities, has been well established at both the SCF and higher levels of theory, for instance, second-order Møller–Plesset (MP2) theory, and used in the evaluation of VCD intensities. The calculation of AATs, on the other hand, at similar levels has posed some difficulties.

Origin Gauges

One difficulty associated with the calculation of AATs is the choice of coordinate system, that is, the origin gauges, because AAT is in general origin dependent, unlike APT, which is origin independent. As a consequence, the evaluated rotatory strengths may be origin dependent. A shift of the origin of the coordinate frame results in an extra term in the expression for the rotatory strength because of the origin dependence of AAT. This extra term can be shown to actually equal zero for exact wavefunctions, yielding an origin independent rotatory strength. However, such exact wavefunctions are not available in practice, and the vanishing of that extra term is not necessarily attained.

It should be mentioned that an alternative expression exists for the electric dipole operator,[37,52] in which the velocities instead of the positions of the particles are involved. The calculated rotatory strength is origin independent for any wavefunctions when the velocity form electric dipole operator is used. However, matrix elements involving velocity (gradient) operators are much more difficult to evaluate accurately, and the errors introduced by using the velocity form of the operator are normally bigger than the uncertainties due to the origin-dependent rotatory strength unless very large basis sets are used.[53] As a consequence, the velocity form of the electric dipole operator is rarely used in practical calculations of rotatory strengths.

Efforts have long been made to seek special coordinate systems, or origin gauges for convenience, to minimize the uncertainties due to the use of inexact electronic wavefunctions. The common origin (CO) gauge is the earliest one adopted in ab initio calculations of rotatory strengths. In the CO gauge, a single origin is chosen for the evaluation of all AATs. Not surprisingly, different choices of the origin in the CO gauge may yield different rotatory strengths. Although some of the arbitrariness may be removed from the choice of origin by selection of a structure-related point such as the molecular center of mass, the possibility remains that magnetic properties may be poorly represented at a point remote from centers of charge density. Stephens has proposed a distributed origin (DO) gauge in conjunction with the MFP approach.[51] The DO gauge in the VCT formalism was subsequently developed and implemented as well.[54] In the DO gauge with origins at nuclei, each AAT is evaluated with its own origin at the equilibrium position of the nucleus corresponding to that AAT. The rotatory strength calculated in the DO gauge is independent of the common origin of the coordinate frame, although it is dependent on the choice of the DOs. The DO gauge has gained wide applications since its appearance. With rare exceptions, the DO results are better than the CO results.

More recently, the so-called London atomic orbitals (LAO)[55] [also termed the gauge-independent atomic orbitals (GIAO)[56]], which have been used widely in calculating magnetic properties of molecules, have been applied in calculating rotatory strengths.[57] In the GIAO method, each atomic orbital is assigned its own gauge origin, defined as the position of the nucleus carrying that orbital. The GIAOs are gauge origin independent and the calculated magnetic properties are therefore also independent of the origin.[58]

Another approach involves the so-called localized orbital/local origin (LORG) scheme in conjunction with the random phase approximation (RPA). It was formulated[59] but not implemented. In the LORG method, each localized (occupied) molecular orbital carries its own gauge origin.

Electronic Wavefunctions and Calculation of Matrix Elements

For most calculations of rotatory strengths to date, each of the APTs and AATs as well as the force field has been evaluated using SCF (uncorrelated) wavefunctions. However, the evaluation of the force field and the APTs has been well established at higher levels of theory, for instance MP2 theory, and some have been applied to VCD intensities.

The first full MP2 VCD calculations, in which the AATs as well as the APTs and the force field are all calculated at the MP2 level of theory using the VCT method (VCT/MP2), appeared only very recently.[60] In the VCT/MP2 approach, the electronic ground state, Ψ_0, for all matrix element evaluations is approximated by the normalized MP2 wavefunction, i.e.,

$$\Psi_0 = N(\Phi_{HF} + \Phi_{MP2}) = N\left(\Phi_{HF} + \sum_{a<b}^{\substack{occ \\ spin}} \sum_{r<s}^{\substack{vir \\ spin}} C_{ab}^{rs} \Phi_{ab}^{rs}\right) \qquad [20]$$

where N is the normalization factor, Φ_{HF} is the HF determinantal wavefunction which represents Ψ_0 in the SCF approach, and Φ_{ab}^{rs} is the doubly excited configuration generated by replacement of the occupied orbitals a and b by the virtual orbitals r and s. The excitation energy, $E_e - E_0$, in Eq. [19] is the electronic energy difference between the HF wavefunction and the singlet-spin-adapted singly excited configuration, namely,

$$E_e - E_0 = \epsilon_r - \epsilon_a - J_{ar} + 2K_{ar} \qquad [21]$$

where J_{ar} and K_{ar} are the Coulomb and exchange integrals, respectively. The definition of excitation energy by Eq. [21] is consistent with a "pure" SCF approach, but is also expected to reflect the actual difference between ground and excited states to a reasonable approximation in the MP2 approach when uncorrelated excited wavefunctions are used. This is because, without introduction of comparable correlation into the excited state wavefunction, which would require triply excited configurations, use of the MP2 energy for the ground state would exaggerate unrealistically the excitation energy.

In addition to the ground state wavefunctions, two different approaches have been used to approximate the excited states, $\{\Psi_e\}$, involved in the summation in Eq. [19] in the VCT approach. In the simplest approach, $\{\Psi_e\}$ is approximated as the set of all singlet-spin-adapted singly excited configurations derived from the HF wavefunction. In the second approach, partially correlated excited state wavefunctions, $\{\Psi_e\}$, are derived from an all singles configuration interaction (SCI) calculation. The calculated energy differences between the HF wavefunction and the SCI wavefunctions are used as the excitation energies in Eq. [19] in both the SCF and the MP2 methods. This procedure, however, has been shown not to lead to significant changes at the SCF level as well as the MP2 level.[60,61]

For the MFP approach, most calculations have used SCF wavefunctions. However, very recently, Bak et al.[57] have implemented the MFP theoretical approach at the correlated multiconfigurational self-consistent-field level by using the complete active space wavefunctions (CASSCF) expressed over conventional and gauge invariant basis sets.

Wavefunction derivatives involved in Eq. [12] and [17] comprise the major part of the evaluations of AATs. In earlier days, a finite difference method was used to obtain these derivatives.[62] The implementation of analytical coupled perturbed Hartree–Fock (CPHF) methods[63–66] has permitted these derivatives to be obtained more efficiently. In the MFP approach, two sets of CPHF calculations are required to obtain $\partial\Psi_0(\vec{\mathbf{R}})/\partial Q$ and $\partial\Psi_0(\vec{\mathbf{R}}^0,\vec{\mathbf{B}})/\partial\vec{\mathbf{B}}$, respectively. In the VCT approach, only the quantities $\partial\Psi_0(\vec{\mathbf{R}})/\partial Q$ are required.

In an alternative approach, the APTs and AATs are calculated by the RPA method.[67] In this method, the APT and AAT are converted from nuclear electromagnetic shielding tensors which are directly related to the APT and AAT.[24,68] These nuclear shielding tensors are expressed as sums over states, which are evaluated by the SCF-RPA method.

Geometry and Force Field

Molecular geometries and force fields are the single most important considerations for VCD calculations. Wherever possible, good geometries and force fields must be used. One common practice to obtain these properties is to perform geometry optimization and evaluate the force field at the optimized equilibrium geometry at the highest available (and feasible) level of theory. The calculated frequencies may be scaled uniformly[69] or selectively[70] for improved agreement with experiment.

The experimental geometries and force fields may be used if available, although this is rarely the case for the force fields. In some cases, "corrected" ab initio geometries,[71] where systematic deviations are known to occur, are used.

If possible, the MP2/6-31G* level, or density functional methods, should be used to obtain geometries and force fields. However, calculations at the SCF level with the basis set denoted 6-31G*$^{(0.3)}$ have been found to yield geometries comparable to MP2 geometries and satisfactory force fields in many cases.[61,72–74] The 6-31G*(0.3) basis set differs from the conventional 6-31G* basis set only in the *d*-orbital exponent, where the conventional exponent, 0.8 for N, O and C atoms, is changed to 0.3.

Basis Sets

Generally, a good description of the electric, and especially the magnetic, dipole transition moments requires large polarized basis sets that include basis functions that are derivatives of nuclear coordinates. However, the size of such basis sets effectively precludes application of VCD theories to all but the smallest molecules. To permit VCT calculations of VCD intensities on larger molecules, smaller basis sets were explored, and the 6-31G*(0.3) basis set in the VCT/SCF method was found to produce VCD intensities in excellent agreement with experimental results.[61,72–74] This basis set was originally designed to mimic rotatory strengths derived from a much larger polarized basis set[75] denoted as 6-31G~, but was found to produce geometries and frequencies closer to experiment.[61,72–74,76] In fact, the success of the 6-31G*(0.3) basis set appears to be due to the improved geometry and force field it yields. Satisfactory results may be obtained for VCD intensities even with quite small basis sets for the calculation of the AATs, provided that geometries and force fields are derived from higher-level calculations. Because APTs are also typically available in conjunction with force field calculations, it is recommended that these also be used.

Ideally, for magnetic properties, gauge-origin independent atomic orbitals (GIAO)[55,56] should be used as discussed earlier. These orbitals reduce the sensitivity of calculated properties to the size of the basis set and therefore permit applications to larger molecules. The use of GIAO for VCD calculations has recently been reported.[57]

Computer Programs

The VCT formalism was first implemented in the FREQ85 program[65] which was designed to interface with an extensively modified Gaussian 82. FREQ85 was subsequently upgraded to FREQ86 in order to interface with Gaussian 86. The most recent version is the VCT90 program,[54,66] which was designed to interface with a slightly modified Gaussian 90.[77]

A typical procedure of performing VCD computations with the VCT method consists of three steps. The first step is geometry optimization, which is normally accomplished by using the Gaussian ab initio program. The second step, also performed by the Gaussian program, is the computation of the force field, molecular orbitals (MOs), and derivatives of the Fock and overlap matrices (which are used in calculating nuclear derivatives of MO coefficients).

These data are written to the read-write file from which, in the final step, they are fed into the VCT90 program in which the actual computation of VCD intensities based on the VCT formalism is accomplished.

The MFP approach was first implemented in a modified Gaussian 80 program, and a finite difference method was used to compute the wavefunction derivatives.[62] The MFP formalism has been implemented in the CADPAC ab initio program[78] and is fully analytical. An MFP implementation at the SCF level has also been carried out by Morokuma and Sugeta.[79] An implementation of MFP using CASSCF wavefunctions and gauge-invariant orbitals has been accomplished by Bak et al.[57]

The RPA calculations of VCD intensities are accomplished by a program mainly used for nuclear shielding tensor calculations.[80]

APPLICATIONS

The few ab initio calculations of VCD spectra that had been carried out up to 1990 have been reviewed.[38] Since then, many more systems have been investigated, and some of the earlier systems have been reinvestigated in a more rigorous fashion. These are briefly discussed below.

(*S*)-NHDT

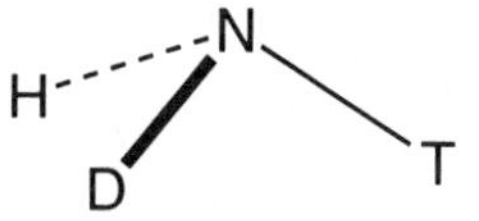

1

The simplest real molecule that can be made chiral by isotropic substitution is ammonia (**1**). The experimental geometry and force field are very well characterized and essentially indistinguishable from those derived by high-level theoretical calculations. The chiral isotopomer, therefore, has served as a benchmark for the performance of all a priori theoretical VCD models, the basis set, and the level of correlation required to produce converged results. NHDT was also used to study extensively the basis set dependence and gauge dependence of the rotatory strengths by the MFP method,[57,81,82] the VCT method,[54] and the RPA method.[67] These studies, together with many calculations on other molecules, have demonstrated that either very large polarized basis sets or modified conventional basis sets have to be used to yield reliable VCD intensities, and DO gauge is preferred to CO gauge. The most rigorous VCD calculations on NHDT have been collected in Reference 60 and are shown in Table 1.

Table 1 Rotatory Strengths[a] of (*S*)-NHDT[b]

		VCT[d]							MFP			
		6-31G*			6-31G*(0.3)		vd/3p(u)		vd/3p[e]	pVTZ(++)[f]		
No	Freq[c]	SCF	MP2	MS[g]	SCF	MP2	SCF	MP2	SCF	SCF	CAS1	CAS2
1	846	−3.36	−3.84	−3.67	−4.22	−5.14	−4.57	−5.44	−4.94	−6.30	−4.72	−5.11
2	1192	6.73	7.66	7.30	8.14	9.89	9.02	10.72	9.89	11.88	9.50	10.18
3	1480	−3.50	−3.92	−3.75	−4.07	−4.90	−4.56	−5.36	−5.10	−5.71	−4.82	−5.14
4	2195	0.46	0.51	0.53	0.68	0.82	0.64	0.82	1.17	0.75	0.61	0.76
5	2632	−0.26	−0.35	−0.37	−0.59	−0.75	−0.58	−0.85	−1.18	−0.67	−0.64	−0.79
6	3603	0.02	0.05	0.05	0.11	0.14	0.11	0.18	0.23	0.13	0.14	0.17

[a]In units of 10^{-44} esu^2 cm^2.

[b]Geometry of NH_3: SCF/vd/3p(u), r_{NH} = 0.998 Å, $<_{HNH}$ = 107.6°; MP2/vd/3p(u), r_{NH} = 1.009 Å, $<_{HNH}$ = 107.0°; expt (Reference 118), r_{NH} = 1.012 Å, $<_{HNH}$ = 106.7°.

[c]MP2/vd/3p(u) force field. Dipole strengths (10^{-40} esu^2 cm^2) for modes 1–6 are 467.0, 35.9, 57.3, 6.2, 7.4, 7.4, respectively (Reference 60).

[d]Reference 60.

[e]Reference 82; Duncan and Mills harmonic force field (Reference 118); at experimental geometry, in the DO gauge.

[f]Reference 57; at experimental geometry, in the CO gauge, with GIAO orbitals, and ab initio force field; CAS1, 924 CSFs; CAS2 71947 CSFs.

[g]MP2–SCI method.

Cyclopropanes

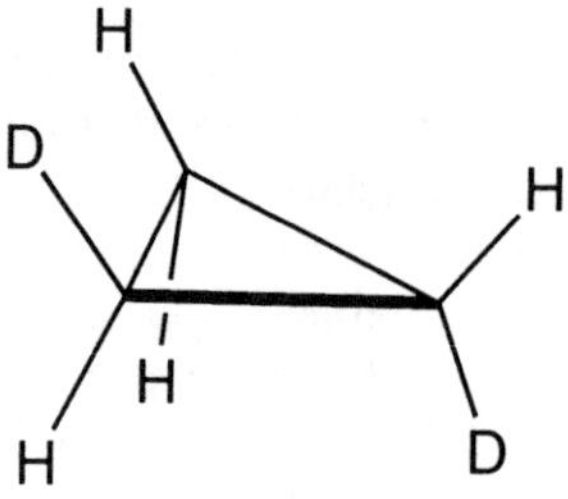

The calculated VCD properties of *trans*-(1*S*,2*S*)-dideuteriocyclopropane[62] **2** are the first ab initio results obtained by any a priori method, that is the MFP method, in 1986. Several geometries, including the 4-31G optimized geometry, and three different force fields were employed. SCF wavefunctions were calculated at the 4-31G level, and wavefunction derivatives were obtained by numerical differentiation using a modified version of the Gaussian 80 program. This molecule also served as a test case to establish proper displacement step sizes in the numerical differentiation approach when calculating the two types of wavefunction derivatives mentioned in the previous section. Compound **2** was reinvestigated by the same method in 1990 with a considerably larger basis set [6-31G(ext)] using the experimental geometry and an empirical force field[83] and again in 1990 with a variety of basis sets and gauge origins.[84] The ab initio VCD spectrum of **2** has also been calculated by the RPA method,[85] using a polarized basis set, 6-31G(pol), at experimental geometry, and a 6-31G(ext) force field. The RPA results were compared with the experimental results, and good agreement was found.

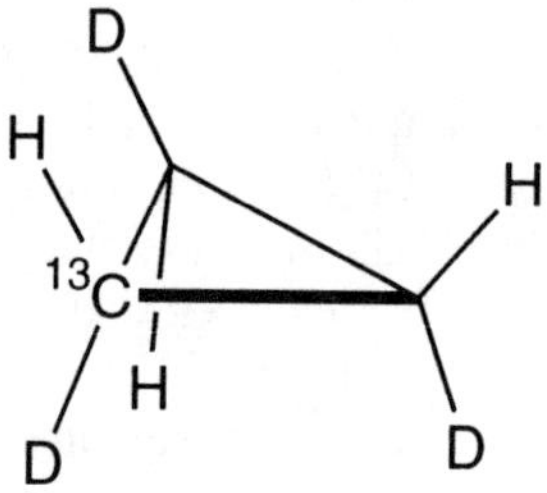

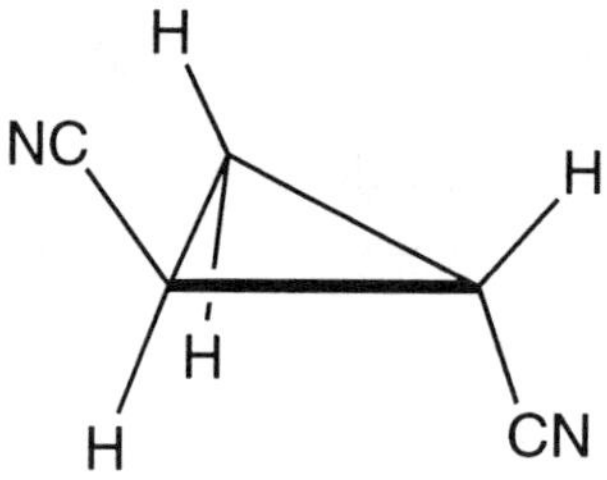

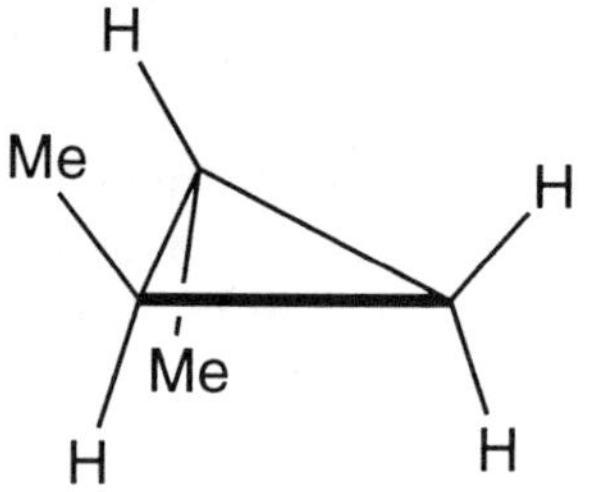

The calculated and measured rotatory strengths of (2*S*,3*S*)-cyclopropane-1-^{13}C-1-d_1-2-d_1-3-d_1 **3** in the C-H and C-D stretching region were published by Freedman and Nafie.[37] The calculations were done by using the MFP method with the 6-31G(ext) basis set at the experimental geometry and a slightly modified empirical force field. Both DO and CO gauges were employed, and the results from the two gauges were found to be similar. All the signs in the C-H and C-D stretching region were correctly reproduced by the calculations, but the absolute intensities were a factor of two and five larger than the observed dipole and rotatory strengths, respectively.

The VCD of *trans*-(1*S*,2*S*)-dicyanocyclopropane[86] **4** was one of the early results obtained by use of the MFP method as implemented in the CADPAC program (i.e., using analytical methods to obtain wavefunction derivatives). The 6-31G** basis set was employed for obtaining APT and AAT. The force field was derived from scaling the ab initio 6-31G* force field at the 6-31G** equilibrium geometry. The results for uniform and nonuniform (6 scale factors) scaling were compared to dipole and rotatory strengths estimated from the experimental spectra. The VCD originating from the skeletal deformations, including those of the cyano groups, were reasonably well reproduced, difficulties being experienced with the CH stretches.

The IR and VCD spectra of (1*R*,2*R*)-dimethylcyclopropane **5** have been measured (in the C—H stretching region) and theoretically calculated by Amos et al.[87] The MFP method implemented in the CADPAC program was used. Amos et al. used a 6-31G* equilibrium geometry and unscaled force field but reported rotatory strengths only with a 3-21G basis set. The authors claimed substantial agreement between the theoretically simulated VCD spectrum and their experimental spectrum which covers the C—H stretching region. The dipole and rotatory strengths of **5** were also calculated by the RPA method,[85] by employing the polarized basis set 6-31G(pol), at the experimental geometry and a 6-31G** force field. The sign patterns from this calculations are the same as the ones calculated by Amos et al. except for a few weak bands.

Oxiranes

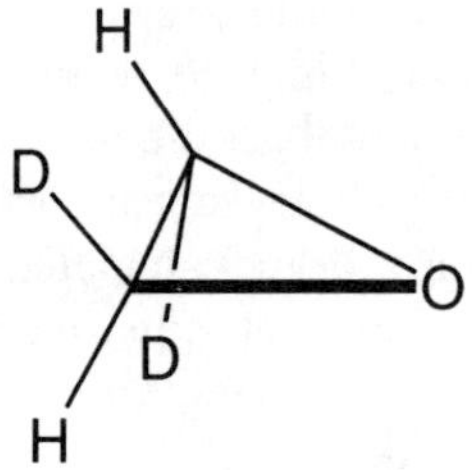

H D D H O

6

H Me H H O

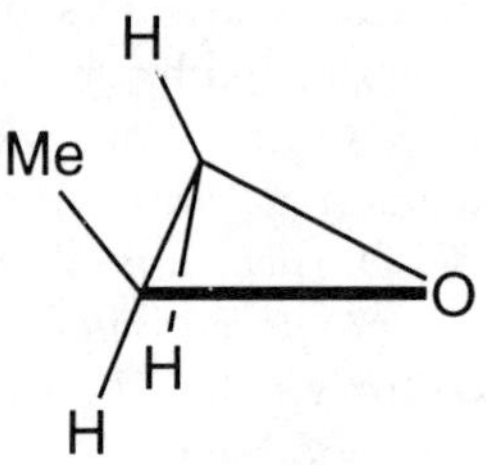

7

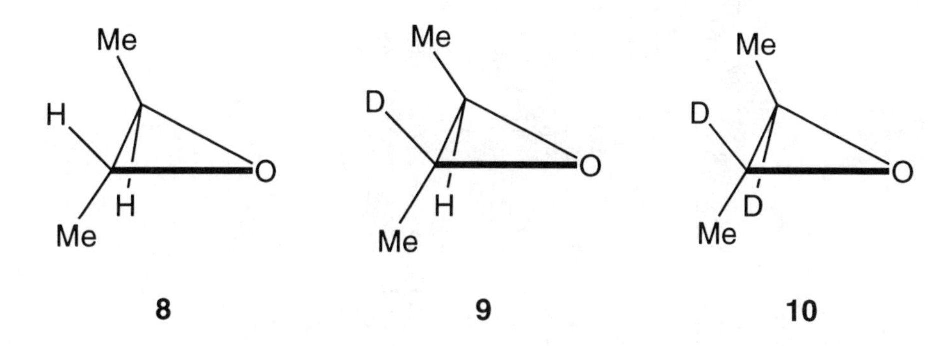

Oxirane can be made chiral by dideuterio substitution. *trans*-2,3-Dideuteriooxirane **6** has been the target molecule for VCD studies with various methods and at different levels of theory because of its small size and conformational rigidity. This is also the only molecule for which the rotatory strengths for the complete vibrational spectral range have been obtained experimentally.[88] The IR and VCD intensities of **6** were first calculated for the C—H and C—D stretching region[89] and then the complete spectral region.[83] The method used in these studies is the MFP method with the 6-31G(ext) basis set at the experimental geometry,[90] which was also used in all subsequent MFP and RPA studies of this molecule. The force field was a scaled ab initio 6-31G(ext) one. Comparison to the experimental spectra[91] in the stretching region shows good agreement between the MFP-DO and the observed VCD intensities.

Molecule **6** was reinvestigated recently[92] with the MFP-DO method at a higher level of theory, namely, the APT and the force field were calculated at the MP2 level, whereas the AAT was calculated at the SCF level by using a near-SCF limit basis set, the vd/3p basis set.[82,93] Excellent agreement with the experimentally obtained rotatory strengths[88] is observed. The large basis set employed in the latter study permits more accurate harmonic force fields and APTs to be obtained, and discrepancies between the calculated and the measured rotatory strengths therefore are mainly attributed to the lack of correlation in the AATs. Neglect of correlation effects is not insignificant in calculating both dipole and rotatory strengths.

Dutler and Rauk[65,75] have published IR and VCD intensities for **6** as well as its mono- and trideuterio isotopomers using the VCT-CO method with the 6-31G~ basis set. The optimized 6-31G~ geometry and scaled 6-31G~ force field were used. The results show that agreement with the experimental VCD spectrum both in the C—H and C—D stretching region and the mid-IR region is excellent.

Recently, the first full MP2 VCD calculations by the VCT/MP2 method have been applied to **6**. The force field, APTs, and AATs were all calculated at the MP2 level. The results together with the experimental and selected MFP rotatory strengths are listed in Table 2. The VCT/MP2 results show that significant improvements are observed for the conventional 6-31G* basis set, and the

Table 2 Rotatory Strengths[a] of (2*S*,3*S*)-Dideuteriooxirane **6** in DO Gauge

		VCT												MFP[b]			
		6-31G*				6-31G*(0.3)				vd/3p				vd/3p		Expt[c]	
No	Sym.	SCF	$\|\Delta\|$[d]	MP2	$\|\Delta\|$[d]	SCF	$\|\Delta\|$[d]	MP2	$\|\Delta\|$[d]	SCF	$\|\Delta\|$[d]	MP2	$\|\Delta\|$[d]	MP2[e]	$\|\Delta\|$[d]	*R*	Freq
1	b	−0.4		−0.1		0.6		0.2		0.9		0.3		0.0			(673)
2	a	2.1		−4.1		4.7		−3.2		11.1		4.4		11.8			(754)
3	b	−4.6	x	−0.3	x	−0.7	x	1.2		−0.7	x	2.7		1.7		(+)	(817)
4	a	8.9	78	13.9	178	6.0	20	8.3	66	0.9	82	5.7	13	8.0	60	(+5.0)	(885)
5	b	7.0	x	−2.7	56	1.8	x	−3.2	48	1.0	x	−6.6	6	−3.7	40	−6.2	914
6	a	−17.6	39	−15.2	48	−18.9	35	−6.8	77	−27.6	5	−19.9	31	−28.2	3	−29.0	948 (961)
7	b	2.0	82	5.2	53	3.6	68	10.7	4	6.7	40	16.8	51	10.1	9	11.1	1102 (1106)
8	a	2.8	x	−1.2	76	0.1	x	−8.8	80	5.9	x	−5.3	9	−6.3	29	−4.9	1109 (1112)
9	a	5.4	78	10.1	58	8.1	66	10.1	58	8.0	67	13.3	45	10.9	55	24.1	1226 (1235)
10	b	−5.2	108	−.2	92	−1.5	41	−1.6	36	−1.7	31	−1.0	61	−1.3	48	(−2.5)	(1339)
11	a	−4.3	71	−9.0	40	−4.9	67	−5.4	64	−7.1	53	−11.1	26	−8.3	45	(−15.0)	(1397)
12	b	−15.6	50	−6.6	37	−8.3	20	−2.3	78	−11.7	12	−4.3	59	−11.2	8	−10.4	2232 (2240)
13	a	13.3	10	7.0	42	9.1	25	3.1	74	11.8	3	5.2	57	12.7	5	12.1	2252 (2254)
14	a	−22.5	153	−10.7	20	−16.2	82	−3.9	56	−20.3	128	−5.5	38	−24.4	174	−8.9	3014 (3015)
15	b	28.8	153	12.0	5	16.7	46	4.1	64	23.1	102	7.6	34	22.8	100	11.4	3027 (3028)
Average $\|\Delta\|$[f]			99		59		58		59		72		36		48		
Wrong signs			3		1		3		0		3		0		0		

[a]In units of 10^{-44} esu^2 cm^2.
[b]Reference 92.
[c]Reference 88; values in parentheses are from the gas-phase spectrum.
[d]$|\Delta| = |R^{exp} - R^{calc}| / |R^{exp}| \times 100\%$, where *R* represents the rotatory strength; x denotes an incorrect sign prediction.
[e]Calculated using MP2 force field, MP2-APTs and SCF-AATs.
[f]The bands with incorrect signs are taken into account.

6-31G*(0.3) basis set produces rotatory strengths that are better in terms of signs but not necessarily better in terms of absolute intensities at the MP2 level. The very large basis set, i.e., vd/3p, produces the best rotatory strengths, which are in agreement with the experimental rotatory strengths both in signs and in absolute intensities. The RPA-DO method has also been applied to calculate VCD intensities of 6, first for the C—H and C—D region[94] using a polarized basis set, 6-31G/pol, and a scaled ab initio 6-31G(ext) force field, and later for the complete region of spectrum[85] using the 6-31G/pol basis set at a 6-31G** force field. Both show good agreement with experiment.[88,91]

trans-1,2-Dideuteriocyclobutane and (2*S*)-methyloxirane (propylene oxide) 7 were the other of the first two molecules[106] to have their predicted VCD properties compared to experiment. The equilibrium geometry and SCF wavefunctions were determined using a modified Gaussian 80 program and the 4-31G basis set, and the ab initio force field was scaled. Wavefunction derivatives were obtained by a numerical differentiation method using the optimum step sizes established in the case of *trans*-(1*S*,2*S*)-dideuteriocyclopropane. Comparison of the simulated VCD spectra from this study to the VCD spectrum of neat (−)-(*S*)-7 measured[95] in the 850–1550 cm^{-1} range was encouraging.

The VCD of 7 was recalculated[96,97] in 1988 with the MFP method using analytical methods as implemented in CADPAC rather than finite difference methods. The DO gauge and substantially large basis sets were employed. The VCD spectrum was also remeasured and rotatory strengths were obtained.[96] The basis set, geometry, and force field dependence of the VCD intensities of this molecule were investigated in those studies.

Molecule 7 was even more thoroughly investigated[98] in 1991 both theoretically (using the MFP method) and experimentally. Results from these studies show that VCD intensities are relatively insensitive to the basis set used in calculating equilibrium geometries and to the distinction between theoretical and corrected geometries used in those studies, but very sensitive to the force fields employed. Excellent agreement for the sign patterns between calculated and measured rotatory strengths are observed when large basis sets are used to calculate equilibrium geometry, force field, APTs, and AATs. The rotatory strengths of 7 were also studied[85] by the RPA-DO method using the 6-31G/pol basis set at the experimental geometry and 6-31G** force field. Most of the signs of the observed VCD bands are reproduced in this study.

Rauk and Yang reported[61,66] the VCD studies on 7 by using the VCT method with the 6-31G*(0.3) basis set. Both the CO and the DO gauges were used. The VCT-CO and the VCT-DO rotatory strengths are in good agreement with the measured ones for the pattern of signs and relative intensities, but the absolute intensities are overall smaller than the observed ones, as has been observed for all the above-mentioned VCD studies of this molecule.

The experimental and theoretical IR and VCD spectra of the dimethyl substituted oxiranes, namely (2*R*,3*R*)-dimethyloxiranes[66,72] 8 (2*R*,3*R*)-dimethyloxiranes-2-d_1[66,72] 9, and (2*R*,3*R*)-dimethyloxiranes-2,3-d_2[66,72] 10, were reported in 1992. The VCT method with the 6-31G*$^{(0.3)}$ basis set was used in this study. All properties, that is, the equilibrium geometry, force field, APTs, and AATs, were obtained at the SCF/6-31G*$^{(0.3)}$ level. The similarities between the simulated VCT/6-31G*$^{(0.3)}$ and experimental[72] VCD spectra are striking, especially for **8**. The simulated and experimental IR and VCD spectra of (2*R*,3*R*)-dimethyloxiranes-2-d_1 are shown in Figure 1. The VCT-CO inten-

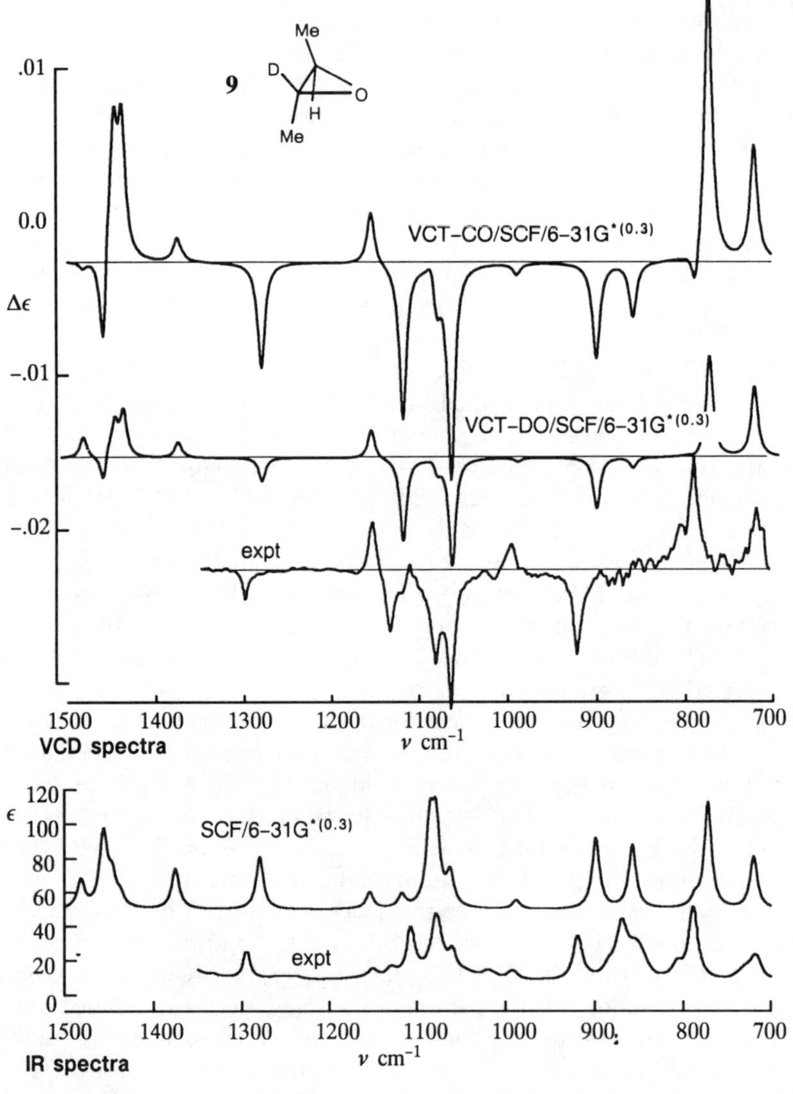

Figure 1 Theoretically simulated, Lorentzian lineshape of 5 cm^{-1} half width at half height, and experimental (CS_2) VCD, and IR spectra of (2*R*,3*R*)-dimethyloxirane-2-d_1 **9** in the region 700–1500 cm^{-1} (Reference 72). SCF frequencies are scaled by 0.9.

sities are uniformly higher than the VCT-DO ones, a phenomenon also observed for all other VCT calculations. Detailed analysis[66] shows that the average errors of the VCT-CO results, compared to the experimental rotatory strengths, are uniformly higher than the VCT-DO results for all three dimethyloxiranes.

Thiiranes

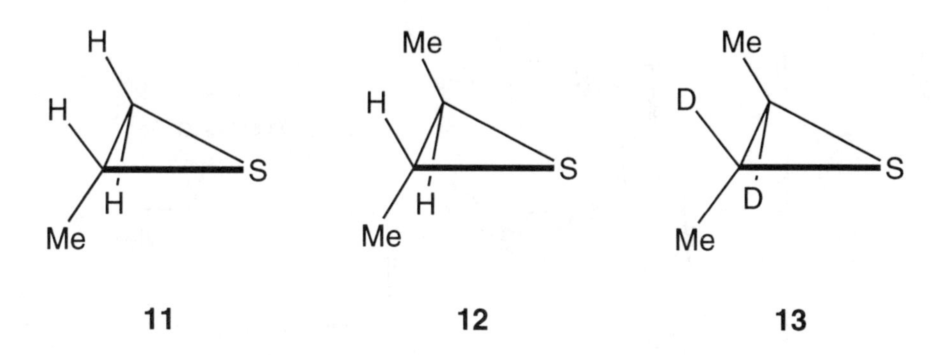

The IR and VCD spectra of (2*R*)-methylthiirane **11** were studied[66,99–101] by both MFP and VCT methods. The experimental VCD spectra of **11** were measured early in 1987[102] and again in 1991.[103] It was first studied theoretically in 1988 by the MFP method with both the CO and the DO gauges.[99] In this study, the wavefunction derivatives were obtained using a numerical differentiation method, a corrected theoretical geometry, a scaled force field, and a 6-31G* basis set. Comparison with experiment[102] indicated that the simulated VCD spectrum in the CO gauge at the center of mass bears a closer resemblance to the experimental VCD spectrum.

As the first VCD study partially at the MP2 level of theory, Amos et al. studied the VCD of **11** using the MFP method with the force field and APTs at the MP2 level, and AATs at the SCF level.[100] The equilibrium geometries and force fields were obtained with a variety of basis sets, and only the results obtained with a DZP basis set were quoted. The authors concluded that the correlation effects improve both the APTs and frequencies, and that correlation affected frequencies more than intensities in the C-H stretching region.

Molecule **11** was also studied by the VCT/SCF/6-31G$^{*(0.3)}$ approach,[66,101] in which all properties were obtained at the SCF/6-31G$^{*(0.3)}$ level. The calculated signs of all rotatory strengths correctly reproduce the available experimental rotatory strengths[103] by both the VCT-CO and the VCT-DO approaches.

The experimental and theoretical IR and VCD spectra of the dimethyl substituted thiiranes, namely (2*R*,3*R*)-dimethylthiirane[66,101] **12** and (2*R*,3*R*)-dimethylthiirane-2,3-d_2[66,101] **13** have been reported. The VCT method with the 6-31G$^{*(0.3)}$ basis set was used in those studies. All properties were obtained at the SCF/6-31G*$^{(0.3)}$ level. The predicted signs of the rotatory strengths in both the CO and the DO gauges for **12** are all in agreement with those of the available measured rotatory strengths.[66,101] The average error[66] of VCT-DO rotatory strengths (35%) is much lower than that of VCT-CO rotatory strengths (112%). A similar trend was observed for **13**, for which the average errors[66] for VCT-DO and VCT-CO rotatory strengths are 32% and 91%, respectively. The simulated and experimental IR and VCD spectra of **12** are illustrated in Figure 2.

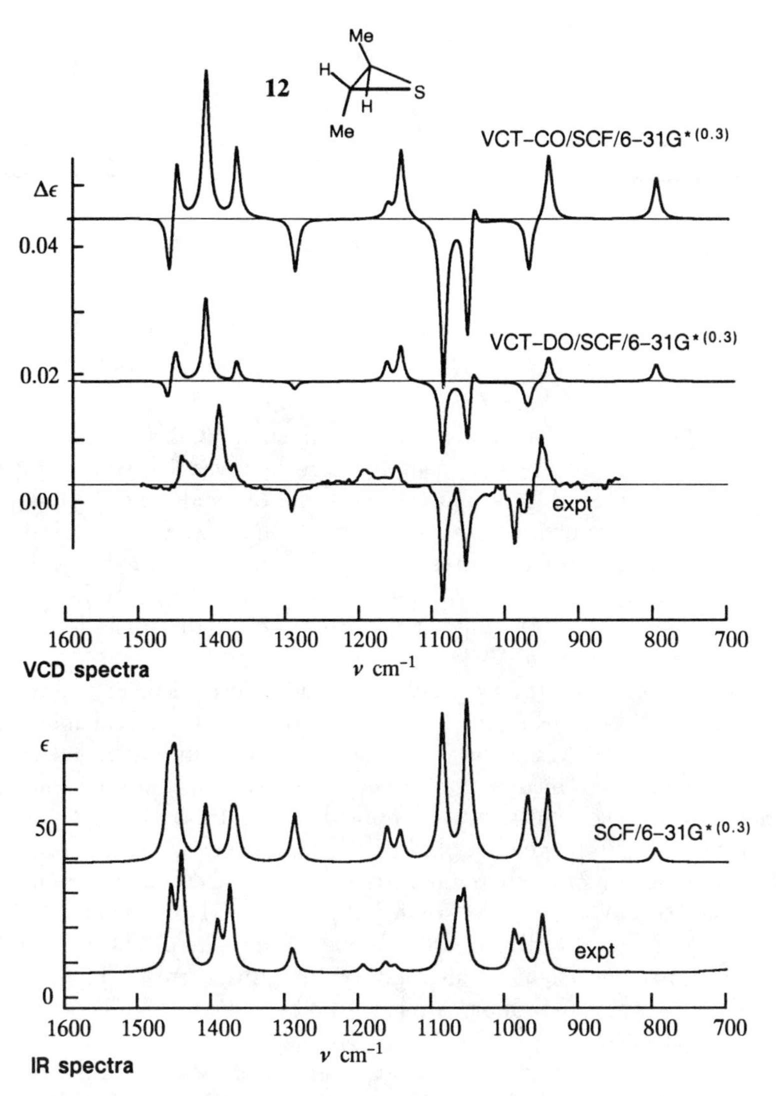

Figure 2 Theoretically simulated, Lorentzian lineshape of 5 cm^{-1} half width at half height, and experimental (CCl_4) VCD and IR spectra of (2*R*,3*R*)-dimethylthiirane **12** in the region 700–1600 cm^{-1} (Reference 101). SCF frequencies are scaled by 0.9.

Aziridines

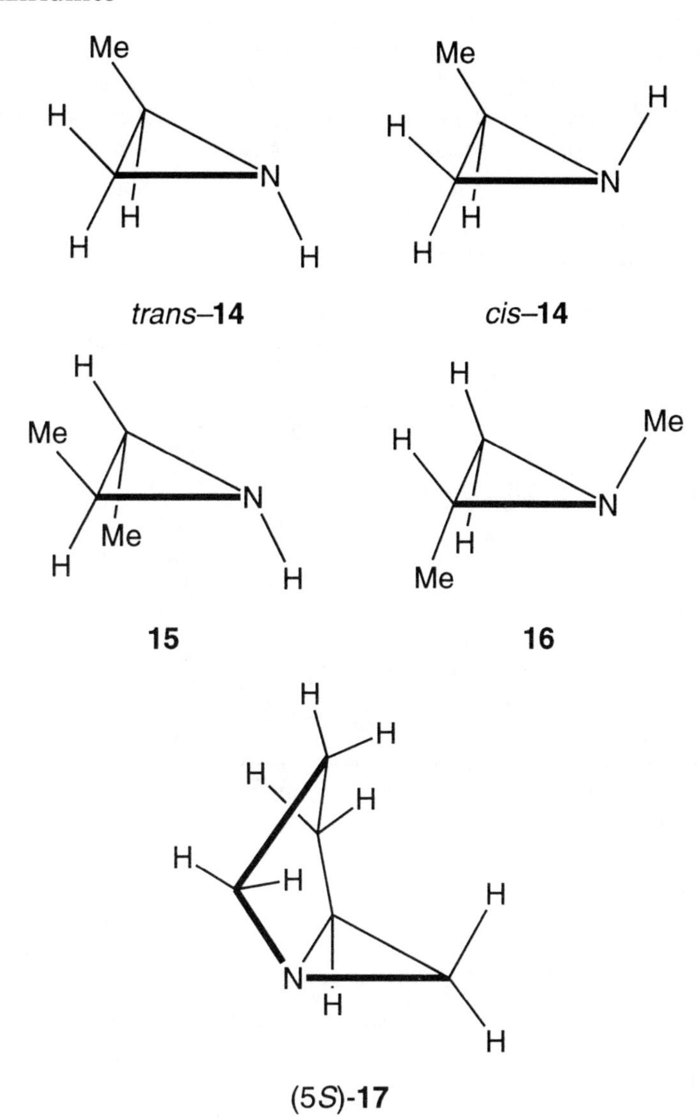

Aziridines are of special interest among the strained three-ring heterocycles, because the tricoordinated nitrogen atom constitutes an additional distinct chiral center, and in many cases, isomerism by N-inversion must be considered in the interpretation of the VCD spectra.

The experimental and theoretical IR and VCD spectra have been reported for (2*R*)-methylaziridine[73] **14**, (2*S*,3*S*)-dimethylaziridine[74] **15**, and (1*R*,2*R*)-dimethylaziridine[74] **16**. The VCT method with the 6-31G$^{*(0.3)}$ basis set was used in the calculations. The equilibrium geometry, force field, APTs, and

AATs were also obtained with 6-31G*$^{(0.3)}$ basis set at the SCF level. In the case of **14,** the MP2/6-31G* energies obtained at the optimized MP2/6-31G* geometries indicate a trans/cis mixture of 0.706:0.294 at 298 K. The force fields used for **14** were either optimally scaled or uniformly scaled, the two being very similar. Some bands of the experimental spectra could be assigned to each invertomer, and those sensitive to the change of configuration at N identified.[73]

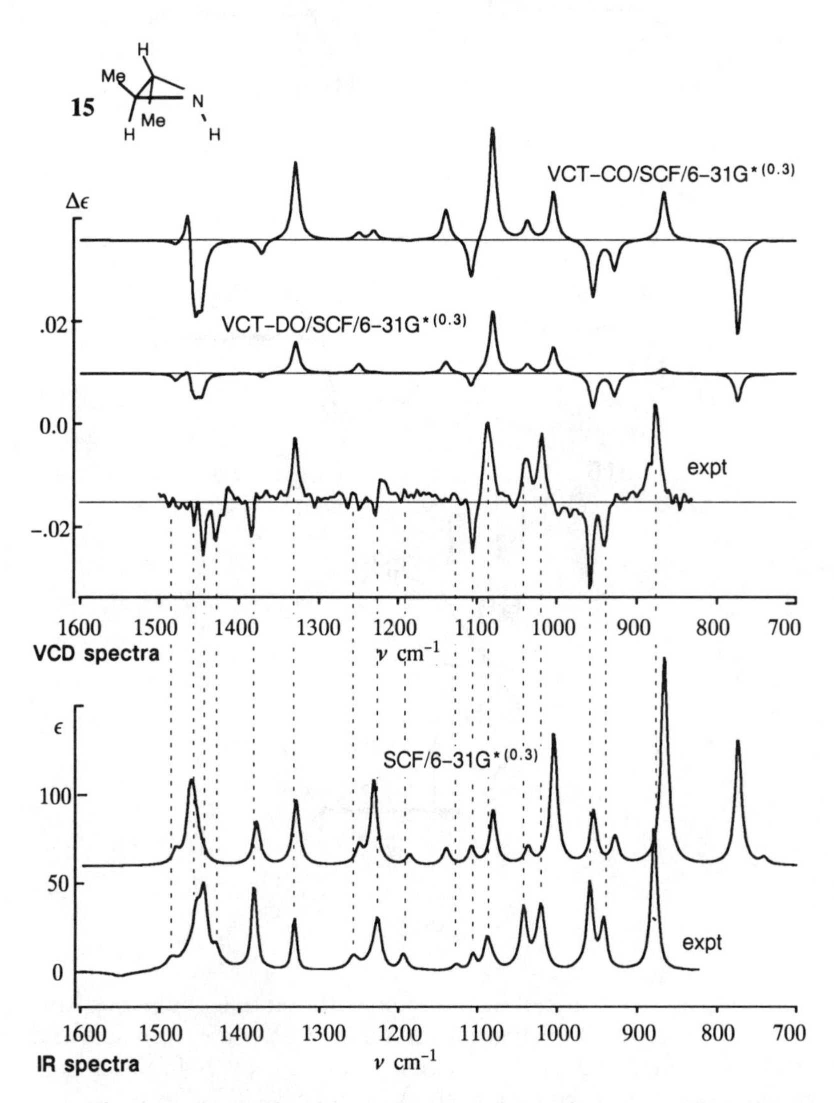

Figure 3 Theoretically simulated, Lorentzian lineshape of 5 cm^{-1} half width at half height, and experimental (CCl_4) VCD and IR spectra of (2*S*,3*S*)-dimethylaziridine **15** in the region 700–1600 cm^{-1} (Reference 74). SCF frequencies are scaled by 0.9.

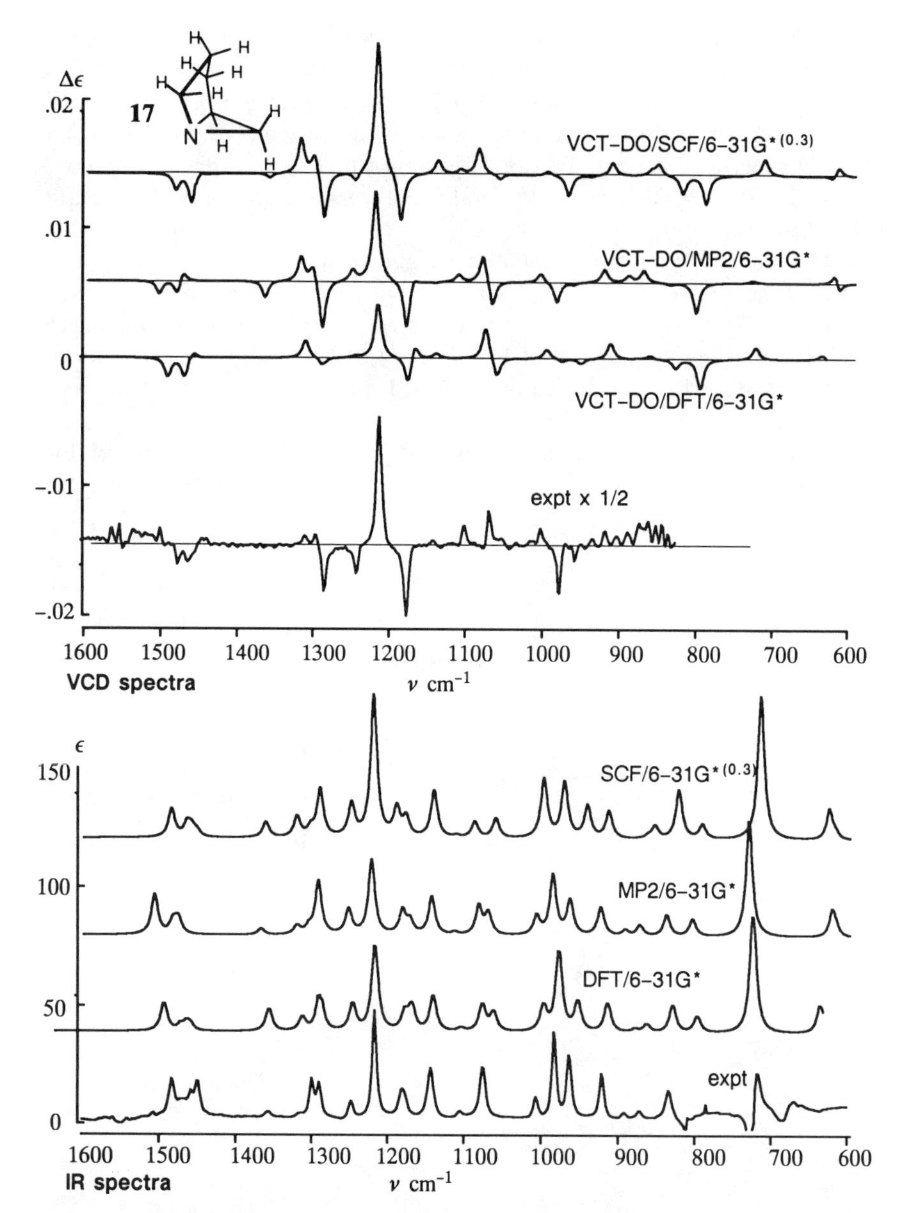

Figure 4 Theoretically simulated, Lorentzian lineshape of 5 cm^{-1} half width at half height, and experimental (Reference 119) VCD and IR spectra of 1-azabicyclo-[3.1.0]hexane **17** in the region 600–1600 cm^{-1}. SCF, MP2, and DFT frequencies are scaled by 0.90, 0.95, and 0.97, respectively.

The spectra of the aziridines were also compared to corresponding oxirane and thiirane spectra, and common patterns in the VCD spectra were found.

The simulated and experimental IR and VCD spectra of **15** are shown in Figure 3 to illustrate the level of agreement that can be achieved. Comparison of the VCD spectra of **15** and **16** suggested several VCD features that could be attributed to the chirality of the skeletal framework independent of the nature of the chiral centers.[74]

The experimental and simulated IR and VCD spectra of 1-azabicyclo-[3.1.0]hexane **17** are shown in Figure 4.[104] The prominent features near 1200 cm^{-1} and 1300 cm^{-1} clearly establish the absolute configuration of the bicyclic ring skeleton. The investigation of **17** represents the first application of density functional theory for the determination of a molecular force field for VCD calculations.[104]

2-Carbomethoxyaziridine **18** exists as an equilibrium mixture of four diastereomeric forms, the prominent conformer (94%) of which has a cis con-

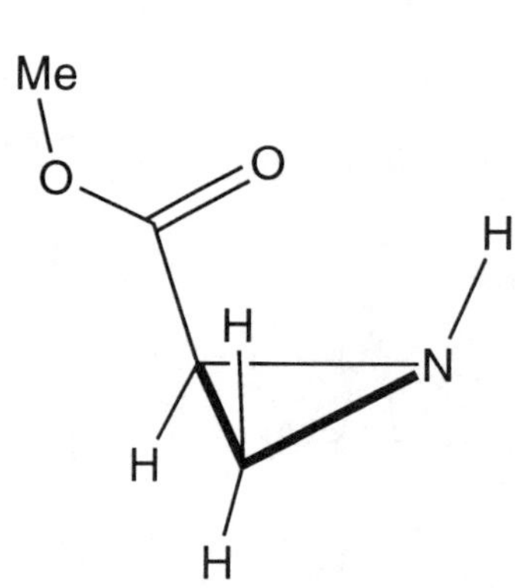

Z-cis-**18**

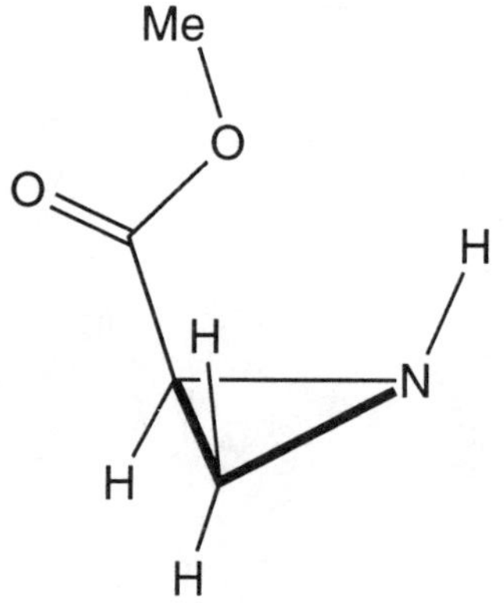

E-cis-**18**

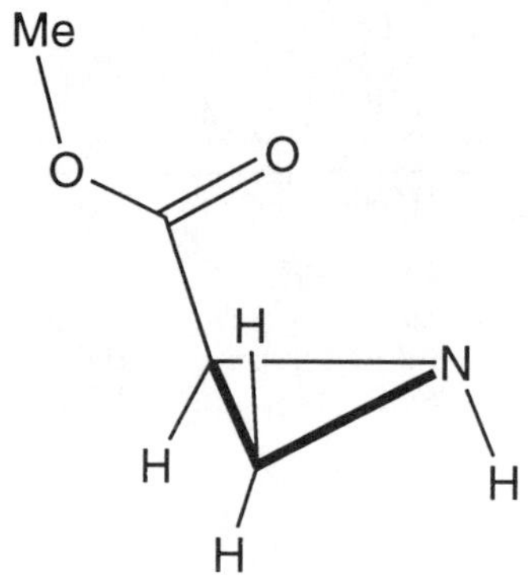

Z-trans-**18**

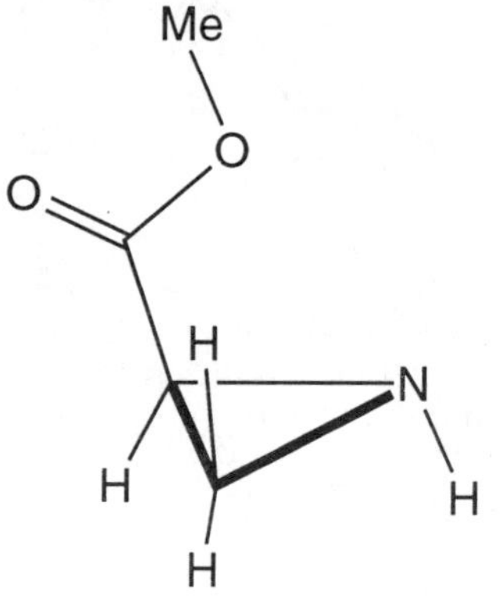

E-trans-**18**

figuration at N and an orientation of the carbomethoxy group that permits formation of a weak N—H to O=C hydrogen bond. The theoretical simulated VCD spectra of the equilibrium mixture are compared with the experimental spectrum in Figure 5. Excellent agreement of the two is evident, particularly with the fully MP2 calculation.[105]

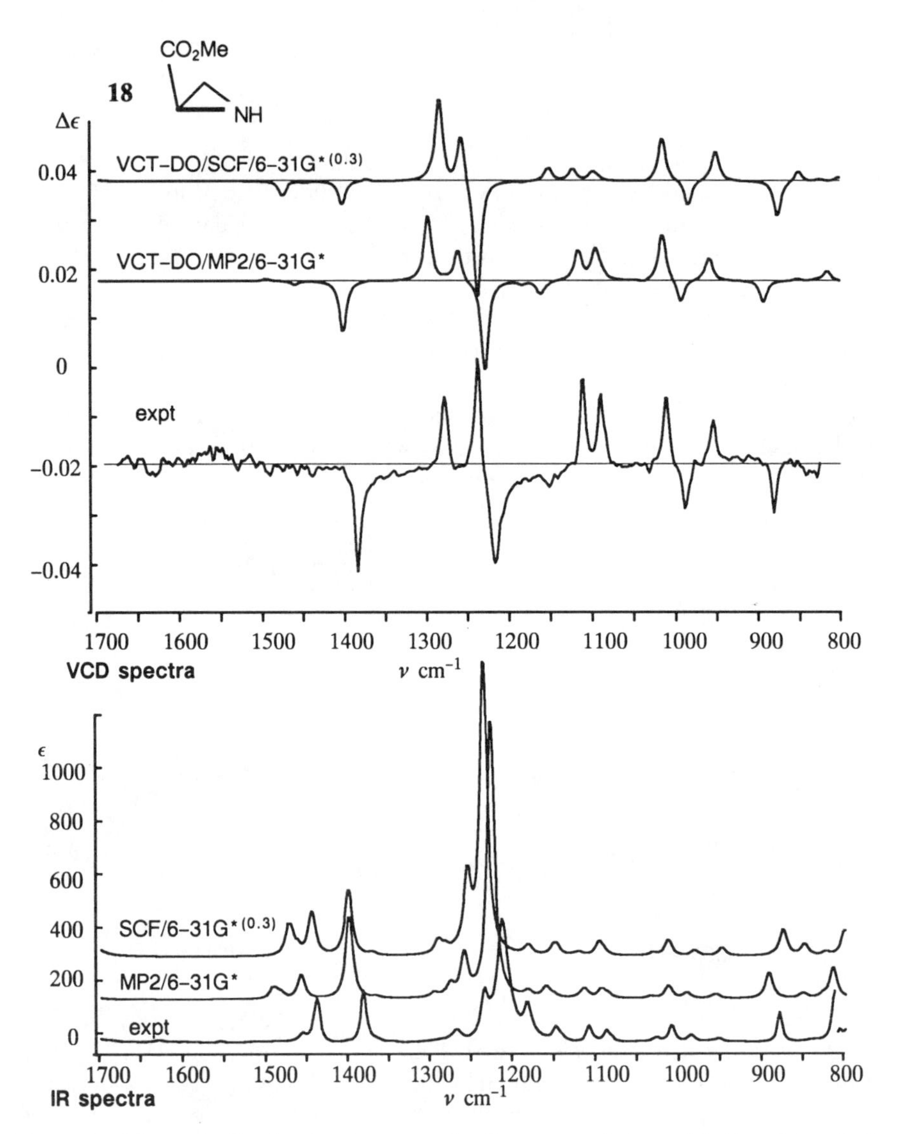

Figure 5 Theoretically simulated, Lorentzian lineshape of 5 cm^{-1} half width at half height, and experimental (Reference 119) VCD and IR spectra of 2-carbomethoxy-aziridine **18** in the region 700–1800 cm^{-1}. The theoretical spectra are for the mixture of the two *cis* isomers: 0.942 Z-cis-**18** and 0.056 E-*cis*-**18**. SCF and MP2 frequencies are scaled by 0.90 and 0.95, respectively.

Cyclobutanes

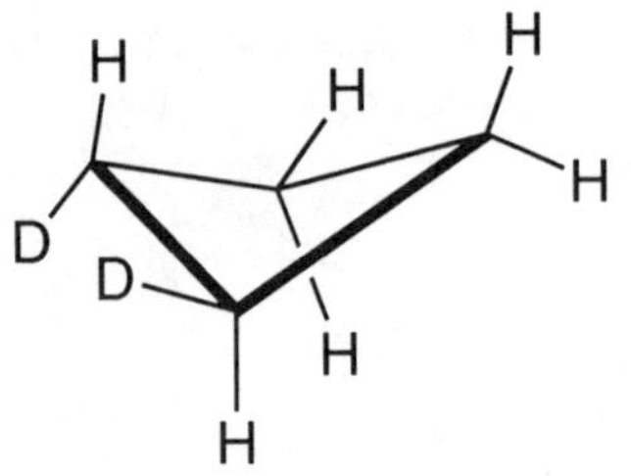

19

The calculated VCD properties of *trans*-1,2-dideuteriocyclobutane[106] **19** were one of the first to be compared with experimental data.[107] The MFP method was used. The force field was that of Annamalai and Keiderling.[108] The equilibrium geometry and SCF wavefunctions were determined using a modified Gaussian 80 program and the 4-31G basis set. Wavefunction derivatives were obtained by numerical differentiation using the optimum step sizes established for *trans*-(1*S*,2*S*)-dideuteriocyclopropane. The calculated VCD intensities were represented as simulated spectra of a 1:1 equilibrium mixture of axial and equatorial isomers. Comparison with experimental spectra available for the C—H and C—D stretching region, and in the mid-IR range, 700 cm^{-1} to 1500 cm^{-1}, was reasonably good.

Lactones

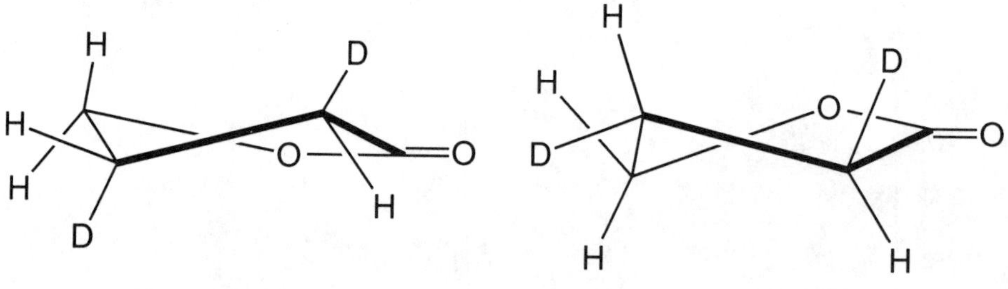

axial-**20** equatorial-**20**

The VCD spectra of the two conformers of (2*S*,3*S*)-dideuteriobutyrolactone **20** were calculated using the MFP method and several basis sets.[109] The ab initio SCF/6-31G** force field yielded good qualitative agreement between the summed spectra and the experimentally measured one. The appearance of the VCD spectrum and the qualitative agreement with experiment were not improved by nonuniform scaling of the ab initio force field. Simpler theoretical models of calculating VCD spectra did not yield satisfactory results even at the qualitative level.[109]

Allenes

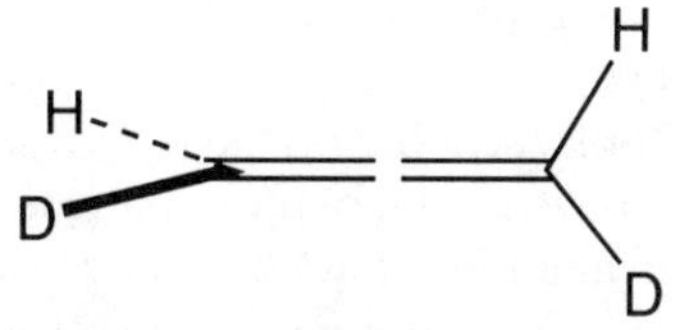

21

Stephens and co-workers have reported[110] a VCD study of 1,3-dideuterioallene **21** using the MFP method and two large basis sets, TZ/2P and 6-31G(ext). The results were compared to various empirical and approximate models. The results for the rotatory strengths obtained with both basis sets agree in sign and are quite close in absolute magnitude, the greatest deviation being a factor of 2 for the C—H stretching vibrations. Experimental optical rotatory strengths were not available, but overall agreement with experimentally determined dipole strengths was also satisfactory.

Alcohols

α-Deuterioethanol 22 was the first system investigated for which account had to be taken of the conformational equilibrium between chiral diastereomeric conformations. The VCD spectra of 22 and several other deuter-

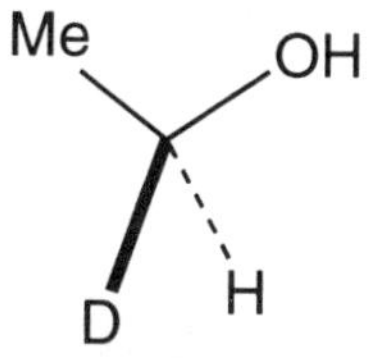

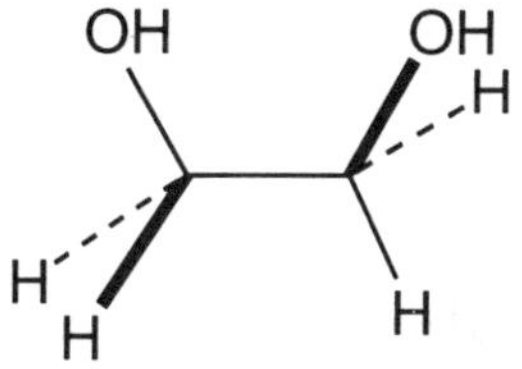

22 23

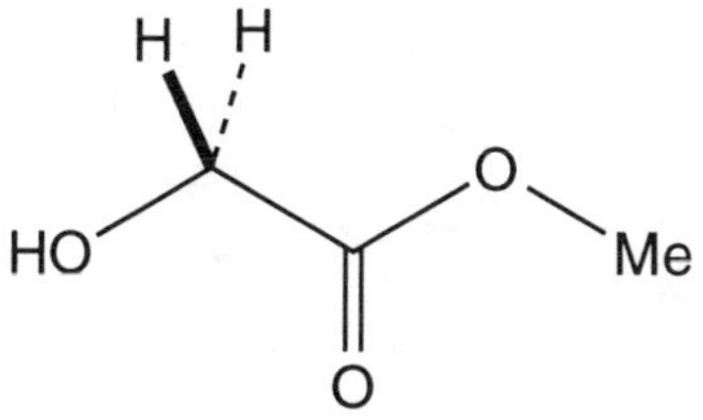

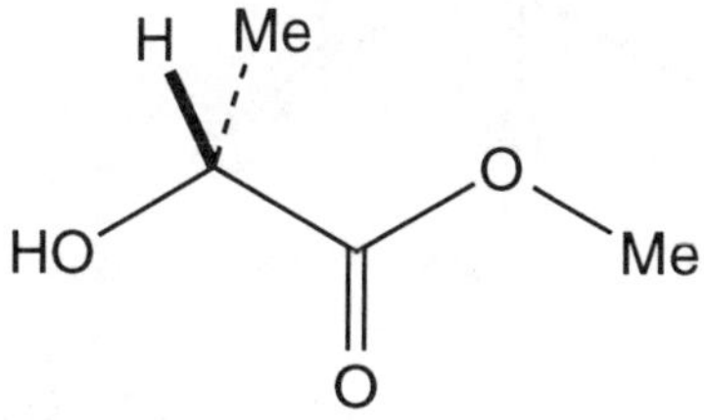

24 25

ated isotopomers were first predicted by Dothe, Lowe, and Alper using the MFP method in the DO gauge and 6-31G** basis set; reasonable agreement was found between measured spectra and predictions primarily for the gauche conformers.[111] The VCD spectrum of **22** was subsequently predicted at the VCT-CO level using the polarized 6-31G~ basis set and nonuniformly scaled force field, with Boltzmann weighting of the conformer populations. Very good agreement of the theoretical and measured spectra was achieved.[75]

Morokuma and Sugeta[79] used a modified Gaussian 82 and the MPROP program[112] together with the code of their own to calculate dipole and rotatory strengths of ethylene glycol **23** using the MFP formalism. In their approach, the wavefunction position derivatives were obtained analytically and the wavefunction field derivatives were obtained numerically. All the calculations were carried out at the HF/3-21G level.

Bursi and Stephens[113] reported VCD studies of deuterated methyl glycolate, **24**-d_1 and **24**-d_4, using the MFP method implemented in the CADPAC program. The α-hydroxy ester **24** can exist in a cyclic hydrogen-bonded conformation that may exhibit evidence of an enhanced magnetic dipole transition moment of the methylene stretch due to the ring current mechanism proposed by Nafie and co-workers.[114] Analysis of the MFP results and comparison with a second conformer in which there is no cyclic structure did not yield evidence of the expected enhancement.[113] The theoretical results support an earlier and similar study on methyl lactate **25**, which also exists in a cyclic conformation and for which similar conclusions were reached, in spite of the fact that both the observed and calculated rotatory strengths of the α-methine stretch do exhibit enhanced rotatory strength.[115]

Larger Molecules

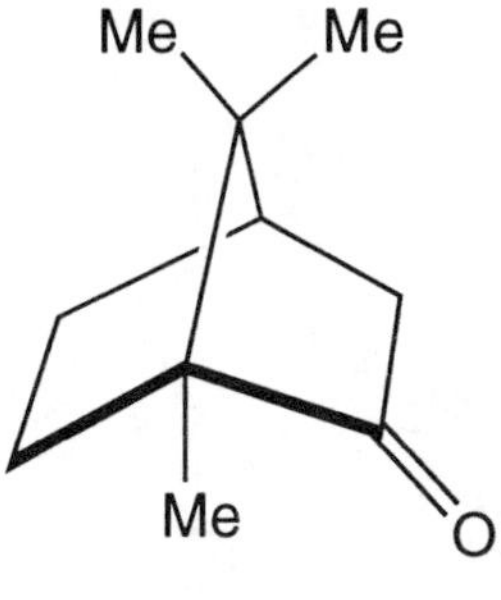

26 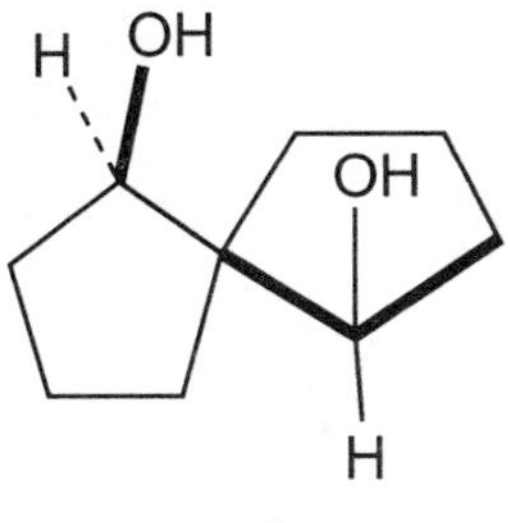 **27**

The largest molecules for which high-level ab initio calculations have been carried out are camphor **26**[116] and the spiro-diol **27**.[117] The VCD spectrum of camphor **26** was calculated by the MFP method in the DO gauge using geometry, force field, and APT generated at the MP2/6-31G* level. The AATs were calculated at the SCF/6-31G* level except in the transformation to the DO gauge, for which MP2 APTs were used. The agreement with the experimental spectrum is good, reinforcing the necessity of using correlated levels of theory to obtain the geometry and force field.

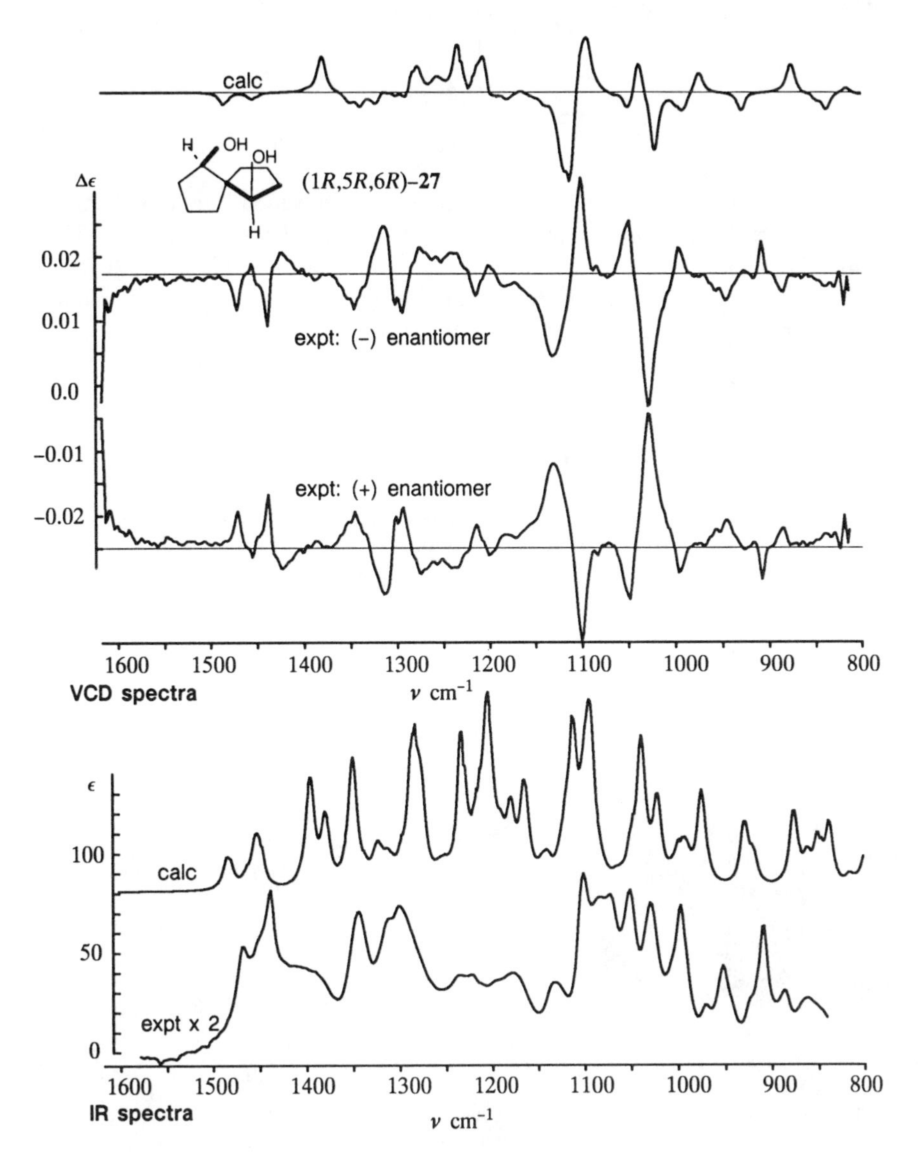

Figure 6 VCD and IR spectra of **27**. Experimental: VCD of both (+) and (−) enantiomers. Calc: Superposition of six conformations of (1*R*,5*R*,6*R*)-**27**, 6-31G$^{*(0.3)}$ geometry, force field, and APTs; 6-31G DO-AATs; calculated frequencies are uniformly scaled by 0.9; assuming Lorentzian lineshape of 5 cm^{-1} half width at half height.

The VCD spectrum of each of the six lowest energy conformers of *cis,cis* spiro[4,4]nonane-1,6-diol **27** was calculated at the VCT/SCF level in the DO gauge using the 6-31G$^{*(0.3)}$ basis set. The simulated spectrum shown in Figure 6 is that of the equilibrium mixture. Comparison with the two experimental VCD spectra of the enantiomers permits unambiguous determination of the absolute configuration as (−)-(1*R*,5*R*,6*R*)-**27**.

CONCLUSIONS

The relatively new chiroptical technique, vibrational circular dichroism spectroscopy, is coming of age. Instrumentation and theoretical models are sufficiently good to establish VCD as the only spectroscopic method for the determination of absolute configurations of molecules in solution and potentially in the gas phase.

ACKNOWLEDGMENTS

We thank the Natural Sciences and Engineering Research Council of Canada for financial support of this work, and the University of Calgary for a Postdoctoral Fellowship to one of us (D. Y.)

REFERENCES

1. L. A. Nafie and D. Che, in *Modern Nonlinear Optics,* Part 3, M. Evans and S. Kielich, Eds., Advances in Chemical Physics Series, Vol. LXXXV, John Wiley & Sons, New York, 1994. Theory and Measurement of Raman Optical Activity.
2. L. D. Barron, *Molecular Light Scattering and Optical Activity,* Cambridge University Press, Cambridge, U.K., 1982.
3. L. Hecht, L. D. Barron, A. Gargaro, Z. Q. Wen, and W. Hug, *J. Raman Spectrosc.,* **23,** 401 (1992). Raman Optical Activity Instrument for Biochemical Studies.
4. E. C. Hsu and G. Holzwarth, *J. Chem. Phys.,* **57,** 4678 (1972). Optical Activity of Vibrational Transitions. Coupled Oscillator Model.
5. G. Holzwarth, E. C. Hsu, H. S. Mosher, T. R. Faulkner, and A. Moscowitz, *J. Am. Chem. Soc.,* **96,** 251 (1974). Infrared Circular Dichroism of Carbon-Hydrogen and Carbon-Deuterium Stretching Modes. Observations.
6. L. A. Nafie, T. A. Keiderling, and P. J. Stephens, *J. Am. Chem. Soc.,* **98,** 2715 (1976). Vibrational Circular Dichroism.
7. P. J. Stephens and R. Clark, in *Optical Activity and Chiral Discrimination,* S. F. Mason, Ed., D. Reidel, 1978, pp. 263–287. Vibrational Circular Dichroism: The Experimental Viewpoint.
8. L. A. Nafie and D. W. Vidrine, in *Fourier Transform Infrared Spectroscopy,* Vol. 3, J. R. Ferraro and L. J. Basile, Eds., Academic Press, New York, 1982, pp. 83–123. Double Modulation Fourier Transform Spectroscopy.

9. T. A. Keiderling, in *Practical Fourier Transform Spectroscopy*, J. R. Ferraro and K. Krishnan, Eds., Academic Press, New York, 1989, pp. 203–284. Vibrational Circular Dichroism: Comparison of Techniques and Practical Applications.

10. P. L. Polavarapu, in *Fourier Transform Infrared Spectroscopy*, Vol. 4, J. R. Ferraro and L. J. Basile, Eds., Academic Press, New York, 1985, pp. 61–96. Fourier Transform Infrared Vibrational Circular Dichroism.

11. P. J. Stephens and M. A. Lowe, *Annu. Rev. Phys. Chem.*, **36**, 213 (1985). Vibrational Circular Dichroism.

12. S. Birkle, M. Moses, B. Kagalovsky, D. Jano, M. Gulotta, and M. Diem, *Biophys. J.*, **65**, 1202 (1993). Infrared CD of Deoxy Oligonucleotides: Conformational Studies of 5′d(GCGC)3′, 5′d(CGCG)3′, 5′d(CCGG)3′, and 5′d(GGCC)3′ in Low and High Salt Aqueous Solution.

13. A. Annamalai and T. A. Keiderling, *J. Am. Chem. Soc.*, **109**, 3125 (1987). Vibrational Circular Dichroism of Poly(ribonucleic acids). A Comparative Study in Aqueous Solution.

14. V. Baumruk and T. A. Keiderling, *J. Am. Chem. Soc.*, **115**, 6939 (1993). Vibrational Circular Dichroism of Proteins in Water Solution.

15. T. Xiang, D. J. Gross, and M. Diem, *Biophys J.*, **65**, 1255 (1993). Strategies for the Computation of Infrared CD and Absorption Spectra of Biological Molecules; Ribonucleic Acids.

16. D. Tsankov, T. Eggimann, G. Liu, and H. Wieser, in *9th International Conference on Fourier Transform Spectroscopy, Proceedings*, Vol. 2089, J. E. Bertie and H. Wieser, Eds., SPIE, Bellingham, WA, 1994, pp. 178–179. FT-IR Vibrational Circular Dichroism (VCD) for Studies of Polymers.

17. C. W. Deutsche and A. Moscowitz, *J. Chem. Phys.*, **49**, 3257 (1968). Optical Activity of Vibrational Origin. I. A Model Helical Polymer.

18. C. W. Deutsche and A. Moscowitz, *J. Chem. Phys.*, **53**, 2630 (1970). Optical Activity of Vibrational Origin. II. Consequences of Polymer Conformation.

19. M. Born and R. Oppenheimer, *Ann. Phys.*, **84**, 457 (1927). Quantum Theory of the Molecules.

20. D. P. Craig and T. Thirunamachandran, *Mol. Phys.*, **35**, 825 (1978). A Theory of Vibrational Circular Dichroism in Terms of Vibronic Interactions.

21. D. P. Craig and T. Thirunamachandran, *Can. J. Chem.*, **63**, 1773 (1985). The Adiabatic Approximation in the Ground-State Manifold.

22. L. A. Nafie and T. B. Freedman, *J. Chem. Phys.*, **78**, 7108 (1983). Vibronic Coupling Theory of Infrared Vibrational Transitions.

23. P. J. Stephens, *J. Phys. Chem.*, **89**, 748 (1985). Theory of Vibrational Circular Dichroism.

24. A. D. Buckingham, P. W. Fowler, and P. A. Galwas, *Chem. Phys.*, **112**, 1 (1987). Velocity-Dependent Property Surfaces and the Theory of Vibrational Circular Dichroism.

25. T. B. Freedman and L. A. Nafie, in *Methods in Enzymology*, Vol. 226, part C, J. F. Riordan and B. L. Vallee, Eds., Academic Press, Inc., San Diego, CA 1994, pp. 306–319. Infrared Circular Dichroism.

26. D. Tsankov, T. Eggimann, and H. Wieser, *Appl. Spectrosc.*, **49**, 132 (1995). FTIR-VCD with Bomem MB100—A New Alternative.

27. M. Diem, P. J. Gotkin, J. M. Kupfer, A. G. Tindall, and L. A. Nafie, *J. Am. Chem. Soc.*, **99**, 8103 (1977). Vibrational Circular Dichroism in Amino Acids and Peptides. 1. Alanine.

28. F. Devlin and P. J. Stephens, *Appl. Spectrosc.* **41**, 1142 (1987). Vibrational Circular Dichroism Measurement in the Frequency Range of 800 to 650 cm^{-1}.

29. M. Diem, G. M. Roberts, O. Lee, and A. Barlow, *Appl. Spectrosc.*, **42**, 20 (1988). Design and Performance of an Optimized Dispersive Infrared Dichrograph.

30. O. Lee and M. Diem, *Anal. Instrum.*, **20**, 23 (1992). Infrared CD in the 6-Micrometer Spectral Region: Design of a Dispersive Infrared Dichrograph.

31. E. D. Lipp and L. A. Nafie, *Appl. Spectrosc.*, **38**, 20 (1984). Fourier Transform Infrared Vibrational Circular Dichroism: Improvements in Methodology and Mid-Infrared Spectra Results.

32. P. L. Polavarapu, *Appl. Spectrosc.*, **43**, 1295 (1989). New Developments in Fourier Transform Infrared Vibrational Circular Dichroism Measurements.

33. P. L. Polavarapu, P. G. Quincey, and J. R. Birch, *Infrared Phys.*, **30**, 175 (1990). Circular Dichroism in the Far Infrared and Millimeter Wavelength Regions: Preliminary Measurements.

34. S. J. Cianciosi, K. M. Spencer, T. B. Freedman, L. A. Nafie, and J. E. Baldwin, *J. Am. Chem. Soc.*, **111**, 1913 (1989). Synthesis and Gas-Phase Vibrational Circular Dichroism of (+)-(S,S)-Cyclopropane-1,2-$^2H_2$.

35. P. L. Polvarapu, *Chem. Phys. Lett.*, **161**, 485 (1989). Rotational-Vibrational Circular Dichroism.

36. L. Rosenfeld, *Z. Physik*, **52**, 161 (1928). Quantum-Mechanical Theory of the Natural Optical Activity of Liquids and Gases.

37. T. B. Freedman and L. A. Nafie, in *Non-Linear Optics*, Part 3. M. Evans and S. Kielich, Eds., Advances in Chemical Physics Series, Vol LXXXV, John Wiley & Sons, New York, 1994, pp. 207–263. Theoretical Formalism and Models for Vibrational Circular Dichroism Intensities.

38. A. Rauk, in *New Developments in Molecular Chirality*, P. G. Mezey, Ed., Kluwer Academic Publishers, Amsterdam, The Netherlands, 1991, pp. 57–92. Vibrational Circular Dichroism Intensities: Ab Initio Calculations.

39. G. Holzwarth and I. Chabay, *J. Chem. Phys.*, **57**, 1632 (1972). Vibrational Circular Dichroism Observed in Crystalline α-$NiSO_4.6H_2O$ and α-$ZnSeO_4.6H_2O$ between 1900 and 5000 cm^{-1}. H. Sugeta, C. Marcott, T. R. Faulkner, J. Overend, and A. Moscowitz, *Chem. Phys. Lett.*, **40**, 397 (1976). Infrared Circular Dichroism Associated with the Carbon-Hydrogen Stretching Vibration of Tartaric Acid.

40. J. A. Schellman, *J. Chem. Phys.*, **58**, 2882 (1973). Vibrational Optical Activity. *Ibid.*, **60**, 343 (1974). Erratum: Vibrational Optical Activity.

41. S. Abbate, L. Laux, J. Overend, and A. Moscowitz, *J. Chem. Phys.*, **75**, 3161 (1981). A Charge Flow Model for Vibrational Rotational Strengths. M. Moskovits and A. Gohin, *J. Phys. Chem.*, **86**, 3947 (1982). Vibrational Circular Dichroism: Effect of Charge Fluxes and Bond Currents.

42. L. A. Nafie and T. B. Freedman, *J. Phys. Chem.*, **90**, 763 (1986). Ring Current Mechanism of Vibrational Circular Dichroism. T. B. Freedman and L. A. Nafie, *Top. Stereochem.*, **17**, 113 (1987). Stereochemical Aspects of Vibrational Optical Activity.

43. C. J. Barnett, A. F. Drake, R. Kuroda, and S. F. Mason, *Mol. Phys.*, **41**, 455 (1980). A Dynamic Polarization Model for Vibrational Optical Activity and the Infrared Circular Dichroism of a Dihydro[5]helicene.

44. L. D. Barron, in *Molecular Light Scattering and Optical Activity*, Cambridge University Press, Cambridge, U.K., 1982, pp. 317–321. P. L. Polavarapu, *Mol. Phys.*, **49**, 645 (1983). A Comparison of Bond Moment and Charge Flow Models for Vibrational Circular Dichroism Intensities. J. R. Escribano, T. B. Freedman, and L. A. Nafie, *J. Phys. Chem.*, **91**, 46 (1987). A Bond-Origin-Independent Formulation of the Bond Dipole Model of Vibrational Circular Dichroism.

45. L. A. Nafie and T. H. Walnut, *Chem. Phys. Lett.*, **49**, 441 (1977). Vibrational Circular Dichroism Theory: A Localized Molecular Orbital Model. T. H. Walnut and L. A. Nafie, *J. Chem. Phys.*, **67**, 1491 (1977). Infrared Absorption and the Born–Oppenheimer Approximation. I. Vibrational Intensity Expression. T. H. Walnut and L. A. Nafie, *J. Chem. Phys.*, **67**, 1501 (1977). Infrared Absorption and the Born–Oppenheimer Approximation. II. Vibrational Circular Dichroism.

46. T. B. Freedman and L. A. Nafie, *J. Phys. Chem.*, **88**, 496 (1984). Molecular Orbital Approaches to the Calculation of Vibrational Circular Dichroism.

47. T. B. Freedman and L. A. Nafie, *J. Chem. Phys.*, **78**, 27 (1983). Vibrational Optical Activity Calculations Using Infrared and Raman Atomic Polar Tensors. *Ibid.*, **79**, 1104 (1983).

Erratum: Vibrational Optical Activity Calculations Using Infrared and Raman Atomic Polar Tensors.

48. T. B. Freedman, L. A. Nafie, and D. Yang, *Chem. Phys. Lett.*, **227**, 419 (1994). Ab Initio Locally Distributed Origin Gauge Calculations of Vibrational Circular Dichroism Intensity: Formulation and Application to (S,S)-Oxirane-2,3-$^2H_2$.

49. T. B. Freedman and L. A. Nafie, *J. Phys. Chem.*, **89**, 374 (1988). Vibrational Coupling Calculations of Vibrational Circular Dichroism Intensities Using Floating Basis Sets.

50. L. A. Nafie, *J. Phys. Chem.*, **96**, 5687 (1992). Velocity-Gauge Formalism in the Theory of Vibrational Circular Dichroism and Infrared Absorption.

51. P. J. Stephens, *J. Phys. Chem.*, **91**, 1712 (1987). Gauge Dependence of Vibrational Magnetic Dipole Transition Moments and Rotational Strengths.

52. C. A. Mead and A. Moscowitz, *Int. J. Quantum Chem.*, **1**, 243 (1967). Dipole Length Versus Dipole Velocity in the Calculation of Infrared Intensities with Born–Oppenheimer Wave Functions.

53. R. D. Amos, K. J. Jalkanen, and P. J. Stephens, *J. Phys. Chem.*, **92**, 5573 (1988). Alternative Formalism for the Calculation of Atomic Polar Tensors and Atomic Axial Tensors.

54. D. Yang and A. Rauk, *J. Chem. Phys.*, **97**, 6517 (1992). Vibrational Circular Dichroism: Ab Initio Vibronic Coupling Theory Using the Distributed Origin Gauge.

55. F. London, *J. Phys. Radium.*, **8**, 397 (1937). Quantum Theory of Interatomic Currents in Aromatic Compounds.

56. R. Ditchfield, *Mol. Phys.*, **27**, 789 (1974). Self-Consistent Perturbation Theory of Diamagnetism I. A Gauge-Invariant LCAO Method for N.M.R. Chemical Shifts. R. Ditchfield, *Chem. Phys. Lett.*, **40**, 53 (1976). GIAO Studies of Magnetic Shielding in FHF^- and HF. R. Ditchfield, *J. Chem. Phys.*, **65**, 3123 (1976). Theoretical Studies of Magnetic Shielding in H_2O and $(H_2O)_2$. D. B. Chesnut, *Chem. Phys.*, **110**, 415 (1986). NMR Chemical Shift Bond Length Derivatives of the First- and Second-Row-Atom Molecules. D. B. Chesnut and C. K. Foley, *J. Chem. Phys.*, **84**, 853 (1986). Chemical Shifts and Bond Modification Effects for Some Small First-Row-Atom Molecules.

57. K. L. Bak, P. Jørgensen, T. Helgaker, K. Ruud, and H. J. A. Jensen, *J. Chem. Phys.*, **98**, 8873 (1993). Gauge-Origin Independent Multiconfigurational Self-Consistent-Field Theory for Vibrational Circular Dichroism.

58. E. Dalgaard, *Chem. Phys. Lett.*, **47**, 279 (1977). Comments on the Use of London's Field Dependent Orbitals.

59. A. E. Hansen, P. J. Stephens, and T. D. Bouman, *J. Phys. Chem.*, **95**, 4255 (1991). Theory of Vibrational Circular Diachroism: Formalisms for Atomic Polar and Axial Tensors Using Noncanonical Orbitals.

60. D. Yang and A. Rauk, *J. Chem. Phys.*, **100**, 7995 (1994). Vibrational Circular Dichroism Intensities: Calculations by Ab Initio Second Order Møller-Plesset Vibronic Coupling Theory.

61. A. Rauk and D. Yang, *J. Phys. Chem.*, **96**, 437 (1992). The VCD and IR Spectra of 2-Methyloxirane and 2,3-Dimethyloxirane: Ab Initio Vibronic Coupling Theory with the 6-31G$^{*(0.3)}$ Basis Set.

62. M. A. Lowe, G. A. Segal, and P. J. Stephens, *J. Am. Chem. Soc.*, **108**, 248 (1986). The Theory of Vibrational Circular Dichroism: *trans*-1,2-Dideuteriocyclopropane.

63. J. A. Pople, R. Krishnan, H. B. Schlegel, and J. S. Binkley, *Int. J. Quantum Chem., Quantum Chem. Symp.*, **13**, 225 (1979). Derivative Studies in Hartree–Fock and Møller–Plesset Theories. J. Gerratt and I. M. Mills, *J. Chem. Phys.*, **49**, 1719 (1968). Force Constants and Dipole-Moment Derivatives of Molecules from Perturbed Hartree–Fock Calculations. Part I. J. Gerratt and I. M. Mills, *J. Chem. Phys.*, **49**, 1730 (1968). Force Constants and Dipole-Moment Derivatives of Molecules from Perturbed Hartree–Fock Calculations. II. Applications to Limited Basis Set SCF-MO Wavefunctions. R. M. Stevens, R. Pitzer, and W. N. Lipscomb, *J. Chem. Phys.*, **38**, 550 (1963). Perturbed Hartree–Fock Calculations. I. Magnetic Susceptibility and Shielding in the LiH Molecule. T. C. Caves and M. Karplus, *J. Chem.*

Phys., **50,** 3649 (1969). Perturbed Hartree–Fock Theory. I. Diagrammatic Double-Perturbation Analysis. R. McWeeny, *Rev. Mod. Phys.*, **32,** 335 (1960). Some Recent Advances in Density Matrix Theory.

64. R. D. Amos, N. C. Handy, K. J. Jalkanen, and P. J. Stephens, *Chem. Phys. Lett.*, **133,** 21 (1987). Efficient Calculation of Vibrational Magnetic Dipole Transition Moments and Rotational Strengths.
65. R. Dutler, Ph. D. Dissertation, The University of Calgary (1988). Theoretical Study of Asymmetric Reactions Induced by Circularly Polarized Light.
66. D. Yang, Ph.D. Dissertation, The University of Calgary (1992). Implementation of Ab Initio Vibronic Coupling Theory to Interpret Vibrational Circular Dichroism Spectra.
67. K. J. Jalkanen, P. J. Stephens, P. Lazzeretti, and R. Zanasi, *J. Chem. Phys.*, **90,** 3204 (1989). Nuclear Shielding Tensors, Atomic Polar and Axial Tensors, and Vibrational Dipole and Rotational Strengths of NHDT.
68. P. Lazzeretti and R. Zanasi, *Chem. Phys. Lett.*, **112,** 103 (1984). Connection between the Nuclear Electric Shielding Tensor and the Infrared Intensities. P. Lazzeretti, R. Zanasi, and P. J. Stephens, *J. Phys. Chem.*, **90,** 6761 (1986). Magnetic Dipole Transition Moments and Rotational Strengths of Vibrational Transitions: An Alternative Formalism.
69. J. A. Pople, H. B. Schlegel, R. Krishnan, D. J. DeFrees, J. S. Binkley, M. J. Frisch, R. A. Whiteside, R. F. Hout, and W. J. Hehre, *Int. J. Quantum Chem., Quantum Chem. Symp.*, **15,** 269 (1981). Molecular Orbital Studies of Vibrational Frequencies.
70. P. Pulay, G. Fogarasi, F. Pang, and J. E. Boggs, *J. Am. Chem. Soc.*, **101,** 2550 (1979). Systematic Ab Initio Gradient Calculation of Molecular Geometries, Force Constants, and Dipole Moment Derivatives. P. Pulay, in *Modern Theoretical Chemistry,* Vol. 4, H. F. Schaefer III, Ed., Plenum, New York, 1977, pp. 153–185. Direct Use of the Gradient for Investigating Molecular Energy Surfaces.
71. C. E. Blom, P. J. Slingerland, and C. Altona, *Mol. Phys.*, **31,** 1359 (1976). Application of Self-Consistent-Field Ab Initio Calculations to Organic Molecules. I. Equilibrium Structure and Force Constants of Hydrocarbons.
72. S. T. Pickard, H. E. Smith, P. L. Polavarapu, T. M. Black, A. Rauk, and D. Yang, *J. Am. Chem. Soc.*, **114,** 6850 (1992). Synthesis, Experimental and Ab Initio Theoretical Vibrational Circular Dichroism, and Absolute Configurations of Substituted Oxiranes.
73. A. Rauk, T. Eggimann, H. Wieser, G. V. Shustov, and D. Yang, *Can. J. Chem.*, **72,** 506 (1994). Vibrational Circular Dichroism Spectra of 2-Methylaziridine: Dominance of the Asymmetric Center at Nitrogen.
74. D. Yang, G. Shustov, T. Eggimann, H. Wieser, and A. Rauk, *Can. J. Chem.*, **71,** 2028 (1993). Framework Stereochemistry and Vibrational Circular Dichroism: 1,2- and 2,3-Dimethylaziridines.
75. R. Dutler and A. Rauk, *J. Am. Chem. Soc.*, **111,** 6957 (1989). Calculated Infrared Absorption and Vibrational Circular Dichroism Intensities of Oxirane and Its Deuterated Analogues.
76. A. Rauk, T. Eggimann, H. Wieser, and D. Yang, *Can. J. Chem.*, **70,** 464 (1992). The Infrared Spectrum of 2-Methylaziridine from Scaled Ab Initio Force Fields.
77. M. J. Frisch, M. Head-Gordon, G. W. Trucks, J. B. Foresman, H. B. Schlegel, K. Raghavachari, M. A. Robb, J. S. Binkley, C. Gonzalez, D. J. DeFrees, D. J. Fox, R. A. Whiteside, R. Seeger, C. F. Melius, J. Baker, R. L. Martin, L. R. Kahn, J. J. P. Stewart, S. Topiol, and J. A. Pople, Gaussian 90, Gaussian, Inc., Pittsburgh, PA (1990). The version at the University of Calgary was modified to write extra information to the Read-Write file (RWF) for VCD calculations.
78. R. D. Amos, *The Cambridge Analytic Derivatives Package 5.0,* Cambridge University, Cambridge, U.K., 1992.
79. K. Morokuma and H. Sugeta, *Chem. Phys. Lett.*, **134,** 23 (1987). Ab Initio Derivative Calculation of Vibrational Circular Dichroism.
80. P. Lazzeretti and R. Zanasi, *J. Chem. Phys.*, **87,** 472 (1987). Electromagnetic Nuclear Shielding Tensors and Their Relation to Other Second-Order Properties. A Study of the

Methane Molecule. P. Lazzeretti and R. Zanasi, *Phys. Rev.* **A, 33,** 3727 (1986). Electric and Magnetic Nuclear Shielding Tensors: A Study of the Water Molecule.

81. K. J. Jalkanen, P. J. Stephens, R. D. Amos, and N. C. Handy, *Chem. Phys. Lett.,* **142,** 153 (1987). Basis Set Dependence of Ab Initio Predictions of Vibrational Rotational Strengths: NHDT.
82. K. J. Jalkanen, P. J. Stephens, R. D. Amos, and N. C. Handy, *J. Phys. Chem.,* **92,** 1781 (1988). Gauge Dependence of Vibrational Rotational Strengths: NHDT.
83. P. J. Stephens, K. J. Jalkanen, and R. W. Kawiecki, *J. Am. Chem. Soc.,* **112,** 6518 (1990). Theory of Vibrational Rotational Strengths: Comparison of A Priori Theory and Approximate Models.
84. K. J. Jalkanen, R. W. Kawiecki, and P. J. Stephens, *J. Phys. Chem.,* **94,** 7040 (1990). Basis Set and Gauge Dependence of Ab Initio Calculations of Vibrational Rotational Strengths.
85. F. Faglioni, P. Lazzeretti, M. Malagoli, and R. Zanasi, *J. Phys. Chem.,* **97,** 2535 (1993). Calculation of Infrared and Vibrational Circular Dichroism Intensities via Nuclear Electromagnetic Shielding.
86. K. J. Jalkanen, P. J. Stephens, R. D. Amos, and N. C. Handy, *J. Am. Chem. Soc.,* **109,** 7193 (1987). Theory of Vibrational Circular Dichroism: *trans*-1*(S)*,2*(S)*-Dicyanocyclopropane.
87. R. D. Amos, N. C. Handy, A. F. Drake, and P. Palmieri, *J. Chem. Phys.,* **89,** 7287 (1988). The Vibrational Circular Dichroism of Dimethylcyclopropane in the C-H Stretching Region.
88. T. B. Freedman, K. M. Spencer, N. Ragunathan, and L. A. Nafie, *Can. J. Chem.,* **69,** 1619 (1991). Vibrational Circular Dichroism of *(S,S)*-[2,3-$^2H_2$]Oxirane in the Gas Phase and in Solution.
89. K. J. Jalkanen, P. J. Stephens, R. D. Amos, and N. C. Handy, *J. Am. Chem. Soc.,* **110,** 2012 (1988). Theory of Vibrational Circular Dichroism: *trans*-2,3-Dideuteriooxirane.
90. C. Hirose, Bull. *Chem. Soc. Jpn.,* **47,** 1311 (1974). Microwave Spectra and r_0, r_s, and r_m Structures of Ethylene Oxide.
91. T. B. Freedman, M. G. Lee, L. A. Nafie, J. M. Schwab, and T. Ray, *J. Am. Chem. Soc.,* **109,** 4727 (1987). Vibrational Circular Dichroism in the Carbon-Hydrogen and Carbon-Deuterium Stretching Modes of *(S,S)*-[2,3-$^2H_2$]Oxirane.
92. P. J. Stephens, K. J. Jalkanen, F. J. Devlin, and C. F. Chabalowski, *J. Phys. Chem.,* **97,** 6107 (1993). Ab Initio Calculation of Vibrational Circular Dichroism Spectra Using Accurate Post-Self-Consistent-Field Force Fields: *trans*-2,3-Dideuteriooxirane.
93. F. B. van Dijneveldt, IBM Res. Rept. RJ945, 35 (1971). Gaussian Basis Sets for the Atoms H-Ne for Use in Molecular Calculations.
94. K. J. Jalkanen, P. J. Stephens, P. Lazzeretti, and R. Zanasi, *J. Phys. Chem.,* **93,** 6583 (1989). Random Phase Approximation Calculations of Vibrational Circular Dichroism: *trans*-2,3-Dideuteriooxirane.
95. P. L. Polavarapu and D. F. Michalska, *J. Am. Chem. Soc.,* **105,** 6190 (1983). Vibrational Circular Dichroism in *(S)*-(−)-Epoxypropane. Measurement in Vapor Phase and Verification of the Perturbed Degenerate Mode Theory. P. L. Polavarapu and D. F. Michalska, *Mol. Phys.,* **52,** 1225 (1984). Mid Infrared Vibrational Circular Dichroism in *(S)*-(−)-Epoxypropane Bond Moment Model Predictions and Comparison to the Experimental Results. P. L. Polavarapu and D. F. Michalska, *Mol. Phys.,* **55,** 723 (1985). Errata: Mid Infrared Vibrational Circular Dichroism in *(S)*-(−)-Epoxypropane Bond Moment Model Predications and Comparison to the Experimental Results. P. L. Polavarapu, B. A. Hess, and L. J. Schaad, *J. Chem. Phys.,* **82,** 1705 (1985). Vibrational Spectra of Epoxypropane.
96. R. W. Kawiecki, F. Devlin, P. J. Stephens, R. D. Amos, and N. C. Handy, *Chem. Phys. Lett.,* **145,** 411 (1988). Vibrational Circular Dichroism of Propylene Oxide.
97. M. A. Lowe and J. S. Alper, *J. Phys. Chem.,* **92,** 4040 (1988). Ab Initio Calculations of Vibrational Circular Dichroism in Propylene Oxide: Geometry and Force Field Dependence.

98. R. W. Kawiecki, F. Devlin, P. J. Stephens, and R. D. Amos, *J. Phys. Chem.*, **95**, 9817 (1991). Vibrational Circular Dichroism of Propylene Oxide.

99. H. Dothe, M. A. Lowe, and J. S. Alper, *J. Phys. Chem.*, **92**, 6246 (1988). Vibrational Circular Dichroism of Methylthiirane.

100. R. D. Amos, N. C. Handy, and P. Palmieri, *J. Chem. Phys.*, **93**, 5796 (1990). Vibrational Properties of *(R)*-Methylthiirane from Møller-Plesset Perturbation Theory.

101. P. L. Polavarapu, S. T. Pickard, H. E. Smith, T. M. Black, A. Rauk, and D. Yang, *J. Am. Chem. Soc.*, **113**, 9747 (1991). Vibrational Circular Dichroism and Absolute Configuration of Substituted Thiiranes.

102. P. L. Polavarapu, B. A. Hess, L. J. Schaad, D. O. Henderson, L. P. Fontana, H. E. Smith, L. A. Nafie, T. B. Freedman, and W. M. Zuk, *J. Chem. Phys.*, **86**, 1140 (1987). Vibrational Spectra of Methylthiirane.

103. P. L. Polavarapu, P. K. Bose, and S. T. Pickard, *J. Am. Chem. Soc.*, **113**, 43 (1991). Vibrational Circular Dichroism in Methythiirane: Ab Initio Localized Molecular Orbital Predictions and Experimental Measurements.

104. A. Rauk, D. Yang, D. Tsankov, H. Wieser, Yu. Koltypin, A. Gedanken, and G. V. Shustov, *J. Am. Chem. Soc.*, **117**, 4160 (1995). Chiroptical Properties of 1-Azabicyclo[3.1.0]hexane in the Vaccum–UV and IR Regions.

105. D. Yang, D. Tsankov, H. Wieser, G. V. Shustov, and A. Rauk, to be published. The Electronic and Vibrational Chiroptical Properties of 2-Carbomethoxyaziridine.

106. M. A. Lowe, P. J. Stephens, and G. A. Segal, *Chem. Phys. Lett.*, **123**, 108 (1986). The Theory of Vibrational Circular Dichroism: *trans*-1,2-Dideuteriocyclobutane and Propylene Oxide.

107. A. Annamalai, T. A. Keiderling, and J. S. Chickos, *J. Am. Chem. Soc.*, **106**, 6254 (1984). Vibrational Circular Dichroism of *trans*-1,2-Dideuteriocyclobutane. A Comparison of Fixed Partial Charge and Localized Molecular Orbital Theories with Different Force Fields. A. Annamalai, T. A. Keiderling, and J. S. Chickos, *J. Am. Chem. Soc.*, **107**, 2285 (1985). Vibrational Circular Dichroism of *trans*-1,2-Dideuteriocyclobutane. Experimental and Calculational Results in the Mid-Infrared.

108. A. Annamalai and T. A. Keiderling, *J. Mol. Spectrosc.*, **109**, 46 (1985). Vibrational Spectra and Normal Coordinate Calculations for *cis*- and *trans*-1,2-Dideuteriocyclobutanes.

109. P. Malon, L. J. Mickley, K. M. Sluis, C. N. Tam, T. A. Keiderling, S. Kamath, J. Uang, and J. S. Chickos, *J. Phys. Chem.*, **96**, 10139 (1992). Vibrational Circular Dichroism Study of (2*S*,3*S*)-Dideuteriobutyrolactone. Synthesis, Normal Mode Analysis, and Comparison of Experimental and Calculated Spectra.

110. A. Annamalai, K. J. Jalkanen, U. Narayanan, M.-C. Tissot, T. A. Keiderling, and P. J. Stephens, *J. Phys. Chem.*, **94**, 194 (1990). Theoretical Study of the Vibrational Circular Dichroism of 1,3-Dideuterioallene: Comparison of Method.

111. H. Dothe, M. A. Lowe, and J. S. Alper, *J. Phys. Chem.*, **93**, 6632 (1989). Calculations of the Infrared and Vibrational Circular Dichroism Spectra of Ethanol and Its Deuterated Isotopomers.

112. H. Nakatsuji, K. Handa, K. Endo, and T. Yonezawa, *J. Am. Chem. Soc.*, **106**, 4653 (1984). Theoretical Study of the Metal Chemical Shift in Nuclear Magnetic Resonance. Ag, Cd, Cu, and In Complexes.

113. R. Bursi and P. J. Stephens, *J. Phys. Chem.*, **95**, 6447 (1991). Ring Current Contributions to Vibrational Circular Dichroism? Ab Initio Calculations for Methyl Glycolate-d_1 and -d_4.

114. L. A. Nafie, M. R. Oboodi, and T. B. Freedman, *J. Am. Chem. Soc.*, **105**, 7449 (1983). Vibrational Circular Dichroism in Amino Acids and Peptides. 8. A Chirality Rule for Methine C* α-H Stretching Modes.

115. R. Bursi, F. J. Devlin, and P. J. Stephens, *J. Am. Chem. Soc.*, **112**, 9430 (1990). Vibrationally Induced Ring Current? The Vibrational Circular Dichroism of Methyl Lactate.

116. F. J. Devlin and P. J. Stephens, *J. Am. Chem. Soc.*, **116**, 5003 (1994). Ab Initio Calculation of Vibrational Circular Dichroism Spectra of Chiral Natural Products Using MP2 Force Fields: Camphor.

117. J. A. Nieman, B. A. Keay, M. Kubicki, D. Yang, A. Rauk, D. Tsankov, and H. Wieser, *J. Org. Chem.*, **60**, 1918 (1995). Determining Absolute Configuration by Spectroscopic Means: The Vibrational Circular Dichroism Spectrum of (+)-(1*S*,5*S*,6*S*)- and (−)-(1*R*,5*R*,6*R*)-Spiro[4.4]nonane-1,6-diol.

118. J. L. Duncan and I. M. Mills, *Spectrochim. Acta*, **20**, 523 (1964). Calculation of Force Constants and Normal Coordinates. IV. XH_4 and XH_3 Molecules.

119. Compounds 1-azabicyclo[3.1.0]hexane **17** and 2-carbomethoxyaziridine **18** were synthesized by G. V. Shustov and A. Rauk, and *cis,cis*-spiro[4,4]nonane-1,6-diol **27** by J. A. Nieman and B. A. Keay in the Department of Chemistry at the University of Calgary. The experimental spectra were measured by D. Tsankov in the laboratory of H. Wieser in the Department of Chemistry at the University of Calgary. IR and VCD spectra were recorded in the CCl_4 solution in a 0.15 mm NaCl cell on a Bomem MB100 spectrometer at 4 cm^{-1} resolution. For 1-azabicyclo[3.1.0]hexane **17** and 2-carbomethoxyaziridine **18**, 250 scans were collected for IR spectra and 5000 scans were collected for VCD spectra. For *cis,cis*-spiro[4,4]nonane-1,6-diol **27**, 500 scans were collected for IR spectra and 10,000 scans were collected for VCD spectra. The concentrations used are 1.28 M for 1-azabicyclo[3.1.0]hexane **17**, 0.1174 M for 2-carbomethoxyaziridine, and 1.29 M for *cis,cis*-spiro[4,4]nonane-1,6-diol **27**.

APPENDIX

Compendium of Software for Molecular Modeling

Donald B. Boyd

Department of Chemistry, Indiana University-Purdue University at Indianapolis (IUPUI), Indianapolis, Indiana 46202-3274

INTRODUCTION

The palette of computational chemistry continues to expand. A breathtaking array of computer programs is at the disposal of chemists to aid them in their research and teaching. This compendium, which has grown rapidly with each volume of *Reviews in Computational Chemistry*,[1] is updated, selectively expanded, and reorganized for easier use. The information herein will link the reader to more than 2500 programs for studying molecules and presenting the results. The aims of the compendium are to provide a ready reference for researchers as well as newcomers to the field, to advance computer-aided chemistry by making the tools widely known, and to benefit both developers and consumers of software.

As used here and expounded on in Volume 1 of *Reviews in Computational Chemistry*, *molecular modeling* refers to the generation, manipulation, and/or representation of realistic molecular structures and associated physicochemical properties.[2] Any way in which computers can expedite or make practical research on molecules is *computational chemistry*. The terms "molecular modeling" and "computational chemistry" are used interchangeably in the context of this compendium.

Not only should many techniques be included under the umbrella of molecular modeling/computational chemistry, but also the scientist who lacks theoretical training must be welcomed to apply the techniques when appropriate. It is obvious that as computational chemistry software packages become easier to use through sophisticated graphical user interfaces with pull-down menus and point-and-click buttons, it is incumbent on developers to make sure their software is foolproof so it can only be appropriately applied. The more elegant the software, the less the user should have to think about software, hardware, and theoretical technicalities, and the more the user can concentrate on the scientific question at hand.

With the ever-increasing array of software available for molecular modeling, it is useful for purposes of this compendium to attempt to categorize the wide array of programs. Software packages listed here have been divided into two broad categories based on the platform on which they run, that is, on an inexpensive personal microcomputer or on a more powerful computer, such as a minicomputer, mainframe, workstation, supercomputer, or massively parallel machine. Within each of these two broad categories, we have further subdivided the software (and the corresponding suppliers) according to the main thrust:

1. General purpose molecular modeling
2. Quantum chemistry calculations
3. Databases of molecular structures
4. Molecular graphics
5. Quantitative structure–property relationships and statistics
6. Other applications

Group 1 includes multifunctional and molecular mechanics programs. Programs intended mainly for molecular orbital or other quantum mechanical calculations are in Group 2. Group 3 features software for storage and retrieval of molecular structure data. Group 4 contains programs to visualize molecules (but not to minimize an energy). Group 5 encompasses programs to study quantitative structure–property relationships and perform chemometric and other statistical analyses. Last, Group 6 includes software for computer-aided organic chemistry, crystallographic structure determination, scientific writing, and other computational tools of interest to chemists.

Programs are organized in each group alphabetically by the name of a program for which the vendor or supplier is well known. Other software products of that vendor are then listed with it. Although this organization is somewhat arbitrary, it helps keep this compendium more compact than it would otherwise be. Individual program names can tracked down with the aid of the subject index for this volume. Some of the more sophisticated molecular modeling packages—really suites of software—span more than one group. Suppliers who offer several strategic products in more than one group are listed in each.

For each software package, we give a brief description, the address and telephone number of the supplier, and other pertinent information (e.g., other programs offered by the vendor). The descriptions are concise overviews, not reviews, and an effort has been made to free the descriptions of commercial embellishment. When possible, we give an electronic mail address and/or an 800 number for toll-free telephone calls from within the United States to make it easier to communicate with the supplier.

Version numbers of the programs are generally not included in this compendium because they are constantly, but irregularly, changing. Version numbers can represent important milestones in the evolution of a program in terms of added functionality or in terms of bugs that have been fixed. The reader is encouraged to check on the latest version number before embarking on a computational research project.

Prices of software, which range from essentially free to more than $100,000 (U.S.), are not included because they are subject to change and to specific conditions of distribution. In some cases, particularly with respect to QCPE (Indiana University, Bloomington), significant software, which has been validated with respect to expected output, can be obtained at minimal cost. With so much excellent software and so many suppliers now vying for attention, the consumer of software products is in a good position to explore all possibilities.

Whereas most of the commercial molecular modeling vendors provide excellent support in terms of quickly answering users' questions and fixing bugs periodically, it should be kept in mind that some publishers sell the software "as is." Also, providing customer support may not be among the top priorities of individual suppliers of software. Obviously, the price of a software product includes the anticipated cost of the level of support to be provided. Almost invariably, a user will encounter a question when using a program, so the availability of an accommodating support service can be reassuring.

Besides the software listed here, other molecular modeling programs are being developed in academic and industrial laboratories around the world. However, because the availability, documentation, and degree of support of these other programs are highly variable, it is impractical to include them all. Some software may be optimized for a particular machine or to take advantage of a machine's unique graphics capabilities, in which case the hardware vendor may be an additional source of information about a program.

There is no warranty expressed or implied as to completeness or accuracy of the material described herein or referred to. Readers are encouraged to pursue further details germane to their own interests. Inclusion in this compendium should not be construed as an endorsement. Product names are the registered symbols or trademarks of their respective organizations. It should be kept in mind that new software products of interest to computational chemists are continually appearing in the marketplace (and some are falling by the wayside).

THE INTERNET

The Internet is a network of computer networks based on the TCP/IP (Transmission Control Protocol/Internet Protocol) protocols. Starting as the U.S. Department of Defense's ARPAnet circa 1970, it has grown to be an almost wholly privatized, global resource connecting more than 11,000 networks with 2 million host machines and 20–25 million computer users in more than 90 countries.[3] It is estimated that Internet traffic has been growing at about 15% per month recently. In the U.S., one-third of all business mail in the last 5 years traveled over telephone wires, including both electronic mail and facsimile.

More and more frequently, computational chemists transfer academic and public domain software via file transfer protocol (ftp) over Internet or other networks connected to Internet, such as Bitnet. Thanks to the speed and accuracy of this method of communication, an individual can receive or download software at essentially no cost.

Information on software and archival resources available over Internet can be obtained with Archie, Gopher, or World Wide Web. For instance, Archie, a system that has been in existence for a few years, permits searching indexes with more than 2 million files on more than 1000 public servers. To run Archie, a user can telnet from a networked workstation to a server such as archie.rutgers.edu, archie.sura.net, archie.unl.edu, archie.ans.net, archie.mcgill.ca, archie.au, archie.funet.fi, or archie.doc.ic.ac.uk. Next the user must login as archie and type "help" for additional information. At the "archie>" prompt, a keyword is typed corresponding to the subject matter of interest. It is also possible to make an inquiry by e-mail to archie@archie.unl.edu, for example. Files of interest can be downloaded by ftp or by Internet Ftpmail (e-mail ftp-mail@decwrl.dec.com).

To use ftp, the user types the "telnet site" command on a networked computer to reach the server (site) with the desired files. At the "ftp>" prompt, typing "help" obtains the ftp commands. Typically, the files that are available for sharing are in a public directory, such as "pub/filename." Files have to be identified by their name; unfortunately, it is not possible with ftp to inspect the contents of a file before it is transmitted.

There are many Gopher sites around the world, mostly at universities and government laboratories. A few Gopher sites of particular interest to computational chemists include those at the Ohio Supercomputer Center maintained by Dr. Jan K. Labanowski and colleagues (infomeister.osc.edu), at the American Chemical Society headquarters in Washington, DC (acsinfo.acs.org), and at Northern Illinois University maintained by Dr. Steven M. Bachrach (hackberry.chem.niu.edu).

The most recent medium for obtaining information on the Internet is through the World Wide Web (WWW or simply Web). A sampling of Universal Resource Locations (URL) of WWW servers of particular interest to computational chemists is given in Table 1. Much information is available on-line through Mosaic, Netscape, or one of the other WWW browsers and viewers. From the sites listed in Table 1, it is possible to obtain additional information and to navigate to other Web servers, which are rapidly proliferating around the world. The latest news about new resources is available at InterNIC's WWW (Table 1) and Gopher addresses (ds.internic.net).

Often scientists wish to contact a colleague by electronic mail to ask about a piece of software or for other information. Several strategies for obtaining an individual's electronic address are possible. If the colleague is at a U.S. or Canadian university, an e-mail address may be listed in the American Chemical Society's *Directory of Graduate Research*. Similar compilations may be available in other nations. If the colleague's institution maintains a Gopher or World Wide Web server, it is possible that this server will contain a faculty or staff directory that can be searched on-line. Another strategy is to telnet to a Netfind server, such as bruno.cs.colorado.edu, mudhoney.micro.umn.edu, redmont.cis.uab.edu, monolity.cc.ic.ac.uk, netfind.if.usp.br, netfind.anu.edu.au, or nic.nm.kr; log on as netfind and follow the instructions. If the electronic address of the institution (site) is known, an e-mail query can be sent to postmaster@site asking for the individual's user identification at that site. Another facility is WHOIS; to use it, send a mail message to service@nic.ddn.mil with "WHOIS last_name, first_name" in the subject line. The Usenet Name Server can be accessed by sending a mail message with the subject line blank and the body containing "send usenet-addresses/last_name" to mail-server@rtfm.mit.edu. The Knowbot Information Service can be accessed by sending a mail message with the body containing "query last_name" to netaddress@sol.bucknell.edu. As a last resort, the moderators of some of the chemistry-related electronic bulletin boards can be contacted.

Of interest to computational chemists are several electronic bulletin boards. The bulletin board at the Ohio Supercomputer Center (OSC) is devoted to computational chemistry in general. It is supervised by Dr. Jan K. Labanowski and is one of the largest and busiest, with more than 2000 subscribers. Other bulletin boards are focused on a single program or set of programs. Table 2 lists electronic addresses of some pertinent bulletin boards. For each topic in the first column, the address given in the second column is that of the administrator of the bulletin board, who can be informed that you wish to subscribe; the address in the last column is for posting notices that go to all current subscribers (mail exploder). The notices can be used to ask questions, exchange useful information, carry on topical discussions, and obtain information about software. Individuals using these services, which are almost always free, are usually asked to follow certain rules set by the administrator, to avoid unnecessary or inappropriate traffic.

Finally Table 3 gives information for contacting some of the high-performance computing centers in the United States. These have many of the common, compute-intensive, computational chemistry codes in production mode.

Table 1 Sampling of Universal Resource Location (URLs) for World Wide Web Server Home Pages of Interest to Computational Chemists

Sponsor or topic	Comments	URL[a,b]
Acrobat Reader Software	Reading ACS supplementary material	http://www.adobe.com/Software.html
AMBER		http://www.amber.ucsf.edu/amber/amber.html
American Chemical Society		http://acsinfo.acs.org/, http://pubs.acs.org/, and http://www.acs.org/
American Crystallographic Association		http://www.hwi.buffalo.edu/ACA/
Apple Computer		http://www.apple.com/
Bionet		http://www.bio.net/
BIOSYM Technologies, Inc.		http://www.biosym.com/biosym
Biotechnology		http://www.inform.umd.edu/Education_Resources/AcademicResources
Birkbeck College	Internet course on protein structure and connections to many Web resources	http://www.cryst.bbk.ac.uk/
Brookhaven Protein Data Bank		http://www.pdb.bnl.gov/
CADPAC		http://www.cray.com/PUBLIC/DAS/files/CHEMISTRY/CADPAC.txt
Cambridge Crystallographic Data Centre (CCDC)		http://csdvx2.ccdc.cam.ac.uk/
Cambridge University Molecular Modeling		http://www.ch.cam.ac.uk/SGTL/home.html
CambridgeSoft Corporation		http://www.camsci.com/
Chemical Abstracts Service	STN International	http://info.cas.org/welcome.html

Chemical Information Sources from Indiana University (CIS-IU)	Web connections to many sites related to molecular modeling, graphics, and information	http://www.indiana.edu/~cheminfo
Cornell Theory Center		http://www.tc.cornell.edu/
Digital Equipment		http://www.digital.com/
Directory of chemistry-related mailing lists		http://bionmr1.rug.ac.be/chemistry/overview.html
Directory of commercial outfits	Includes pharmaceutical and chemical companies	http://www.engr.iupui.edu/~ho/interests/
Hewlett-Packard		http://www.dmo.hp.com/computing/main.html
Human Genome Project		http://gdbwww.gov.org/
IBM		http://www.ibm.com/
Imperial College (Dr. Henry Rzepa)		http://www.ch.ic.ac.uk/rzepa.html and http://www.ch.ic.ac.uk/chemical_mime_first.html
Indiana University-Purdue University at Indianapolis Chemistry Department	Includes *Reviews in Computational Chemistry*	http://chem.iupui.edu/
International Union of Crystallography	Includes World Directory of Crystallographers	http://www.iucr.ac.uk/welcome.html
InterNIC	AT&T directory and database services	http://www.internic.net/

(continued)

Table 1 *(continued)*

Sponsor or topic	Comments	URL[a,b]
Kinemage software for *Protein Science*		gopher://orion.oac.uci.edu:1071/11/protein/Kinemage
Klotho Biochemical Compounds Database		http://ibc.wustl.edu/klotho/
Lawrence Livermore National Laboratory		http://www.llnl.gov/
Los Alamos National Laboratory		http://www.lanl.gov/
MacroModel		http://www.cc.columbia.edu/~chempub/mmod/mmod.html
Mathematica		http://www.wri.com/mathsource.html
MDL Information Systems		http://www.mdli.com/
MedChem/BioByte QSAR Database		http://fox.pomona.claremont.edu/chem/qsar-db/index.html
Microsoft Corp.		http://www.microsoft.com/
Molecular biology sequence database		http://www.nih.gov/molbio/
Molecular graphics programs at the Lawrence Livermore National Laboratory		http://www-dsed.llnl.gov/documents/tests/chem.html
Molecular Model Types and Rendering Techniques		http://scsg9.unige.ch/eng/toc.html
Molecular Simulations Inc.		http://www.msi.com/
National Center for Atmospheric Research		http://http.ucar.edu/metapage.html

National Center for Supercomputing Applications	Mosaic	http://www.ncsa.uiuc.edu/
National Institutes of Health		http://www.nih.gov/
National Institutes of Health 3D Drug Structure Databank		http://www.nih.gov/molecular_modeling/drugbank.html
National Institutes of Health molecular modeling		http://www.nih.gov/molecular_modeling/mmhome.html
National Institute of Standards and Technology		http://www.nist.gov/
National Library of Medicine		http://www.nlm.nih.gov/
National Science Foundation		http://www.nsf.gov/
NetBiochem	Medical biochemistry	http://www.hahnemann.edu/Heme-Iron/NetWelcome.html
Netscape	Netscape Navigator	http://home.netscape.com/
NewsPage	Includes some news about computers and software	http://www.newspage.com/
North Carolina Supercomputing Center		http://tfnet.ils.unc.edu/
Northern Illinois University Chemistry (Dr. Steven Bachrach)	Information for computational chemists on employment opportunities	http://hackberry.chem.niu.edu/

(continued)

Table 1 *(continued)*

Sponsor or topic	Comments	URL[a,b]
	conferences, journal submissions, software, and directory of chemists	
NRL_3D Database	Protein sequences and structures	http://www.gdb.org/Dan/proteins/nrl3d.html
Oakridge National Laboratory		http://www.ornl.gov/
Ohio Supercomputer Center	Computational Chemistry List (CCL) with archives of messages, programs, job and conference information	http://www.osc.edu/chemistry.html
Pittsburgh Supercomputing Center		http://pscinfo.psc.edu/
PMD	Parallel molecular dynamics	http://tincan.bioc.columbia.edu/pmd/
Protein Science		http://www.prosci.uci.edu/
QCPE		http://www.indiana.edu/
Royal Society of Chemistry	Chemical Web connections in UK	http://chemistry.rsc.org/rsc/
San Diego Supercomputer Center		http://www.sdsc.edu/
SCOP	Software for searching and using PDB files	http://scop.mrclmb.cam.ac.uk/scop
Silicon Graphics		http://www.sgi.com/

Software for Science magazine		http://www.scitechint.com/scitech
Starting points for Internet exploration		http://www.ncsa.uiuc.edu/SDG/Software/Mosaic/StartingPoints/NetworkStartingPoints.html
Sun Microsystems		http://www.sun.com/
SWISS-PROT	Protein sequence database	http://expasy.hcuge.ch/sprot/sprot-top.html
SYBYL SPL scripts		http://www.nih.gov/molecular_modeling/mmhome.html
UNICHEM		http://www.cray.com/apps/UNICHEM/Mainpage.html
University of California, San Diego (Dr. Kent R. Wilson)	Instructional physical chemistry	http://www-wilson.ucsd.edu/education/samplegateway.html
University of Sheffield, Department of Information Studies		http://www2.shef.ac.uk/
Wavefunction, Inc.		http://wavefun.com/
WebCrawler	Searching and general navigation on Web	http://www.webcrawler.com/
WHAT IF		http://swift.embl-heidelberg.de/whatif/
World Wide Web Project		http://info.cern.ch/hypertext/WWW/TheProject.html
WPDB	Software for analyzing PDB files	http://cuhhca.hhmi.columbia.edu/wpdb.html
XMol		ftp://ftp.msc.edu/pub/xmol/
X-PLOR		http://xplor.csb.yale.edu/
XtalView		http://www.sdsc.edu/0h/CCMS/Packages/xtalview

[a]http indicates a HyperText Transfer Protocol. Other protocols include Gopher and ftp.

[b]Developers do change the organization of their WWW files from time to time. If a URL does not work, it sometimes helps to truncate the subdirectories (the parts of the address after the third slash), so as to access a higher level directory of the host. The URLs are case sensitive. A Beginner's Guide to HyperText Markup Language can be found at URL http://www.ncsa.uiuc.edu/General/Internet/WWW/HTMLPrimer.html.

Table 2 Electronic Mail Addresses of Bulletin Boards and Information Exchanges of Interest to Computational Chemists

Topic	To join	To post
AMBER	amber-request@cgl.ucsf.edu	amber@cgl.ucsf.edu
BIOSYM	dibug-request@comp.bioz.unibas.ch	dibug@comp.bioz.unibas.ch
Brookhaven Protein Data Bank	listserv@pdb.pdb.bnl.gov	pdb-l@pdb.pdb.bnl.gov
CAChe	majordomo@pacificu.edu, jim_currie@unixmail.pacificu.edu	CAChe@pacificu.edu
Cerius2	listproc@msi.com (send message subscribe ⟨list name⟩ ⟨Person's Name⟩	c2-l@msi.com
CHARMm	charmm-bbs-request@emperor.harvard.edu	charmm-bbs@emperor.harvard.edu
Chemical education	listserv@uwf.bitnet, whalpern@uwf.bitnet	chemed-l@uwf.cc.uwf.edu
Chemical information courses	listserv@iubvm.ucs.indiana.edu, wiggins@indiana.edu	cicourse@iubvm.usc.indiana.edu
Chemical information science	listserv@iubvm.ucs.indiana.edu, wiggins@indiana.edu	chminf-l@iubvm.ucs.indiana.edu
Chemical structure indexing	listserv@derwent.co.uk (with message subscribe chemind-l ⟨first_name⟩ ⟨last_name⟩), btown@derwent.co.uk	chemind-l@derwent.co.uk
Chemistry conferences	listserv@umdd.umd.edu	chemconf@umdd.umd.edu
Chemometrics	listserv@umdd.umd.edu, to2@umail.umd.edu	ics-l@umdd.umd.edu

Computational biology, sequence databases, GCG software, protein crystallography, etc.	biosci@net.bio.net, biosci-server@net.bio.net, kristoff@genbank.bio.net	comp-bio@net.bio.net, genbankb@net.bio.net, info-gcg@net.bio.net, x-tal-log@net.bio.net, etc., for the respective new groups
Computational chemistry (global, all topics)	chemistry-request@osc.edu, jkl@osc.edu	chemistry@osc.edu
Crystallography	helpdesk@concise.level-7.co.uk	—
Crystallography	lachlan@dmp.csiro.au (for further information)	sci.techniques.xtallography in Usenet
Genetic algorithms	ga-list-request@aic.nrl.navy.mil, ga-list-request@aic.nrl.navy.mil	ga-list@aic.nrl.navy.mil
Genetic algorithms	ga-molecule-request@tammy.harvard.edu	ga-molecule@tammy.harvard.edu
Genetic algorithms in chemistry	ga-molecule-request@interval.com, listmgr@interval.com, ngo@interval.com	ga-molecule@interval.com
GROMOS	gromos-request@igc.ethz.ch	gromos@igc.ethz.ch
High-performance computing	more@hpcwire.ans.net (with 601 in subject line)	—
HyperChem	hyperchem-request@hyper.com	hyperchem@hyper.com
ISIS	mailserv@mdli.com, admin-isis@mdli.com	isisforum-l@mdli.com
LabVIEW and LabWindows	info-labview-request@pica.army.mil and listserv@tamvm1.tamu.edu	—
MDL programming language	isispl@mdli.com	—
Minitab	mailbase@mailbase.ac.uk	—

(continued)

Table 2 *(continued)*

Topic	To join	To post
Molecular modeling and computational chemistry (U.K.)	mailbase@mailbase.ac.uk, cstajs@staffs.ac.uk	chem-mod@mailbase.ac.uk, chem-com@mailbase.ac.uk
Neural networks	neuron-request@psych.upenn.edu	neuron-request@psych.upenn.edu
O	o-info-request@kaktus.kemi.aau.dk (with subscribe in subject line)	o-info@kaktus.kemi.aau.dk
Organic chemistry	orgreq@quant.chem.rpi.edu, breneman@xray.chem.rpi.edu	orgchem@quant.chem.rpi.edu
Organometallic chemistry	cabku01@mailserv.zdv.uni-tuebingen.de	—
Parallel molecular dynamics	pmd-request@cumbnd.bioc.columbia.edu, windemut@cumbnd.bioc.columbia.edu	pmd-request@cumbnd.bioc.columbia.edu
Polymer physics	listserv@hearn.nic.surfnet.nl, peter@fenk.wau.nl	polymerp@hearn.nic.surfnet.nl
QUANTA/CHARMm	listproc@msi.com (send message subscribe ⟨list name⟩ ⟨Person's Name⟩, bobf@msi.com	quant-l@msi.com
SAS	listserver@vm.sas.com	—
Simulated annealing	anneal-request@cs.ucla.edu	anneal@cs.ucla.edu
SPASMS	spasms-users-request@cgl.ucsf.edu	spasms-users@cgl.ucsf.edu
Statistics	list-serve@julia.math.ucla.edu	—
SYBYL	sybylreq@quant.chem.rpi.edu	sybyl@quant.chem.rpi.edu
World Association of Theoretical Organic Chemists	listserver@ic.ac.uk, rzepa@ic.ac.uk	watoc@ic.ac.uk
X-PLOR	listproc@msi.com (send message subscribe ⟨list name⟩ ⟨Person's Name⟩, bobf@msi.com	xplor-l@msi.com

Table 3 Electronic Mail Addresses of Some High-Performance Computing Centers

Site	Telephone	E-mail address
Cornell Theory Center	607-254-8610	consult@ctc.edu
National Center for Atmospheric Research	303-497-1225	scdinfo@ncar.ucar.edu
National Center for Supercomputing Applications	217-244-0072	consult@ncsa.uiuc.edu
Ohio Supercomputer Center	614-292-9248	oscinfo@osc.edu
Pittsburgh Supercomputing Center	412-268-6350	remarks@psc.edu
San Diego Supercomputer Center	619-534-5000	consult@sdsc.edu

REFERENCES

1. D. B. Boyd, in *Reviews in Computational Chemistry,* Vol. 1, K. B. Lipkowitz and D. B. Boyd, Eds., VCH Publishers, New York, 1990, pp. 383–392. Compendium of Software for Molecular Modeling.
2. D. B. Boyd, in *Reviews in Computational Chemistry,* Vol. 1, K. B. Lipkowitz and D. B. Boyd, Ed., VCH Publishers, New York, 1990, pp. 321–354. Aspects of Molecular Modeling.
3. E. Krol, *The Whole Internet User's Guide and Catalog,* O'Reilly & Associates, Sebastopol, CA, 1992.

SOFTWARE FOR PERSONAL COMPUTERS

Apple; Compaq; Gateway; IBM; other PC clones.

General Purpose Molecular Modeling

Alchemy III
Tripos, Inc.
1699 South Hanley Road, Suite 303
St. Louis, MO 63144-2913, U.S.A.
Tel. 800-323-2960, 314-647-1099, fax 314-647-9241, e-mail support@tripos.com (U.S.A.), tel. 44-344-300144, fax 44-344-360638 (U.K.), tel. 81-3-5228-5617, fax 81-3-5228-5581 (Japan)

Structure building, manipulation, comparison. SYBYL energy minimization of organic and biomolecules. Stick, space-filling, or cylinder (thick bonds) display. Interface to Chemical Abstracts Service registry files. Molfile transfer to SYBYL and LabVision. ChemPrint (under Windows) for 2D chemical structure drawing. PowerSearch for systematic and Monte Carlo conformational searching in Windows environment. Macintosh and PCs (DOS and Windows).

CAChe WorkSystem
CAChe Scientific (Oxford Molecular Group)
P.O. Box 4003
Beaverton, OR 97076, U.S.A.
Tel. 800-544-6634, 503-526-5000, fax 503-526-5099
e-mail info@cache.com

Graphical pre- and postprocessor for semiempirical molecular orbital calculations: extended Hückel, MOPAC, and ZINDO. Structure building from library of fragments and molecules; crystal modeling; manipulation. Stick, ball-and-stick, and space-filling display. Orbital, electron density, and electrostatic maps; reaction energy surfaces; IR and UV spectra prediction; MM2 energy minimization and dynamics. BLOGP and BLOGW for prediction of octanol/water partition coefficient and water solubility. ProjectLeader for automating calculations and organizing results for QSAR. WorkSystem runs on Tektronix-enhanced Macintosh with coprocessor and stereoscopic graphics. Personal CAChe runs a standard Macintosh. GroupServer runs on UNIX servers. IBM's Mulliken for performing molecular mechanics, semiempirical, density functional, and ab initio calculations through the ProjectLeader graphical user interface. Structure and other files are available via Gopher from nsgopher.pacificu.edu. Macintosh, PCs (Windows), Power Macintosh, and IBM RISC (UNIX) workstations.

Chem3D Pro
CambridgeSoft Corporation
875 Massachusetts Avenue, Suite 61
Cambridge, MA 02139, U.S.A.
Tel. 800-315-7300 (ext. 1200), 617-491-2200 (ext. 1612), fax
617-491-8208, e-mail info@camsci.com, support@camsci.com

Structure building, manipulation, MM2 energy minimization and molecular dynamics. Molfile conversion. Chem3D for stick, ball-and-stick, and space-filling display. 2D-to-3D conversion. ChemDraw for 2D chemical structure drawing. ChemDraw Pro for color 2D chemical structure drawing. ChemFinder and ChemFinder Pro for managing libraries of 2D and 3D structures. ChemOffice and ChemOffice Pro for integrated combinations of the other programs. Client-server link to Tripos's UNITY. Crossfire Client to Beilstein Handbook database. CS MOPAC for a graphical interface to MOPAC 93. World Wide Web URL http://www.camsci.com/and ftp site ftp.camsci.com for downloadable demonstration programs. Macintosh, PCs (Windows), Sun, Silicon Graphics, Digital workstations.

ChemMod II
Fraser Williams Scientific Systems, Ltd.
London House, London Road South
Poynton, Cheshire, SK12 1YP, England, U.K.
Tel. 44-625-871126, fax 44-625-871128

Structure building, manipulation. Force field minimization. Stick, ball-and-stick, and space-filling display. Sabre for managing a database of 2D chemical structures. Macintosh. PC-Chemmod for structure building, manipulation, energy minimization of molecules with up to 2000 atoms. Stick and space-filling display. PCs.

Chem-X
  Chemical Design Ltd.
  Dr. Keith Davies
  Roundway House, Cromwell Park
  Chipping Norton, Oxon, OX7 5SR, England, U.K.
  Tel. 44-1608-644000, fax 44-1608-644244 (U.K.), tel. 81-3-3345-1411,
    fax 81-3-3344-3949 (Japan)
  and
  Chemical Design Ltd.
  200 Route 17 South, Suite 120
  Mahwah, NJ 07430, U.S.A.
  Tel. 201-529-3323, fax 201-529-2443
An integrated, molecular system for molecular visualization and computation of organic, inorganic, peptide, and polymeric compounds. Stick, ball-and-stick, and space-filling representations. Chem-X/DRAW for 2D and 3D structure drawing. PCs (DOS and Windows) and Macintosh.

Desktop Molecular Modeller
  Oxford Electronic Publishing
  Oxford University Press
  Walton Street
  Oxford, OX2 6DP, England, U.K.
  Tel. 44-865-56767, ext. 4278, fax 44-865-56646 (U.K.), tel.
    212-679-7300, fax 212-725-2972 (U.S.A.)
Structure building, manipulation. Energy minimization. Stick, ball-and-stick, and space-filling display. PCs.

HAMOG
  P.O. Box 1247
  Birkenstrasse 1A
  Schwerte, D-5840, Germany
Halle Molecular Graphics software for building, manipulation. Electrostatic potentials based on electronegativities; interfaces to ECEPP and MM2P. Stick, ball-and-stick, and space-filling display. PCs.

HyperChem
  Hypercube Inc.
  Dr. Neil S. Ostlund
  419 Phillip Street, #7

Waterloo, Ontario N2L 3X2, Canada
Tel. 800-960-1871, 519-725-4040, fax 519-725-5193,
e-mail info@hyper.com

Model building, manipulation. Stick, ball-and-stick, space-filling, and dot surface display. Semiempirical calculations by extended Hückel, CNDO, INDO/1, INDO/S, MINDO/3, MNDO, AM1, PM3, ZINDO/1, and ZINDO/S. UV, IR, electrostatic potential, and molecular orbital plots. 2D-to-3D conversion. Protein and DNA fragment libraries. MM+, BIO+ (implementations of MM2 and CHARMM, respectively), OPLS, and AMBER molecular mechanics and dynamics. Solvent box. ChemPlus for structure fitting, 3D rendering, crystal building, conformational searching, sequence editing, and computing log *P* and other QSAR properties. HyperNMR for quantum mechanical computational prediction of chemical shifts and coupling constants. Demonstration software is available by ftp from ftp.hyper.com in directory /pub/demo. URL is http://www.hyper.com/. PCs (DOS and Windows) and client to UNIX servers.

MacMimic
InStar Software AB
IDEON Research Park
S-223 70 Lund, Sweden
Tel. 46-46-182470, fax 46-46-128022, e-mail sundinkc@dna.lth.se,
ok2aps@gemini.lcd.lu.se

Structure building, manipulation, comparison. Energy minimization by authentic MM2 force field, dihedral angle driver for structures with up to 200 atoms. Stick and ball-and-stick display, multiple structures in multiple windows, structures with up to 32,000 atoms. Macintosh.

MicroChem
Chemlab, Inc.
1780 Wilson Drive
Lake Forest, IL 60045, U.S.A.
Tel. 312-996-4816

Structure building, manipulation; energy minimization of organic, inorganic, and polymer units. Stick, ball-and-stick, and space-filling display. Craig plots for QSAR. PCs.

MOBY
Springer-Verlag New York, Inc.
Electronic Media Services
175 Fifth Avenue
New York, NY 10010, U.S.A.
Tel. 800-777-4643, ext. 653, 212-460-1653, fax 212-473-6272
and
Springer-Verlag GmbH
Dr. Rainer Stumpe

Tiergartenstrasse 17
D-6900 Heidelberg, Germany
Tel. 49-6221-487-406, fax 49-6221-487-288, e-mail
stumpe@spint.compuserve.com (Germany), 81-03-812-0331 (Japan)
Structure building, manipulation. Stick, dot surface, and orbital display. Geometry optimization and dynamics of 150 atoms by AMBER parameters. MNDO and AM1 semiempirical calculations. Demonstration version available via anonymous ftp from benny.bsc.mass.edu (134.241.41.5). ARGOS for conversion of connection table to 2D structure drawings. AUTONOM for computerized assignment of chemical nomenclature to structures from graphical input. AUTONOM Batch for PCs and VAX. C-Design for 2D chemical structure drawing. PCs (DOS).

Molecular Modeling Pro
WindowChem Software
420-F Executive Court North
Fairfield, CA 94585, U.S.A.
Tel. 800-536-0404, 707-864-0845, fax 707-864-2815
Model building, manipulation. Stick, ball-and-stick, space-filling, and dot surface display. Energy minimization with the MOLY force field. Estimation of surface area, volume, log *P*, and other QSAR descriptors. Molecular Analysis Pro for QSAR analysis and multiple regression. GPMAW (General Protein Mass Analysis for Windows) for sequence analysis and hydrophobicity prediction. Peptide Companion to predict HPLC behavior. Enzyme Kinetics. Fit-Regression Analysis. CD-ROM with Properties of Organic Compounds from the CRC Press Database. pK_a Database with 2000 values. Dictionary of Chemistry and other dictionaries. PCs (Windows) and Macintosh.

MOLGEN
Dr. Milan Hudecek
CHERS
P. Horova 18
SK-841 07 Bratislava, Slovakia
Fax 42-70492-2855-9, e-mail hudeck@fns.uniba.sk
and
J. Eric Slone Consulting Services
5500 Holmes Run Parkway, No. 501
Alexandria, VA 22304-2851, U.S.A.
Tel. 703-461-7078, fax 703-451-6639, e-mail eslone@masoni.gmu.edu
Structure building, manipulation. Stick, ball-and-stick, space-filling, and dot surface display. MM2 geometry optimization. Charge, log *P*, and molar refraction calculations. Interfaces to AMPAC and other modeling programs. Databases of octanol/water partition coefficients for 2500 compounds and 1500 3D structures. 2D-to-3D structure conversion. Structural database and searching.

Demonstration copy on infomeister.osc.edu in directory /pub/chemistry/software/MS-DOS/MOLDEN-demo. PCs.

Nemesis
Oxford Molecular Ltd.
The Magdalen Centre, Oxford Science Park
Oxford, OX4 4GA, England, U.K.
Tel. 44-865-784600, fax 44-865-784601, e-mail omlemail@vax.ox.ac.uk (U.K.), tel. 81-33-245-5004, fax 81-33-245-5009 (Japan)
and
Oxford Molecular Inc.
7005 Backlick Court, Suite 200
Springfield, VA 22151-3903, U.S.A.
Tel. 703-658-4854, fax 703-658-4887, e-mail syazdi@presto.ig.com
Structure building, manipulation. Stick, ball-and-stick, space-filling, and dot surface display. Energy calculations with the COSMIC force field. Partial charges from Charge-2. PC-PROT+ (sequence analysis), PC-TAMMO+ (protein-lipid modeling), and MASCA (statistics). URL is http://www.oxmol.co.uk/. Macintosh and PCs (Windows).

PCMODEL
Serena Software
Dr. Kevin E. Gilbert
P.O. Box 3076
Bloomington, IN 47402-3076, U.S.A.
Tel. 812-333-0823, 812-855-1302/9415, fax 812-332-0877, e-mail gilbert@indiana.edu
Structure building, manipulation. Energy minimization by MMX (an extension of MM2 and MMP1). Stick and dot surface display for organic, inorganic, organometallic, hydrogen-bonded, pi-bonded, and transition state systems. Solvent dynamics. Structure files can be read and written for MM2, MOPAC, X-ray crystal data, and others. GMMX for conformational searching, MOPAC, PCDISPLAY for molecular display with ORTEP and PLUTO, BKM, MDLFMTS, MDLCMP, PCM-OrbDraw, PCM-Vibrate, and PCM-NMR programs. PCs (DOS), Macintosh, Silicon Graphics, Sun, and IBM RS/6000 versions.

Quantum Chemistry Calculations

ATOM
Project Seraphim
Department of Chemistry

University of Wisconsin
Madison, WI 53706, U.S.A.
Also, ATOMPLUS, H2ION, and GAUSS2 for educational uses.

CACAO
Dr. Davide M. Proserpio
Istituto di Chimica Strutturistica Inorganica dell'
Università di Milano
Via Venezian 21
20133 Milan, Italy
Tel. 39-2-70635120, 39-55-2346653, fax 39-2-70635288,
39-55-2478366, e-mail stinch@imisiam.mi.cnr.it,
mealli@cacao.issecc.fi.cnr.it
Computer-Aided Composition of Atomic Orbitals. Molecular orbital calculations with extended Hückel method, orbital plots, and Walsh diagrams. Also available via anonymous ftp from cacao.issecc.fi.cnr.it (149.139.10.2). PCs.

HMO
Trinity Software
Campton Common, Unit No. 3
P.O. Box 960
Campton, NH 03223, U.S.A.
Tel. 800-352-1282, 603-726-4641, fax 603-726-3781 (U.S.A.), tel.
44-734-787917, fax 44-734-773867 (U.K.)
Graphics-based Hückel molecular orbital calculator of energies and orbitals of pi electrons. Diatomic Molecular Mechanics and Motion. EnzymeKinetics for fitting Michaelis–Menten kinetics parameters. ESP (Experimental Section Processor) for organizing synthetic procedures in publication format. LabSystant for evaluating quantitative lab data. PC-Mendeleev for studying periodic table. SynTree for synthesis planning. TAPP (Thermodynamic and Physical Properties) database with physical and thermodynamic data on more than 10,000 compounds. SRS II (Scientific Reference System) for managing bibliographies. MassSpec for identifying peaks in mass spectra. PCs and Macintosh.

MOPAC
QCPE
Creative Arts Building 181
Indiana University
840 State Highway 46 Bypass
Bloomington, IN 47405, U.S.A.
Tel. 812-855-4784, fax 812-855-5539, e-mail qcpe@ucs.indiana.edu
Semiempirical molecular orbital package for optimizing geometry and studying reaction coordinates. Extensive library of more than 100 programs for quantum mechanics, molecular mechanics, and molecular graphics, including

AMPAC, ARVOMOL (surface areas and volumes of molecules), CHEMICALC-2 (log *P*), CNINDO/D, DISMAP (distance maps from PDB files), DRAW (a graphical complement to MOPAC), EXTOIN (coordinate conversion to *Z* matrix), FORTICON8 (extended Hückel), FORTICONMAC, HAM/3, MLDC8 (NMR analysis), MM2, MNDO, MOLDRAW (chemical shifts), MOLSVMAC (molecular volume and surface areas), MOLVIB, MOLVIEW, MOLYROO, (structure drawing) MOPC (orbital plots), NAMOD (molecular graphics), NorMode, PCILO/2, POLYATOM (ab initio), PROTEAN II (structure determination), RING (puckering), SCIBASE (management of references), SILMUT (DNA sequence analysis), SIMVOL/PLTSIM (molecular similarity), STERIMOL (substituent steric parameters), VIBMOL and VIBRAM (normal mode visualization), VISION3 (contour plots from MM2 output), and VSSMAC (electrostatic potential energy mapping). Current catalogs and information are available by anonymous ftp from qcpe6.chem.indiana.edu (129.79.74.206). PCs and Macintosh.

Databases of Molecular Structures

Accord
Synopsys Scientific Systems Ltd.
175 Woodhouse Lane
Leeds, LS2 3AR, England, U.K.
Tel. 44-113-245-3339, fax 44-113-243-8733, e-mail
support@synopsys.co.uk, sales@synopsys.co.uk
and
Synopsys Scientific Systems Ltd.
116 Village Boulevard, Suite 200
Princeton Forrestal Village
Princeton, NJ 08540-5799, U.S.A.
Tel. 609-734-7431, fax 609-520-1702
Makes chemical structure drawings from ChemDraw and ISIS/Draw understandable to Microsoft Excel spreadsheets. Library of QSAR parameters for 250 substituents. Protecting Groups database with more than 20,000 reactions, Methods in Organic Synthesis (MOS) database with about 10,000 reactions, and the BioCatalysis database with biomolecule-mediated reactions. PCs, Macintosh, and Power Macintosh.

Aldrich Catalog
Aldrich Chemical Company, Inc.
Data Products
P.O. Box 355
Milwaukee, WI 53201, U.S.A.
Tel. 800-231-8327, 414-273-3850, fax 800-962-9591, 414-273-4979
(U.S.A.), tel. 44-747-822211, fax 44-747-823779 (U.K.)

Menu-driven software to search database of chemical products in the catalog. On CD-ROM for PCs. Material Safety Data Sheets on CD-ROM.

ATLAS of Protein and Genomic Sequences
Protein Identification Resource (PIR)
National Biomedical Research Foundation
3900 Reservoir Road, NW
Washington, DC 20007, U.S.A.
Tel. 202-687-2121, e-mail garavelli@nbrf.georgetown.edu
Retrieval system and protein sequence database. PCs and VAX.

CD-React
InfoChem GmbH
Springer-Verlag Group
Landsberger Strasse 408
81241 Munich, Germany
Tel. 49-89-583002, fax 49-89-5803839
Database of 1.8 million reactions. ChemReact database with 370,000 reactions. ChemSynth with 80,000 reactions. ChemSelect database of 9,900 reactions searchable by MDL and Chem-X software; a demonstration version is available on svserv@dhdspri6.bitnet. Lysis reaction database. MOLKICK for setting up queries to STN, Dialog, and Questel. Beilstein database with structures, chemical, and physical data. CD-ROMs for PCs. PC-Search and ChemReact32 for database searching. PCs (DOS and Windows).

CHCD Dictionary of Natural Products
Chapman & Hall
Scientific Data Division
2-6 Boundary Row
London, SE1 8HN, England, U.K.
Tel. 44-071-865-0066, fax 44-071-522-9623
Database of more than 80,000 alkaloids, terpenoids, antibiotics, peptides, carbohydrates, lipids, steroids, flavinoids, and related compounds. *Dictionary of Organic Compounds. Dictionary of Inorganic Compounds. Dictionary of Analytical Reagents*. On CD-ROM for PCs.

ChemQuery
Maxwell Online
8000 Westpark Drive
McLean, VA 22102, U.S.A.
Tel. 703-442-0900, fax 703-356-4023 (U.S.A.), tel. 44-081-992-3456 (U.K.)
Substructure searching of ORBIT chemical files. PCs.

CIPSLINE PC
Prous Science Publishers
Apartado de Correos 540
08080 Barcelona, Spain
Tel. 34-3-459-2220, fax 34-3-258-1535
Database of almost 20,000 2D structures and biological activities. PC diskette. *Drugs of the Future* on CD-ROM and *Drug Data Report* on CD-ROM for PCs.

Current Facts in Chemistry
Beilstein Information Systems, Inc.
15 Inverness Way East
P.O. Box 1154
Englewood, CO 80150-1154, U.S.A.
Tel. 800-275-6094, 303-792-2652, fax 303-792-2818 (U.S.), 49-69-7917-633, fax 49-69-7917-636, e-mail helpdesk@beilstein.com (Germany)
Database of 2D structures, data, and reactions. URL is http://www.beilstein.com. CD-ROM for PCs.

DARC-CHEMLINK
Questel
83-85 Boulevard Vincent Auriol
75646 Paris Cédex 13, France
Tel. 33-144-23-64-64, fax 33-144-23-64-65
Preparation and transfer of queries for DARC database searches. PCs. DARC Inhouse for maintaining databases of 2D structures on VAX.

GenBank
National Center for Biotechnology Information
National Library of Medicine
National Institutes of Health
8600 Rockville Pike
Bethesda, MD 20894, U.S.A.
Tel. 301-496-2475
Gene sequence database. More than 20 other databases including SWISS-PROT, ACeDB, FlyBase, Eukaryotic Promoter Database, Restriction Enzyme Database, Transcription Factor Database, and CarbBank/CCSD. Databases and software development tools for sequence data available for anonymous ftp ncbi.nlm.nih.gov (130.14.25.1).

Generic TOPFRAG
Derwent, Inc.
1313 Dolley Madison Boulevard
McLean, VA 22101, U.S.A.
Tel. 800-451-3451, 703-790-0400, fax 703-790-1426 (U.S.A.), tel.

44-071-242-5823, fax 44-071-405-3630 (U.K.), tel. 81-03-3581-7711, fax 81-03-3505-0513 (Japan)
Preparation of queries for Derwent's on-line database of the patent literature. PCs. World Patents Index and other databases searchable by DERPICT software. CD-ROM for PCs.

geNMR
IvorySoft Scientific Software
Dr. P. H. M. Budzelaar
Amerbos 330
1025 ZV Amsterdam, The Netherlands
Tel. and fax 31-20-6326913
Iterative simulations to do fitting of observed and calculated NMR spectra. PCs (DOS) and Macintosh.

Inorganic Crystal Structure Database
Fachinformationszentrum (FIZ) Karlsruhe
P.O. Box 2465
D-7514 Eggenstein-Leopoldshafen, Germany
Tel. 49-7247-808-253, fax 49-7247-808-666, e-mail pls0d@fizvax.kfl.de (Germany), cansnd@vm.nrc.ca (Canada)
and
Scientific Information Service Inc.
7 Woodland Avenue
Larchmont, NY 10538, U.S.A.
Tel. 914-834-8864, fax 914-834-8903
ICSD has 3D structures of more than 35,000 inorganic substances. PC, IBM and VAX versions. Available on-line through STN International. On CD-ROM for PCs. RETRIEVE for searching CD-ROM and CRYSTAL VISUALIZER for displaying data.

Marinlit
Dr. John W. Blunt
Department of Chemistry
University of Canterbury
Christchurch, New Zealand
E-mail j.blunt@csc.canterbury.ac.nz
Database of 6000 marine products searchable by Cambridge Scientific's ChemFinder.

NIST Crystal Data File
International Center for Diffraction Data
1601 Park Lane
Swarthmore, PA 19081-2389, U.S.A.
Tel. 215-328-9400

Crystallographic data on more than 170,000 crystalline materials. EDD (Electron Diffraction Database) with crystallographic data on more than 71,000 crystalline materials. PC-PDF (Powder Diffraction File) on CD-ROM for PCs.

NIST Structures and Properties Database
National Institute of Standards and Technology
U.S. Department of Commerce
Gaithersburg, MD 20899, U.S.A.
Tel. 301-975-2208, fax 301-926-0416, e-mail rdj3@enh.nist.gov
Thermodynamic data for almost 5000 gas phase compounds. Estimation of structures drawn into program using Benson's additivity rules. IVTANTHERMO database with enthalpies of formation and other thermodynamic properties for 2300 substances. NIST/NASA/CARB Biological Macromolecule Crystallization Database with crystal growth conditions. PCs.

Nucleic Acid Database (NDB)
Dr. A. R. Srinivasan
Rutgers University
Chemistry Department
Brunswick, NJ 08903, U.S.A.
Tel. 201-932-4619, e-mail ndbadmin@helix.rutgers.edu
Relational database with data on more than 200 DNA, RNA, and tRNA structures, including coordinates for more than 100 of them.

PsiBase
Hampden Data Services, Ltd.
9 Peachcroft Centre
Peachcroft Road
Abingdon, Oxon, OX14 2NA, England, U.K.
Tel. 44-235-559559, fax 44-235-559585
Management of databases of 2D chemical structures. PsiGen for 2D chemical structure drawing. PCs (Windows).

Sadtler Molecular Structure Search Software
Bio-Rad, Sadtler Division
Sadtler Research Laboratories
3316 Spring Garden Street
Philadelphia, PA 19104, U.S.A.
Tel. 215-352-7800, fax 215-662-0585
Drawing 2D structure queries. IR Search to retrieve vibrational spectra. Carbon-13 Search to retrieve NMR spectra. PCs.

Softron Substructure Search System
Gesellschaft für Technisch-Wissenschaftliche Software mbH
Rudolf Diesel Strasse 1

W-8032 Gräfelfing, Germany
Tel. 49-089-855056, fax 49-089-852170
Management of databases of 2D chemical structures. PC, IBM, and VAX.

STN EXPRESS
STN International
Chemical Abstracts Service
2540 Olentangy River Road
P.O. Box 3012
Columbus, OH 43210-0012, U.S.A.
Tel. 800-753-4227, 800-848-6533, 614-447-3600, fax 614-447-3713, e-mail help@cas.org (U.S.A.), tel. 49-7247-808-555, fax 49-7247-808-131 (Germany), tel. 81-3-3581-6411, fax 81-3-3581-6446 (Japan)
Preparation of structure queries, transfer of queries for on-line database searches, and capture of hits. Imports structures from ChemDraw, Alchemy III, and ISIS/Draw. Structures can be used to search BEILSTEIN, GMELIN, MARPAT (the CAS Markush patent file), REGISTRY, SPECINFO, CASREACT, and ChemInformRX databases. SciFinder for graphical user interface to STN databases. Internet address stnc.cas.org (134.243.5.32). *12th Collective Index of Chemical Abstracts* on CD-ROM. CASurveyor with recent CAS abstracts on CD-ROM. PCs (DOS and Windows) and Macintosh.

SYNLIB
Distributed Chemical Graphics, Inc.
1326 Carol Road
Meadowbrook, PA 19046, U.S.A.
Tel. 215-885-3706, fax 215-355-0954
Synthesis library with about 70,000 searchable reactions from the literature. Macintosh, Silicon Graphics, and VAX.

TOXNET
National Library of Medicine/Environmental Protection Agency
National Institutes of Health
8600 Rockville Pike
Bethesda, MD 20894, U.S.A.
Tel. 301-496-6531, 301-496-1131
Toxicology Data Network with data files on hazardous, carcinogenic, and other substances. Medical Literature Analysis and Retrieval System (MEDLARS) for access to databases from the Specialized Services department of the National Library of Medicine.

WPDB
Drs. Ilya Shindyalov and Phil E. Bourne
Department of Biochemistry and Molecular Biophysics

Columbia University
630 W. 168th Street
New York, NY 10032, U.S.A.
Tel. 212-305-3456, fax 212-305-7379, e-mail
wpdb@cuhhca.hhmi.columbia.edu
Macromolecular structure interrogation and analysis of Brookhaven Protein Data Bank. Described in http://cuhhca.hhmi.columbia.edu/wpdb.html and available by anonymous ftp from cuhhca.hhmi.columbia.edu (128.59.98.1) in directory /pub/programs/WPDB/doc. Compressed databases of random selection of 100 structures and 320 unique structures. PCs (Windows).

Molecular Graphics

Ball & Stick
Cherwell Scientific Publishing
The Magdalen Centre, Oxford Science Park
Oxford, OX4 4GA, England, U.K.
Tel. 44-865-784800, fax 44-865-784801, e-mail csp@sable.ox.ac.uk,
k360171%edvz.uni-linz.ac.bitnet
and
Cherwell Scientific Publishing
15 Auburn Place
Brookline, MA 02146, U.S.A.
Tel. 617-277-4200, fax 617-739-4836, e-mail csp@netcom.com
Molecular graphics of structures imported from molecular modeling packages, rotation. A demonstration version is available via anonymous ftp from ftp.unilinz.ac.at. Macintosh. ModelMaker for data analysis. PCs (Windows).

chemVISION
Molecular Arts Corporation
1532 East Katella Avenue, Suite 1000
Anaheim, CA 92805-6627, U.S.A.
Tel. 800-431-5222, 714-634-8100, fax 714-634-1999, e-mail
info@molecules.com
Stick, ball-and-stick, and space-filling displays. chemDIAGRAM for 2D chemical structure drawing. chemEXHIBIT for molecular graphics. chemSAVER is screen saver. chemCLIPART. chemSCENES. PCs (Windows). Molecules-3D for drawing, building, cleanup, and graphics of molecular structures. PCs (Windows) and Macintosh.

Imdad
Molecular Applications Group
445 Sherman Avenue, Suite T
Palo Alto, CA 94306, U.S.A.

Tel. 800-229-7382, 415-473-3030, fax 415-473-1795, e-mail info@mag.com
Interactive model building, molecular graphics, and animation of small molecule and macromolecule displays. Macintosh, PowerMac, and PCs. LOOK for protein modeling, homology building, and related literature retrieval. SegMod (Segment Match Modeling) module for multiple sequence alignment. Silicon Graphics, Digital Alpha, and Sun workstations.

Kinemage
Protein Science
University of Washington, SJ-70
Seattle, WA 98195, U.S.A.
Tel. 206-685-1039, fax 206-685-2674, e-mail prosci@u.washington.edu
PREKIN and MAGE by Dr. David C. Richardson (Duke University) for visualization of structures in Brookhaven Protein Data Bank format from the journal *Protein Science*. Available via Gopher from gopher://orion.oac.uci.edu:1071/11/protein/Kinemage. Macintosh.

Modeler
COMPress Division of Queue, Inc.
338 Commerce Drive
Fairfield, CT 06430, U.S.A.
Tel. 800-232-2224, 203-335-0906, fax 203-336-2481
Display and manipulation of 3D models using keyboard input. Molecular Animator for creating and displaying 3D models. Molecular Graphics for display and manipulation of atomic coordinate data. ChemFile II for creating databases of 2D chemical structures with associated text. PCs and Macintosh.

Molecular Animator
Kalmia Company Inc.
71 Dudley Street
Cambridge, MA 02140, U.S.A.
Tel. 617-864-5567
Model building and vibration. Modeler: Molecular Design Editor for building and displaying ball-and-stick representations. Molecular Graphics for wire, ball-and-stick, and space-filling display. Diatomic Molecular Mechanics and Motion. Educational software, including General Chemistry. PCs (DOS). SynTree for synthesis planning. SRS II for scientific reference system. Macintosh.

Molecules
Atlantic Software
P.O. Box 299
Wenham, MA 01984, U.S.A.
Tel. 508-922-4352

Builds and plots 3D structures. DNA/RNA Builder. Protein Predictor (for secondary structure) and N.N.Charge (partial charges) both based on neural network approach. Macintosh.

NanoVision
ACS Software
American Chemical Society
Distribution Office
P.O. Box 57136, West End Station
Washington, DC 20037, U.S.A.
Tel. 800-227-5558, 202-872-4363, fax 202-872-6067
A 3D visualization program capable of rotating molecules with up to 32,000 atoms, for the Macintosh. Stick and space-filling displays. Analytical Chemistry by Open Learning (ACOL) for computer-aided teaching. ChemStock for managing database of laboratory chemicals. DESIGN-EASE for factorial design. DESIGN-EXPERT for multiple response optimization. EndLink to capture bibliographic data from on-line services. EndNote and EndNote Plus for creating databases of bibliographic information in conjunction with word processing. EQUIL for aqueous solution equilibrium calculations. Gardner's Chemical Names and Trade Names database. Grant Tracker. Instant EPA's Air Toxics, Integrated Risk Information System, and Pesticide Facts databases. IR Mentor for spectral interpretation. IR SearchMaster to manage data libraries. LabADVISOR information database on regulated chemicals. Origin for scientific graphics and data analysis (under Windows). PCNONLIN for pharmacokinetic models. PeakFit for deconvolution. SpecTool for managing spectral data. TableCurve for curve fitting. Un-Scan-It for entering *x, y* data into PCs and Macintosh. UniVersions for unit conversion and physical constant database. ACS *Directory of Graduate Research, College Chemistry Faculties,* and *Chemical Sciences Graduate School Finder* on CD-ROM. *Journal of the American Chemical Society* and *Biochemistry* on CD-ROM. The ACS is second party distributor of about 70 software products such as Alchemy III, CAChe, Chem3D Plus, DRUGIDEA, ISIS, SciWords, and NIST Structures and Properties Database and Estimation Program. PCs and Macintosh. Information about products and ACS publicators is available through the World Wide Web at URLs http://www.acs.org/ and http://pubs.acs.org/.

ORTEP_for_Mac
Ohio Supercomputer Center
1224 Kinnear Road
Columbus, OH 43212-1163, U.S.A.
Tel. 614-292-9279, fax 614-292-7168, e-mail jkl@osc.edu,
jkl@ohstpy.bitnet
Molecular graphics. RasMol, Babel, Chemfont, Periodic_Table_for_Mac, xvibs, and Vibratio. Available via anonymous ftp from infomeister.osc.edu and gopher site infomeister.osc.edu (port 70 or 73).

Protein Expert
BioSoftware Marketing
4151 Middlefield Road, Suite 109
Palo Alto, CA 94303-4743, U.S.A.
Tel. 800-465-4276, 415-858-0522, 415-858-0521, e-mail sunger@crl.com
Protein structure visualization and analysis. Macintosh. Second party distributor for software products of CambridgeSoft, SciVision, Megalon, Hypercube, and other software.

SCHAKAL
Dr. E. Keller
Kristallographic Institut der Universität Freiburg
Hebelstrasse 25
W-7800 Freiburg, Germany
Tel. 49-761-203-4279, fax 49-761-203-3362, e-mail kell@ruf.unifreiburg.de
Generation and display of stick, ball-and-stick, and space-filling representations. PCs and UNIX workstations.

Quantitative Structure-Property Relationships and Statistics

CHEMEST
Technical Database Services, Inc.
135 West 50th Street
New York, NY 10020-1201, U.S.A.
Tel. 212-245-0044, fax 212-247-0587
Estimation of molecular properties, such as pK_a, solubility, and viscosity. MicroQSAR for property estimation, including environmental partitioning and toxicity risk assessment, and statistical analysis. LOGKOW database with partition coefficients and other data for 14,000 organic compounds. PCs (DOS). TDS Numerica on-line databases and software tools for chemistry, toxicology, and environmental sciences.

JMP
SAS Institute Inc.
SAS Campus Drive
Cary, NC 27513, U.S.A.
Tel. 919-677-8000, fax 919-677-8123, e-mail support@unx.sas.com (U.S.A.), tel. 49-6221-4160, fax 49-6221-474-850 (Germany)
Statistical analysis using spreadsheet format; experimental design. Demonstration version available. URL is http://www.sas.com/. Macintosh. Anonymous ftp at ftp.sas.com. SAS statistical and data management system on IBM mainframes, VAX, PCs (Windows), Sun, Hewlett-Packard and IBM UNIX workstations.

LOGKOW

Syracuse Research Corporation
Environmental Sciences Center
Merrill Lane
Syracuse, NY 13210-4080, U.S.A.
Tel. 315-426-3350, fax 315-426-3429

Octanol–water partition coefficient estimation. DERMAL for estimating dermal permeability coefficient. WS/KOW for estimating water solubility. Atmospheric Oxidation Rate Program. MPBPVP for estimating melting point, boiling point, and vapor pressure. Henry's Law Constant Program. Biodegradation Probability Program. PC-HYDRO for estimating hydrolysis rates. PC-KOC for estimating soil adsorption coefficient. Databases including SMILES Notations, SOLV-DB, Chemical Pointer File, Physical Properties Data Base. Environmental Fate Data Bases (DATALOG, CHEMFATE, BIOLOG, and BIODEG). PCs (DOS and Windows).

MINITAB

Minitab Inc.
3081 Enterprise Drive
State College, PA 16801-3008, U.S.A.
Tel. 814-238-3280, fax 814-238-4383 (U.S.A.), tel. 44-21-471-4199, fax 44-21-471-5169 (U.K.)

Statistical analysis using spreadsheet format; experimental design. Index of available macros and other statistical information can be obtained by sending the message "send index" to statlib@lib.stat.cmu.edu. PCs (DOS and Windows), Macintosh, VAX, and mainframes.

MOLCONN-X

Hall Associates Consulting
Dr. Lowell H. Hall
Department of Chemistry
Eastern Nazarene College
Quincy, MA 02170, U.S.A.
Tel. 617-773-6350, ext. 280, fax 617-773-6324

Computes topological indexes from molecular structures for use in QSAR analysis. PCs, Macintosh, and VAX.

OncoLogic

LogiChem
P. O. Box 357
Boyertown, PA 19512, U.S.A.
Tel. 213-367-1636

Expert system to predict carcinogenicity of compounds.

OREX ExpertSystem
Scientific Services
Bergstrasse 15
8405 Donaustauf, Germany
Tel. 49-9403-8149, fax 49-9412-8123
Knowledge-based prediction of the type of pharmacological activity a structure will exhibit based on a database of more than 8500 drugs with their biological activities and other data. PCs.

Pirouette
InfoMetrix, Inc.
2200 Sixth Avenue, Suite 833
Seattle, WA 98121, U.S.A.
Tel. 206-441-4696, fax 206-441-0841, e-mail infomtrx@halcyon.com
Chemometric analysis based in part on the ARTHUR pattern recognition program. InStep for routine application of statistical models. EinSight for multivariate and visual data exploration. PCs (DOS).

PSI-Plot
Poly Software International
P.O. Box 526368
Salt Lake City, UT 84152, U.S.A.
Tel. 801-485-0466, fax 801-485-0480
Statistical analysis. PCs (DOS).

QSAR-PC
Biosoft
22 Hills Road
Cambridge, CB2 1JP, England, U.K.
Tel. 44-223-68622, fax 44-223-312873
Regression analysis for quantitative structure–activity relationships. FIG.P for scientific graphics. PCs.

SciLogP
SciVision
128 Spring Street
Lexington, MA 02173, U.S.A.
Tel. 800-861-6274, 617-861-6660, fax 617-861-6252, e-mail scivision@delphi.com, info@scivision@terranet.com
Calculation of octanol/water partition coefficients using N. Bodor's method. SciLogW for aqueous solubility prediction. SciPredictor (SciProtein) for protein secondary structure and homology modeling. SciPolymer for polymer property estimation. PCs (Windows). CAD Gene for gene construction. Macintosh.

SigmaStat
Jandel Scientific Software
2591 Kerner Boulevard
San Rafael, CA 94901, U.S.A.
Tel. 800-452-6335, 415-453-6700, fax 415-453-7769, e-mail sales@jandel.com (U.S.A.), tel. 49-2104-36098, fax 49-02104-33100 (Germany)
Statistical modeling. SigmaPlot for graphing. SigmaScan/Image for image processing. TableCurve 2D for curve fitting. TableCurve 3D for fitting surfaces to *x*, *y*, *z* data. Mocha for image analysis. PeakFit for deconvoluting curves. URL is http://www.jandel.com/. PCs (Windows).

STATGRAPHICS Plus
Manugistics, Inc.
2115 East Jefferson Street
Rockville, MD 20852-4999, U.S.A.
Tel. 800-592-0050, 301-984-5000, fax 301-984-5094
Statistical modeling. PCs (Windows).

STATISTICA
StatSoft
2325 East 13th Street
Tulsa, OK 74104, U.S.A.
Tel. 918-583-4149, fax 918-583-4376
Statistical modeling. PCs (DOS and Windows) and Macintosh.

StatMost
DataMost Corporation
P.O. Box 65389
Salt Lake City, UT 84165, U.S.A.
Tel. 801-484-3860, fax 801-484-3870
Statistical modeling and graphing. PCs (Windows).

TOPMOST
Health Designs, Inc.
183 East Main Street
Rochester, NY 14604, U.S.A.
Tel. 716-546-1464, fax 716-546-3411 (U.S.A.), tel. 44-379-644122, fax 44-379-651165 (U.K.), tel. 81-3-536-4604, fax 81-3-536-4768 (Japan)
Calculation of electronic charges and related parameters by quick methods based on electronegativity. TOPKAT programs for calculating QSAR descriptors and statistical modeling. TOPKAT/TPS for using statistical models, such as for carcinogenicity, mutagenicity, skin and eye irritation, teratogenicity; TOPKAT/EA for environmental assessment, such as rat and minnow toxicity.

TOPDRAW for graphical input. CONKAT for file input/output. PROGNOSYS for construction of TOPKAT-compatible statistical models. PCs (DOS) and Digital VAX.

Other Applications

BIOPROP
Office of Technology Licensing
University of California, Berkeley
2150 Shattuck Avenue, Suite 510
Berkeley, CA 94704, U.S.A.
Tel. 510-643-9525, fax 510-642-4566, e-mail
domino@uclink2.berkeley.edu
Dr. Steven M. Muskal's program for neural network simulator for protein conformation prediction and other data analysis. BLSS for statistics. Miscellaneous programs available by anonymous ftp from berkeley.edu in directory /pub. PCs (DOS), Macintosh, UNIX workstations.

BioWorld Online
Cartermill Inc.
1615 Thames Street
Baltimore, MD 21231-3445, U.S.A.
Tel. 410-563-5382, fax 410-563-5389, e-mail info@bestpl.hcf.jhu.edu
BioWorld Today daily fax biotechnology newsletter. BEST North America database with research profiles, patents, and facilities of more than 125 universities and research institutions and five U.S. federal agencies. Available on CD-ROM using FOLIO Views software. URL is http://www.cartermill.com/.

Bookends Pro
Westing Software
2960 Paradise Drive
Tiburon, CA 94920, U.S.A.
Management of references. Macintosh.

BrainMaker Professional
California Scientific Software
10024 Newtown Road
Nevada City, CA 95959, U.S.A.
Neural network package. PCs (Windows).

CA-Cricket Graph
Computer Associates International, Inc.
One Computer Associates Plaza
Islandia, NY 11788-7000, U.S.A.

Tel. 800-225-5224, fax 516-342-5734
Data display. Macintosh.

CAD/Chem
AI Ware
11000 Cedar Avenue
Cleveland, OH 44106, U.S.A.
Tel. 800-537-3338, 216-421-2380
Neural networks optimization of formulations.

CHEM-CALC
Dr. Paul Haberfield
1666 52nd Street
Brooklyn, NY 11203, U.S.A.
Unit conversion, formula weights, rate constants.

Chemeleon
Exographics
P.O. Box 655
West Milford, NJ 07480-0655, U.S.A.
Tel. 201-728-0188, fax 201-728-0735
Cut and paste structures between different applications. PCs (Windows). Con-Systant for interconverting chemical structure files. PCs (DOS).

ChemIntosh
SoftShell International
715 Horizon Drive, Suite 390
Grand Junction, CO 81501, U.S.A.
Tel. 800-240-6469, 303-242-7502, fax 303-242-6469, e-mail info@softshell.com
2D chemical structure drawing on Macintosh. ChemWindow for 2D chemical structure drawing. C-13 NMR module for predicting chemical shifts. MS Calculator for obtaining empirical formulas from masses. SciWords for spell checking. Art of Science clip art. Entropy for chemical database management. PCs (DOS and Windows) and Macintosh.

CHEMiCALC
Chemical Concepts Corporation
Dr. Bert Ramsay
912 Main Street, Suite 6
Ann Arbor, MI 48104, U.S.A.
Tel. 800-243-6023, 313-741-1192, fax 313-663-7937, e-mail chm_ramsay@emuvax.emich.edu
Molecular weight and other chemical mathematics. PCs (DOS and Windows) and Macintosh.

Chemistry Citation Index
Institute of Scientific Information, Inc.
3501 Market Street
Philadelphia, PA 19104, U.S.A.
Tel. 800-523-1850, 215-386-0100, fax 215-386-6362 (U.S.A.), tel. 44-895-270016, fax 44-895-256710 (U.K.)

Database of cited papers as reported in ISI's *Science Citation Index* (SCI), plus abstracts and keywords. *Biotechnology Citation Index. Neuroscience Citation Index. Materials Science Citation Index. Index Chemicus* database with data on 200,000 new compounds each year. *Current Contents*. On CD-ROM for PCs and Macintosh.

Chemistry 4D-Draw
ChemInnovation Software
8190-E Mira Mesa Boulevard
San Diego, CA 92126, U.S.A.
Tel. 619-566-2846, fax 619-566-4138, e-mail cis@cheminnovation.com

Chemical structure drawing (formerly ChemNameStru); the NameExpert module converts IUPAC names to 2D structure drawings. URL is http://www.cheminnovation.com/. PCs (Windows) and Macintosh.

ChemStructure
Megalon S.A.
10, rue St. Honoré
CH-2000 Neuchatel, Switzerland
Tel. 41-38-24-75-24, fax 41-38-24-76-16
and
Megalon
400 Bel Marin Keys Boulevard, Suite 102
Novato, CA 94949, U.S.A.
Tel. 415-884-3002, fax 415-884-2279, e-mail info@megalon.com

2D chemical structure drawing. Compounds database with chemical structure of 27,000 organic compounds. Unistat for statistical analysis. Helix Systems' ResearchStation for electronic notebook integration of data from multiple applications. Demonstration version via anonymous ftp from ftp.megalon.com, via Gopher from gopher.megalon.com, and via World Wide Web at URL http://megalon.megalon.com/. PCs (Windows).

ChemWord
Laboratory Software Ltd.
2 Ivy Lane
Broughton, Aylesbury, Bucks, HP22 5AP, England, U.K.
Tel. 44-296-431234, fax 44-296-397231

2D chemical structure drawing for PCs.

ChemWords
Scientific Software
17 Squire Court
St. Louis, MO 63146, U.S.A.
Tel. 314-993-8586
Spelling checker with 30,000-word dictionary. Macintosh.

CHIRON
Dr. Stephen Hanessian
Department of Chemistry
Université de Montréal
P.O. Box 6128, Station A
Montréal, Québec H3C 3J7, Canada
Tel. 514-343-6738, fax 514-343-5728, e-mail hanessia@ere.umontreal.ca
Analysis and perception of stereochemical features in molecules and selection from a database of chiral precursors for total syntheses. 2D-to-3D structure conversion. Macintosh, Silicon Graphics, VAX. ChemProtect for selection of appropriate protective groups in more than 150 reaction conditions. Macintosh.

CORINA
Dr. Johann Gasteiger
Computer-Chemie-Centrum
Universität Erlangen-Nürnberg
Nägelsbackstrasse 25
D-91052 Erlangen, Germany
Tel. 49-09131-85-6570, fax 49-09131-85-6566, e-mail gasteiger@eros.ccc.uni-erlangen.de
Converts 2D-to-3D structures. EROS for predicting reactions and reaction products. F3D for molecular modeling. MAGIC for drawing 2D chemical structures. MEDTOOL for drawing 2D chemical structures with valence check. PETRA for empirical evaluation of charges and bond energies for use in QSAR. VEGAS for 2D drawing of structural formulas. WODCA for synthesis planning. URL is http://schiele.organik.uni-erlangen.de/services/3d.html. PCs and Sun.

Data Desk
Data Description, Inc.
P.O. Box 4555
Ithaca, NY 14852, U.S.A.
Tel. 800-573-5121, 607-257-1000, fax 607-257-4146
Exploratory data analysis and visualization. Macintosh.

DATAMARINER
Logica UK, Ltd.
Betjeman House, 104 Hills Road

Cambridge, CB2 1CQ, England, U.K.
Rule induction and neural networks data exploration.

DeltaGraph
DeltaPoint, Inc.
2 Harris Court, Suite B-1
Monterey, CA 93940, U.S.A.
Tel. 800-446-6955, 408-648-4000, fax 408-648-4020
Data display. Macintosh.

DESIGN-EASE
STAT-EASE, Inc.
2021 East Hennepin Avenue, Suite 191
Minneapolis, MN 55413, U.S.A.
Tel. 612-378-9449, fax 612-378-2152
Experimental design. DESIGN-EXPERT for 3D experimental design. PCs.

Enriching Quantum Chemistry with Mathcad
Journal of Chemical Education: Software
Department of Chemistry
University of Wisconsin-Madison
1101 University Avenue
Madison, WI 53706-1396, U.S.A.
Tel. 608-262-5253, fax 608-262-0381, e-mail jcesoft@macc.misc.edu
Exercises for teaching quantum theory. Periodic Table Stack. Molecular Dynamics of the F + H_2 Chemical Reaction. About 70 other programs for instruction in chemistry, such as KinWorks, Quantum Barrier, and Animated Demonstrations II. Also 650 programs for classroom use distributed by Project Seraphim. PCs, Macintosh, and other microcomputers.

Expressionist
Prescience
939 Howard Street, No. 110
San Francisco, CA 94103
Tel. 800-827-6284, fax 415-882-0530
Equation editor. Macintosh.

GAUSS
Aptech Systems, Inc.
23804 S.E. Kent-Kangley Road
Maple Valley, WA 98038, U.S.A.
Tel. 206-432-7855, fax 206-432-7832, e-mail info@aptech.com
Mathematical and statistical modeling. PCs and UNIX workstations.

GeneWorks
IntelliGenetics, Inc. (Oxford Molecular Group)
700 East El Camino Real, Suite 300
Mountain View, CA 94040 U.S.A.
Tel. 800-876-9994, 415-962-7300, fax 415-962-7302, e-mail dhelmrei@presto.ig.com (U.S.A.), tel. 32-3-219-5352, fax 32-3-219-5354 (Belgium), tel. 81-45-661-3414, fax 81-45-661-3429 (Japan)

Nucleic acid and protein sequence analysis, including secondary structure prediction. GenBank, EMBL, SWISS-PROT, PIR, PROSITE, NASITE, and VectorBank databases on CD-ROM. Macintosh. PC/GENE for sequence analysis on PCs. IntelliGenetics Suite Sequence Analysis Software and GENESEQ database with patented protein and nucleic acid sequences. Sun and VAX.

gluplot
Dr. Jan De Leeuw
Room 8118 MSB
University of California, Los Angeles
Los Angeles, CA 90024, U.S.A.
Tel. 310-205-9550, e-mail deleeuw@math.ucla.edu

Gopher server with software related to statistics; powerlisp on ftp site ftp.stat.ucla.edu. PCs, Macintosh, and UNIX workstations.

HiQ
National Instruments
6504 Bridge Point Parkway
Austin, TX 78730, U.S.A.
Tel. 800-433-3488, 512-794-0100, fax 512-794-8411

Mathematical modeling. Macintosh. LabWindows and LabVIEW for instrument control. PCs (Windows).

Kekulé
PSI INTERNATIONAL, Inc.
810 Gleneagles Court, Suite 300
Baltimore, MD 21286, U.S.A.
Tel. 410-821-5980, fax 410-296-0712

Converts computer-scanned chemical structure drawings into molfiles and SMILES. CIS with chemical, environmental, and drug databases. PCs (Windows) and Macintosh.

MacDNASIS
Hitachi Software Engineering America, Ltd.
1111 Bayhill Drive, Suite 395
San Bruno, CA 94066, U.S.A.
Tel. 800-624-6176, 302-658-8444, fax 415-615-7699

DNA, RNA, and protein sequence analysis and prediction of secondary structure. CD-DATA for GenBank, EMBL, NBRF-PIR, PROSITE, and SWISS-PROT databases on CD-ROM. Macintosh and PCs (Windows).

Maple
Waterloo Maple Software
450 Phillip Street
Waterloo, Ontario N2L 5J2, Canada
Tel. 800-267-6583, 519-747-2373, fax 519-747-5284, e-mail info@maplesoft.on.ca (Canada), tel. 49-6221-487-180, fax 49-6221-487-184, e-mail 100275.163@compuserve.com (Germany)
Mathematical modeling. THEORIST for symbolic algebra and graphing. PCs (Windows), Macintosh, and Power PC.

Mathcad
Mathcad, Inc.
P.O. Box 290
Buffalo, NY 14207-0120, U.S.A.
Tel. 800-685-5624, fax 716-873-0906
Scientific calculations, graphing, and equation solver. Simple quantum chemistry calculations can be set up. PCs and Macintosh.

MATLAB
The MathWorks, Inc.
24 Prime Park Way
Natick, MA 01760-1500, U.S.A.
Tel. 508-653-1415, fax 508-653-6284, e-mail info@mathworks.com (U.S.A.), tel. 44-223-301303, fax 44-223-301128, e-mail cewen@mathworks.com (U.K.)
Mathematical modeling. SIMULINK for simulating nonlinear dynamic systems. Toolboxes for symbolic math, neural networks, statistics, spline analysis, signal processing, and image processing. PCs (Windows) and Macintosh.

Mathematica
Wolfram Research, Inc.
100 Trade Center Drive
Champaign, IL 61820-7237, U.S.A.
Tel. 800-441-6284, 217-398-0700, fax 217-398-0747, e-mail info@wri.com (U.S.A.), tel. 44-1993-883400, fax 44-1993-883800, e-mail support-euro@wri.com (U.K.)
Mathematical modeling. Applications can be obtained by anonymous ftp or gopher from mathsource.wri.com or World Wide Web at URL http://www.wri.com/mathsource.html. PCs (DOS and Windows), Macintosh, and UNIX workstations.

MathType
  Design Science, Inc.
  4028 Broadway
  Long Beach, CA 90803, U.S.A.
Equation editor. PCs (Windows).

METABOLEXPERT
  CompuDrug North America, Inc.
  P.O. Box 23196
  332 Jefferson Road
  Rochester, NY 14692-3196, U.S.A.
  Tel. 716-292-6830, fax 716-292-6834
  and
  CompuDrug Chemistry Ltd.
  Hollán Ernó utca 5
  H-1136 Budapest, Hungary
  Tel. 36-1-112-4874, fax 36-1-132-2574
Prediction of possible metabolic products based on a library of known transformations. AGRO-METABOLEXPERT for agrochemicals. HPLC-METABOLEXPERT for chromatographic properties of metabolites. DRUGIDEA for QSAR regression analysis. ELEUX suggesting mobile phase in HPLC. HAZARDEXPERT for prediction of toxicity of organic compounds. LABSWARE for general computations and statistics. MOLIDEA for molecular modeling, molecular mechanics; CNDO/2 and CNDO/S calculations. PKALC for estimation of acidity and basicity. PROLOGP for lipophilicity calculation using R. F. Rekker method. PROLOGD for distribution coefficients based on predicted log P(octanol/water) and pK_a values. PCs (DOS). METABOLEXPERT also runs on VAX.

MLab
  Civilized Software, Inc.
  7735 Old Georgetown Road, Suite 410
  Bethesda, MD 20814, U.S.A.
  Tel. 301-652-4714, fax 301-656-1069
Curve fitting and solution of differential equations. PCs (DOS), Macintosh, Sun, and Silicon Graphics.

Molecular Presentation Graphics (MPG)
  Hawk Scientific Systems
  170 Kinnelon Road, Suite 8
  Kinnelon, NJ 07405, U.S.A.
  Tel. 201-838-6292, fax 201-838-7102 (U.S.A.), tel. 44-734-787917, fax 44-734-773867 (U.K.)
Drawing 2D chemical structures. ChemWhere for searching for MPG struc-

tures in word processing documents. Datalyst II for management of databases of chemical structures in dBASE-compatible files. Exographics' ConSystant and Chemeleon for structure data file conversions. PCs.

mpp
  Dr. Peter Gedeck
  Institut für Physikalische Chemie I
  Egerlandstrasse 3
  8520 Erlangen, Germany
  Tel. 49-9131-85-7335, e-mail gedeck@pctc.chemie.uni-erlangen.de
MOPAC Pre-Processor constructs *Z*-matrix from structural fragments.

NeuRun for Applications
  Neural Computer Sciences
  Unit 3, Lulworth Business Centre, Nutwood Way
  Totton, Southampton, SO4 2WW, U.K.
  Tel. 44-703-667775, fax 44-703-663730
Neural network package. PCs (Windows).

Outokumpu HSC Chemistry
  ARSoftware
  8201 Corporate Drive, Suite 1110
  Landover, MD 20785, U.S.A.
  Tel. 800-257-0073, 301-459-3773, fax 301-459-3776, e-mail
    arsoftware@arclch.com
Calculations related to chemical reactions, equilibria, and phase diagrams. CAD Crystallography for calculation and representation of crystal structures. PCs (Windows). Reseller for PCMODEL, MOPAC, Kekulé, and other programs for PCs and Macintosh.

PHARMSEARCH
  O'Hara Consulting, Inc.
  215 12th Street, SE
  Washington, DC 20003-1427
  Tel. 800-949-5120, 202-543-5120, fax 202-544-7159
Database of more than 54,000 patents and 70,000 structures searchable through Questel.

PLTCOR
  Integrated Graphics
  312 Nevada Street
  Northfield, MN 55057, U.S.A.
  Tel. 507-663-3107, fax 507-646-3107
Maintenance of bibliographies. PCs.

Prism
GraphPad Software
10855 Sorrento Valley Road, Suite 203
San Diego, CA 92121, U.S.A.
Tel. 800-388-4723, 619-457-3909, fax 619-457-8141
Data analysis and visualization. InPlot for curve fitting. InStat for statistics. InTend for laboratory chemical calculations. PCs and Macintosh.

Pro-Cite
Personal Bibliographic Software, Inc.
P.O. Box 4250
Ann Arbor, MI 48106, U.S.A.
Tel. 313-996-1580, fax 313-996-4672
Maintenance of bibliographies. Biblio-Links and Biblio-Link II for connection to on-line services and CD-ROMs. PCs (Windows) and Macintosh.

QUASAR
International Union of Crystallography
E-mail sendcif@iucr.ac.uk
Program for processing X-ray structures in the format of Crystallographic Information Files (CIF). CYCLOPS for dictionary validation.

RAIN
Dr. Eric Fountain
Institut für Organische Chemie
Technische Universität München
Lichtenbergstrasse 4
W-8046 Garching, Germany
Tel. 49-89-3209-3378, fax 49-89-3209-2727
Proposes pathways between reactants and products. PCs.

Reference Manager
Research Information System
Camino Corporate Center
2355 Camino Vida Roble
Carlsbad, CA 92009, U.S.A.
Tel. 800-722-1227, 619-438-5526, fax 619-438-5573, e-mail ruinfo@ris.risinc.com
Maintenance of bibliographies. PCs (DOS), NEC, and Macintosh.

Scientific Word
TCI Software Research
1190 Foster Road

Las Cruces, NM 88001, U.S.A.
Tel. 800-874-2383, 505-522-4600, fax 505-522-0116, e-mail sales@tcisoft.com
Technical word processing. Scientific WorkPlace for technical word processing, 2D and 3D graph creation, and BibTeX bibliography creation. LaTeX documents. PCs (Windows).

SciWords for Chemistry
Pool, Heller, and Milne, Inc.
9520 Linden Avenue
Bethesda, MD 20814, U.S.A.
Tel. 301-493-6595, fax 301-897-3487, e-mail pool@phm.com
Dictionary with more than 75,000 organic and inorganic names and technical words in chemistry, physics, and biology. SciWords for Agriculture. SciWords for Biology. SciWords for Environment. InfoChem's products. PCs (DOS and Windows) and Macintosh.

SYBASE
Sybase, Inc.
6475 Christie Avenue
Emeryville, CA 94608, U.S.A.
Tel. 800-792-2731, 510-596-3500, fax 510-658-9441
Relational database management system. UNIX workstations.

Virtual Notebook System
Forefront Group
1360 Post Oak Boulevard, Suite 1660
Houston, TX 77056, U.S.A.
Tel. 713-961-1101, fax 713-961-1149, e-mail sarar@ssg.com
Electronic lab notebook. PCs (Windows), Macintosh, and UNIX platforms.

SOFTWARE FOR MINICOMPUTERS, SUPERMINICOMPUTERS, WORKSTATIONS, AND SUPERCOMPUTERS

Alliant; AT&T; Convex; Cray; Digital; Evans & Sutherland; Fujitsu; Hewlett-Packard; Hitachi; IBM; Intel; Kendall Square; Kubota; NEC; Silicon Graphics; Star; Sun; Thinking Machines, and others.

General Purpose Molecular Modeling

AMBER

Dr. Peter A. Kollman
Department of Pharmaceutical Chemistry
University of California
San Francisco, CA 94143, U.S.A.
Tel. 415-476-4637, fax 415-476-0688, e-mail pak@cgl.ucsf.edu

Assisted Model Building using Energy Refinement. Energy minimization, molecular dynamics, and free energy perturbation (FEP) calculations. SPASMS (San Francisco Package of Applications for the Simulation of Molecular Systems). tLEaP (nongraphical) and xLEaP (graphical) interfaces to AMBER and SPASMS. GIBBS for free energy. SANDER for NMR fitting. VAX, Cray, and UNIX workstation versions.

AM2000

American Molecular Technologies, Inc.
14785 Omicron Drive, Texas Research Park
San Antonio, TX 78245, U.S.A.
Tel. 210-677-6000, fax 210-677-6070

Molecular simulations, including energy minimizations, dynamics, and free energy perturbation calculations on macromolecules. Quest for ab initio molecular orbital calculations. Ultimol for molecular visualizations. Galaxy for Hansch-type and 3D QSAR and molecular modeling.

ANNEAL-RING

Mr. Patrick Franc
NYU Industrial Liaison Office
NYU Medical School
New York, NY 10003, U.S.A.
E-mail franc@cs.nyu.edu

Simulated annealing to be used in conjunction with MULTIC conformational search routine of MacroModel.

AutoDock

Dr. Arthur Olson
Department of Molecular Biology MB5
Research Institute of Scripps Clinic
10666 North Torrey Pines Road
La Jolla, CA 92037, U.S.A.
Tel. 619-554-9702, fax 619-554-6860, e-mail olson@scripps.edu

Monte Carlo docking of ligands into receptors. GRANNY for molecular graphics in conjunction with GRAMPS. UNIX workstations.

CHARMM
Dr. Martin Karplus
Department of Chemistry
Harvard University
12 Oxford Street
Cambridge, MA 02138, U.S.A.
Tel. 617-495-4018, fax 617-495-1792, e-mail karplus@huchel.bitnet, brbrooks@helix.nih.gov

Molecular dynamics package using Chemistry at Harvard Macromolecular Mechanics force field. Extensive scripting language for molecular mechanics, simulations, solvation, electrostatics, crystal packing, vibrational analysis, free energy perturbation (FEP) calculations, quantum mechanics/molecular mechanics calculations, stochastic dynamics, and graphing data. Convex, IBM, Intel, Silicon Graphics, Sun, and VAX.

Chem-X
Chemical Design Ltd.
Dr. Keith Davies
Roundway House, Cromwell Park
Chipping Norton, Oxon, OX7 5SR, England, U.K.
Tel. 44-1608-644000, fax 44-1608-644244 (U.K.), tel. 81-03-3345-1411, fax 81-03-3344-3949 (Japan)
and
Chemical Design Ltd.
200 Route 17 South, Suite 120
Mahwah, NJ 07430, U.S.A.
Tel. 201-529-3323, fax 201-529-2443

An integrated, modular system for molecular visualization and computation in all areas of chemistry. Chem-X/MODEL provides for molecular building and displays, geometry and energy calculations, fitting and conformational analysis, and crystal symmetry calculations. ChemQM for quantum mechanical calculations using ICON8, CNINDO, MOPAC, AMPAC, Gaussian, GAMESS, and ORACLE. Interfaces to AMBER, DGEOM, PSI77, and VSS. Chem-Inorganic for modeling organometallic and inorganic systems. LAZY for calculation of powder diffraction spectra. ChemPolymer for modeling polymers. ChemProtein for protein modeling. Pharmacophore Identification for finding structural features related to bioactivity. ChemNovel for de novo ligand builder and prediction of synthetic feasibility. ChemStat for QSAR, statistical analysis, and 3D QSAR. Substituent Database for QSAR descriptors. ChemLib interfacing files to other applications. Silicon Graphics IRIS, IBM RS/6000, and Digital VAX and Alpha.

Customized Polymer Modeling Program
Higher Dimension Research, Inc.

7650 Currell Boulevard, Suite 340
St. Paul, MN 55125, U.S.A.
Tel. 612-730-6205, fax 612-730-6206
Monte Carlo and molecular dynamics simulations on polymers. Silicon Graphics, Hewlett-Packard, and IBM RS/6000.

ECEPP/3
Dr. Harold A. Scheraga
Baker Laboratory of Chemistry
Cornell University
Ithaca, NY 14853-1301, U.S.A.
Tel. 607-255-4034, fax 607-255-4137, e-mail mwv@cornellc.bitnet
Empirical energy calculations on peptides and proteins.

GRID
Molecular Discovery Ltd.
Dr. Peter Goodford
West Way House, Elms Parade
Oxford, OX2 9LL, England, U.K.
Tel. 44-993-830385, fax 44-993-830966
Nonbonded force field probing for sites of interaction between small molecules/functional groups and 3D protein structures. GRIN helps create input. Output from GRID can be prepared in formats suitable for display in molecular modeling packages or for analysis in statistical programs. VAX, Evans & Sutherland, and UNIX workstations.

GROMOS
Biomos B.V.
Laboratory of Physical Chemistry
University of Groningen
Nijenborgh 16
9747 AG Groningen, The Netherlands
Tel. 31-50-63-4329/4323/4320, fax 31-50-63-4200
Groningen Molecular Simulation system for batch processing. SPC solvation model. PCMCAD for polymer/biopolymer mechanics. UNIX workstations.

Insight/Discover
BIOSYM Technologies, Inc.
9685 Scranton Road
San Diego, CA 92121-3752, U.S.A.
Tel. 619-458-9990, fax 619-458-0136, e-mail rcenter@biosym.com (U.S.A.), tel. 44-1256-817577, fax 44-1256-817600 (U.K.), tel. 81-0473-53-6997, fax 81-0473-53-6330 (Japan)

Insight II, an interactive graphics program for building, loop searching, manipulating, and analyzing molecules. Discover for molecular mechanics and dynamics on single processors or in parallel. Insight Xpress is subset for bench chemists. DeCipher for analysis of simulations. DelPhi for calculation and visualization of Poisson-Boltzmann electrostatic potentials. Analysis for treatment of trajectory data. Apex-3D for statistically fitting 3D structural features to bioactivities. Biopolymer for building proteins and nucleic acids. Converter for 2D-to-3D conversion of structures in MACCS databases. Homology for construction of proteins by sequence homology. Ludi for de novo ligand design from receptor site geometry and a library of structural fragments. NMRchitect suite consisting of Felix Assign, Felix Model, and other modules for processing and displaying multidimensional NMR data, NMR Database for managing data, DG II and Simulated Annealing for structure generation, IRMA for refinement, and NMR Analysis for evaluation. Felix for Windows. TurboNMR for predicting chemical shifts. Profiles-3D for analyzing protein structure and folding. Search/Compare for conformational searches. Sketcher for 2D-to-3D conversion using distance geometry. Amorphous Cell for polymers in glassy melts. Crystal Cell for thermodynamic and mechanical properties. Interphases for monolayers and bilayers. Networks for properties of elastomers and gels. Phase Diagram for polymer solutions. Polymerizer for modeling polymers. Prism for phase equilibria. QSPR for structure–property relationships of polymers. RIS for statistical properties of chains. Synthia for quantitative structure–property relationships of polymers. Viscoelasticity for polymers in dilute solution. SolidState suite for structures and properties of materials, including Characterize for comparison to experimental data, Solids Adjustment for studying derivatives, Solids Builder for building models of metals and glasses, Solids Docking for Monte Carlo, Solids Refine for dynamic refinement, Solids Simulation for dynamics, Structure Image for lattices, and Structure Solve for solving crystal structures. Xsight for interface to XtalView, REPLACE, and ProLSQ; X-ray structure validation and visualization module ProStat. VAX, Cray, and Silicon Graphics, IBM, and other UNIX workstations. CONSISTENT for interconverting molfiles on PCs.

MacroModel
Dr. W. Clark Still
Department of Chemistry
Columbia University
New York, NY 10027, U.S.A.
Tel. 212-280-2577, fax 212-678-9039, e-mail
mmod@still3.chem.columbia.edu, still@cuchmc.chem.columbia.edu
Graphical molecular modeling package. MULTIC for molecular mechanics, molecular dynamics, and conformational searching of organic molecules, proteins, nucleic acids, and carbohydrates. AMBER-, MM2-, and MM3-like and

OPLS force fields; implicit solvation model. Reads Cambridge and Brookhaven PDB files. VAX, Convex, Alliant, Cray, and UNIX workstations.

MM3

Technical Utilization Corporation, Inc.
235 Glen Village Court
Powell, OH 43065, U.S.A.
Tel. 614-885-0657, e-mail bender@osc.edu

N. L. Allinger's molecular mechanics program for energy minimization of organic molecules. Includes CRSTL for crystal lattices, MINP for keyboard input, MEDIT for interactive editing, and VIBPLT for vibrational animation. MM2 for molecular mechanics. Stochastic conformational searching. Source code. VAX and UNIX workstations.

MODEL

Dr. Kosta Steliou
Department of Chemistry
Boston University
590 Commonwealth Avenue
Boston, MA 02215, U.S.A.
Tel. 617-353-2480, fax 617-353-2497, e-mail steliou@chem.bu.edu

Molecular modeling with AMBER-like and MM2 force fields. Batch conformational searching with BAKMDL. Interfaces to AMPAC, MacroModel, GAUSSIAN86, SYBYL, PCMODEL, and CHEM-3D. VAX.

MOIL

Dr. Ron Elber
Department of Physical Chemistry
and the Israel Center for Structural Biology
The Hebrew University
Givat Ram
Jerusalem 91904, Israel
Tel. 972-2-585-484, fax 972-2-513-742, e-mail ron@ala.fh.huji.ac.il, ron@pap.chem.uic.edu

Molecular dynamics and minimization of proteins; locally enhanced sampling and free energy calculations along reaction paths by perturbation or thermodynamic integration. Hewlett-Packard, IBM, Silicon Graphics, and Stardent workstations. moil-view for visualization of shaded spheres and sticks on Silicon Graphics. Available by anonymous ftp from 128.248.186.70.

Moldy

Dr. Keith Refson
Department of Earth Sciences, Parks Road
Oxford University

Oxford, OX1 3PR, U.K.
Tel. 44-865-272026, fax 44-865-272072, e-mail keith@earth.ox.ac.uk
Molecular dynamics of solvated mineral surfaces. Available by anonymous ftp from earth.ox.ac.uk in the /pub directory. UNIX and VAX.

Prometheus
Proteus Molecular Design Ltd.
Proteus House, Lyme Green Business Park
Macclesfield, Cheshire, SK11 OJL, England, U.K.
Tel. 44-625-500555, fax 44-635-500666
Protein model building based on artificial intelligence and energy minimization.

PROPHET
BBN Systems and Technologies Corporation
10 Moulton Street
Cambridge, MA 02238, U.S.A.
Tel. 617-873-2669, fax 617-873-3776, e-mail prophet-info@bbn.com
Molecular building, molecular mechanics, simulations, and graphics. Statistical and mathematical modeling and display. Sequence analysis. Structural and sequence database retrieval. Information is available via anonymous ftp and Gopher from www-prophet.bbn.com (192.1.100.150) and via World Wide Web at URL http://www-prophet.bbn.com/. UNIX workstations, such as Sun and VAX (Ultrix), and X-emulators on Macintosh (A/UX) or PCs.

QUANTA/CHARMm
Molecular Simulations Inc.
16 New England Executive Park
Burlington, MA 01803-5297, U.S.A.
Tel. 800-756-4674, 617-229-9800, fax 617-229-9092, e-mail support@msi.com (U.S.A.), tel. 44-734-568056, 44-223-413300, fax 44-223-413301, 44-223-421591, e-mail support@msicam.co.uk (U.K.), tel. 81-3-3818-6511, fax 81-3-3818-6550 (Japan)
Structure building, manipulation, energy minimization, molecular dynamics, Boltzmann jump Monte Carlo conformational searching, and protein homology building. QUANTA molecular graphics system is integrated with the CHARMm molecular dynamics software using force fields derived from the Chemistry at Harvard Macromolecular Mechanics force field and the Merck Modeling Force Field. Cluster CHARMm and FARMm CHARMm for simulations in clustered computing environments. X-PLOR for X-ray and NMR structure refinement and simulated annealing. Crystal Workbench combining Quanta and X-PLOR. NMR Workbench with NMR·Pipe and NMR·Compass for multidimensional spectral processing and assigning peaks to proteins. QUANTA/NMR for protein construction. QUANTA/Bones for fitting elec-

tron density maps, X-FIT for model fitting, QUANTA/StAR for X-ray model building. QSPR-Polymer for property estimation. Enzymix for enzymatic reaction energies. Polaris for modeling and solvation free energies. UHBD for Brownian dynamics. BIOGRAF for modeling of drugs, proteins, carbohydrates, lipids, and DNA/RNA. Polygraf for modeling polymers, materials, and solvents. Professional Polygraf for polymer modeling and structure–property analysis. Cerius^2 suite of programs for modeling of polymeric, small molecular, and inorganic materials: C^2·Polymorph for predicting crystal polymorphs, C^2·Powder Indexing, C^2·X-GEN for X-ray data processing, C^2·EXAFS for extended X-ray absorption fine structure refinement, C^2·Mechanical Properties for predicting elastic moduli, C^2·MFA for molecular field analysis, C^2·CAVEAT to search databases of template structures, C^2·DBAcess for searching CAVEAT and Catalyst databases of 3D structures, and C^2·Amorphous Builder for Monte Carlo polymer packing. Drug Discovery Workbench (DDW) for spreadsheet QSAR with modules C^2·Alignment, C^2·Analysis, C^2·Conformers, C^2·Genetic Algorithms, C^2·MSA, C^2·QSAR+, C^2·Receptor, and C^2·Visualizer. MCSS for identifying ligand binding sites; GROUPBUILD and HOOK for de novo ligand design. MODELER to build proteins by homology. Quantum Mechanics Workbench with modules C^2·MOPAC, C^2·Gaussian, and C^2·CASTEP for plane wave pseudopotential calculations of solids and surfaces. C^2·SDK for software developer's kit. Products of Polygen, Molecular Simulations Inc., Cambridge Molecular Design, and BioCAD. UniChem for Cray Research, Inc. Silicon Graphics, Cray, Sun, Convex, Digital, Hewlett-Packard, IBM, and other UNIX workstations, as well as Macintosh and PCs (X-Windows). AVS ChemistryViewer for visualization and analysis of data input for and output from programs, such as Gaussian, MOPAC, etc. Anonymous ftp from avs.ncsc.org. UNIX workstations.

SPARTAN
  Wavefunction, Inc.
  Dr. Warren J. Hehre
  18401 Von Karman Avenue, Suite 370
  Irvine, CA 92715, U.S.A.
  Tel. 714-955-2120, fax 714-955-2118, e-mail sales@wavefun.com, support@wavefun.com
Model building, molecular mechanics (SYBYL, MM2, MM3), and ab initio (Hartree–Fock, Møller–Plesset, direct HF) and semiempirical (MNDO, AM1, PM3) molecular orbital calculations with and without solvent effects. Graphical front-end and postprocessor of the output. Electron density and electrostatic plots. Interface to Gaussian 92. Convex, Digital, Hewlett-Packard, IBM, and Silicon Graphics versions.

SCARECROW
  Dr. Leif Laaksonen
  Center for Scientific Computing

P.O. Box 405
FIN-02101 Espoo, Finland
Tel. 358-0-4572378, fax 358-0-4572302, e-mail leif.laaksonen@csc.fi

Analysis of molecular simulation trajectories from CHARMM, Discover, YASP, MUMOD, GROMOS, and AMBER. Interface to ICON8 for extended Hückel calculations and to VSS for electrostatic potentials. 2D graphics of surfaces and electron density and orbitals. Silicon Graphics.

SYBYL

Tripos, Inc.
1699 South Hanley Road, Suite 303
St. Louis, MO 63144-2913, U.S.A.
Tel. 800-323-2960, 314-647-1099, fax 314-647-9241, e-mail support@tripos.com (U.S.A.), tel. 44-344-300144, fax 44-344-360638 (U.K.), tel. 81-3-5228-5617, fax 81-3-5228-5581 (Japan)

An integrated molecular modeling package with capabilities for molecular mechanics, conformational searching, minimization, semiempirical and ab initio molecular orbital calculations, molecular graphics, active analog approach, and molecular dynamics. Tripos, AMBER-, and MM2-like force fields. Components for handling small molecules, biomolecules, and polymers. A programming language for macros. Interfaces to Cambridge Structural Database, Brookhaven Protein Data Bank, and QCPE programs. LabVision is a subset of SYBYL for bench chemists using ESV workstations. QSAR based on Comparative Molecular Field Analysis (CoMFA) and interface to Daylight's CLOGP and CMR; Partial Least Squares; SIMCA. Molecular Spreadsheet for management and analysis of structures and data. N. L. Allinger's MM3 molecular mechanics programs for industrial customers. R. S. Pearlman's CONCORD knowledge-based model builder for rapid generation of 3D databases from connectivity databases. T. Blundell's COMPOSER for building proteins by homology. POSSUM and PROTEP for searching databases for secondary structure motifs. W. L. Jorgensen's BOSS (Biochemical and Organic Simulation System) for Monte Carlo simulations. Molecular Silverware for solvating molecules. R. Dammkoehler's RECEPTOR for constrained conformational searching. TRIAD Base for analyzing 1D and 2D spectral data. TRIAD NMR for multidimensional data processing and structure determination. K. Wüthrich's DIANA for generating conformations in torsional space subject to geometrical constraints. T. James's MARDIGRAS+ for refining nuclear Overhauser effect distances. NMR1 and NMRZ of New Methods Research Inc. CAPRI for Computer-Assisted Peak Resonance Identification in multidimensional NMR data. Flex-Model for polymer modeling and solubility prediction. J. Brickmann's MOLCAD for visualization with Gourard-shaded and transparent surfaces on Silicon Graphics. LEAPFROG for generating new ligand structures based on CoMFA model or receptor structure. DISCO for finding structural features related to bioactivity. MatchMaker for inverse protein folding. ProTable for protein analysis. NITRO terminal emulator for Macintosh and PCs. X-

Windows for Macintosh, PC, and X terminals. VAX, Silicon Graphics, Evans & Sutherland, and Sun versions. User-written SYBYL Programming Scripts (SPL) are available via anonymous ftp from extreme.chem.rpi.edu and can be browsed through the NIH Molecular Modeling Home Page at URL http://www.nih.gov/molecular_modeling/mmhome.html under the Network-Based Services button.

WHAT IF
Dr. Gerrit Vriend
European Molecular Biology Laboratory
Meyerhofstrasse 1
6900 Heidelberg, Germany
Tel. 49-6221-387-473, fax 49-6221-387-517, e-mail vriend@embl-heidelberg.de
Protein modeling package with molecular graphics, homology building, database searches, and options for NMR- and X-ray-related work. Interfaces to GROMOS, GRID, RIBBONS, and DSSP. Documentation available by anonymous ftp from swift.emb-heidelberg.de in directory /pub/whatif/writeup. URL is http://swift.embl-heidelberg.de/whatif/. VAX/Evans & Sutherland, NeXT, Silicon Graphics, Sun, Bruker, and PCs (DOS and Linux).

Yeti
Dr. Angelo Vedani
Biographics Laboratory
Swiss Institute for Alternatives to Animal Testing
Aeschstrasse 14
CH-4107 Ettingen, Switzerland
E-mail vedani@czheth5a.bitnet
Molecular mechanics with special treatment of hydrogen bonding, solvation, and metal ions. Also Yak for receptor modeling based on directionality of potential binding points on a ligand. Digital and Silicon Graphics.

Quantum Chemistry Calculations

ACES II
Dr. Rodney J. Bartlett
Quantum Theory Project
362 Williamson Hall
University of Florida
Gainesville, FL 32611-2085, U.S.A.
Tel. 904-392-1597, fax 904-392-8722, e-mail aces2@qtp.ufl.edu
Ab initio molecular orbital code specializing in the evaluation of the correlation energy using many-body perturbation theory and coupled-cluster theory. Ana-

lytic gradients of the energy available at MBPT (2), MBPT (3), MBPT (4), and CC levels for restricted and unrestricted Hartree–Fock reference functions. MBPT (2) and CC gradients. Also available for ROHF reference functions. UNIX workstations.

ADF
    Dr. E. J. Baerends
    Department of Theoretical Chemistry
    Vrije Universiteit
    De Boelelaan 1083
    1081 HV Amsterdam, The Netherlands
    Tel. 31-020-548-2978, fax 31-020-646-1479, e-mail tevelde@chem.vu.nl
Amsterdam Density Functional program including X-alpha parameterization.

AIMPAC
    Dr. Richard F. W. Bader
    Department of Chemistry
    McMaster University
    Hamilton, Ontario L8S 4M1, Canada
    Tel. 416-525-9140, ext. 3499, fax 416-522-2509, e-mail
        bader@mcmail.cis.mcmaster.ca
Calculation of electron density, its gradient and Laplacian, and related properties.

AMPAC
    Semichem
    Dr. Andrew J. Holder
    7128 Summit
    Shawnee, KS 66216, U.S.A.
    Tel. 913-268-3271, fax 913-268-3445, e-mail aholder@vax1.umkc.edu
Semiempirical molecular orbital calculations with M. J. S. Dewar's SAM1 parameterization, including d orbitals for transition metals, and a graphical user interface. MINDO/3, MNDO, MNDOC, AM1, and PM3 methods. Convex, Digital VAX and Alpha, Cray, Silicon Graphics, Hewlett-Packard, IBM RS/6000, Sun, and PCs (X-Windows).

Argus
    Dr. Mark A. Thompson
    Pacific Northwest Laboratory
    P.O. Box 999, Mail Stop K1-87
    Richland, WA 99352, U.S.A.
    Tel. 509-375-6734, fax 509-375-6631, e-mail d3f012@pnlg.pnl.gov
Semiempirical (EHT, INDO1, INDO1/S, and NDDO1) and SCF calculations for spectroscopic properties. C language. Available via anonymous ftp from

pnlg.pnl.gov (130.20.64.11). Cray, DECstation, Hewlett-Packard, IBM RS/6000, Sun, and PCs.

ASTERIX
Computer Physics Communications (CPC) Program Library
Queen's University of Belfast
Belfast BT7 1NN, North Ireland, U.K.
Fax 44-232-239182, e-mail cpc@qub.ac.uk, cpc@queens-belfast.ac.uk, cpcserver@daresbury.ac.uk
and
Dr. Marie-Madeleine Rohmer
Laboratoire de Chimie Quantique
Institut Le Bel
4, rue Blaise Pascal
F-67000 Strasbourg, France
Tel. 33-88-41-61-42, fax 33-88-61-20-85, e-mail rohmer@frccsc21.bitnet
Ab initio calculations for large organometallic and other compounds. FORTRAN programs designed for high-performance vector and parallel processing machines.

CADPAC
Lynxvale WCIU Programs
Dr. Roger Amos
20 Trumpington Street
Cambridge, CB2 1QA, England, U.K.
Tel. 44-223-336384, e-mail cadpac@theory.chemistry.cambridge.ac.uk
Cambridge Analytical Derivatives Package. General purpose ab initio calculations. Cray and other versions. Spectro for analyzing anharmonic force fields and calculating positions and intensities of lines, including Fermi resonance effects. A description is available at World Wide Web URL http://www.cray.com/PUBLIC/DAS/files/CHEMISTRY/CADPAC.txt.

CHELPG
Dr. Curt M. Breneman
Department of Chemistry
Rensselaer Polytechnic Institute
Troy, NY 12180, U.S.A.
Tel. 518-276-2678, e-mail breneman@quant.chem.rpi.edu
Computes electrostatic-potential-derived charges from ab initio wavefunctions generated by one of the Gaussian packages. UNIX and OpenVMS machines.

COLUMBUS Program System
Dr. Isaiah Shavitt
Dr. Russell M. Pitzer

Department of Chemistry
Ohio State University
Columbus, OH 43210, U.S.A.
Tel. 614-292-1668, fax 614-292-1685, e-mail shavitt@mps.ohio-state.edu, pitzer@neon.mps.ohio-state.edu, shepard@tcg.anl.gov, lischka@itc.univie.ac.at

Modular FORTRAN programs for performing general ab initio, multireference single and double excitation configuration interaction (CI), averaged coupled-pair functional, and linearized coupled-cluster method calculations. Available via anonymous ftp from ftp.tcg.anl.gov (146.137.200.2). Cray and other versions.

DeFT
Dr. Alain St-Amant
Department of Chemistry
University of Ottawa
10 Marie Curie
Ottawa, Ontario K1N, 6N5, Canada
Tel. 613-564-2234, fax 613-564-6793, e-mail st-amant@theory.chem.uottawa.ca

Density functional theory calculations.

DMol
BIOSYM Technologies, Inc.
9685 Scranton Road
San Diego, CA 92121-3752, U.S.A.
Tel. 619-458-9990, FAX 619-458-0136, e-mail rcenter@biosym.com (U.S.A.), tel. 44-1256-817577, fax 44-1256-817600 (U.K.), tel. 81-0473-53-6997, fax 81-0473-53-6330 (Japan)

Density functional theory calculations. deMon for density functional calculations. Turbomole for Hartree–Fock and MP2 ab initio calculations. ZINDO for extended Hückel, PPP, CNDO, and INDO semiempirical molecular orbital calculations and prediction of electronic spectra. Plane Wave for band structures of semiconductors. ESOCS for electronic structure of solids. DSolid for density functional theory calculations of periodic solids. Silicon Graphics and IBM workstation versions.

GAMESS
Dr. Michael Schmidt
Department of Chemistry
Iowa State University
Ames, IA 50011, U.S.A.
Tel. 515-294-9796, fax 515-294-5204, e-mail mike@si.fi.ameslab.gov, theresa@si.fi.ameslab.gov

General Atomic and Molecular Electronic Structure System. Ab initio calculations with analytic energies and first derivatives and numerical second derivatives for effective core potential (ECP) calculations. Parallel execution on UNIX workstations. VAX, Hewlett-Packard, IBM, Cray, and UNIX workstations.

Gaussian
Gaussian, Inc.
Dr. Michael Frisch
Carnegie Office Park, Building 6
Pittsburgh, PA 15106, U.S.A.
Tel. 412-279-6700, fax 412-279-2118, e-mail info@gaussian.com
Gaussian 92 for ab initio molecular orbital calculations (Hartree–Fock, direct HF, Møller–Plesset, CI, reaction field theory, electrostatic potential-derived charges, vibrational frequencies, etc.). Gaussian 92/DFT for density functional theory calculations with Slater (X-alpha) and Becke's gradient-corrected exchange functionals. Self-consistent reaction field and polarized continuum solvation models. Browse Quantum Chemistry Database System for archival storage of computed results. Processing on parallel machines. Newzmat to interconvert molfile formats. Interface to SPARTAN. Convex (UNIX), Cray (UniCOS), DEC-Alpha (OpenVMS), DECstation (ULTRIX), Fujitsu (UXP/M), Hewlett-Packard (HP-UX), Hitachi (HI-OSF, VOS3), IBM (AIX, VM, MVS), KSR1 (UNIX), Kubota Titan (UNIX). Multiflow (UNIX), NEC (SUPER-UX), Silicon Graphics (IRIX), Sun (SunOS, Solaris), and VAX (OpenVMS, OSF1). Guassian 92 for PCs (Windows).

GRADSCF
Polyatomics Research Institute
Dr. Andrew Komornicki
1101 San Antonio Road, Suite 420
Mountain View, CA 94043, U.S.A.
Tel. 415-964-4013, e-mail 0394@rhea.cray.com
Ab initio calculations. Cray and other versions.

HONDO
Dr. Michel Dupuis
IBM
Department 48B, Mail Stop 428
Kingston, NY 12401, U.S.A.
Tel. 914-385-4965, fax 914-385-4965, e-mail
michel@kgnvma.vnet.ibm.com
Ab initio calculations of RHF, UHF, GVB, MCSCF, Möller–Plesset, and CI wavefunctions. Geometry optimization and transition state location. Force constants, vibrational spectra, and other properties. IBM 3090 and other models. IBM.

KGNMOL

Dr. Enrico Clementi
Centro di Ricerca, Sviluppo e Studi Superiori in Sardegna
Casella Postale 488
09100 Cagliari, Italy
Tel. 39-70-279-62-231, fax 39-70-279-62-220, e-mail enrico@crs4.it

Ab initio calculations. ATOMSCF, ALCHEMY-II (direct CI and MCSCF), ATOMCI, BNDPKG2 (bands in solids), BROWIAN, GDFB, GDFMOL/GDFMD, HONDO-8, HYCOIN (Hylleraas-CI), MELD (CI), MOLCAS-1, PHOTO (excited states), PLH-91 (band structure of polymers), QMDCP (Kohn-Sham orbitals), REATOM, SIRIUS (complete and restricted active space MCSCF calculations), VEH-91 (Valance Effective Hamiltonian method). KGNGRAF for interactive computer graphics. Molecular dynamics by KGNMCYL, KGNMD, and KGNNCC. PRONET for prediction of backbone conformations of proteins with a neural network procedure, and other programs described in the book series *MOTECC: Modern Techniques in Computational Chemistry,* E. Clementi, Ed., 1989–1991, ESCOM, Leiden, The Netherlands. Information available via anonymous ftp from malena.crs4.it (156.148.7.12). IBM machines under VM, MVS, and AIX operating systems.

MOLCAS

Dr. Bjorn O. Roos
University of Lund, Sweden
Department of Theoretical Chemistry
P.O. Box 124, Chemical Center
S-221 00 Lund, Sweden
Tel. 46-46-108251, fax 46-46-104543, e-mail teobor@garm.teokem.lu.se

CASSCF ab initio calculations.

MOLDEN

CAOS/CAMM Center
University of Nijmegen
Nijmegen 6525 ED, The Netherlands

Displaying molecular density using output from GAMESS-UK and Gaussian. Available by anonymous ftp from camms1.caos.kun.nl (131.174.82.237). UNIX workstations.

MOPAC

QCPE
Creative Arts Building 181
Indiana University
840 State Highway 46 Bypass
Bloomington, IN 47405, U.S.A.
Tel. 812-855-4784, fax 812-855-5539, e-mail qcpe@ucs.indiana.edu

Semiempirical molecular orbital package for optimizing geometry and studying reaction coordinates. The latest version, MOPAC 93, has additional capabilities to recognize symmetry point groups, compute nonlinear optical properties, handle polymers, layers, and solids, and model solvent effects. Extensive library of about 800 other programs from academia and industry for quantum mechanics, molecular mechanics, structure generation from NMR data, quantitative structure–activity relationships, and molecular graphics, including ABACUS (regression analysis), AMSOL (semiempirical molecular orbital calculations including solvation effects), BIGSTRN3, CHEMICALC-2 (log *P*), CNDO/S, CNINDO, CPKPDB, CRYSTAL (solid state quantum chemistry), DGEOM, DISGEO (distance geometry), DNMR (NMR analysis), DNMR6 (spectra calculations), DRAW, ECEPP2 (Empirical Conformational Energy Program for Peptides), ESTAR (electrostatics), FORTICON8 (extended Hückel), FACTANAL (factor analysis), GEPOL (surface areas), GRID (charges from electrostatic potentials), INTERCHEM (comprehensive molecular modeling package, including PIFF for molecular mechanics with pi electrons, PROTEINS for accessing Brookhaven Protein Data Bank files, CONVERT for interchanging molfile formats, THREEDOM for 3D database searching, PRESTO for aligning protein and nucleic acid sequences, and interfaces to MOPAC and GAMESS-UK), MAXWELL (electrostatic energies), MD Display (animation and Ramachandran plots), MLDC8 (NMR spectra), MNDO, MNDOC, MOLDEN (electron density maps from Gaussian and GAMESS output), MOLFIT (superposition), MOLY-86 (modeling package), MOS (molecular orbitals calculations for spectroscopy), MS (molecular surfaces), MSEED (solvent-accessible surface areas), NM Display (normal modes), NOEL (molecular similarity), NOEMOL (NMR analysis), PAP (protein analysis and graphics), PCILO3 (Perturbative Configuration Interaction using Localized Orbitals), PDM88 (Point charges), PEPA (population analysis), POCKETT (holes in proteins), PPP-MO, PRODEN (electron density), PSDD (neural network simulator for drug design), PSI77 (orbital plots), QCFF/PI (molecular mechanics), ROVI (vibrational fine structure), RRKM, SAMPLS (partial least squares), SEAL (structure superposition), SIBFA (intermolecular interactions), SIMVOL (molecular volumes, areas, and similarity), SYBST (STERIMOL parameters), VIBRATE (normal modes), VOID (protein packing), and mdXvu (visualization of AMBER trajectory files). QCPE has also accepted the responsibility to distribute "semicommercial" academic software, such as MM2, MM3 (94), MOPAC 93, POLYRATE, PEFF, and COMPARE-CONFORMER. MM3 (94) includes dynamics and force field parameter estimation. Most programs in library are in FORTRAN and are available as source code. Current catalogs and information are available by anonymous ftp from qcpe6.chem.indiana.edu (129.79.74.206). Many of the programs run on several hardware platforms, including Digital, IBM, Silicon Graphics, Stardent, Sun, Fujitsu, and Cray.

MOPLOT2
  Dr. Dennis L. Lichtenberger

Department of Chemistry
University of Arizona
Tucson, AZ 85721, U.S.A.
E-mail dlichten@ccit.arizona.edu

Orbital and density plots from linear combinations of Slater or Gaussian type orbitals. Available by anonymous ftp on infomeister.osc.edu (128.146.36.5) in directory /pub/chemistry/software/SOURCES/FORTRAN/moplot2. Silicon Graphics, Convex, and VAX.

MORATE
Computer Physics Communications (CPC) Program Library
Queen's University of Belfast
Belfast, BT7 1NN, Northern Ireland, U.K.
Fax 44-232-239182, e-mail cpc@qub.ac.uk, cpc@queens-belfast.ac.uk, cpcserver@daresbury.ac.uk

Dynamics calculations of reaction rates by semiempirical molecular orbital theory. POLYRATE for chemical reaction rates of polyatomics. POLYMOL for wavefunctions of polymers. HONDO for ab initio calculations. RIAS for configuration interaction wavefunctions of atoms. FCI for full configuration interaction wavefunctions. MOLSIMIL-88 for molecular similarity based on CNDO-like approximation. JETNET for artificial neural network calculations. More than 1350 other programs, most written in FORTRAN for physics and physical chemistry.

NCSAdisco
Dr. Harrell Sellers
Department of Chemistry
South Dakota State University
Brookings, SD 57007, U.S.A.
Tel. 605-688-6374, fax 605-688-5822, e-mail hsellers@ncsa.uiuc.edu

Ab initio calculations on metal surfaces and other systems. Mainframes and workstations.

PDM93
Dr. Donald E. Williams
Department of Chemistry
University of Louisville
Louisville, KY 40292, U.S.A.
Tel. 502-852-5975, fax 502-852-8149, e-mail dew01@uxray5.chem.louisville.edu

Electric Potential Derived Multipoles method to find optimized net atomic charges and other site multipole representations. Accepts input from Gaussian. UNIX workstations and VAX.

PS-GVB
Schrödinger, Inc.
80 South Lake Avenue, Suite 735
Pasadena, CA 91101, U.S.A.
Tel. 818-568-9392, fax 818-568-9778, e-mail info@psgvb.com
Pseudospectral Generalized Valence Bond calculations on molecules, clusters, and crystals. Convex, Cray, Hewlett-Packard, IBM RS/6000, and Silicon Graphics. The graphical user interface runs on Hewlett-Packard, Silicon Graphics, and Sun workstations. Also Cerius^2 interface.

PSI88
Dr. William L. Jorgensen
Yale University
P.O. Box 6666
New Haven, CT 06511, U.S.A.
Tel. 203-432-6278, fax 203-432-6144, e-mail bill@doctor.chem.yale.edu
Plots of wavefunctions in three dimensions from semiempirical and popular ab initio basis sets. Silicon Graphics, Sun, VAX, Cray, and others.

UniChem
Cray Research, Inc.
Cray Research Park, 655 Lone Oak Drive
Eagan, MN 55121, U.S.A.
Tel. 612-683-3688, fax 612-683-3099, e-mail sgustaf@rhea.cray.com
A package with a graphics front-end for structure input and visualizations of electron density, electrostatic potentials, molecular orbitals, and molecular surfaces from quantum applications. DGauss for density functional theory calculations with nonlocal SCF corrections and geometry optimization. CADPAC for ab initio calculations. MNDO93 for semiempirical molecular orbital calculations. Interface to Gaussian 92 quantum chemistry program. Information is available at World Wide Web URL http://www.cray.com/apps/UNICHEM/Mainpage.html. Silicon Graphics and Macintosh (X-Windows) networked to a Cray.

Databases of Molecular Structures

Cambridge Structural Database
Cambridge Crystallographic Data Centre
12 Union Road
Cambridge, CB2 1EZ, England, U.K.
Tel. 44-1223-336408, fax 44-1223-336033, e-mail
software@crystallography.chemistry.cambridge.ac.uk
Database with more than 130,000 X-ray structures of low-molecular-weight organic and organometallic compounds. QUEST for data retrieval using con-

nectivity and text searching. QUEST3D for 3D structure searching. PLUTO for stick, ball-and-stick, and space-filling plots of retrieved 3D structures. GSTAT for statistical summaries of molecular geometrical data retrieved by QUEST, and VISTA for interactive statistical analysis. BUILDER converts structures to CSD format. The CSD is also available in MACCS and UNITY formats. CD-ROM with the 3D graphics version of the CSD system. URL is http://csdvx2.ccdc.cam.ac.uk/. VAX, Silicon Graphics, and Sun.

CAST-3D
STN International
Chemical Abstracts Service
2540 Olentangy River Road
P.O. Box 3012
Columbus, OH 43210, U.S.A.
Tel. 800-753-4227, 800-848-6533, 614-447-3600, fax 614-447-3713, e-mail help@cas.org (U.S.A.), tel. 49-7247-808-555, fax 49-7247-808-131 (Germany), tel. 81-3-3581-6411, fax 81-3-3581-6446 (Japan)
CAS 3D Structure Template file is a subset of structures from the Chemical Abstracts Service registry file. More than 370,000 compounds with limited conformational flexibility have been three-dimensionalized with CONCORD. The MDL SDfile and SYBYL molfiles are searchable by UNITY, ISIS, ChemDBS-3D, and other 3D database management programs. NEWCRYST database produced by FIZ Karlsrugh (Germany) with structural data on inorganic and organic crystal structures. STN provides on-line access to many databases with chemical, physical, thermodynamic, toxicological, pharmaceutical, biomedical, and patent data.

Catalyst
Molecular Simulations Inc.
16 New England Executive Park
Burlington, MA 01803-5297, U.S.A.
Tel. 800-756-4674, 408-522-0100, e-mail support@msi.com, support@biocad.com
Catalyst/Hypo for building molecular models, systematic conformational searching, and statistical fitting of 3D structural features to bioactivities. Pharmacophoric hypotheses can be used to search for matches in databases of 3D structures. Catalyst/Info for managing databases of 1D, 2D, and 3D data. Silicon Graphics and networked Macintosh and PCs.

CAVEAT
Office of Technology Licensing
University of California, Berkeley
2150 Shattuck Avenue, Suite 510
Berkeley, CA 94704, U.S.A.

Tel. 510-643-9525, fax 510-642-4566, e-mail
domino@uclink2.berkeley.edu
Programs of Dr. Paul A. Bartlett to convert Cambridge Structural Database to one with bond vectors and to search the latter for specified vector relationships. UNIX workstations. TRIAD database of more than 400,000 energy-minimized tricyclic structures for automated design and ILIAD database of more than 100,000 energy-minimized linking structures in MacroModel, CAVEAT, MDL, SYBYL, and PDB formats. Silicon Graphics, IBM RS/6000, and Sun.

Chem-X
Chemical Design Ltd.
Dr. Keith Davies
Roundway House, Cromwell Park
Chipping Norton, Oxon, OX7 5SR, England, U.K.
Tel. 44-1608-644000, fax 44-1608-644244 (U.K.), tel. 81-03-3345-1411, fax 81-03-3344-3949 (Japan)
and
Chemical Design Ltd.
200 Route 17 South, Suite 120
Mahwah, NJ 07430, U.S.A.
Tel. 201-529-3323, fax 201-529-2443
ChemCore module to three-dimensionalize 2D structures, interfaces to reformat MACCS, SMILES, or DARC-2D databases, ChemDBS-1 module to build 3D databases, and ChemDBS-3D module to search 3D databases. Chem-X/BASE for entry level, structural database management. 3D database searching with Flexifit accounts for conformational flexibility while storing only one conformation. ChemDiverse for measuring molecular diversity in structure libraries. Chapman & Hall's *3D Dictionary of Drugs* (12,000 medicinally interesting compounds), *3D Dictionary of Natural Products* (50,000 antibiotics, alkaloids, and terpenoids), and *3D Dictionary of Fine Chemicals* (105,000 organics). Derwent's *Standard Drugs File* (30,000 compounds). ChemRXS for reaction database searching. InfoChem's database of 370,000 reactions. Synopsys's database of 10,000 reactions for protecting groups. SPECS and BioSPECS databases (20,000 drugs). Biobyte Medchem Masterfile database (26,000 compounds). Chem-X/INVENTORY and Chem-X/REGISTRY for managing inventories of compounds. UNIX workstations, PCs (Windows), and Macintosh.

Cobra
Oxford Molecular Ltd.
The Magdalen Centre, Oxford Science Park
Oxford, OX4 4GA, England, U.K.
Tel. 44-865-784600, fax 44-865-784601, e-mail omlemail@vax.ox.ac.uk (U.K.), tel. 81-33-245-5004, fax 81-33-245-5009 (Japan)
and

Oxford Molecular Inc.
7005 Backlick Court, Suite 200
Springfield, VA 22151-3903, U.S.A.
Tel. 703-658-4854, fax 703-658-4887, e-mail syazdi@presto.ig.com
Constructs multiple conformers from a library of 3D fragments and rules; accepts SMILES notation input. Iditis is a relational database of protein structures and related data from the Brookhaven Protein Data Bank. Serratus is a nonredundant database of amino acid sequences from NBRF-PIR, SWISS-PROT, and GenBank. Antheprot for sequence analysis. Asp (Automated Similarity Package) for comparisons of molecular electrostatic fields. Anaconda for gnomonic projection and alignment of properties of molecules. Constrictor for distance geometry. Cameleon for multiple protein sequence alignment. AbM for modeling and humanizing variable fragments of antibodies and energy refining them with EUREKA force field. MAD (Molecular Advanced Design) for general molecular modeling. PIMMS molecular graphics. Pro-Explore for sequence analysis and biomolecular modeling. Pro-Simulate for graphical interface to molecular simulations with GROMOS, AMBER, and BOSS. Pro-Quantum for graphical interface to semiempirical (MOPAC, extended Hückel) and ab initio (CADPAC) calculations. Rattler for semiempirical charge calculation. Tsar for spreadsheet QSAR based on molecular properties, connectivity, and substituent property database. Vamp for semiempirical molecular orbital calculations, including ESR hyperfine coupling constants and ^{13}C NMR shifts. AMBER with LEAP, INTERFACE, GIBBS, SANDER, and SPASMS programs. VAX, Hewlett-Packard, IBM RS/6000, and Silicon Graphics workstations.

Daylight Toolkits
DAYLIGHT Chemical Information Systems, Inc.
18500 Von Karman Avenue, Suite 450
Irvine, CA 92715, U.S.A.
Tel. 714-476-0451, fax 714-476-0654, e-mail info@daylight.com
Chemical information platform for integration of chemical software tools including nomenclature (SMILES), 2D and 3D structural database management, similarity searching, display, geometry, and modeling. The modules include Depict for 2D molecular display, Fingerprint for chemical structure characterization, GEMINI for molfile conversions, GENIE/SMARTS for generalized substructure searching, Merlin for searching databases, SMARTS for representing substructure patterns, SMILES for structure representation, Thor interface to chemical information databases, and X-WIDGETS for building user interfaces. CASTOR for managing a catalog of compounds. Clustering Package for mining compound databases. Dervish for 3D molecular display. Merlin and MerlinServer for searching large databases of structures. PCmodels for predicting lipophilicity (CLOGP) and molar refractivity (CMR). Printing Package for I/O. Rubicon for distance geometry generation of 3D structures. Thor and ThorServer for interface to database management. ThorMan for database management functions. Maybridge93 database of 35,000 commercial compounds.

MedChem94 database of 26,000 compounds and their properties. Spresi93 database of 2.2 million compounds catalogued by institutes in Moscow and Berlin. TSCA93 database of 100,000 compounds affected by U.S. EPA Toxic Substance Control Act. WDI93 database with 42,000 drugs in the DERWENT World Drug Index. ALADDIN for geometric database searching on a VAX. Unix workstations, VAX, and PCs and Macintosh running X-Windows.

DOCK

Dr. Irwin D. Kuntz
Department of Pharmaceutical Chemistry
School of Pharmacy
University of California
San Francisco, CA 94143-0446, U.S.A.
Tel. 415-476-1397, fax 415-476-0688, e-mail kuntz@picasso.ucsf.edu

Samples the six degrees of freedom involved in the relative placement of two 3D rigid structures and scores their fit. SPHGEN searches for concave regions on a protein surface and defines cavity in terms of overlapping spheres. DOCK2 and DOCK3.0 for searching 3D databases to find ligands of appropriate shape. Companion programs CLUSTER, DISTMAP, and CHEMGRID. Silicon Graphics and VAX.

ISIS

MDL Information Systems, Inc.
14600 Catalina Street
San Leandro, CA 94577, U.S.A.
Tel. 800-635-0064, 800-326-3002, 510-895-1313, 510-895-2213, fax 510-895-6092, 510-483-4738, e-mail techsupp@mdli.com
and
Molecular Design MDL AG
Mühlebachweg 9
CH-4123 Allschwil 2, Switzerland
Tel. 41-61-4812656, fax 41-61-4812721, e-mail hotline@mdlag.ch (Switzerland), tel. 44-276-681777, fax 44-276-681724 (U.K.), tel. 0120-177-007, 81-43-299-3211, 81-3-3419-9171, fax 81-43-299-3077, 81-3-3419-9179 (Japan)

Integrated Scientific Information System for management of databases of 2D and 3D structures and associated properties on multiple platforms. PC (Windows), Macintosh, Fujitsu FMR, and NEC terminal support of ISIS/Draw and ISIS/Base; ISIS/Desktop. Academic version of ISIS/Draw available from http://www.mdli.com/. ISIS SAR Table for outputting structure-activity data to Microsoft Excel. MACCS-II for managing and searching databases of 2D and 3D structures on a single platform. Fixed conformation and conformationally flexible searching of 3D databases. Databases of structures three-dimensionalized by CONCORD, including CMC-3D of known pharmaceutical agents mentioned in *Comprehensive Medicinal Chemistry* (5000 medicinally interesting compounds; C. Hansch, et al., Pergamon Press, Elmsford, NY, 1990), ACD-3D from the Available Chemicals Directory (330,000 commercial chemicals), and MDDR-3D from the *Drug Data Reports* (12,000 drugs under development). REACCS managing and searching databases of reactions from the literature, including THEILHEIMER, REACCS-JSM, ORGSYN, Current Literature File, CHIRAS, and METALYSIS. Xenobiotic Metabolism database of known metabolic transformations. ORAC for managing databases of chemical reactions. OSAC for managing databases of 2D chemical structures. Digital VAX, IBM RS/6000, and others.

Protein Data Bank
Protein Data Bank
Chemistry Department, Building 555
Brookhaven National Laboratory
Upton, NY 11973-5000, U.S.A.
Tel. 516-282-3629, fax 516-282-5751, e-mail pdb@bnl.gov
Database of more than 3000 sets of atomic coordinates of proteins and other macromolecules, such as nucleic acids and carbohydrates, derived from X-ray crystallography, NMR spectroscopy, and modeling. Procheck for evaluation of stereochemical quality of protein structures. BLDKIT for model building. BENDER for bent wire models. CONECT generates full connectivity from atomic coordinates in Brookhaven database. DGPLOT for diagonal plots on printer. DIHDRL for torsional angles. DSTNCE for interatomic distances. FISIPL for phi/psi plots. Getentry for searching data bank entries and addresses of crystallographers. NEWHEL92 for helix parameters. PDB-Browse for handling indexes of data bank entries on workstations. STEREO to extract *x, y, z* atomic coordinates from printed stereo molecular graphics. Database available via anonymous ftp and Gopher from pdb.pdb.bnl.gov (130.199.144.1). Atomic coordinates and graphic images are available from URL http: //www.pdb.bnl.gov/. Also on this server is a file crystlist.adr with electronic mail addresses of crystallographers. Atomic coordinate files are also available on CD-ROM.

SCOP
Dr. Tim Hubbard
Centre for Protein Engineering
Medical Research Council Centre, Hills Road
Cambridge, CB2 2QH, U.K.
Tel. 44-1223-402131, fax: 44-1223-402140, e-mail
th@mrccpe.cam.ac.uk
Structural Classification of Proteins database with information on folds and families. URL of World Wide Web server at http://scop.mrc-lmb.cam.ac.uk/scop/. UNIX workstations and Macintosh.

SpecInfo
Chemical Concepts
Boschstrasse 12
P.O. Box 100202
D-6940 Weinheim, Germany
Tel. 49-6201-606-435, fax 49-6201-606-430
Database of 350,000 compounds for which IR, NMR, and mass spectral data are available. VAX.

UNITY
Tripos, Inc.
1699 South Hanley Road, Suite 303
St. Louis, MO 63144-2913, U.S.A.
Tel. 800-323-2960, 314-647-1099, fax 314-647-9241, e-mail support@tripos.com (U.S.A.), tel. 44-344-300144, fax 44-344-360638 (U.K.), tel. 81-3-5228-5617, fax 81-3-5228-5581 (Japan)
Combines 2D and 3D searching and storage with other molecular design tools. Searches Cambridge Structural Database, Chemical Abstracts Service registry file, or any MACCS database. 3D searches account for conformational flexibility. CONCORD for rapid generation of a single, high-quality conformation from connectivity of a small molecule (2D-to-3D conversion). Chapman & Hall's *CHCD Drugs, Fine Chemicals,* and *Natural Products* databases of 2D and 3D structures and the 2D *Organometallic Compound Database.* InfoChem's SPRESI database of 2.2 million organic compounds. Current Drugs Ltd.'s Agrochemical Patent Fast Alert and ID Patent Fast Alert databases. Sadtler's CSEARCH database with carbon-13 data for 70,000 compounds. Legion, Selector, and UNISON for molecular diversity management of libraries of compounds in combinatorial chemistry. BioSTAR for data handling from high throughput screening. VAX, UNIX workstations, Macintosh, and PCs (X-Windows). UNISON for Windows.

Molecular Graphics

AVS
Advanced Visual Systems, Inc.
300 Fifth Avenue
Waltham, MA 02154, U.S.A.
Tel. 617-890-4300, fax 617-890-8287
Graphical user interface. UNIX workstations.

FLEX
Dr. Michael Pique
Research Institute of Scripps Clinic

10666 North Torrey Pines Road
La Jolla, CA 92037, U.S.A.
E-mail mp@scripps.edu
Molecular graphics of molecular models and MD trajectories. Available via anonymous ftp from perutz.scripps.edu (137.131.152.27). Sun, Digital, and Stardent.

GEMM
Dr. B. K. Lee
National Institutes of Health
Room 4B15, Building 37
Bethesda, MD 20892, U.S.A.
Tel. 301-496-6580, fax 301-402-1344, e-mail bkl@helix.nih.gov
Generate, Edit, and Manipulate Molecules system for graphics. Silicon Graphics.

GRAMPS
Dr. T. J. O'Donnell
1307 West Bryon Street
Chicago, IL 60613, U.S.A.
Tel. 312-327-9390, e-mail tj@bert.eecs.uic.edu
General purpose graphics and animation toolkit for molecular models, such as stick figures, ball-and-stick, CPK, dot surfaces, wire-mesh surfaces, and fully shaded polygon surfaces. Animation of molecular dynamics trajectories. Silicon Graphics.

GRASP
Dr. Anthony Nicholls
Department of Biochemistry and Molecular Biophysics
Columbia University
630 W. 168th Street
New York, NY 10032, U.S.A.
Tel. 212-305-3882, fax 212-305-7932, e-mail nicholls@cuhhca.hhmi.columbia.edu
Molecular graphics. A demonstration version is at ftp site cumbnd.bioc.columbia.edu (128.59.96.103) in directory /grasp.

LIGHT
Dr. Bernard Brooks
National Institutes of Health
Bethesda, MD 20892, U.S.A.
Tel. 301-496-0148, fax 301-496-2172, e-mail brbrooks@helix.nih.gov
Ray trace graphics program for CHARMM files. Plotting package PLT2.

MidasPlus
Dr. Tom Ferrin
Department of Pharmaceutical Chemistry
University of California
San Francisco, CA 94143, U.S.A.
Tel. 415-476-1540, fax 415-476-0688, e-mail tef@cgl.ucsf.edu, midas@cgl.ucsf.edu
Real-time interactive stick, ball-and-stick, space-filling, and ribbon displays. Silicon Graphics.

MOLSCRIPT
Dr. Per Kraulis
Karolinska Institute
Stockholm, Sweden
E-mail pjk19@mbug.bio.cam.ac.uk
Black and white postscript molecular graphics. Silicon Graphics.

OpenMolecule
Andataco Computer Peripherals
9550 Waples Street, Suite 105
San Diego, CA 92121, U.S.A.
Tel. 800-334-9191, 619-453-9191, fax 619-453-9294, e-mail inquire@andataco.com
Molecular graphics for a Sun SPARCstation.

RasMol
Dr. Roger Sayle
Glaxo Research and Development
Greenford Road
Greenford, Middlesex, UB6 0HE, U.K.
Tel. 44-081-966-3567, fax 44-081-966-4476, e-mail ras32425@ggr.co.uk, ros@dcs.ed.ac.uk
Molecular graphics visualization tool for proteins and nucleic acids. Available by anonymous ftp from ftp.dcs.ed.ac.uk (129.215.160.5) in directory /home/ftp/pub/rasmol. Sun, Digital Alpha, IBM RS/6000, and Silicon Graphics. The Macintosh version RasMac is available from ftp.dcs.ed.ac.uk in /home/ftp/ pub/rasmol/rasmac.sit.hqx. URL is ftp://ftp.dcs.ed.ac.uk/export/rasmol/.

RASTER3D
Dr. Ethan A. Merritt
Department of Biological Structure SM-20
University of Washington
Seattle, WA 98195, U.S.A.

Tel. 206-543-1421, fax 206-543-1524, e-mail merritt@xray.bchem.washington.edu

Raster rendering of proteins and other molecules in stick, ball-and-stick, space-filling, and ribbon representations. Available via anonymous ftp from stanzi.bchem.washington.edu (128.95.12.38). Silicon Graphics and Sun.

Ribbons

Dr. Mike Carson
Center for Macromolecular Crystallography
University Station, Box THT-79
Birmingham, AL 35294, U.S.A.
Tel. 205-934-1983, fax 205-934-0480, e-mail carson@gtx.cmc.uab.edu

Display of proteins as rendered images showing secondary structure and other features. Silicon Graphics and Evans & Sutherland.

SciAn

Dr. Eric Pepke
Supercomputer Computations Research Institute
Florida State University
Tallahasse, FL 32306-4052, U.S.A.
E-mail pepke@scri.fsu.edu

Scientific visualization and animation program that can be applied to molecules. Available via anonymous ftp from ftp.scri.fsu.edu. Silicon Graphics and IBM RS/6000 workstations with Z-buffer capability.

XMol

Minnesota Supercomputer Center, Inc.
1200 Washington Avenue South
Minneapolis, MN 55415, U.S.A.
Tel. 612-337-0200, fax 612-337-3400, e-mail xmol@msc.edu, doherty@msc.edu

Molecular graphics from molfiles produced in Alchemy, Brookhaven PDB, CHEMLAB-II, Gaussian, MOLSIM, MOPAC, and MSCI's XYZ formats. Available by anonymous ftp from ftp.msc.edu in directory/pub/xmol. World Wide Web URL is http://www. arc.umn.edu/GVL/software/xmol/XMol.html. DECstation, Silicon Graphics, and Sun. Availably by anonymous from ftp.msc.edu.

Xpdb

Mr. Vinod T. Nair
Center for High Performance Computing
University of Texas
Austin, TX 78712, U.S.A.
E-mail vtn@almach.chpc.utexas.edu

Molecular graphics. Sun.

Quantitative Structure–Property Relationships and Statistics

ADAPT

Dr. Peter Jurs
152 Davey Laboratory
Department of Chemistry
The Pennsylvania State University
University Park, PA 16802, U.S.A.
Tel. 814-865-3739, fax 814-865-3314, e-mail pcj@psuvm.psu.edu

Adapted Data Analysis using Pattern recognition Toolkit. Generates molecular descriptors and applies metric methods to find structure–property relationships. VAX and Sun.

CLOGP

BioByte Corp.
P.O. Box 517
Claremont, CA 91711-0517, U.S.A.
Tel. 909-982-6645, fax 909-624-5509, e-mail al@iris.claremont.edu

Prediction of octanol/water partition coefficients from SMILES structure representation using methodology of Drs. Corwin Hansch and Al Leo. CMR for calculating molar refractivities. THOR-R/W for molecular database management. MERLIN for searching THOR databases. GENIE for substructure searching. QREG for regression analysis. C-QSAR for interactive data analysis of multivariate chemical and biological information. VOWT/XMR for computing molecular weight and volume. MASTERFILE database of 43,000 experimental log $P_{o/w}$ values and experimental 9,400 pK_a values; STARLIST database of preferred experimental log $P_{o/w}$ and pK_a values. BIOEQ/PHYSEQ databases with 3000 QSAR equations dealing with biological and physical properties. SIGFILE with 14,000 QSAR parameters for 3200 substituents. Products originally developed in Pomona College Medchem project. VAX. MacLogP for the Macintosh.

EXPOD

American InterFace Computer Inc.
One Westlake Plaza, Suite 200
1705 Capital of Texas Highway South
Austin, TX 78746, U.S.A.
Tel. 512-327-5344, fax 512-327-5176 (U.S.A.), fax 81-3-3277-0567 (Japan)

EXpert system for POlymer Design. Knowledge-based system for predicting physical properties from structure–property relationships for more than 2000 polymers. Hewlett-Packard, Sony, and Sun.

HINT
eduSoft LC
P.O. Box 1811
Ashland, VA 23005, U.S.A.
Tel. 619-566-1127, fax 619-586-1481, e-mail haney@netcom.com
Hydropathic INTeraction by empirical calculation of atomistic hydrophobicity of molecules. Grid points based on energetics of hydrophobic and hydrophilic fields can be used for contouring a hydrophobic space and for scoring ligand–macromolecule and macromolecule–macromolecule interactions. Interfaced to SYBYL (CoMFA), Insight II, and Chem-X. URL is http://www.i2020.net/edusoft/hint.html. Silicon Graphics and Evans & Sutherland ESV workstations.

POLLY
Dr. Subhash Basak
Center for Water and the Environment
University of Minnesota
5013 Miller Trunk Highway
Duluth, MN 55811, U.S.A.
Tel. 218-720-4279, fax 218-720-4219, e-mail sbasak@ua.d.umn.edu
Generation of connectivity and other molecular descriptors for use in QSAR and similarity/dissimilarity analysis. Silicon Graphics and PCs.

SIMCA-R
Umetri AB
Box 1456
S-901 24 Umea, Sweden
Tel. 46-90-196890, fax 46-90-197685
and
UMETRICS, Inc.
371 Highland Road
Winchester, MA 01890, U.S.A.
Tel. 617-756-0433, fax 617-721-2652
Data handling, statistical modeling (projection of latent structures, principal components analysis), and plotting for QSAR. SIMCA-IPI for integrated process intelligence. VAX and PCs. MODDE for experimental design with multiple regression analysis and partial least squares. PCs (Windows).

Other Applications

CAMEO
Dr. William L. Jorgensen
Department of Chemistry
Yale University
P.O. Box 6666

New Haven, CT 06511, U.S.A.
Tel. 203-432-6278, fax 203-432-6144, e-mail bill@doctor.chem.yale.edu

Computer-Assisted Mechanistic Evaluation of Organic reactions and prediction of products. pK_a and reaction enthalpy predictions. VAX.

CCP4
Dr. David Love
SERC Daresbury Laboratory
Warrington, WA4 4AD, England, U.K.
Tel. 44-925-603528, fax 44-925-603100, e-mail ccp4@daresbury.ac.uk

Suite of almost 100 protein crystallography programs for data processing, scaling, Patterson search and refinement, isomorphous and molecular replacement, structure refinement, such as PROLSQ, phase improvement (solvent flattening and symmetry averaging), and presentation of results, such as SURFACE for accessible surface area. Available via anonymous ftp from gserv1.dl.ac.uk in directory /pub/ccp4 and mail server info-server@daresbury.ac.uk. VAX and UNIX platforms.

Comprehensive Chemistry CD Package
Falcon Software
1 Hollis Street
Wellesley, MA 02181, U.S.A.
Tel. 617-235-1767

Courseware for chemistry teaching.

FRODO
Dr. Florante A. Quiocho
Howard Hughes Medical Institute
Baylor College of Medicine
One Baylor Plaza
Houston, TX 77030, U.S.A.
Tel. 713-798-6565, fax 713-797-6718, e-mail
faq@dino.qci.bioch.bcm.tmc.edu

Modular graphics and crystallographic applications. Evans & Sutherland. CHAIN is a newer, supported program for electron density fitting and molecular graphics that runs on Evans & Sutherland (PS300 and ESV) and Silicon Graphics.

GCG Package
Genetics Computer Group, Inc.
University Research Park
575 Science Drive, Suite B
Madison, WI 53711, U.S.A.

Tel. 608-231-5200, fax 608-231-5202, e-mail help@gcg.com, beers@gcg.com

A suite of more than 110 programs for analysis of nucleic acid and amino acid sequences in molecular biology and biochemistry. Formerly called the Wisconsin Sequence Analysis Package. BESTFIT for optimal alignment of similarity of two sequences. PileUp for multiple sequence alignment. PEPTIDESTRUCTURE for predictions of secondary structure of peptide sequences based on Chou-Fasman and Garnier-Osguthorpe-Robson rules. PepPlot+ for predictions of Chou-Fasman's secondary structure, Eisenberg's hydrophobic moment, and Kyte-Doolittle's hydropathy. HelicalWheel for plots of distribution of hydrophobic residues. Includes latest versions of sequence databases from GenBank, EMBL, PIR, SWISS-PROT, and VecBase. Runs on VAX (OpenVMS), Digital (Ultrix), Silicon Graphics, and Sun platforms. Macintosh interface.

Gopher

Internet Gopher Developers
100 Union Street, Suite 190
Minneapolis, MN 55455, U.S.A.
Fax 612-625-6817, e-mail gopher@boombox.micro.umn.edu

Internet client/server for a distributed information delivery system. Client program for workstations is available by anonymous ftp from boombox.micro.umn.edu (132.84.132.2). UNIX workstations, VAX, IBM, Macintosh, PCs, and NeXT.

Latticepatch

Dr. Alexandra Lee Klinger
Department of Biochemistry
Health Sciences Center
Charlottesville, VA 22920, U.S.A.
E-mail alk2p@fermi.clas.virginia.edu

Analysis of diffraction data to be collected by area detectors. Silicon Graphics.

LHASA

Dr. Alan K. Long
Department of Chemistry
Harvard University
12 Oxford Street
Cambridge, MA 02138, U.S.A.
Tel. 617-495-4283, fax 617-496-5618, e-mail long@layla.harvard.edu
and
LHASA/UK
School of Chemistry
University of Leeds

Leeds, LS2 9JT, England, U.K.
Tel. 44-532-336531, fax 44-532-336565, e-mail jan@vax.bio.leeds.ac.uk
Logic and Heuristics Applied to Synthetic Analysis for retrosynthetic analysis of organic compounds using reactions stored as transforms in knowledge base. APSO, a teaching version of LHASA that displays successful and unsuccessful reactions. DEREK (Deductive Estimation of Risk from Existing Knowledge) for predicting toxicology of organic structures. PROTECT database of protecting groups in reactions. VAX.

MORASS
Dr. David Gorenstein
University of Texas Medical Branch at Galveston
Galveston, TX 77555, U.S.A.
E-mail bruce@dggin1.utmb.edu
Hybrid matrix NOE structural refinement. A World Wide Webb server on NMR and computational chemistry is available at URL http://www.nmr.utmb.edu/.

Mosaic
National Center for Supercomputing Applications
University of Illinois at Urbana-Champaign
Champaign, IL 61820, U.S.A.
Tel. 217-244-4130, fax 217-244-1987, e-mail orders@ncsa.uiuc.edu, alpha@ncsa.uiuc.edu
Communication of text and graphics files between computers using World Wide Web (WWW) servers. UNIX, PC, and Macintosh versions available via NCSA anonymous ftp from ftp.ncsa.uiuc.edu (141.142.20.50). Alternative anonymous ftp sources for Mosaic include sunsite.unc.edu (all versions), ftp.sunet.se (all versions), and ftp.luth.se (Macintosh version). Alvis (Alpha Shape Visualizer) for construction and rendering of objects from x, y, z coordinates in 3D space. Polyview for graphics rendering. NCSA ImageTool for data visualization. Silicon Graphics. NCSA Contours for data rendering. NCSA Image for image data manipulation. Macintosh.

MSP
Dr. Michael L. Connolly
2269 Chestnut Street, Suite 279
San Francisco, CA 94123, U.S.A.
Tel. 415-346-3505, e-mail connolly@netcom.com
Suite of programs for computing dotted, analytical, and polyhedral molecular surfaces. Omega for curvature of polyhedral surface.

O
Dr. T. Alwyn Jones
Blueberry Hill, Dalby

S-755 91 Uppsala, Sweden
Tel. 46-18-174982, fax 46-18-536971, e-mail alwyn@xray.bmc.uu.se, gerard@xray.bmc.uu.se

Molecular graphics for crystallographic determination of protein structures. URL of World Wide Web site is http://kaktus.kemi.aau.dk/. Silicon Graphics, Evans & Sutherland ESV, and Digital Alpha/OSF1. DEJAVU databases for the program "O" with about 3000 proteins. With license agreement, DEJAVU can be obtained by anonymous ftp from rigel.bmc.uu.se.

ORACLE
Oracle Corporation
500 Oracle Parkway
Redwood Shores, CA 94065, U.S.A.
Tel. 415-506-7000, fax 415-506-7200

Relational database management system for distributed data. VAX and other computers.

PROLSQ
Dr. Wayne A. Hendrickson
Department of Biochemistry and Molecular Biophysics
Columbia University
630 W. 168th Street
New York, NY 10032, U.S.A.
Tel. 212-305-3456, fax 212-305-7379

PRotein Least SQuares for refinement of X-ray diffraction data. VAX, Cray, and others.

QCLDB
Japan Association for International Chemical Information
Nakai Building, 6-25-4 Honkomagome, Bunkyo-ku
Tokyo 113, Japan
Tel. 81-3-5978-3601, 81-3-5978-3622, fax 81-3-5978-3600

Quantum Chemistry Literature Database with bibliographic information on more than 25,000 papers reporting ab initio atomic and molecular calculations since 1978. Distributed on magnetic tape; searching software runs on FACOM, HITAC, Hitachi, Sony, Hewlett-Packard, IBM, Silicon Graphics, Sun, and VAX. NQRS database with nuclear quadrapole resonance spectra; NQRS PC SEARCH software. PCs.

SQUASH
Molecular Structure Corporation
Software Department
3200 Research Forest Drive

The Woodlands, TX 77381-4238, U.S.A.
Tel. 800-543-2379, fax 713-292-2472
Macromolecular phase refinement.

XtalView
Dr. Duncan E. McRee
Computational Center for Macromolecular Science
Research Institute of Scripps Clinic
10666 North Torrey Pines Road
La Jolla, CA 92037, U.S.A.
Tel. 619-534-5100, e-mail ccms-request@sdsc.edu, dem@scripps.edu
Complete package for solving X-ray crystal structures such as those of proteins. SHAPE for analyzing molecular surfaces. FLEX for displaying and animating molecular graphics. Ftp site ftp.sdsc.edu. Sun, Silicon Graphics, and Digital (ULTRIX) workstations.

Author Index

Subject Index

Computer programs are denoted in boldface; databases and journals are in italics.